Micro/Nano Structures and Systems: Analysis, Design, Manufacturing, and Reliability

Micro/Nano Structures and Systems: Analysis, Design, Manufacturing, and Reliability

Editor

Stelios K. Georgantzinos

MDPI • Basel • Beijing • Wuhan • Barcelona • Belgrade • Manchester • Tokyo • Cluj • Tianjin

Editor
Stelios K. Georgantzinos
Department of Aerospace
Science and Technology
National and Kapodistrian
University of Athens
Psachna
Greece

Editorial Office
MDPI
St. Alban-Anlage 66
4052 Basel, Switzerland

This is a reprint of articles from the Special Issue published online in the open access journal *Micromachines* (ISSN 2072-666X) (available at: www.mdpi.com/journal/micromachines/special_issues/Micro_Structures_Systems).

For citation purposes, cite each article independently as indicated on the article page online and as indicated below:

LastName, A.A.; LastName, B.B.; LastName, C.C. Article Title. *Journal Name* **Year**, *Volume Number*, Page Range.

ISBN 978-3-0365-6545-3 (Hbk)
ISBN 978-3-0365-6544-6 (PDF)

Contents

About the Editor

Stelios K. Georgantzinos

Dr. Stelios K. Georganzinos is an Assistant Professor in the Department of Aerospace Science and Technology (2021) of the School of Sciences of the University of Athens. He holds a Ph.D. in Mechanical and Aerospace Engineering, University of Patras (2010); an MBA, University of Patras (2009); MSc in Mechanical Engineering, University of Thessaly (2007); and a Diploma in Mechanical and Aerospace Engineering, University of Patras (2005). He teaches postgraduate courses and has served and is serving as a supervisor in numerous postgraduate theses and doctoral dissertations. His research interests focus on the design, analysis, and simulation of advanced materials, structures, and systems at multiple scales (nano–micro–macro), as well as additive manufacturing and digitization. His scientific work has received significant international recognition; has published more than 110 papers (most as a principal investigator) in international journals, books, and conferences; and has received numerous citations (2022). He has participated as a researcher and scientific supervisor in numerous international and nationally competitive research projects. He is a regular Editorial Board Member and Guest Editor of international journals with high impact factors. He has served as an organizer of international symposia, as well as a member of scientific committees, chair of thematic areas, and invited speaker at international conferences. He is a reviewer for a significant number of scientific journals, as well as an evaluator of national and international research proposals.

Preface to "Micro/Nano Structures and Systems: Analysis, Design, Manufacturing, and Reliability"

The subject of the book is the analysis, design, manufacturing, and reliability of micro- and nanostructures and systems, with a focus on the use of modern computational and analytical methods in conjunction with experimental techniques.

The book is intended for researchers and engineers working in the field of micro- and nanosystems, including those in aerospace, automobiles, and biosensors, as well as those studying nano- and micro-sized systems and devices, such as nanoactuators, nanoprobes, and micro/nano-electromechanical systems.

The authors of the book's chapters are experts in the field of micro- and nanosystems, and they have made significant contributions to the understanding of the physical and mechanical behavior of micro- and nanostructures. They have collaborated to create a comprehensive and valuable resource for those working in this field.

This book is the result of the collective efforts of many individuals and organizations, and the authors would like to acknowledge and express their gratitude to all those who have contributed to the success of this special issue.

In conclusion, the book *Micro/Nano Structures and Systems: Analysis, Design, Manufacturing, and Reliability* presents a valuable collection of articles for researchers and engineers working in the field of micro- and nanosystems. It aims to provide a complete understanding of the physical and mechanical behavior of micro- and nanostructures and to advance the field through the development and application of modern computational and analytical methods.

Stelios K. Georgantzinos
Editor

micromachines

Editorial

Editorial for the Special Issue on Micro/Nano Structures and Systems: Analysis, Design, Manufacturing, and Reliability

Stelios K. Georgantzinos

Laboratory for Advanced Materials, Structures and Digitalization, Department of Aerospace Science and Technology, National and Kapodistrian University of Athens, 34400 Psachna, Greece; sgeor@uoa.gr

Abstract: The advancement of fundamental sciences in recent decades has led to an increased focus on the prediction of phenomena occurring at the micro and nano scales. Micro- and nanostructures have a wide range of applications in various fields, such as aerospace and automobiles, and are widely used in nano- and micro-sized systems and devices, such as biosensors, nanoactuators, and nanoprobes. The design of these structures relies on a complete understanding of their physical and mechanical behaviors. Mechanics plays a crucial role at the micro- and nanoscales, from the generation of nanostructures to the properties of nanocomposite materials and the manufacturing and design of machines, structures, sensors, actuators, fluidics, and more. This Special Issue aims to bring together high-quality papers that advance the field of micro- and nanostructures and systems through the use of modern computational and analytical methods, in conjunction with experimental techniques, for their analysis, design, manufacture, maintenance, quality, and reliability.

Keywords: microstructures; nanostructures; mechanics; additive manufacturing; quality; design; analysis; finite element analysis

Citation: Georgantzinos, S.K. Editorial for the Special Issue on Micro/Nano Structures and Systems: Analysis, Design, Manufacturing, and Reliability. *Micromachines* **2023**, *14*, 253. https://doi.org/10.3390/mi14020253

Received: 16 January 2023
Accepted: 18 January 2023
Published: 19 January 2023

Micro- and nanostructures and systems have become increasingly vital in various fields, such as aerospace, automobiles, and biomedical engineering. These structures and systems, which are typically on the scale of micro/nanometers, possess unique mechanical and physical properties that make them valuable for a broad range of applications, such as biosensors, nanoactuators, nanoprobes, and micro/nano-electromechanical systems. However, the design, manufacturing, and reliability of these structures and systems necessitate a comprehensive understanding of their behavior at the micro- and nanoscales. Therefore, the analysis of micro- and nanostructures and systems is an active field of research, with the objective of developing reliable methods for the prediction and control of phenomena at these scales.

The residual stress is defined as the presence of stress in a thin-film material layer without any externally applied forces. It can be compressive or tensile, vary over extremely large ranges of values, and even exhibit changes in the sign of the stress state. The residual stress is highly dependent on a number of factors, including processing conditions used during the deposition, type of material system (thin-films and substrate materials), and other processing steps performed after the thin-film layer has been deposited, particularly those involving exposure to elevated temperatures. The origins of residual stress can involve several complex and interrelated factors. Therefore, there is still no generally applicable theory to predict residual stresses in thin films. This lack of information can make device design more time consuming, expensive, and risky. Huff [1] reviewed the topic of residual stresses in deposited thin-film material layers and their effect on device behavior in micro- and nano-system development. The review discusses the origins of residual stress, its impact on device behavior, and methods to measure, control, and mitigate the impact of residual stress in micro- and nano-system device design and fabrication.

The fabrication of three-dimensional (3D) nanostructures has gained considerable attention in various fields, such as physics, chemistry, engineering sciences, and biology, due to their superior functionalities in comparison to planar nanostructures. However, the fabrication of 3D nanostructures remains a challenging task. To further advance their applications in commercial devices, there is a need for reliable fabrication methods, improved control, and enhanced ability to integrate multiple functions. Geng et al. [2] reviewed in depth a powerful method for realizing 3D nanostructures through atomic layer deposition-assisted 3D assembly using various sacrificial templates. The aim is to provide a comprehensive understanding of the use of atomic layer assembly (ALA) in a variety of sacrificial templates for 3D nanostructures and to highlight recent advancements. The ultimate goal is to unlock the full potential of this method for use in nanodevice applications.

Research and development of additive manufacturing (AM) technology has been ongoing for nearly three decades, and microscale AM has emerged as one of the most rapidly growing areas within the field. Significant progress has been made in the development and commercialization of new and innovative microscale AM processes, as well as in the application of these processes in a wide range of fields. However, there are still major challenges in terms of design, materials, processes, and ability to fabricate true three-dimensional structures and systems at a microscale. For example, microscale AM fabrication technologies are subjected to certain limitations and constraints due to their small scale, which may require the use of specialized design methodologies to overcome them. Rogkas et al. [3] reviewed the main processes, materials, and applications of current microscale AM technology. The aim is to identify future research needs for this technology and to discuss the need for the introduction of a design methodology. This study focuses on the design aspects, highlighting the advantages and limitations of AM at the microscale, as well as the selection of processes and materials.

Despite being a promising technology, fifth-generation (5G) wireless communication has yet to be fully implemented. To achieve the goals of 5G standards, such as higher data rates and ultrahigh-definition video streaming, the use of the millimeter wave (mmWave) band is crucial. However, the mmWave spectrum poses several challenges, including high connection losses, short wavelength, and restricted bandwidth, as well as path-loss challenges. To overcome these challenges, an antenna with wide bandwidth, high gain, narrow steerable beam, high isolation, low side-lobe levels, and multiband features is required. To achieve this, researchers have employed various strategies and techniques in the traditional antenna design process to improve performance in terms of bandwidth, gain, and efficiency, and to reduce mutual coupling between closely co-located antenna elements in MIMO and arrays. The latest state-of-the-art techniques, such as metamaterials, parasitic patches, hybrid feeding, EBG structure, and the impact of slots with different geometrical shapes in the radiator, are discussed by Ahmad et al. [4]. This study briefly reviews mutual coupling reduction techniques and focuses on the role of reconfigurability. Finally, it reviews the challenges in the field of antenna design and potential solutions to solve these challenges.

Deployable structures have the ability to significantly alter their geometric shapes by switching between different lattice configurations. One way to achieve this is by using compliant mechanisms as the lattice units, which can help prevent wear and friction among multi-part mechanisms. Liu and Hao [5] introduced two unique deployable structures that are based on a programmable compliant bistable lattice. Several novel parameters are introduced in the bistable mechanism to better control its behavior. By adjusting these parameters, the programmable bistable lattice can be optimized for specific goals, such as a larger deformation range or higher stability. The first structure is designed for 1D deployable movement and consists of multiple series-connected bistable lattices. The second structure, a cylindrical deployable mechanism, is designed based on the curved double tensural bistable lattice in order to explore the 3D bistable characteristics. This study mainly focuses on four types of bistable mechanisms, which are obtained by dividing the long segment of traditional compliant bistable mechanisms into two equal parts and setting a series of angle data to them. The results of the experiments and the FEA simulations confirm the feasibility of these compliant deployable structures.

Zhang et al. [6] presented a new refractive index sensor structure, which consists of a metal-insulator-metal (MIM) waveguide with two rectangular baffles and a U-shaped Ring Resonator (USRR). The sensor's transmission characteristics were theoretically investigated using the finite element method. The simulation results indicate that Fano resonance is a sharp asymmetric resonance generated by the interaction between a discrete narrow-band mode and a successive wide-band mode. This study further explored the formation of broadband and narrowband, and identified key factors that affect the sensor's performance. The best sensitivity of this refractive-index sensor is 2020 nm/RIU and the figure of merit (FOM) is 53.16. The proposed sensor is promising for use in nanophotonic sensing applications.

Cesmeci et al. [7] evaluated the performance of five different magnetorheological micropump designs, including two of their proposed designs and three from existing designs found in the literature. Comparisons were made using physics-based simulations, utilizing the fully coupled magneto-solid-fluid interaction simulations in the COMSOL Multiphysics software. To ensure a fair and meaningful comparison, all designs were given the same material and geometric properties, and the simulations were run for one complete pumping cycle. The results revealed that the proposed flap and duckbill valve models were able to pump 1.09 µL and 1.16 µL, respectively, in 1 s, which was a higher output than the other existing micropump models. Furthermore, when the magnetic flux density was at its maximum at 0.5 s, the flap and duckbill valve models could pump almost twice as much fluid as some of the existing valve models. The proposed models also had response times that were nearly five times faster than some of the existing models. Finally, the proposed micropump models demonstrated improved performance compared to existing designs, with higher net fluid volume output, low leakage during the contraction and expansion phase, and faster response times. These findings have significant implications for a wide range of applications, including insulin dosing systems for T1D patients, artificial organs, organ-on-chip applications, and micro-cooling systems.

Zhang et al. [8] investigated the changes in residual stress that occur during the heating and solidification of SiCp/Al composites. A one-way Fluid-Structure Interaction (FSI) model was developed to simulate the solidification process of molten material. By using process parameters, the model was able to predict the temperature distribution, liquid- and solid-state material transformation, and residual stress. Additionally, this study analyzed the crack that initiates from the thermal stress in the recast layer and proposed a mathematical model for the crack tip stress. The results revealed a wide range of residual stresses, ranging from 44 MPa to 404 MPa. The model was validated by comparing its results to the experimental data from three points on the surface layer.

Ye et al. [9] presented a study on the use of deep-narrow grooves (DNGs) of nickel-based alloy GH4169 in the aerospace industry. The electrochemical milling (EC-milling) process is a popular method for manufacturing special structures, such as DNGs, as it is highly efficient and does not result in residual stresses, burrs, or tool wear. However, the current EC-milling process has the disadvantage of poor removal of electrolytic by-products in the inter-electrode gap (IEG), which affects the machining accuracy and surface quality. To address this issue, a novel tube tool with a wedged end face was designed to create a pulsating flow field in the IEG, which enhances the removal of electrolytic by-products and improves the machining quality of DNGs. The flow field simulations and the experimental results showed that the tube tool with a wedged angle of 40 degrees was most suitable for the EC-milling process and was able to produce DNGs with a width of 1.49 mm $\pm$ 0.04 mm, a taper of $1.53^\circ \pm 0.46^\circ$, and a surface roughness of 1.04 µm, along with a milling rate of 0.42 mm/min. Additionally, increasing the spindle speed and the feed rate further improved the machining quality of DNGs. This study concludes that the use of the novel tube tool with a wedged end face can significantly improve the machining quality of DNGs in the aerospace industry.

To understand the impact of thermodynamic conditions on hydromechanical cavitation, a modified Singhal cavitation model was developed by Wu et al. [10] that considers thermodynamic effects. Using this model, numerical simulations were performed on a full flow field of an automotive electronic water pump at different temperatures (25 °C, 50 °C, and 70 °C). The results revealed that the trend of the simulation and experimental values was consistent across all flow rates and met the requirements for cavitation analysis. As the temperature increased, the low-pressure area inside the impeller of the electronic pump decreased, resulting in a decrease in NPSHr and an increase in cavitation resistance. Additionally, as cavitation developed, the maximum pressure pulsation amplitude in the impeller channel gradually increased, leading to increased vibration in the pump.

Lee et al. [11] demonstrated a flip-chip μ-bump bonding technology for a millimeter-wave wireless communication application that uses indium phosphide (InP) and silicon carbide (SiC) substrates. The proposed process includes a SiO_2-based dielectric passivation process, a sputtering-based pad metallization process, an electroplating bump process for creating a flat-top μ-bump shape, a dicing process without peeling off the dielectric layer, and a SnAg-to-Au solder bonding process. This process enables the fabrication of 10 mm long InP-to-SiC coplanar waveguide lines with 10 daisy chains interconnected by one hundred μ-bumps. The performance of the CPW lines is uniform with an insertion loss deviation within ±10% along with an average insertion loss of 0.25 dB/mm, while achieving return losses of more than 15 dB at a frequency of 30 GHz. Additionally, a resonant tunneling diode device was fabricated and its DC and RF characteristics were investigated.

Liu and Hao [12] proposed a new type of linear guiding mechanism to address the common issues associated with traditional slide rail guides. The proposed mechanism utilizes compliant flexible members to transfer motion, force, and energy and is designed in a cylindrical shape, with a central platform that moves along its axis. The mechanism is comprised of several in-parallel curved compound double parallelogram mechanisms (CDPMs) connected by decoupling parallelogram mechanisms. Nonlinear finite element analysis (FEA) was used to analyze the stiffness of the mechanism, and it was found that the decoupling mechanisms can significantly improve stiffness in the undesired movement directions while maintaining original stiffness along the axis of motion. The proposed mechanism was also tested using a 3D printed prototype, and the results showed good agreement with the FEA analysis with a maximum error of 9.76%.

Bhoi et al. [13] evaluated the effects of different parameters on commonly used coating materials for micro-milling applications on high-speed steel substrates. Four different coating materials were chosen: titanium nitride (TiN), diamond-like carbon (DLC), aluminum titanium nitride (AlTiN), and titanium silicon nitride (TiSiN). A 3D finite element model was created in Abaqus to analyze the Hertzian normal stress when a 4 N normal load was applied using a rigid ball with a radius of 200 μm. The maximum normal stress found in the model was 12,109 MPa, with a deviation of 2.63% from the analytical results. The results indicated that the TiSiN coating with a thickness of 3 microns is the most optimal for micro-milling applications as it has the least plastic equivalent strain in the substrate.

Tseni et al. [14] presented a novel technique for optimizing inspection, warranty, and maintenance policies for micromachines that experience wear degradation. Their model takes into account the costs of quality control, warranty periods, and maintenance, and determines the optimal combination of these factors to improve product reliability and quality. The proposed optimization tool provides a comprehensive solution for determining the best approach to the maintenance and quality management of micromachines. To the best of the authors' knowledge, this is the first time a warranty period has been incorporated into an optimization model for micromachines in the open literature, and it can bring significant benefits to their quality promotion strategy.

In recent years, micro-annular beams have become more widely used, expanding the possibilities for laser processing. However, the current methods for generating these beams have limitations, such as a spot of energy at the center of the beam. Sun et al. [15] presented a new method for generating an annular beam using a Fresnel zone plate with an annular structure, which was machined using a femtosecond laser. The results showed that by focusing the beam, an annular beam without a spot in the center can be obtained, and the radius and focal length of the annular beam can be easily adjusted. Additionally, by using two annular Fresnel zone plates that are concentrically connected, a concentric double-ring beam in the same focal plane can be obtained. The simulation and experimental results were consistent, demonstrating the potential of this method for applications requiring nontraditionally shaped laser beams.

In conclusion, the analysis and fabrication of micro- and nanostructures and systems is an active field of research, with a focus on developing reliable methods for the prediction and control of various phenomena at these scales. The residual stress, 3D nanostructures, and microscale additive manufacturing are some examples of the challenges that need to be addressed in order to advance the applications of micro- and nanostructures and systems in various fields, such as aerospace, automobiles, and biomedical engineering. The studies discussed in this editorial have provided valuable insights and perspectives on the current state of the field and have highlighted the need for further research in order to unlock the full potential of micro- and nanostructures and systems for commercial use.

Conflicts of Interest: The author declares no conflict of interest.

References

1. Huff, M. Review Paper: Residual Stresses in Deposited Thin-Film Material Layers for Micro- and Nano-Systems Manufacturing. *Micromachines* **2022**, *13*, 2084. [CrossRef] [PubMed]
2. Geng, G.; Zhang, Z.; Li, C.; Pan, R.; Li, Y.; Yang, H.; Li, J. Atomic Layer Assembly Based on Sacrificial Templates for 3D Nanofabrication. *Micromachines* **2022**, *13*, 856. [CrossRef] [PubMed]
3. Rogkas, N.; Vakouftsis, C.; Spitas, V.; Lagaros, N.D.; Georgantzinos, S.K. Design Aspects of Additive Manufacturing at Microscale: A Review. *Micromachines* **2022**, *13*, 775. [CrossRef] [PubMed]
4. Ahmad, I.; Tan, W.; Ali, Q.; Sun, H. Latest Performance Improvement Strategies and Techniques Used in 5G Antenna Designing Technology, a Comprehensive Study. *Micromachines* **2022**, *13*, 717. [CrossRef] [PubMed]
5. Liu, T.; Hao, G. Design of Deployable Structures by Using Bistable Compliant Mechanisms. *Micromachines* **2022**, *13*, 651. [CrossRef] [PubMed]
6. Zhang, X.; Yan, S.; Liu, J.; Ren, Y.; Zhang, Y.; Shen, L. Refractive Index Sensor Based on a Metal-Insulator-Metal Bus Waveguide Coupled with a U-Shaped Ring Resonator. *Micromachines* **2022**, *13*, 750. [CrossRef] [PubMed]
7. Cesmeci, S.; Hassan, R.; Baniasadi, M. A Comparative Evaluation of Magnetorheological Micropump Designs. *Micromachines* **2022**, *13*, 764. [CrossRef] [PubMed]
8. Zhang, W.; Chang, H.; Liu, Y. Study on Solidification Process and Residual Stress of SiCp/Al Composites in EDM. *Micromachines* **2022**, *13*, 972. [CrossRef] [PubMed]
9. Ye, Z.; Qiu, G.; Chen, X. Electrochemical Milling of Deep-Narrow Grooves on GH4169 Alloy Using Tube Electrode with Wedged End Face. *Micromachines* **2022**, *13*, 1051. [CrossRef] [PubMed]
10. Wu, K.; Ali, A.; Feng, C.; Si, Q.; Chen, Q.; Shen, C. Numerical Study on the Cavitation Characteristics of Micro Automotive Electronic Pumps under Thermodynamic Effect. *Micromachines* **2022**, *13*, 1063. [CrossRef] [PubMed]
11. Lee, J.; Lee, J.Y.; Song, J.; Sim, G.; Ko, H.; Kong, S.H. Implementation of Flip-Chip Microbump Bonding between InP and SiC Substrates for Millimeter-Wave Applications. *Micromachines* **2022**, *13*, 1072. [CrossRef] [PubMed]
12. Liu, T.; Hao, G. Design of a Cylindrical Compliant Linear Guide with Decoupling Parallelogram Mechanisms. *Micromachines* **2022**, *13*, 1275. [CrossRef] [PubMed]
13. Bhoi, S.; Kumar, A.; Prasad, A.; Meena, C.S.; Sarkar, R.B.; Mahto, B.; Ghosh, A. Performance Evaluation of Different Coating Materials in Delamination for Micro-Milling Applications on High-Speed Steel Substrate. *Micromachines* **2022**, *13*, 1277. [CrossRef] [PubMed]
14. Tseni, A.D.; Sotiropoulos, P.; Georgantzinos, S.K. Optimization of Quality, Reliability, and Warranty Policies for Micromachines under Wear Degradation. *Micromachines* **2022**, *13*, 1899. [CrossRef] [PubMed]
15. Sun, X.; Zhou, F.; Duan, L. An Annular Fresnel Zone Plate without Central Spots Fabricated by Femtosecond Laser Direct Writing. *Micromachines* **2022**, *13*, 1285. [CrossRef] [PubMed]

micromachines

MDPI

Review

Review Paper: Residual Stresses in Deposited Thin-Film Material Layers for Micro- and Nano-Systems Manufacturing

Michael Huff

Founder and Director of the MEMS and Nanotechnology Exchange, Corporation for National Research Initiatives, Reston, VA 20191, USA; mhuff@mems-exchange.org

Abstract: This review paper covers a topic of significant importance in micro- and nano-systems development and manufacturing, specifically the residual stresses in deposited thin-film material layers and methods to control or mitigate their impact on device behavior. A residual stress is defined as the presence of a state of stress in a thin-film material layer without any externally applied forces wherein the residual stress can be compressive or tensile. While many material properties of deposited thin-film layers are dependent on the specific processing conditions, the residual stress often exhibits the most variability. It is not uncommon for residual stresses in deposited thin-film layers to vary over extremely large ranges of values (100% percent or more) and even exhibit changes in the sign of the stress state. Residual stresses in deposited layers are known to be highly dependent on a number of factors including: processing conditions used during the deposition; type of material system (thin-films and substrate materials); and other processing steps performed after the thin-film layer has been deposited, particularly those involving exposure to elevated temperatures. The origins of residual stress can involve a number of complex and interrelated factors. As a consequence, there is still no generally applicable theory to predict residual stresses in thin-films. Hence, device designers usually do not have sufficient information about the residual stresses values when they perform the device design. Obviously, this is a far less than ideal situation. The impact of this is micro- and nano-systems device development takes longer, is considerably more expensive, and presents higher risk levels. The outline of this paper is as follows: a discussion of the origins of residual stresses in deposited thin-film layers is given, followed by an example demonstrating the impact on device behavior. This is followed by a review of thin-film deposition methods outlining the process parameters known to affect the resultant residual stress in the deposited layers. Then, a review of the reported methods used to measure residual stresses in thin-films are described. A review of some of the literature to illustrate the level of variations in residual stresses depending on processing conditions is then provided. Methods which can be used to control the stresses and mitigate the impact of residual stresses in micro- and nano-systems device design and fabrication are then covered, followed by some recent development of interest.

Keywords: material properties; residual stress; test structures; stress gradients; thin-film material properties; micro- and nano-systems (MNS) fabrication and manufacturing

Citation: Huff, M. Review Paper: Residual Stresses in Deposited Thin-Film Material Layers for Micro- and Nano-Systems Manufacturing. *Micromachines* **2022**, *13*, 2084. https://doi.org/10.3390/mi13122084

Academic Editor: Stelios K. Georgantzinos

Received: 8 October 2022
Accepted: 15 November 2022
Published: 26 November 2022

1. Introduction

The material property of deposited thin-film layers used in the manufacturing of micro- and nano-systems (MNS) (The abbreviation "MNS" is used in this text for "micro- and nano-systems.") often having the considerable interest is the residual stress [1]. There a several reasons for this, including: many MNS devices employ one or more thin-film material layers as mechanically or electro-mechanically functional layers in the design; the performance of MNS devices having mechanical or electro-mechanical functionality can be significantly impacted by the values of the residual stresses in these layers; the behavior of purely electronic devices are also impacted by residual stresses, increasingly so as the device critical dimensions are scaled downwards; the residual stresses in thin-film layers can vary over very large ranges of values depending on the specific processing conditions used during deposition; and many MNS devices employ stacks of thin-film layers thereby involving complex materials systems wherein each of the materials has a different residual stress value [2,3].

The impact of residual stresses on MNS device behavior, performance, workability and reliability can be very large. For example, a mechanically compliant element of a MNS device behaves differently if a significant residual stress is present in the material layer(s) composing the device as shown in a later section. A tensile residual stress results in the element exhibiting a higher mechanical stiffness than otherwise, while a compressive residual stress will lower the stiffness and can also result in Euler buckling of the element. Residual stresses also impact the behavior of electronic devices [4]. A residual stress can alter the crystal lattice of the semiconductor material thereby changing the electronic energy bands and affecting the electron and hole mobilities. At worst, if the residual stresses in thin-film layers are sufficient large, the layer can crack and/or delaminate from the substrate surface thereby rendering the device useless [5]. Residual stresses can also decrease the reliability of micro- and nano-systems [6]. Not all of the effects of residual stresses are negative however; in some circumstances it may be desirable to tailor the residual stress in an important material layer of a device. In any case, it is critically important for MNS designers and manufacturers to be able to evaluate the presence of residual stresses in the thin-film layers used in their device designs and have tools for controlling these stresses and/or mitigating negative impact of these stresses on their device behavior.

Residual stresses can also impact the manufacturing processes of MNS devices. For example, the presence of a residual stress in a thin-film on one side of the substrate can result in significant bowing of the substrate thereby exhibiting either a concave or convex shape [3,5]. This can result in the substrate not being completely flat during subsequent processing steps thereby interfering with the resultant quality of processing steps performed after the thin-film layer has been deposited. For example, a substrate having a bow will not have the correct focal distance across the entire substrate surface during photolithography and can result in device features being out-of-focus and having reduced resolution [7]. Excessive bowing of the substrate can cause the gas flow dynamics across the substrate to be altered sufficiently that it impacts the uniformity of a deposition or etching process [3]. If the residual stresses in the thin-films are sufficiently large it can even cause the thin-films to crack or de-laminate or even result in the fracture of the substrate rendering every die useless [5,6].

2. Origins of Residual Stresses

Residual stresses are defined as the existence of a state of stress in a material in the absence of any externally applied forces [5,8,9]. Residual stresses are believed to be the result of one or more causes, and in most situations, there are a multiplicity of complex and interrelated causes that determine the resultant residual stress in a deposited thin-film layer [5,9]. It is known that the type of process used for deposition (e.g., chemical-vapor deposition, physical vapor deposition, etc.); the specific process parameters (e.g., chemistry, temperature, plasma power, etc.); the material type deposited; the substrate type the layer is deposited onto; and the thickness of the deposited layer all have an impact on the residual stress since there is a wealth of experimental evidence collected over several decades of semiconductor manufacturing showing how much the resultant residual stresses vary based on these effects [2,3,5,6,8,9].

Explaining residual stresses is still a matter of active research and there exists no general theory to enable the prediction of the resultant residual stress in most situations. This section provides a brief overview of the present knowledge of the origins of residual stress. There are several excellent reviews, which go into more detail about the origins of residual stresses that the interested reader is encouraged to review [5,9,10].

Residual stresses are commonly separated into two different components: extrinsic residual stresses due exposure to an external environmental media (e.g., temperature changes, chemical reactions, moisture absorption, etc.) and intrinsic residual stresses that are caused by the thin-film layer's internal structural properties that are a function of the deposition conditions [9]. Each of these is explained in more detail below.

The most common cause of extrinsic residual stress is the result of unequal thermal expansion coefficients (TCE) in different materials [5,9]. Most materials have differences in their respective TCE values. Differences in the TCEs can exist between different material layers in a thin-film layer stack as well as one or more material layers and the substrate. These types of residual stresses are commonly observed in thin-film material deposition processes performed at elevated temperatures, such as chemical vapor depositions (CVD) described below.

For example, if a thin-film material layer having a different thermal expansion coefficient is deposited at an elevated temperature onto a substrate, the different thermal expansion coefficients of the material layer and substrate will result in either the layer or substrate contracting more than the other after the deposition is completed and the substrate and material layer are cooled to room temperature [5,9].

This is illustrated in Figure 1 wherein a substrate is shown in Figure 1a and undergoes a heating during deposition, which causes the substrate to expand as shown by the dotted lines outside of the unheated boundaries of the substrate in Figure 1b. As shown in Figure 1c, a thin-film layer is then deposited onto the top surface of the substrate. Importantly, the deposition occurs when the substrate and deposited thin-film layer are both at an elevated temperature. Compatibility requires the substrate and thin-film layer have the same length. After deposition, the substrate and thin-film layer are cooled to room temperature. Therefore, once the substrate and thin-film obtain an equilibrium state at room temperature, the thin-film having a higher TCE than the substrate will attempt to contract more than the substrate thereby resulting in the situation shown in Figure 1d wherein the curvature of the substrate with thin-film layer on the top is concave-shaped. In this situation, the thin-film layer will be in a state of tensile residual stress and is given a positive sign (+) of stress by convention. If the thin-film layer had a TCE less than the substrate, it would contract less upon cooling and result in the substrate with the thin-film layer having the opposite curvature, that is, a convex shape similar to a dome. In this case, the residual stress would be compressive and given a negative sign (−) by convention.

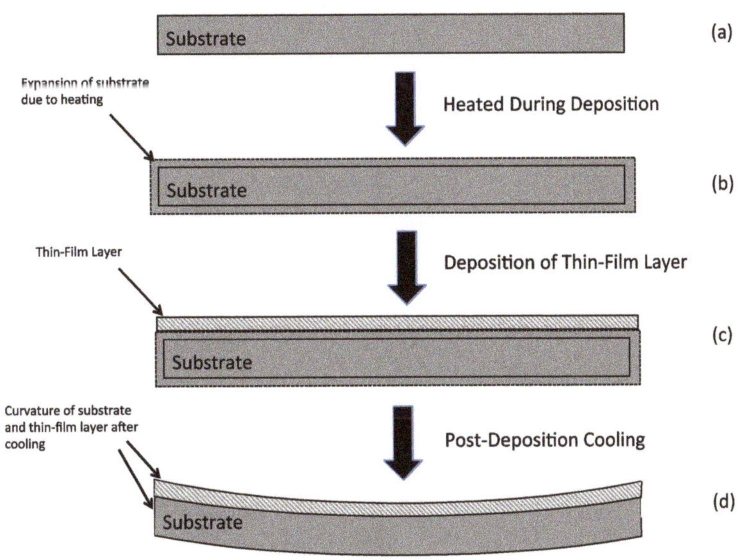

Figure 1. An illustration of residual stresses in thin-film layers deposited onto substrates. In this case, the thin film material layer would be in a state of tensile stress (with a positive sign).

The causes of intrinsic residual stresses are more complicated and are due to the imperfect structural configurations within the deposited material thin-film layer. The structural configurations causing intrinsic residual stresses are often categorized according to their size scale: the first are at the microstructural level and include imperfections between the boundaries of grains and grain columns, voids between grains, and other similar micro-scale defects in the thin-film; and the second are at the atomic level and include atomic point defects, misfits, dislocations, impurity incorporation, etc. All of these structural issues are the result of non-equilibrium growth conditions and/or the incorporation of impurities into the growing thin-film layer. Atomic level defects in the lattice structure from its equilibrium state and imperfections in the microstructure can both cause elastic deformations of the thin-film material layer thereby resulting in an intrinsic residual stress [9,10].

One atomic level cause for intrinsic residual stresses results from lattice mismatches between the substrate and the thin-film material. This is illustrated in Figure 2 wherein a single crystal material layer is epitaxially grown on the surface of a crystalline substrate that is of a different material. The deposited atoms of the thin-film layer chemically bond to the exposed atoms of the substrate. However, the deposited materials layer has a different lattice constant from the substrate and therefore the bonds between the atoms are strained. This causes a compressive residual stress in the deposited thin-film layer [5,9,10]. This type of residual stress occurs when the thin-film layer deposited onto a substrate is of a different material type. For example, a silicon thin-film layer epitaxially grown onto the surface of a clean silicon substrate would not have a lattice mismatch and therefore no intrinsic residual stress due to lattice mismatch.

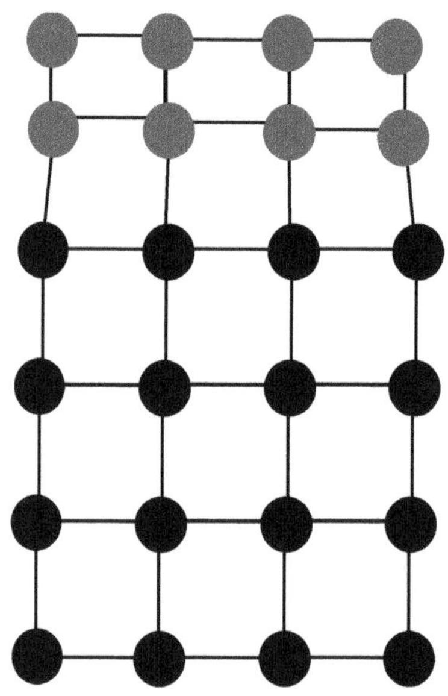

Figure 2. Illustration of a strained thin-film layer epitaxially grown on a crystalline substrate.

Another source of intrinsic residual stresses in deposited polycrystalline thin-film layers is the presence of crystalline grains. Often thin-film layers are deposited at temperatures where it is thermodynamically favorable for the material layer to form micro-crystalline grain structures during growth. These grains can grow in size during the thin-film deposition processes and often exhibit irregular-shaped columnar structures with diameters that increase through the thickness of the material layer thereby resulting in a residual stress [5,9,10].

Intrinsic residual stresses in deposited thin-film layers can result from the incorporation of impurities, such as dopants, into the material layer [5,9,10]. Doping is commonly done in semiconductor thin-film layers, such as polycrystalline silicon (i.e., polysilicon), in order to make these layers sufficiently electrically conductive for useful applications. Impurities introduced into the thin-film material layer have atomic diameters that differ from the host material thereby resulting in a strain in the crystal lattice. For example, single-crystal silicon moderately doped with boron having a smaller atomic diameter causes the doped silicon to contract more compared to un-doped silicon with all else being equal. This would result in a tensile stress in the material layer.

The entrapment of gases into a thin-film layer during deposition can also be a source of intrinsic residual stresses in thin-film layers [9]. Some deposition processes are performed at pressures wherein some background gases may be incorporated into the deposited layer.

Similarly, voids present in thin-film layers and special arrangements of dislocations can also cause a built-in intrinsic stresses [5,9]. The absorption or desorption of moisture by the thin-film layer during deposition can lead to an intrinsic residual stress [9].

Atomic peening whereby ion bombardment onto a thin-film layer, such as during a sputter deposition, can impart sufficient kinetic energy to the thin-film layer surface atoms to result in interstitials causing a built-in state of compressive intrinsic residual stresses [5,9].

If a material layer undergoes a plastic deformation during processing, then this can also result in a residual stress in the layer [11]. Electromigration or a phase transitions can cause a residual stress as well [9].

The thickness of the deposited layer also has a significant impact on the residual stresses. For example, a large number of deposited types of thin-film material layers exhibit a changing residual stress as the thickness increases. Many materials show a slightly compressive stress initially when the thickness is tens of Angstroms, then changing to a net tensile residual stress as the layer thickness increases to around 100 Angstroms, and then changing back to a compressive residual stress when the layer thickness is several hundreds of Angstroms, and finally plateauing to a stable compressive stress as the layer thickness continues to increase [9]. The explanations for these changes are attributed the mechanisms by which the adatoms configure onto the substrate surfaces over time [9].

In short, there are a large number of causes for the resultant value of the residual stress in a thin-film layer. As a practical matter, it is typically not possible to disentangle the causes as well as their relative contributions to the total residual stress in a thin-film [5,9]. Fortunately, separating the contributing factors of the residual stress is usually not a concern to MNS developers.

3. Impact of Residual Stresses

It was stated above that the effects of residual stresses in thin-film layers could significantly impact the behavior of MNS devices. Obviously, cracking, buckling and delamination represent catastrophic events, but the impact of residual stresses in thin-film material layers in the MNS device behavior is often far subtler. A representative example provided.

Consider a conventional simple beam resonator microsystems device (Figure 3). This is a widely used MNS device technology and consists of a double-ended clamped beam that has a resonant frequency given by [12]:

$$f_{r,i} = \frac{i^2\pi}{2L^2}\left(\frac{EI}{\rho A}\right)^{1/2}\left(1 + \frac{SL^2}{iEI\pi^2}\right)^{1/2}, \quad (1)$$

where E is the Young's modulus of the beam material, I is moment of inertial, A is the cross sectional area of the beam (beam thickness, t, times beam width, w), ρ is the density of the beam material, L is the beam length, i is the mode index of the resonance and is an integer number (shown as i = 1 in figure), and S is the tensile force on the beam due to the residual stress, σ, or S = σ A.

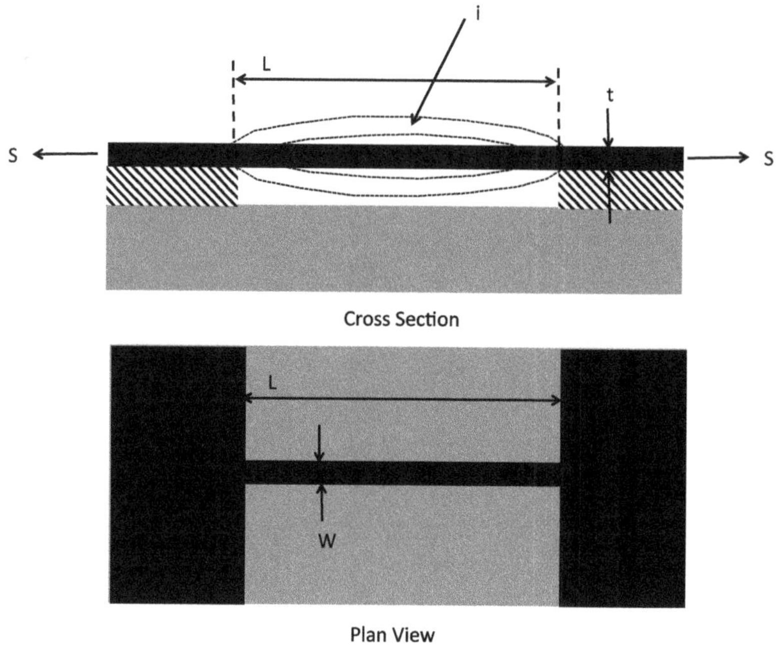

Figure 3. Illustration of Beam Resonator.

Assume the beam is made from polysilicon having a Young's modulus of 160 GPa [13] and a density of 2330 kg/m^3. The beam has a moment of inertia given by:

$$I = \frac{1}{12}w^4, \tag{2}$$

where it is assumed the beam has a thickness equal to the width that is equal to w. Further assume that the beam has a thickness and width of 10 microns, and therefore A = 1.0×10^{-10} m^2 and I = 8.3×10^{-23} m^4. Additionally, assume the length of the beam is 100 microns. Substituting these values into Equation (1) above and simplifying, then the following can be written for the resonant frequency:

$$f_{r,i} = \sqrt{1.4 \times 10^{12} + 10774\sigma}. \tag{3}$$

Figure 4 illustrates how the resonant frequency varies as the tensile residual stress in the beam increases from 0 Pa to 200 MPa. It is worth noting that a residual stress of 200 MPa is not an unusually high value of residual stress in a thin-film layer.

As can be seen from the results of these calculations, the resonant frequency of the beam resonator varies quite significantly as the residual stress increases thereby demonstrating the impact residual stress has on the resonant frequency. In short, the effect of a tensile residual stress in the beam is to make the mechanical stiffness of the beam appear to be higher than if no residual stress were present.

Similar calculations can be performed for the impact of compressive residual stresses on the beam resonator where it can be shown that a compressive residual stress will have the effect of decreasing the resonant frequency. However, the impact of a compressive residual stress must be further evaluated in terms of the onset of Euler buckling. Specifically, a compressive residual stress of sufficient magnitude will result in the beam to buckle and once buckling has occurred, it greatly changes the resonant frequency behavior of the beam structure. In general, buckling is avoided in resonator device designs.

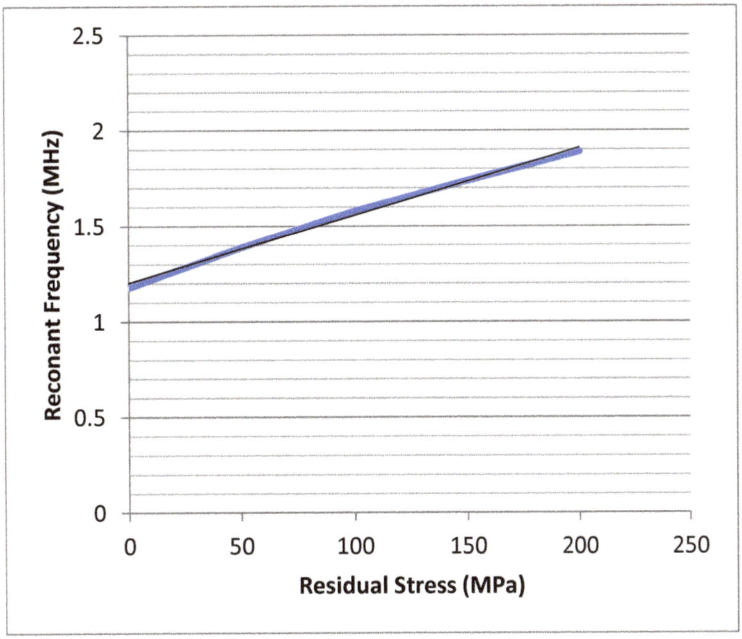

Figure 4. Plot of Resonant Frequency (MHz) verses Residual Stress (MPa) in a Beam Resonator.

4. Impact of Thin-Film Deposition Methods on Residual Stress

Deposition of thin-film material layers is a fundamental capability used in the manufacturing of MNS devices. Some of the many uses of these layers include: implementation of electronic, photonic, electro-mechanical devices, microfluidic, and many other device types. More specifically, in MNS devices thin-films are often used as mechanically functional device layers, as sacrificial layers for formation of free-standing, suspended structures and mechanically movable elements, for making electrical interconnections between devices, as an electrical insulator material, and many others [1–3,14–18].

There are a variety of methods used to deposit thin-films in microsystems fabrication, including: chemical vapor deposition (CVD); physical vapor deposition (PVD); atomic layer deposition (ALD); spin-casting; and electrochemical deposition (It could be asserted that spray resist coatings and thermally conductive tapes are also a form of thin-film deposition, but these are only used for short periods of time to conduct a processing step and are not left as a thin-film layer on the substrate.). This section provides a review of the most commonly used deposition methods. This section will discuss how each of the commonly used deposition processes may effect the residual stress in thin-films. Appendix A provides more information on the deposition methods.

4.1. Thermal Oxidation

Thermally grown silicon dioxides have excellent electrical properties, specifically as electrical insulating layers [14–17]. Thermal oxidation of silicon to form silicon dioxide (SiO_2) is a relatively simple process: an oxide layer is formed on the silicon surface when it is exposed to oxidizing agents and this process is accelerated if the silicon is exposed to an oxidizing agent at elevated temperatures. Deal and Grove provided an accurate analytical model to predict the oxide thickness based on the processing parameters [19].

The oxidation of silicon proceeds by a silicon dioxide layer growing on the surface of the silicon substrate with the interface advancing into the depth of the silicon material. The silicon dioxide formed on the surface has a thickness about 2.27 times that of the thickness of the consumed silicon material [15]. This volume expansion is a potential source of a residual stress in the grown SiO_2 layer and as noted in a later section, the residual stresses in thermal oxides tends to be compressive [5].

4.2. *Chemical Vapor Deposition (CVD)*

Chemical vapor deposition (CVD) is widely used in MNS manufacturing for the deposition of amorphous and polycrystalline thin-films, and under certain special conditions the deposition of single-crystal material layers [15,20]. CVD processes involve the following stages: physical transport of the chemical precursors (A chemical precursor is a chemical compound that participates in a chemical reaction which produces another chemical compound.) to the substrate surface; absorption of the precursors onto the surface; dissociation of the precursors into their chemically reactive components; migration of the chemical reactive components to chemical reaction sites; chemical reaction and the active species involved in the layer growth; and desorption of reaction byproducts from the substrate surface. CVD involves complex chemical reactions and therefore the process parameters including temperature, gas flow rates, and pressure should be accurately controlled. Most CVD processes use elevated temperatures for the deposition. Among the common material layers deposited using CVD include: doped and undoped silicon (amorphous; polycrystalline; and single-crystal); doped and undoped silicon dioxides; and silicon nitride. "In situ doped CVD" allows for the incorporation of dopants (e.g., boron, phosphorous, etc.) into the layer during growth by introducing into the process chamber specific gases that contain the desired dopant species.

There are a number of sub-categories of chemical vapor deposition including: atmospheric chemical vapor deposition (ACVD); low-pressure chemical vapor deposition (LPCVD); plasma-enhanced chemical vapor deposition (PECVD); and atomic layer deposition (ALD), which are reviewed below.

4.2.1. Atmospheric Chemical Vapor Deposition (ACVD)

Atmospheric chemical vapor deposition (ACVD) is performed at atmospheric pressure and the mass transfer-controlled region. ACVD is mostly used for the epitaxial deposition material layers such as silicon that is performed at temperature of 1000 °C or higher [21]. ACVD typically uses a single-wafer process tool configuration. The deposition rates of high-temperature ACVD can be relatively high, typically around 1 micron per minute or more. One major cause of residual stress for single-crystal depositions is any lattice mismatch that exists between the layer being deposited and the substrate and this usually occurs when the deposited layer is of a different material type from that of the substrate. Temperature is another process parameter having a major impact on the deposited material layer properties, particularly when the deposited material layer has a different TCE from that of the substrate.

4.2.2. Low-Pressure Chemical Vapor Deposition (LPCVD)

Low-pressure chemical vapor deposition (LPCVD) is a commonly used type of CVD process for the deposition of common thin-film material layers including: various doped and un-doped forms of silicon dioxide (e.g., phosphosilicate glass [PSG], borophosphosilicate glass [BPSG], low-temperature oxide [LTO], etc.); doped and undoped forms of polycrystalline silicon (polysilicon); silicon nitride; and oxy-nitride [14–17,21–24]. Residual stresses in LPCVD thin-film material layers can be large in magnitude, vary over large ranges of values and have different signs (i.e., tensile or compressive), and have many potential causes. The major process parameters used to modify the materials properties of LPCVD deposited layers is the deposition temperature, chemistry and gas pressure.

4.2.3. Plasma-Enhanced Chemical Vapor Deposition (PECVD)

Plasma-enhanced chemical vapor deposition (PECVD) is a CVD process where energy from a plasma is used to enable the process reactions to take place at lower temperatures, even 200 °C or lower [25]. This is especially useful for substrates having previously deposited materials or previously fabricated elements that are temperature sensitive. The plasma in PECVD can also be used to modify the material properties of layers during deposition [26]. The plasma has two important effects in this process: it provides a non-thermal energy to the reactant gases to allow the precursor gases to dissociate at far lower temperatures; and the free ions bombard the substrate surface to impart non-thermal energy to the absorbed adatoms allowing them to have sufficient surface mobility to find lower energy states (and thereby lower residual stress), as well as improved conformality and uniformity of the deposited thin-film.

Residual stresses in PECVD thin-films vary over large ranges and can exhibit different signs (i.e., compressive (−) or tensile (+)). The major process parameters used to modify the materials properties of PECVD deposited layers are the deposition temperatures, gas pressures, and plasma energy. As a general rule, PECVD thin-films usually have more flexibility in modifying the residual stress and other material properties compared to other CVD processes.

4.2.4. Atomic Layer Deposition (ALD)

Atomic layer deposition (ALD) uses sequentially timed chemical processes to deposit material layers. In the first cycle, the first reactant gas is introduced and reacts with the substrate surface, followed by the next cycle where the second reactant gas is introduced and reacts with the reactants on the substrate surface from the first cycle [27–30]. ALD can also be done at lower temperatures using a plasma enhanced atomic layer deposition (PEALD) [31–33]. The major process parameter to modify the material properties of layers deposited using ALD is the substrate temperature during deposition and the chemistry used for the precursor gases. For PEALD, the plasma power is another parameter that can be used to modify the material properties.

4.3. *Physical Vapor Deposition (PVD)*

Physical vapor deposition (PVD) use physical methods, such as heating or ion bombardment, to produce a vapor phase of the source material that condenses onto the substrate surface. One of the major advantages of PVD deposited materials are they can be deposited at relatively low temperatures in comparison to CVD processes. There are two main types of PVD, evaporation and sputtering.

4.3.1. Evaporation

Evaporation is a thermal process whereby a target material is heated in a high vacuum environment and the material from the target changes its phase state from a solid to a vapor, transports to the substrate surface(s), and then precipitates onto the substrate surface resulting in a deposited layer [34,35].

The major process parameters controlling the material properties of deposited layers using evaporation are the substrate heating and deposition rate.

4.3.2. Sputtering

Sputtering uses the bombardment of a target material with ions generated by a plasma to displace atoms from the target creating a vapor of the target material that condenses onto the substrate surface thereby forming a thin-film layer [36].

DC bias sputtering is used to deposit electrically conductive materials, such as metals. RF and magnetron sputtering can be used for a wider variety of materials types.

The materials properties of deposited layers using sputter deposition are affected by several process parameters, including: pressure; substrate temperature; plasma power; and RF and/or DC biasing powers. By applying a voltage bias to the substrate, ions from the plasma can be made to impact the surface of the substrate and thereby impart energy to the adatoms on the surface. This property of sputtering can be effectively used to modify the material properties of the deposited thin-film layer [8].

4.4. Spin Casting

Spin casting is the deposition of a solution onto a wafer and then spinning it at a specific speed to obtain a uniform coating. Often the wafer after the layer has been spin coated is heated to drive off any solvents used to control the viscosity of the solution and harden the deposited layer. There are several materials that are spin casted.

4.4.1. SU-8

SU-8 is a negative polarity resist sensitive to near-ultraviolet radiation [37,38]. Cross-linking requires a post-bake at temperatures ranging from 150 to 250-degree Celsius resulting in significant shrinkage and a resultant residual stress. The major parameter for impacting the resultant residual stress is the post-bake temperature.

4.4.2. PDMS

Polydimethylsiloxane (PDMS) is a viscoelastic silicon-based organic polymer used in microsystems fabrication, particularly for microfluidic applications [39–41]. PDMS is a viscoelastic material with a low modulus and typically does not exhibit any appreciable residual stresses.

4.4.3. Polyimide

Polyimide is a polymer commonly used in the microelectronics industry for packaging applications [42,43]. It is an electrical insulating material resistant to heat. Polyimides exhibit either thermoset or thermoplastic behaviors. Some of the attractive material properties of polyimide for microsystems applications include: a low mechanical stiffness; biocompatibility, chemical and biological inertness, and low cytotoxicity. Other important characteristics of polyimides include: a high glass transition temperature, high thermal and chemical stability, low dielectric constant, high mechanical strength, low moisture absorption, and high solvent resistance. The processes for use of liquid forms of polyimide are very simple. Polyimide exhibits shrinkage of between 40 to 50% during curing. Therefore, there will be a significant amount of residual stress in the deposited layers [43].

4.4.4. Sol-Gel PZT

Sol-gel PZT is a process whereby the constituents for a piezoelectric material, including lead, zirconate, and titanate (PZT) are in a polymer solution that can be spin casted onto substrates to form thin-film layers [44]. Since the spin-casted material shrinks during curing to drive off the solvents, the deposited material layers develop some amount of residual stress that depends on the thickness and the number of spin-casted layers [44].

4.5. Electrochemical Deposition

Electrochemical deposition methods are wet chemical processes, which involve the reduction of metal ions from an electrolyte solution resulting in the deposition of thin-film layer of metal atoms onto a surface [45–47]. There are two different types of electrochemical deposition processes: the first is electroplating where an electric current is passed between two electrodes placed in the electrolyte solution to supply the charges for the oxidation and reduction reactions; and, electro-less deposition wherein a reducing agent in the electrolyte solution provides the charges for these reactions. Electroless plating is rarely used due to challenges of reproducibility. A variety of metals can be electroplated including: Au; Ag; Cu; Cr; Pt; Ni; Zn; S; Cd; and Pb; as well as number of different alloys composed of metals that can be electrochemically deposited.

The material properties of electroplated layers are impacted by various process parameters including: chemistry solution used, temperature, stirring effectiveness, plating rate and electrical parameters used in plating including type of electrical waveforms used.

5. Thin-Film Residual Stress Measurement

The measurement of the residual stress is very important since, as noted above, it is very dependent on the processing conditions. Therefore, unless there are specific prior measurements taken on the material layer that was deposited with exactly the same process parameters, the residual stress will not be known. Further, the use of lookup tables and/or literature to estimate the value of the residual stresses in deposited layers have limited to no value. Consequently, MNS device developers need to perform measurements of the residual stresses for their specific processing conditions and process sequence used to implement the MNS device. However, the techniques used to measure residual stresses at the macro-scale dimensional domain are not suitable for use at the micro- or nano-scale dimensional scale and more specialized techniques for measuring the films stresses are needed. Fortunately, there are a number of methods to reasonably conduct the accurate measurement of the residual stresses during development and manufacturing. The most commonly used methods for measurement of residual stresses are discussed below and separated based on a classification scheme as follows: wafer curvature; fabricated test structures; and methods requiring the use of sophisticated instruments.

5.1. Wafer Curvature

The easiest and most widely used technique for measuring the residual stresses in thin-film layers is to determine the resultant curvature of the substrate before and after the deposition of the thin-film on one side of the substrate [3,5,9]. This scenario was portrayed in Figure 5 wherein a laser scans across the top surface of the substrate before and after the thin-film deposition.

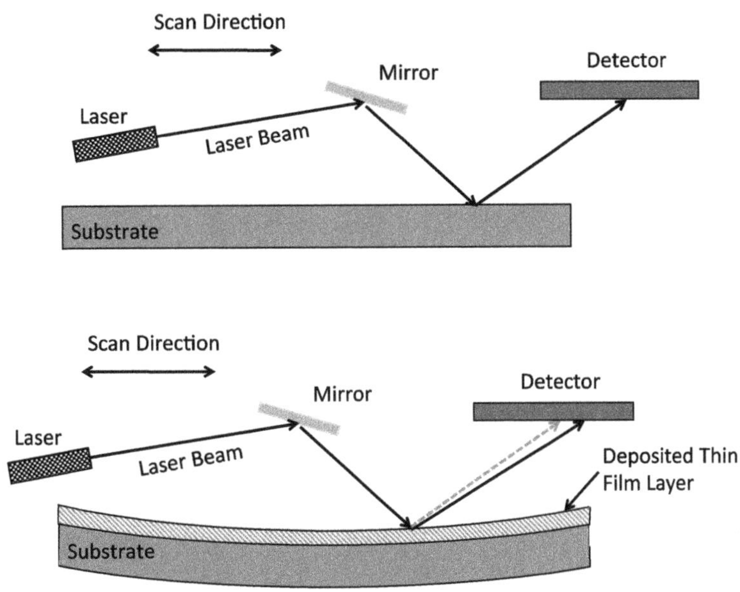

Figure 5. Illustration on the operation of the wafer curvature metrology tool.

The measurement technique shown in Figure 5 operates as follows. As shown at the top, a first scan is performed on the substrate prior to the thin-film layer deposition. This allows a measurement of any wafer bow that may be present before the thin-film deposition. Wafers may exhibit bow due to the manufacturing processes used to make the substrates or previous processing steps performed on the wafers and not compensating for the existing bow would make the measurement inaccurate. A laser scans across the surface along with a mirror and a detector. The detector is usually an electronic imaging array (i.e., CCD or CMOS) that is sensitive to the laser radiation wavelength. The laser radiation emitted from the laser impinges on the mirror at an angle of incidence and bounces off of the mirror surface at an equal angle of reflection. The reflected laser beam then strikes the substrate surface at an incident angle and reflects at an equal angle, and then impinges onto the detector array. The laser, mirrors and detector array are scanned across the substrate surface approximately along the middle of the substrate to measure the wafer radius of curvature. After the thin-film layer has been deposited a second scan is performed across the substrate as shown in the bottom of Figure 5. In this illustration, the thin-film layer has a tensile residual stress present causing the substrate to form a convex or bowl shape. As the laser scans across the surface, the reflection from the substrate surface has a different angle of reflection compared to the previous scan thereby indicating a different radius of curvature. Using this data, the residual stress of the thin-film layer is then calculated and displayed.

The radius of curvature is directly related to the residual stress by knowing the elastic mechanical properties of the substrate material and the thin-film layer using the Stoney equation, given as follows [5,9]:

$$\sigma_f = \frac{\left(E_s t_s^2\right)}{\left[6t_f(1-\nu_s)R\right]} \tag{4}$$

where the subscripts "f" and "s" refer to the thin-film and substrate, respectively, E_s is the substrate Young's modulus, t_s is the substrate thickness, t_f is the thin-film thickness, ν_s is Poisson's ratio of the substrate, and R is the radius of curvature of the substrate with the thin-film on the surface. The Stoney equation assumes the stresses in the film and substrate have isotropic elastic properties, the film thickness is uniform, the temperature is uniform, the stress is uniform throughout the film thickness, the deflection is in the small deflection regime, and there is no cracking or occurrence of other stress relief mechanisms.

Perhaps the assumption of most concern is the substrate exhibits isotropic elastic mechanical properties. In most situations, the thin-films are deposited onto semiconductor substrates, which are single crystals and therefore exhibit anisotropic mechanical properties. A modified form of the Stoney equation that incorporates the anisotropic substrate properties for a thin-film deposited onto a single-crystal silicon substrate with a <100> orientation is given as:

$$\sigma_f = \frac{\left(t_s^2\right)}{\left[6t_f\left(s_{11}^{si}-s_{12}^{si}\right)R\right]} \tag{5}$$

where s_{11}^{si} and s_{12}^{si} are values from the compliance tensor of silicon. The factor $\frac{1}{\left(s_{11}^{si}-s_{12}^{si}\right)}$ represents the biaxial modulus of <100> silicon that has a numerical value of 180 GPa. Other modifications of the Stoney equation are available for other silicon crystallographic orientations as well as for orientations of other anisotropic semiconductor materials [48].

The commercially available residual stress metrology tools perform the calculations to determine the residuals stress automatically as part of the measurement. Typically, a residual stress measurement tool will have a dynamic range of residual stresses that it can measure between 1 MPa to 4 GPa, with both a repeatability and resolution on the measurement of approximately 1 MPa. Calibration standards are available. The accuracy of these systems is typically less than 2.5% or 1 MPa depending on which is larger [49]. Most thin-film stress measurement systems have the capability for varying the temperature of the substrate over a range of −65 °C to 500 °C. This allows the residual stresses to be measured at various temperatures of operation and enables the measurement of the thermal coefficient of expansion (TCE) of thin-film layers.

While the laser scanning method is fast, simple and non-destructive, it does have some disadvantages. First, as noted above, it only measures the average stress over the substrate. That is, this method does not measure the specific values of the residual stresses at different locations across the substrate. It is known the material properties, in general, and the residual stress in particular, can vary across the substrate surface [3]. The reason for the variation in residual stress in deposited thin-films across wafers is because the processing condition parameters vary across the wafer, including: chemical reactive species concentrations; chemical reactions; gas flow rate; plasma power; and temperature variations. No processing parameter can be absolutely controlled and even with the best controls, there will be some amount of random variation. Even small variations in temperature can result in significant variations in the residual stress. Second, the method is based on the assumption the elastic properties of the thin-film are already known. The Young's modulus of thin-films is also process dependent, but usually not to the degree as the residual stress. Young's modulus can be measured separately using various techniques reported in the literature [2]. Third, although the laser scanning method includes the effects of the presence of a stress gradient through the thickness of the thin-film layer in the measured value of the residual stress, it does not allow the stress gradient to be separated out and quantified. The gradient in the residual stress can also have a significant impact on the behavior of thin-film layers and, therefore, should be determined along with the residual stress [50]. The reason for this is if there is a stress gradient in a material layer wherein the neutral axis of the layer is aligned with the mid-point in the layer thickness, the average residual stress will be measured as being zero and yet there can be a significant stress gradient.

5.2. Residual Stress Test Structures

There are a number of test structures, which can be used to measure the residual stresses at locations across the wafer. However, one issue with these methods is they require the fabrication of dedicated test structures. The implementation of test structures means the mask layout design must include dedicated test structures or dedicated substrates would have to be run through the process sequence and then tested. Consequently, there is an appreciable cost associated with the use of test structures.

Additionally, the designs of the test structures must be developed such that they are suitable for measuring the "expected" state of residual stress in the thin-film material layer. This is not a trivial task since the residual stress may not be known. In most circumstances, the MNS developer will use an estimated value of the residual stress, possibly taken from a reference source, as a starting point based on as identical of processing conditions as possible and then perform the designs of the test structures such that the dimensions are varied so as to have test structures that will enable the residual stress to be measured, or an estimate of the residual stress is available from wafer curvature measurements. With an estimate of the residual stress, the dimensions of the test structures can then be developed. Most of the test structures reported in the literature and reviewed herein are made using surface micromachining techniques wherein the thin-film material layer of interest is deposited and patterned on a sacrificial material layer, followed by the removal of the sacrificial layer to release the test structure so that the residual stress measurements can be taken. Since these are thin-film layers, the thicknesses are generally less than a few microns and the width will be several times the thickness, and the lengths will be multiples of the width. Additionally, some of the test structures are suitable for measuring compressive residual stresses, other for measuring tensile residual stresses and some for measuring both compressive and tensile stresses. Lastly, the test structures are generally useful for measuring residual stresses in the ranges from a few tens of MPa to a few GPa.

5.2.1. Buckling Beam Test Structure

One of the first reported test structures for measurement of the residual stress in thin-film layers is the buckling beam test structure shown in Figure 6 [51]. It should be noted that this test structure is only used for thin-film layers exhibiting compressive residual stresses. This method employs an array of beams fabricated from the deposited thin-film layer(s) wherein each beam has a slightly different length, with the other dimensions (i.e., beam width and beam thickness) held constant. Each beam has anchor points attached to the substrate at both ends and the beam is freely suspended between the anchors. This is commonly referred to as the "doubly clamped beam" configuration [52]. The presence of a compressive residual stress in the thin-film layer causes the beam to attempt to expand to a longer length and the beam being attached to the anchors at each end prevents any appreciable expansion. If the beam length is sufficiently long, the residual stress in the beam is relieved by the onset of buckling in the beam. The design of the beams is made so that buckling occurs perpendicular to the substrate surface.

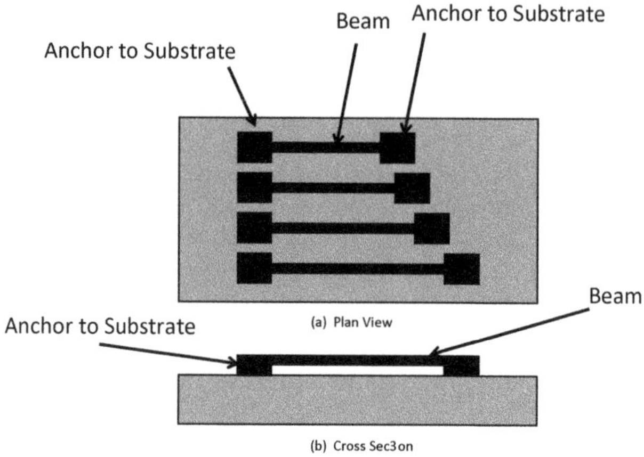

Figure 6. (**a**) Plan view of the fixed-fixed beam array having different lengths used to determine the critical length. (**b**) cross-section of the longest beam.

The residual stress is determined as follows: After the array of beams has been fabricated and released (the process of removing the sacrificial layer under the beams to make them free-standing) and the beam with a length exhibiting the first onset of buckling is observed. The length of the beam at the onset of buckling is called the "critical beam length, or L_{cr}." The critical strain, ε_{cr}, of the thin-film layer at the onset of buckling for a fixed-fixed beam is found from the following equation:

$$\epsilon_{cr} = \frac{\pi^2 t^2}{3L_{cr}^2}, \tag{6}$$

where t is the thickness of the thin-film material layer and the beam [52,53].

Once the critical strain has been calculated in the thin-film using Equation (6), the residual stress in the material layer is determined by multiplying the strain by the Young's modulus of the beam material. The modulus of the thin-film material layer must be known to determine the residual stress.

Figure 7 is a scanning electron micrograph (SEM) of an array of doubly clamped test structures made from low-pressure chemical vapor deposited (LPCVD) polysilicon. The array is composed of 10 separate mini-arrays of beams (5 on the left and 5 on the right) each composed of 10 beams of slightly increasing lengths. The length of the beams increases from the top left of the image towards the bottom left and then from the top right to the bottom right according to the overlaid numbering. As can be observed, the beams in the first (numbered "1") and second (numbered "2") mini-arrays are not buckled and the onset of buckling appears to occur at the first (top) beam in the third (numbered "3") mini-array on the left side. Using the length of the beam at the onset of buckling, Equation (3), and the modulus, allows the residual stress in the thin-film layer to be determined.

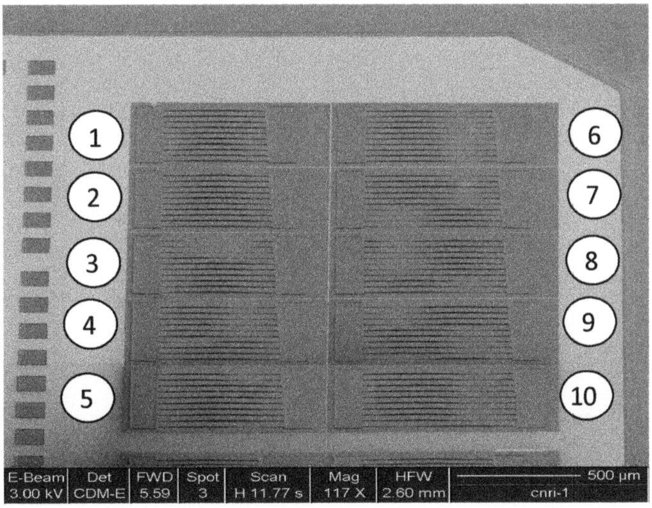

Figure 7. SEM image of an array of fixed-fixed beam test structures used to determine the compressive residual stress in thin-film layers [3].

The doubly clamped beam test structures are useful for measuring the residual stress at specific locations across the substrate, but do have several important shortcomings. First, as noted above, this method can only be used to measure compressive residual stresses. Second, it can be challenging to determine the critical length at the onset of buckling. Van Drieenhuizan [54] has shown that the amount of deflection of the center of a buckled beam can be a small percentage of the thickness of the beam. Differential interference contrast (DIC) or scanning electron microscopy (SEM) can be useful to help detecting the onset of buckling, but even this is prone to some uncertainty. Third, another common issue with this kind of test structure is the presence of stiction (Stiction is a phenomena where a free-standing thin-film layer attaches to the substrate surface. It is mainly due to Van der Waals attraction forces.) effects between the beam and the substrate surface. Stiction may be reduced by the use of a suitable anti-stiction surface coating after the release of the beams [55]. Fourth, the method requires a relatively large array of test structures in order to determine the buckling threshold and this will consume a considerable amount of the substrate area. Fifth, the boundary conditions of the anchors are very important to the calculation of the strain at the buckling onset. If the anchors do not behave as a fixed-fixed type of support, this will make the determination of the residual stress inaccurate [54].

5.2.2. Guckel Ring Test Structures

The Guckel ring test structure (Figure 8) is a buckling type of test structure developed for determining the magnitude of a tensile residual stress [56]. This structure uses a free-standing ring that is clamped on two sides of the ring and with a central beam across the diameter of the ring located 90-degrees from the ring anchors. A tensile residual stress in the material layer the rings are made from causes the ring to take a more oval shape thereby converting the tensile stress into a compressive stress onto the ends of the ring center beam. An array of Guckel rings of varying diameters and center beams width will be used to determine the ring diameter for the onset of buckling. The residual stress for the onset of buckling is found from the maximum displacement of the center beam using:

$$\sigma = \frac{E}{(1-\nu)} \frac{0.515h^2}{R_c^2}, \quad (7)$$

where h is the material layer thickness and R_c is the critical radius for the inset of buckling [57].

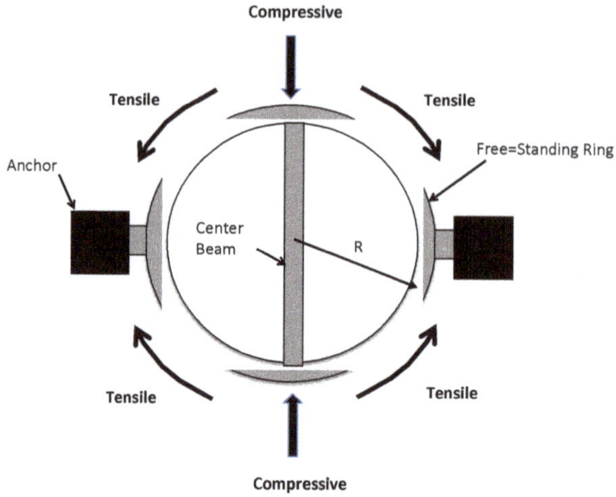

Figure 8. A Guckel ring residual stress test structure.

The Guckel ring test structures, that are a variation of a buckling test structure, are prone to the same shortcomings as the buckling beam test structures. Furthermore, the Guckel rings when buckled can also result in a torque on the ring at the connection points and this can cause uncertainty of the anchor behavior thereby rendering this method less accurate [54].

A similar type of structure, called the diamond structure, shown in Figure 9 is designed for the measurement of both tensile and compressive residual stresses. In this test structure, the cross beams convert the tensile strain in the material layer into a compressive strain acting on the center beam. If a compressive stress is present in the material layer, it can cause the outer beams to buckle. The determination of the residual stresses in this structure is performed using finite-element analysis (FEM) techniques since no analytical equation is possible. These test structures suffer from the same shortcomings as the buckling beam test structures discussed above. Additionally, it has been shown that the stress in the material layer is not converted effectively thereby meaning that large structures are required to measure small strains [54].

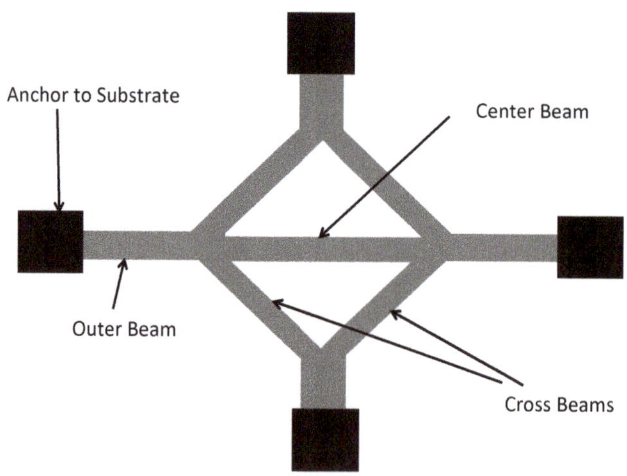

Figure 9. The diamond-shaped test structure.

While the buckling type of test structures can be used to provide a measurement of the residual stress in thin-film layers, it is important to know the shortcomings of these structures. These test structures do not require expensive or sophisticated apparatus to measure, a microscope preferably with DIC, can provide useful data.

5.2.3. Strain-Based Test Structures

Test structures designed to enable the material layer to strain (i.e., deflect) and be measured have been around for a long while. There are a number of different configurations of these types of test structures, of which two will be discussed here.

The first is called the T-structure and is shown in Figure 10 [58]. It is designed to measure a tensile residual stress, but can in some situations be used to measure compressive stresses as well. The design is simple and consists of a free-standing thin-film material layer patterned into the shape of a T. The main element has length L_A and width W and is connected to a perpendicular cross support element having length L and width b. When the layer is released, the residual stress in the main element strains the cross support element by an amount δ. The strain in the main element is given as [58]:

$$\varepsilon = \left(\frac{1}{LA} + \frac{32b^3}{W(2L^3 - 2W^2L + W^3)}\right)\delta. \quad (8)$$

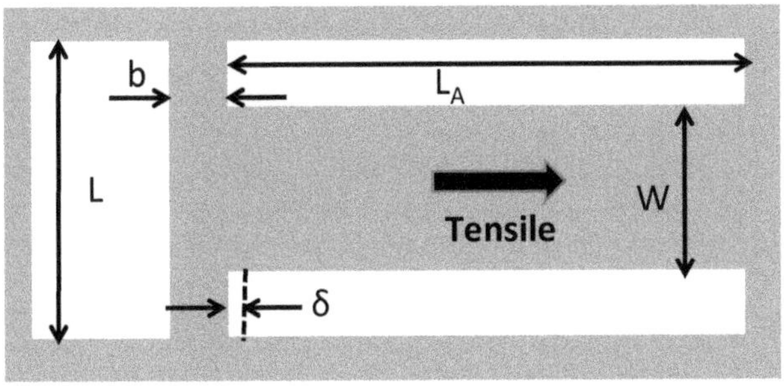

Figure 10. Illustration of the T-structure for measuring residual stress in thin-film layers.

Using the calculated value of the strain and the Young's modulus allows the residual stress to be determined.

The other strain-based test structure is called the H-structure and is shown in Figure 11. This structure is mostly used for measuring tensile residual stresses. The strain in this structure is given by [58]:

$$\varepsilon = \left(\frac{W_1L_2 + W_2L_1}{L_1L_2(W_1 - W_2)}\right)\delta. \quad (9)$$

As before, using the calculated value of the strain and the Young's modulus allows the residual stress to be determined.

A major challenge with the strain-based test structures is that in most circumstances the actual displacements are very small and therefore difficult to measure using an ordinary microscope. This makes these structures relatively inaccurate unless the modulus of the thin-film material layer is low [54]. The use of a SEM can be useful for detecting the displacement.

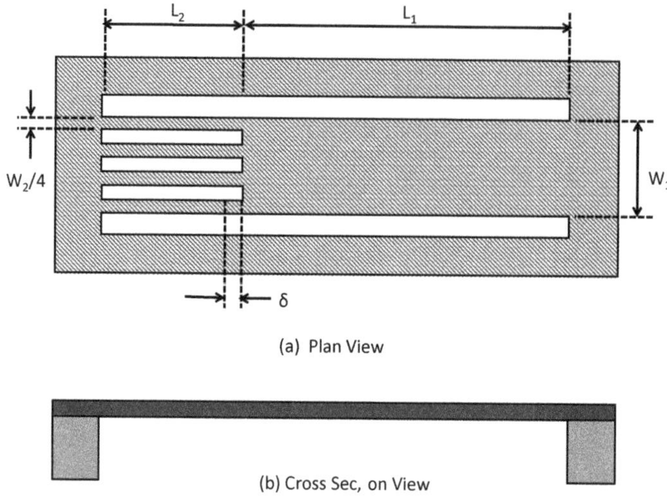

Figure 11. Illustration of the H-structure for measuring residual stress in thin-film layers.

5.2.4. Electrostatic Pull-In Test Structures

The electrostatic pull-in test structures employ a free-standing cantilever, beam or other type of mechanically compliant element fabricated over a ground plane on the substrate surface with a free-space gap between the mechanically compliant element and the ground plane over which a voltage potential is applied. If the voltage is sufficiently large, the mechanically compliant element will pull-in towards the ground plane. Since the pull-in voltage is dependent on the dimensions, material type of the mechanical element, as well as residual stress in the material layer the mechanical element is made from, the residual stress can be determined. The pull-in voltage of a beam type of mechanical element fixed on both ends as a function of the dimensions and the residual stress was developed by Osterberg [59]. The main issue with use of these types of test structures is the large dependency of the pull-in voltage to both the gap and the thickness of the material layer. If these variables are not known with considerable accuracy, the accuracy of the residual stress measurement can be compromised.

5.2.5. Pointer Beam Test Structures

Another commonly used test structure for the determination of the residual stress in thin-film layers is shown in Figure 12. The test structure can be used to determine both tensile and compressive residual stresses and is called the pointer test structure. A free-standing pointer beam having width, w, and length, L, is made from a thin-film material layer. The pointer beam has two supporting struts offset from one another with one end attached to an anchor to the substrate and the other end attached to the pointer beam to cause a rotational motion of the pointer beam in the presence of a residual stress in the material layer. The pointer beam has a scale fabricated on one end where the measurement is taken and another measurement scale made in close proximity to the pointer beam scale. This allows the measurement of the amount of deflection based on the dimensions of the elements in the scales.

Depending on the type and magnitude of the residual stress the layer, the offset supporting struts will either attempt to expand or contract. An expansion of the supporting struts due to compressive residual stresses will cause the pointer beam to move towards the right and a tensile residual stress will result in the pointer beam moving to the left.

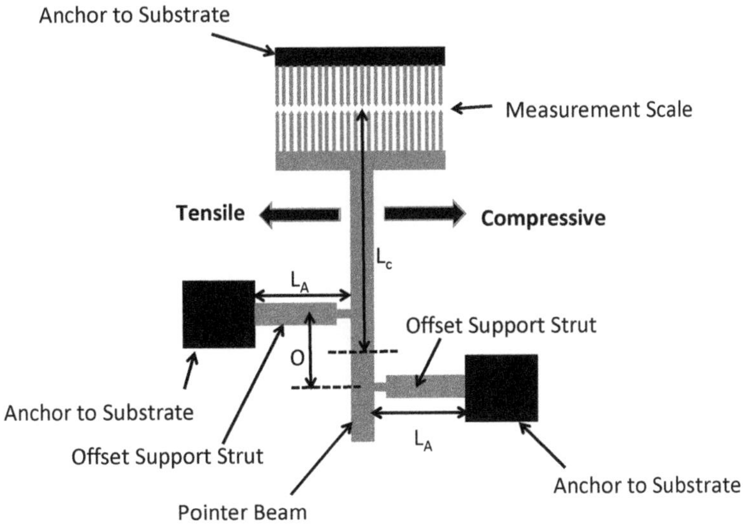

Figure 12. Pointer beam test structure.

The mathematical equation used to calculate the residual stress magnitude as a function of the rotation is given in Van Drieenhuizen [54] wherein the strain, ε, is found as follows:

$$\varepsilon = \frac{O_y}{(L_A + L_B)(L_C + 0.5O)}, \tag{10}$$

where O is the offset distance between the two supporting struts, y is the deflection of the end of the pointer beam (shown not deflected in Figure 12), L_A is the length of one of the supporting struts, L_B is the length of the other supporting strut, and L_C is the length of the pointer beam from the midpoint between the centers of the two offset struts to the end of the pointer beam. Once the strain has been measured from the pointer beam test structure, the residual stress is found by multiplying by the modulus of the beam material.

An alternative form of a pointer test structure is shown in Figure 13. As with the previous test structure, once the strain is measured, and assuming the modulus of the material of the thin-film is known, the stress in the film can be determined. The strain in the thin-film is found using the following equation:

$$\varepsilon = \frac{2L_s\delta}{3L_iL_tC'} \tag{11}$$

where L_s, L_i, and L_t, are the length of the slope, pointer and test beams, respectively, δ, is the displacement of the pointer beam on the gauge scale, and C is a corrective factor [60].

The pointer test structures have certain advantages compared to the buckling type of test structures. First, the pointer test structures can be used to measure both compressive and tensile residual stresses. Second, only a single test structure is needed to perform the measurement as opposed to the buckling test structures wherein an array is needed that will likely consume significantly more die area.

However, the pointer test structure still can suffer from two issues. The first is that stiction can cause the pointer test structure to be attached to the substrate rendering it useless. Again, anti-stiction surface coatings may reduce or eliminate this problem. Another potential issue is the onset of buckling in the test structure. If the compressive stress is sufficiently large, the entire structure can buckle out of place thereby providing a mechanism to relieve the residual stress, but rendering the capability for measurement of the residual stress useless. The probability of this occurring can be reduced by proper selection of the pointer test structure dimensions, but this requires a reasonably good prior knowledge of the value of the residual stress in order to size the dimensions correctly and this may not be possible. While it is possible to fabricate a number of pointer beam test structures having different dimensions so that some of the test structures may not be buckled, this again consumes considerably more substrate area.

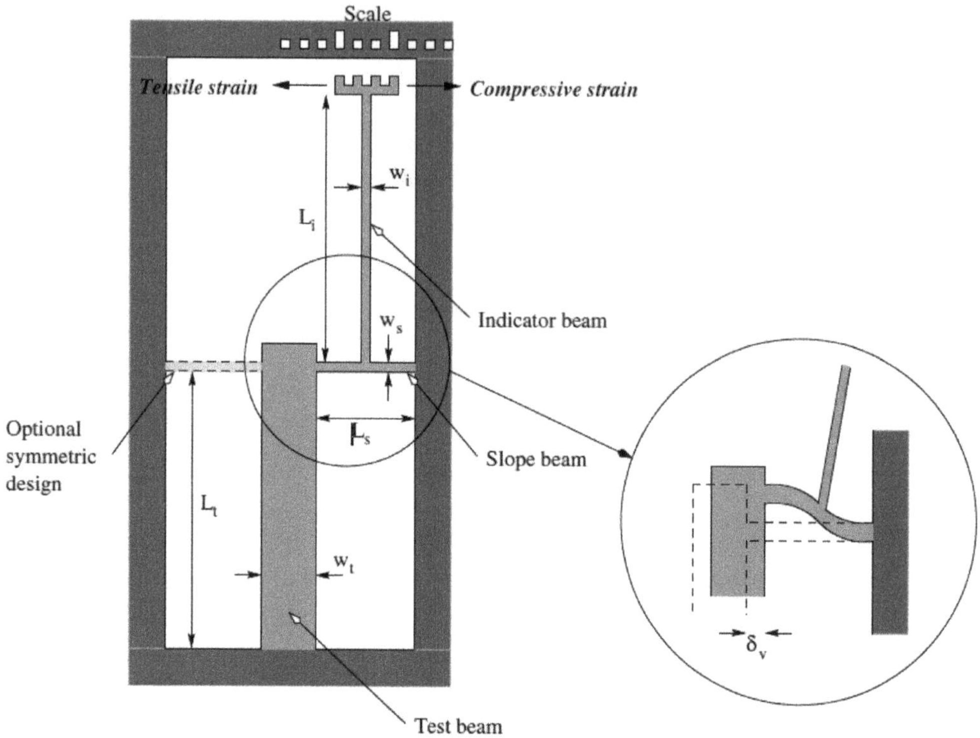

Figure 13. Micro-scale pointer test structure.

5.2.6. Resonator Test Structures

The simple fixed-fixed beam resonator test structure was described earlier in the section on the impact of residual stresses. A resonator test structure can be used to measure the residual stress in thin-film material layer. The key issue in using a resonator to determine the residual stress in a material layer is that the resonant frequency is shifted due to the presence of the residual stress compared to the same resonator without a residual stress. Using Equation (1) enables a very accurate determination of the residual stress as long as the dimensions are accurately known since the resonant frequency can be measured very accurately.

5.2.7. Bulge Test Structures

The bulge test involves micromachining the substrate to form a free-standing membrane of the thin-film material layer that is suspended from the surface of the substrate as illustrated in Figure 14. A uniform pressure loading is applied to the membrane causing it to deflect and the deflection is measured as a function of the applied pressure [58].

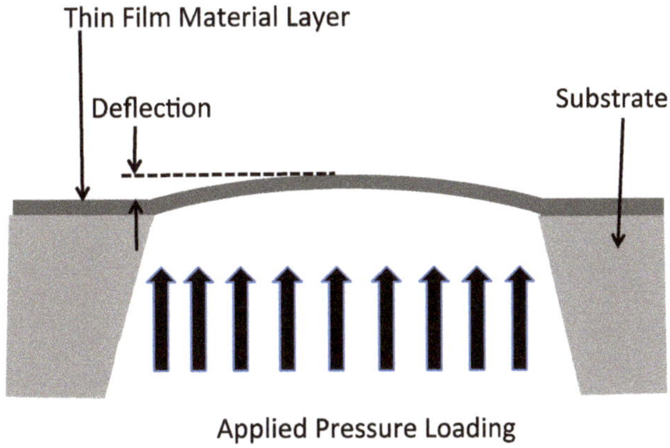

Figure 14. Bulge test structure.

The residual stress, σ_o, in the material layer is related to the membrane deflection, d, the membrane thickness, t, Young's modulus, E, the square membrane edge length, a, and the applied pressure from the following [58]:

$$\left(\frac{Et}{a^4}\right)d^3 + \left(\frac{1.66t\sigma_o}{a^2}\right)d = 0.547p. \tag{12}$$

The bulge test structure can be made using a circular membrane as well and will require a different equation than Equation (12). This test structure can be used both compressive and tensile residual stresses, although if a compressive stress is present the structure is not useful if the membrane is buckled. The deflection of the membrane can be measured using a microscope or interferometer. The major shortcomings of this test structure is that it requires bulk micromachining of the substrate to make the membranes and the measurement of the deflection of the membrane is challenging unless the thin-film material layer has a low modulus value or the membrane is made very large.

5.3. Stress Gradient Test Structures

A stress gradient is when the residual stress varies through the thickness of the film from one side to the other in the direction normal to the substrate surface. It is common for polycrystalline materials, such as polysilicon, to have significant stress gradients since the grains are columnar with the grain size increasing through the thickness [5,9]. Like a residual stress, the presence of stress gradients can also have a significant impact on the resultant device behavior. For example, the fabrication of cantilevers composed of material layers having stress gradients could result in the cantilevers not being straight and flat as would be desired for most applications [50,61]. It is important to note, yet sometimes misunderstood, that it is possible for the average state of stress in a thin-film to be near or at zero, but the film still have a significant stress gradient. Even for devices that are clamped at both ends or around the periphery can have their performance impacted by the presence of a stress gradient.

The test structures described so far are not suitable for determining the gradient of the stress through the material layer. An appropriate test structures for measuring the stress gradients in material layers are the cantilevers as shown in Figure 15. The left side shows a plan view and cross section of a cantilever wherein the unclamped free end of the cantilever has undergone a displacement due to the presence of a stress gradient. This type of curved displacement of the free end of the cantilever would be due to a stress gradient that is more tensile on the top (or more compressive on the bottom) compared to the lower (upper) part of the material layer.

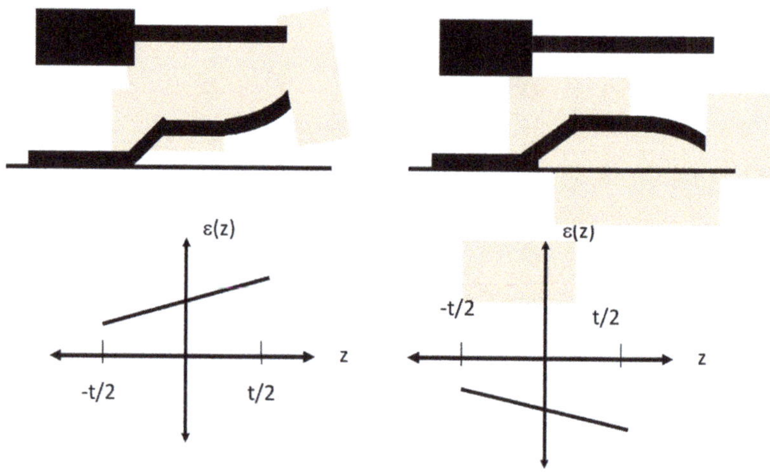

Figure 15. Illustration of stress gradients in cantilevers exhibited as strain gradients that can be used to measure the stress gradient in material layers. The plan and cross-sectional views of the cantilevers are at the top and the strain as a function of the cantilever thickness is given at the bottom.

Once the structure is released the stress gradient is allowed to relax as the cantilever deforms and exhibits a strain. That is, as the cantilever relaxes the stress goes to zero. The gradient in the strain is plotted in the left bottom, which show the strain as a function in the normal direction, z, as a function of the thickness of the layer given by t. The strain has a larger magnitude in the top half of the material layer thickness and a lower magnitude in the lower half. The right side illustrates the effect of a stress gradient where it is compressive in nature and once the cantilever relaxes, the stress goes to zero and the end of the cantilever curving downward as shown in the bottom right exhibits the strain.

There may be both an average stress in the film and a stress gradient. For example, the average stress could be due to the mismatch in the thermal expansion coefficients of the material layer and the substrate. The stress gradient could be due to the columnar structure of the grains of a polycrystalline material layer. Once the cantilever is relaxed, however, both of these stresses will relax and be exhibited as strains. The relaxation of the uniform average stress will be exhibited by a uniform strain that would be an expansion if the average stress is compressive, or contraction if the average stress is tensile. However, the stress gradient will result in a curvature of the released cantilever. This curvature can be used to measure the stress gradient.

The bending moment, M that causes the deflection of the cantilever is given by [52]:

$$M = \int_{-t/2}^{t/2} \sigma(z) z L dz, \tag{13}$$

where σ(z) is the stress gradient, L is the length of the cantilever, and t is the thickness. The deflection of the end of the cantilever is given as:

$$z = \frac{ML^2}{2EI}, \tag{14}$$

where L is the length of the cantilever, E is the modulus of the material of the cantilever, and I is the moment of inertia of the cantilever given by $I = wt^3/12$.

5.4. Other Techniques for Measuring the Residual Stress

In addition to wafer curvature and test structures, there are other methods used to determine the residual stress in thin-film material layers. These methods are based on techniques primarily developed by the material science community and often require the use of sophisticated experimental apparatus and specialized technical staff to perform these experiments and interpret the results.

5.4.1. X-ray Diffraction

X-ray diffraction is a sophisticated technique used to measure the composition of materials, their crystal structure and phases, presence of strain and strain gradients, impurities, and defects in materials [9]. It uses X-rays to irradiate a sample and observes the X-rays that are reflected from or transmitted through the sample undergoing diffraction effects. The accuracy of X-ray diffraction depends on what is being measured, but for determination of crystal structures, the accuracy is typically in the range of $\Delta 2\Theta < \pm 0.04°$ [62–66]. This is mainly due to the precision at which the angles of the sample and system can be aligned, the spread in the diffraction beam on the imaging plane, and other factors. X-ray diffraction can be used on crystalline forms of materials [9].

X-ray diffraction is based on the Bragg equation. The Bragg diffraction condition requires that the wavelength of the radiation used to examine the sample is comparable to the atomic spacing in the material. The X-rays impinge on a sample, penetrate it, and are scattered by the atoms of the crystal whereupon some of the scattered X-rays undergo constructive interference as shown in Figure 16. Constructive interference is based on the X-rays being in phase when the path lengths from two or more waves scattered from lattice planes with inter-planar separation distance of d is equal to an integer multiple of the wavelength. The path length difference between scattered waves undergoing interference is given by 2d sin Θ where Θ is the scattering angle. The result of constructive interference is that the maximum amplitude of the scattered waves is indicative of the crystallographic planes of the solid. In general, Bragg's law is given by:

$$2dsin\theta = n\lambda, \tag{15}$$

where n is an integer and λ is the wavelength of the incident X-rays. The diffraction pattern is the intensity of the scattered waves as a function of scattering angle. The amplitude maximums of the X-ray diffraction intensity $I(\Theta)$ versus angle plot are known as the Bragg peaks and represent the locations where the scattering angles satisfy the Bragg condition for constructive interference.

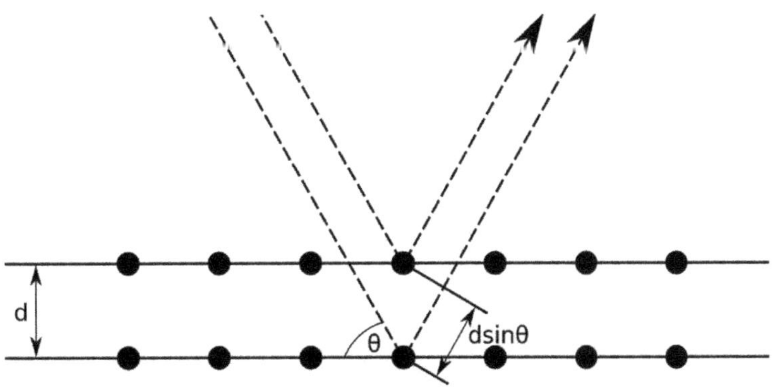

Figure 16. An illustration of Bragg reflection where the dots represent the atoms of a crystal lattice with inter-planar separation distance of d, and the dotted lines represent the impinging and scattered X-rays after interacting with the lattice.

The inter-plane distance, d, of a lattice having index plane (hkl) depends on the lattice parameters of the material. Therefore, an unstressed crystal of a material will have a specific identifiable diffraction pattern. Further, under either a tensile or compressive stress that causes the lattice to strain can also be identified and measured since the values of d are changed as a result of the strain. The measurement of the strain in the material is conducted as a measurement of the shift in the lattice parameters wherein the strain can be expressed as [9]:

$$\varepsilon = \frac{d_{(hkl)} - d_{o(hkl)}}{d_{o(hkl)}}, \tag{16}$$

where $d_{o(hkl)}$ is the value of d of the (hkl) plane under no strain and $d_{(hkl)}$ is the value the d spacing of the (hkl) plane under strain. From Equation (16), the residual stress can be determined by multiplying the strain by the modulus of the material. Similar formulations of this expression are available for materials that are anisotropic.

Typically, a number of X-ray diffraction measurements are taken at different tilt angles and a plot of the intensity I(Θ) of the diffraction angle is made wherein there will be a peak intensity of width 2Θ [9].

While X-ray diffraction methods have been used for decades, a more recent variation of this method was reported called cross-sectional nano-diffraction. This technique uses a cross section sample of the material combined with a pencil-like X-ray beam generated from a synchrotron. This method is able to obtain X-ray beams with diameters of 50 nm or less and can be used in either the reflection or diffraction geometries. This overcomes an issue with conventional X-ray diffraction wherein the average state of residual stress is measured and stress gradients cannot be resolved [67,68].

A strain present in a material layer will result in a slight shift in the atomic spacing of the atoms that the material is made from thereby allowing the residual stress to be determined. Since the spatial resolution of cross-sectional X-ray nano-diffraction can be very good, diffraction measurements can be taken at points through the thickness of a material layer to enable stress gradients to be determined as well [68].

X-ray diffraction is a powerful and accurate method for measuring both residual stress and stress gradients, and can be used to provide information about thin-film layer spatial stress variations across the substrate. There are a few issues that users should be aware of regarding this technique. First, exposure to X-rays can be damaging to materials. Second, this technique is not suitable for use on amorphous material layers. Additionally, some of the X-ray diffraction measurement techniques are destructive to the substrate. For nano-diffraction techniques, a synchrotron is required which is something that access to can be very difficult and costly to obtain. In most instances, performing an X-ray diffraction is a costly experimental method that is mostly used for development purposes.

5.4.2. Raman Spectroscopy

Raman spectroscopy is an analysis technique that uses monochromatic light in the visible, near infrared or ultraviolet regions of the spectrum to induce inelastic scattering, called Raman scattering, with the material sample. The incident laser radiation interacts with the molecular vibrations and phonons in the material that interacts with the incident light resulting in energy shifts in the photons and these energy shifts can provide information about the vibrational modes that are indicative of the residual stress present in the material [69,70].

Raman spectroscopy is performed on an unstressed material sample in order to get a baseline measurement and then performed on the stressed material sample. The Raman measurements will indicate a shift in the wavenumber of the stressed sample compared to the wavenumber of the unstressed sample. For example, single crystal silicon shows a Raman signal at 520.7 cm^{-1} [70] and when the sample is stressed this wavenumber shifts to lower values under tensile stresses and higher values under compressive stresses. Raman spectroscopy is not useful for material layers that are amorphous [69].

5.4.3. Nano-Indentation

Nano-indentation is a commonly used method for measuring a number of important material properties. This approach uses a tip made from a very hard material, such as diamond, having material properties that are known, and pressing the tip into the material being analyzed. The force applied to the tip and into the sample is increased over a range of values. The area of the indentation is measured to determine the hardness, H, according to the following relationship:

$$H = \frac{P_{max}}{A}, \quad (17)$$

where P_{max} is the maximum load applied and A is the indentation area after the load has been removed. The resultant indentation area can be relatively small, such as a few microns or nanometers, thereby making measurement of the residual indentation area difficult. Therefore, it is common to use SEM imaging or AFM for these measurements [5].

Nano-indentation can also be used to measure the modulus of a thin-film material layer (See Figure 17). This is done by forcing the indentation tool into top surface of the thin-film layer so that it penetrates into the layer, and then backing the indentation tool out from the sample and measuring the displacement and force [71].

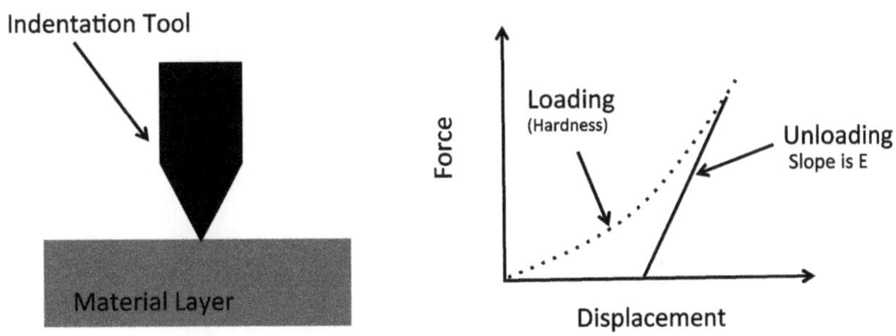

Figure 17. Nano-indentation used to measure modulus of material layer.

Nano-indentation can be used to measure the residual stress [72,73]. Specifically, it has been noticed that the penetration displacement of the indentation tool varies depending on the residual stress in the thin-film layer as illustrated in Figure 18. This phenomenon can be used to measure the residual stress assuming a baseline unstressed force versus displacement is available. The calculation of the residual stress using this technique is complicated. This method can be used at locations across the substrate.

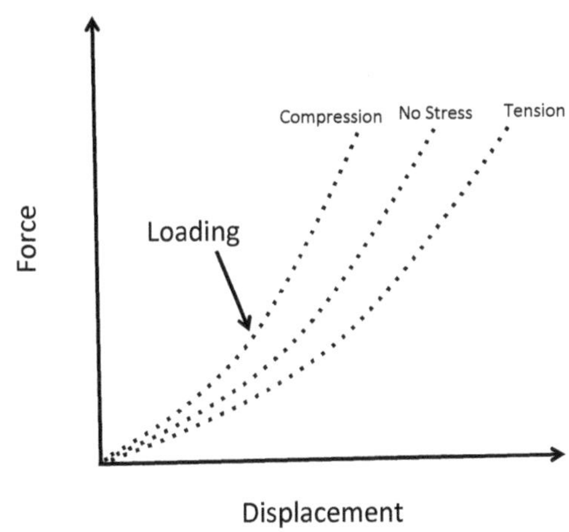

Figure 18. Nano-indentation used to measure residual stress.

5.4.4. Focused-Ion Beam Strain Relief

Focused-Ion Beam (FIB) technology is an extremely useful tool for cross-sectioning and imaging portions of device structures that would otherwise be impossible to image. A FIB performs nanometer dimensional-scale machining using a liquid-metal ion source, such as Gallium, whereby the source is heated causing ionization and resulting in field emission of the Gallium ions. These ions are accelerated to energies usually between 5 and 50 KeV and focused to a small spot size using an electrostatic lens. Material on the substrate surface is sputtered as the ions impinge the material.

Most FIBs include scanning electron microscopy (SEM) imaging capability. FIBs can also be outfitted to perform ion-induced deposition. Conventional FIB technology can machine features down to about 5 to 10 nm and can remove material at rates up to nearly 100 μ^3/s. Newer technology using a helium ion source has recently been introduced into the commercial market and has a resolution below 1 nm [74]. Helium is also less damaging to the surface material than Gallium ions. Using a FIB system, the user can input a 3-D CAD solid model of the desired topology of the machining process and the computer-controlled stage allows very precise registration of sample with the ion beam with submicron positional accuracy. FIBs can be used to machine both conductive and nonconductive materials.

The FIB deposition, imaging and machining capabilities are combined with digital image correction (DIC) to perform residual stress measurements [75]. The approach is illustrated in Figure 19 wherein at the top portion of the figure a very thin material layer is deposited using the FIB over the top of a very small portion of the material layer that the residual stress is to be measured. The FIB deposited layer may have a specific grid pattern as a guide for determining the strain relief in the material layer. The material layer is then machined using the FIB as shown in the bottom. This type of machining pattern is called a ring and leaves a pillar of the material layer surrounded by a ring of removed material. As can be seen, the center pillar expands due to strain relief. It is this strain relief that is a consequence of the residual stress in the material layer. It has been shown that if the depth of the FIB machined ring, given by h, is about the same as the diameter of the center pillar, given by d, the normalized strain relief approaches a value of 1. This means the strain relief is complete at h approaches or exceeds d. When complete strain relief is obtained, the residual stress, σ, in the material layer is given by [75]:

$$\sigma = \frac{E\Delta\epsilon}{(1-\nu)}, \tag{18}$$

where $\Delta\varepsilon$ is the measured strain relief and ν is Poisson's ratio assuming a equi-biaxial residual stress. Other equations can be used for non-equi-biaxial residual stresses.

This technique can also be used to measure the stress gradient in a material layer. The process involves removing material to various incremental depths and then measuring the strain relief that occurs at each of the incremental depths. For this type of analysis finite element modeling (FEM) must be used to determine the residual stresses as a function of depth into the material.

Other types of machined structures can also be used in this technique, including slots, squares, etc. However, these other types of structures require FEM analysis to determine the residual stresses and therefore do not afford themselves to simple analysis.

This technique has a high level of accuracy and can be performed at various locations on substrates in order to determine how the residual stress various across the substrate surface. Moreover, this technique is in theory able to be performed on device wafer samples; those used in manufacturing for MNS production since the size of the areas machined are very small; on the order of material layer thicknesses and thereby typically around 1 micron or less [10,75].

The major disadvantages of FIB are it has slow machining rates and FIBs are very expensive instruments that require some amount of skill to operate effectively. Most MNS foundries do not have an in-house FIB capability.

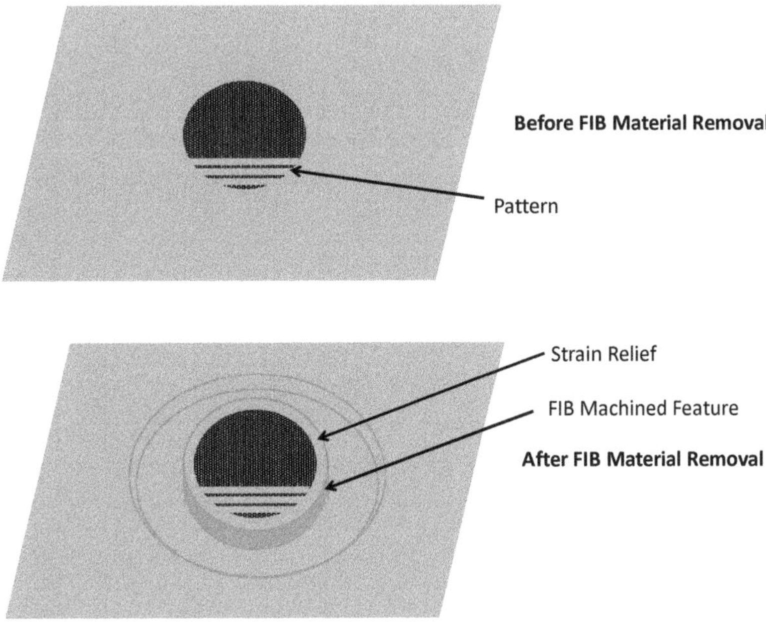

Figure 19. Illustration of FIB machining to measure residual stress in a thin-film material layer.

6. Review of Reported Residual Stresses in Deposited Thin-Film Layers

This section reviews some reported values of the residual stresses of commonly used thin-film material layers in MNS. As noted previously, residual stresses exhibit significant dependency on the processing conditions including temperature, process gases, and method of deposition. These reported values are intended to illustrate the amount of variability in the residual stresses. Importantly, unless the residual stress is known, the MNS designers should always measure the residual stresses in the materials to be used in their MNS device based on the process conditions, equipment and process sequence in their particular situation.

6.1. Thermal SiO_2

As noted above, there is a significant volume expansion as silicon is converted into silicon dioxide that would be expected to result in a compressive residual stress. However, the values of the residual stress at higher growth temperatures can be far lower than would be predicted based on this expansion alone. The explanation for this is based on silicon dioxide exhibiting viscous flow at elevated temperatures, thereby allowing for stress relaxation [5]. Some of the reported values of residual stress in oxide layers are provided in Table 1.

Table 1. Residual stresses measured in silicon dioxide layers grown using wet thermal oxidation.

Deposition Temperature (°C)	Process Gases	Thickness (microns)	Residual Stress (MPa)	Refs.
1000	O_2/H_2	0.430	−331	[76]
950–1050	O_2/H_2	0.5 to 1	−258	[77]

6.2. Low-Pressure Chemical Vapor Deposition (LPCVD)

6.2.1. LPCVD Polysilicon

LPCVD Polysilicon is a commonly used material in the manufacturing of MNS devices and is the most commonly employed material as a structural layer in surface micromachined MNS devices. There is a strong relationship between the processing conditions used during deposition, the microstructure of the deposited thin-film layers, and the residual stress [78]. As shown in Figure 20, the residual stresses in as-deposited polysilicon layers at a differing pressures and temperatures are compressive for deposition temperatures below 580 °C, but at a deposition temperature of 605 °C, the stress transitions to tensile, and at a deposition temperature of 620 °C the stress changes back to being compressive [79].

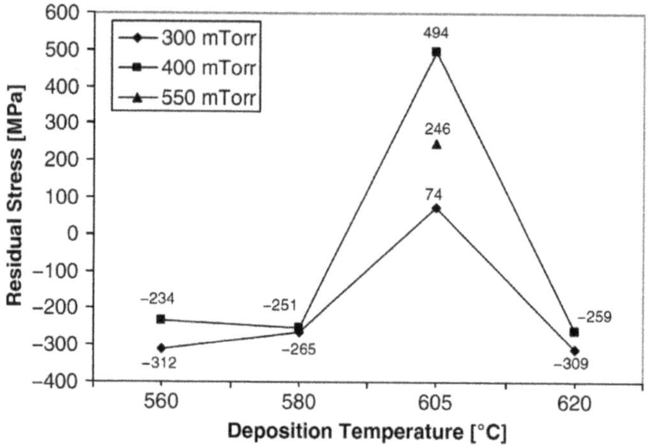

Figure 20. The residual stress in as-deposited polysilicon thin-films with varying deposition pressures and temperatures at a silane flow rate of 30 sccm [79].

Table 2 lists a number of different reported values of residual stress in LPCVD undoped polysilicon thin-film material layers as a function of the processing conditions. As can be seen, there is a huge amount of variability in these values of residual stress. The effects of anneals on doped and undoped polysilicon is provided in a later section.

Table 2. Residual Stresses in As-Deposited, Undoped LPCVD Polysilicon.

Deposition Temperature (°C)	Silane Flow Rate (sccm)	Pressure (mtorr)	Thickness (microns)	Residual Stress (MPa)	References
560 to 630	30	300 to 550	2	−340 to 1750	[79]
570	80	150	1.3	82	[80]
570	100	300	2	270	[81]
600	125	550	0.1	12	[82,83]
615	100	300	2	−200	[81]
620	70	100	0.46	−350 ± 12	[84]

6.2.2. LPCVD Silicon Dioxide (SiO_2)

The residual stress in LPCVD oxides are process dependent. The process conditions and resultant residual stresses in deposited LPCVD oxide layers are shown in Table 3 for as-deposited thin-films. As can be seen, the residual stress in LPCVD oxides tends to become more tensile with increasing deposition temperatures.

Table 3. Resultant residual stresses in as-deposited SiO_2 thin-films deposited using LPCVD.

Deposition Temperature (°C)	Source Gases	Flow Rate Ratio	Pressure (mtorr)	Residual Stress (MPa)	Refs.
425	SiH_4/O_2	0.5	200	−10	[85]
700	$TEOS/O_2$ $TEOS/PH_3$	0.81 5.3	-	200	[86]

6.2.3. LPCVD Silicon Nitride (Si_3N_4)

The residual stress in stoichiometric Si_3N_4 thin-film layers has been reported to have a tensile value of approximately 1 GPa [87]. This is a very high value of residual stress and makes these layers predisposed to cracking if the thickness is more than a few hundred nanometers. Table 4 shows some of the reported values of the residual stresses of stoichiometric Si_3N_4 films. No inference between deposition temperature and residual stress can be drawn from this data.

Table 4. Material Property Values of as-deposited LPCVD SiN thin-films layers.

Temperature (°C)	Gases	Ratio	Pressure (mtorr)	Residual Stress (MPa)	Refs.
850	SiH_2Cl_2/NH_3	0.33	150	967	[88]
785	SiH_2Cl_2/NH_3	0.33	368	1020	[89]

6.2.4. LPCVD Silicon-Germanium (SiGe)

LPCVD SiGe thin-films have been reported to have an as-deposited residual stress that varies from slightly tensile to slightly compressive depending on the germanium content and deposition temperature [90,91]. In one report, the residual stress of LPCVD polycrystalline SiGe deposited at 450 °C ranged from 31 MPa compressive when the germanium content was 64% to 160 MPa compressive with the germanium content was reduced to 47% [91]. Table 5 lists some of reported measured values of residual stresses of LPCVD polycrystalline SiGe thin-film layers. As can be seen, the residual stress tends to reduce from moderately compressive to near zero at deposition temperature of around 450 °C.

Table 5. Reported Residual Stresses of LPCVD Polycrystalline Silicon-Germanium (SiGe) for various process recipes.

Temp. (°C)	Gases	Flow Rates (sccm)	Press (mtorr)	Thickness (microns)	Residual Stress (MPa)	Refs.
400	GeH_4 PH_3 (50% in SiH_4)	219 5	300	5.1	−100	[92]
410 to 440	SiH_4 GeH_4 BCl_3 (in He)	104 to 120 50 to 70 6 to 18	600	1.7 to 2.6	−70 to −228	[93]
450	GeH_4 SiH_4 B_2H_6 (10% in SiH_4)	90 50 85	600	3.1	−10	[92]
550	Si_2H_6 GeH_4 PH_3 (50% in SiH_4)	15 185 5	300	2	−50	[94]

6.2.5. LPCVD
Silicon Carbide (SiC)

A correlation between the deposition pressure and residual stress for un-doped LPCVD polycrystalline SiC thin-films using dichlorosilane and acetylene source gases at a deposition temperature of 900 °C was reported, that included a recipe having near zero residual stress [95]. It was also reported that there was a range of pressures where the residual stress varied significantly with pressure. This range was from 0.5 to 5 torr. At pressures around 0.5 torr the residual stresses highly tensile (i.e., about 700 MPa) and decreased to about 50 MPa at a pressure of 2.5 torr. At pressures higher than 3 torr the residual stress was compressive with a value of about −100 MPa. The correlation between pressure and as-deposited residual stresses in the LPCVD polycrystalline SiC layers was also reported with doped films using NH_3 as a doping gas [96] wherein the minimum residual stress was found to be about 30 MPa at a deposition pressure of 5 torr. Table 6 shows some of the reported residual stress data for polycrystalline SiC deposited using LPCVD for various processing conditions.

Table 6. Reported Residual Stresses for LPCVD Polycrystalline SiC thin-film layers for Various Processing Conditions.

Temp. (°C)	Gases	Gas Flow Rate (sccm)	Press. (torr)	Residual Stress (MPa)	Refs.
900	SiH_2Cl_2 C_2H_4 (5% in H2)	35 180	2	56	[97]
900	SiH_2Cl_2 C_2H_4 (5% in H2)	54 180	2.75	26.9	[98]
900	SiH_2Cl_2 C_2H_4 (5% in H2)	35 180	4	59	[96]
900	SiH_2Cl_2 C_2H_4	10 10	0.15	30 to 250	[99]
930 to 1150	$Si(CH_3)_4$ H_2	10 1000	0.4	−176 to 145	[100]
1010	SiH_4 C_2H_4 HCl H_2 B_2H_6	15 7 Trace 10,000 1.5	2.48	300	[101]
1200	SiH_4 C_2H_4	- -	40	192 to 347	[102]

6.3. *Plasma-Enhanced Chemical Vapor Deposition (PECVD)*

Plasma-enhanced chemical vapor deposition processes employ a plasma as an energy source to facilitate the process and as a result can often perform deposition at much lower temperatures than are possible using LPCVD. The presence of a plasma also provides more capability for modifying the material properties.

6.3.1. PECVD Silicon Dioxide (SiO_2)

The main attraction to PECVD SiO_2 for MNS manufacturing is the low deposition temperatures. The residual stress of PECVD SiO_2 can vary over a large range of values for different process condition as exhibited in Table 7.

Table 7. Reported process conditions for PECVD silicon dioxide layers.

Temp. (°C)	Gases	Gas Flow Rate (sccm) *	Residual Stress (MPa)	Refs.
300	SiH_4 N_2O	430 710	−25	[103]
350	TEOS O_2	2.3 mL/min 9500	−45	[104]
400	SiH_4 N2O N_2	300 9500 1500	−80	[104]

* A dash in the Table means that the data was not given in the source.

6.3.2. PECVD Silicon Nitride (SiN)

Residual stresses in PECVD SiN was reported to be less than 30 MPa at 125 °C and 205 °C [105]. Table 8 is a listing of some of the reported values of residual stresses in PECVD SiN layers as a function of the processing conditions. Based on this data, the transition temperature where the residual stress is nearly zero is about 157 °C.

Table 8. Reported residual stresses for PECVD silicon nitride thin-film layers for various process recipes.

Temp. (°C)	Gases	Gas Flow Rate (sccm)	Pressure (torr)	Power (W)	Residual Stress (MPa)	Refs.
50 to 300	SiH_4 NH_3 N_2	5 45 100	0.45	100	−225 to 300	[106]
55 to 330	SiH_4 NH_3 N_2	5 45 100	0.88	75	−75 to 375	[106]
125 to 300	SiH_4 NH_3 N_2	- 10 2020	0.2 to 0.6	40 to 200	−250 to 250	[107]
300	SiH_4 NH_3 N_2	600 55 1960	0.9	100	178	[108]
300	SiH_4 N_2	15 535	0.5	-	110	[109]

6.3.3. PECVD Silicon

PECVD deposited silicon material layers are amorphous in microstructure. Deposition of silicon at temperatures as low as 100 °C have been reported using PECVD [110], however most of the literature reports deposition temperatures in the range of 250 to 300 °C. Generally, the layers exhibit a fairly high compressive residual stress (i.e., −130 to −575 MPa) depending on the deposition temperature (See Table 9).

Table 9. Reported Residual Stress of PECVD Silicon as a Function of Processing Conditions.

Temp. (°C)	Gases	Gas Flow Rate (sccm)	Press (torr)	Power or Power Density	Residual Stress (MPa)	Refs.
150	SiH_4 Ar	10 7	0.5 to 0.75	77 to 114 mW/cm^2	−370	[111]
250	SiH_4/H_2 PH_3	10 -	0.1	50 mW/cm^2	−130	[112]
300	SiH_4 Ar	0.125 to 0.25 -	0.8	100 to 300 W	−360 to −380	[113]
300	SiH_4 Ar	40 20	0.38 to 0.53	50 to 215 W	−520 to −575	[114]

6.3.4. PECVD Silicon Germanium (SiGe)

The reported residual stresses in PECVD SiGe layers ranges from −225 MPa to about 100 MPa depending on the process conditions. Table 10 lists some of the reported measured data of material properties of PECVD silicon germanium.

Table 10. Reported Residual Stresses of PECVD Silicon Germanium as a Function of Processing Conditions.

Temp. (°C)	Gases	Gas Flow Rate (sccm)	Press (torr)	Power or Power Density	Residual Stress (MPa)	Refs.
300	SiH_4 GeH_4	1.8 ratio	1	203	−175	[115]
350	SiH_4 GeH_4	1.8 ratio	1	203	2	[115]
400	SiH_4 GeH_4	1.8 ratio	1	203	−9	[115]
520–610	SiH_4 GeH_4	42 3	0.45	-	−18 to −225	[116]
520	SiH_4 GeH_4 PH_3 (1% in SiH_4)	30 166 40	0.2	30	19	[117]
590	SiH_4 GeH_4 PH_3 (1% in SiH_4)	30 166 80	0.2	30	79	[117]
590	SiH_4 GeH_4 B_2H_6 (1% in H_2)	30 166 40	0.2	30	100	[117]

6.3.5. PECVD Silicon Carbide (SiC)

Silicon carbide deposited using PECVD can be performed at temperatures below 400 °C using source gases of SiH_4 and CH_4. The as-deposited SiC thin-films are amorphous in microstructure and can exhibit a relatively high compressive residual stress, depending on the process conditions as well as the substrate material type of the SiC is deposited onto. Table 11 provides some of the reported mechanical properties of PECVD SiC thin-films as a function of the processing conditions.

Table 11. Reported Residual Stresses of PECVD Silicon Carbide as a Function of Processing Conditions.

Temp. (°C)	Gases	Gas Flow Rate (sccm)	Press * (torr)	Power * (W)	Residual Stress (MPa)	Refs.
300	$C_6H_{18}Si_2$	-	-	-	−750	[118]
320	SiH_4 CH_4	3.6 8.4 to 32.4	-	-	−93 to −356	[119]
350	SiH_4 (2% Ar) CH_4	2840 1440	1.6	HF: 100 LF: 100 to 150	−80 to 16	[120]
350	$(CH_3)_3SiH$ in He	38	160	HF: 400 LF: 100	−150	[121]

* A dash in the Table means that the data was not given in the source.

6.4. Epitaxial Deposition

Epitaxial Polysilicon

If single-crystal silicon is deposited onto a silicon substrate using epitaxy there is virtually no residual stress in the layer. However, silicon is deposited at a lower temperature (e.g., below 1000 °C) or onto a non-single-crystal silicon substrate, the layer will be polycrystalline in microstructure and its residual stress is dependent on the processing conditions. The major advantage for using an epitaxial process to deposit a polycrystalline layer is that the growth rates for epitaxial are much higher than for LPCVD. Deposition rates are about 1 micron/min, compared to deposition rate of about 10 nm/min or less for LPCVD [122]. This method is mostly used when very thick material layers of silicon are needed in MNS devices. Silicon deposited directly onto a silicon dioxide surface will be polycrystalline. Importantly, epitaxial silicon deposited directly onto silicon dioxide can exhibit adhesion problems. Therefore, a pre-deposition of a thin-layer of LPCVD polycrystalline silicon is often performed to improve adhesion of the epitaxial film. Epitaxial polysilicon can be in situ doped using either PH_3 or B_2H_6. The residual stress in epitaxial polysilicon can range from compressive [123,124] to tensile [125]. Some of the reported material properties as a function of processing conditions for epitaxial deposited polycrystalline material layers are shown in Table 12.

Table 12. Reported Residual Stresses of Epitaxial Thin-Films of Polycrystalline Silicon as a Function of Processing Conditions.

Temp. (°C)	Gases	Gas Flow Rate (sccm)	Deposition Rate (micron/min)	Thickness * (microns)	Residual Stress (MPa)	Refs.
1000	$SiHCl_2$	750 to 1050	0.55 to 0.75	10	−25 to 3	[125]
1000	$SiHCl_2$ PH_3	1050 5%	0.5	10	3	[123]
1050	$SiHCl_2$	360	1	4	Low Tensile	[122]
1080	$SiCl_4$ H_2	15 g/min 200 slm	1	-	Low Tensile	[126]

* A dash in the Table means that the data was not given in the source.

6.5. Evaporative Physical Vapor Deposition (PVD)

The material properties of PVD evaporative deposited thin-films depends on a number of factors including: processing conditions during deposition; material type being deposited; substrate type; material surface onto which the material is deposited; thickness of the deposited layer; and more. While each situation is different some general trends using evaporation can be made as follows [3,5]:

Metal thin-films deposited using evaporation usually exhibit tensile residual stress in the range from 10 MPa to 1 GPa. Dielectric films exhibit both tensile and compressive residual stresses. There appears to be no strong dependence between the residual stress in a thin-film deposited using evaporation and the type of substrate. The magnitude of the residual stress exhibited in non-metallic thin-films layers tend to be small.

A simplistic rationale for the difference seen in the behaviors of metals and non-metals deposited using evaporation has been offered based on the fact that in general metals are strong in tension, but not in compression, whereas dielectrics and semiconductors are strong in compression, but tend to be weak in tension. Table 13 provides some of the measured values of residual stress for metals and non-metal deposited thin-film layers using evaporation from various published references.

Table 13. Residual for a Number of Different Materials Deposited Using Evaporation Physical Vapor Deposition [3,5].

Material Type	Residual Stress (MPa)
Metals	
Ag (Silver)	20
Al (Aluminum)	−74
Au (Gold)	260
Cu (Copper)	60
Cr (Chromium)	850
In (Indium)	0
Mo (Molybdenum)	1080
Pd (Palladium)	60
Ti (Titanium)	0
Non-Metals	
C (Carbon)	−400
Ge (Germanium)	230
Si (Silicon)	300
ZnS (Zinc Sulfide)	−190
MgF2 (Magnesium Fluoride)	300 to 700
SiO (Silicon Oxide)	10

The residual stress has also been found to depend on the thickness of the evaporated layer. Metals with high melting temperature and hard refractory metals both tend to exhibit higher residual stresses than softer metals and metals with lower melting temperatures. It has been reported that the residual stress in thin-film deposited using evaporation only rises to appreciable levels after the thickness has reached 100 Angstroms thick of material. The residual stress increases to large values up to a thickness of about 600 Angstroms, after which the residual stress does not change significantly [3,5].

The temperature during deposition also has a considerable impact on the residual stress in thin-films deposited using evaporation. Specifically, heating of the substrate during deposition modifies the residual stresses by increasing the rates of defect annealing, recrystallization, and growth of grains. The stresses associated with the growth of a thin-film material later decrease rapidly with increasing temperature. Additionally, the diffusion of impurities in and out of the material layer substantially increases with temperature and also has an impact on the residual stress. As a result of all these temperature-related effects, the residual stress may reach a minimum value or exhibit a reversal in sign [3,5].

6.6. Sputter Physical Vapor Deposition (PVD)

As a result of the interactions of the plasma and the gases, there is more complexity associated with sputtering processes compared to evaporation. Therefore, it is harder to develop useful general conclusions about residual stresses in layers deposited using sputtering processes. In any case, sputtering also allows much more freedom in obtaining desired material properties.

Some general tendencies about sputtering: when the substrate temperature during deposition is not elevated, the residual stress tends to be compressive in nature regardless of the material type [3,5]. It is also known that the amount of gas trapped in the thin-film during deposition directly relates to the compressive stress exhibited in the layers and it is suspected this may partially explain the stresses seen in film deposited at lower relative temperatures.

One of the more comprehensive studies of the effects of processing conditions on the residual stresses in sputter deposited thin-films layers was performed by Hoffman and Thornton [127]. Their study was performed using magnetron sputtering which made it possible to deposit the layers over a wide range of pressures and deposition rates without the effects of plasma bombardment and substrate heating. This work reported two distinct regions with an almost discontinuous change in the material properties. Specifically, at low sputtering pressures, using lighter mass sputtering gases, targets with higher masses, and lower deposition rates, the deposited layers exhibited compressive intrinsic stresses and also exhibited values of electrical resistivity and optical reflectance that were very similar to the bulk values of these properties. These films tended to entrap more gases species into the films during deposition. Conversely, tensile stresses were observed in sputtered thin-film layers deposited at higher pressures, using heavier sputtering gases, lighter mass target materials, and oblique angle of incidences. These films tended to incorporate less trapped gases into the layers during deposition.

It is known that elevated working pressures induce columnar grain growth having inter-crystalline voids, which is the so-called Zone 1 developed by Hoffman and Thornton and such layers tend to exhibit tensile stress. At lower operating pressures the zone 1 structure is suppressed and the energetic particle bombardment of the sputtered atoms leads to films having a compressive residual stress that is thought to be due to the atomic peening mechanism during deposition.

6.6.1. Sputter Deposited Silicon

The advantage of sputtering deposition processes for the deposition of silicon layers is the low temperatures these processes can be performed at compared to other methods such as LPCVD and PECVD. Sputter-deposited silicon thin-film layers on silicon dioxide surfaces have been demonstrated with low residual compressive stresses at practical deposition rates and smooth surface roughness [128].

Silicon layers deposited using sputtering are amorphous unless the substrate is heated substantially during deposition or the substrate is annealed after the deposition. It has been reported that annealing temperatures of at least 800 °C are needed in order to induce crystallization of the deposited silicon layers [129]. The silicon can be doped or undoped based on the dopant levels present in the target materials. Since the sputtering efficiency varies with element atomic mass, most silicon layers are sputtered using undoped source targets. Table 14 provides some of the reported residual stresses of sputter deposited silicon layers onto different substrates and processing conditions.

Table 14. Residual Stresses of Sputter-Deposited Silicon Layers on Various Substrates and Processing Conditions. All Depositions were performed at Room Temperature [130].

Substrate	Power (kW)	Pressure (mtorr)	Deposition Rate (nm/min)	Residual Stress (MPa)
Silicon	1.5	8	23	34
Silicon	1.5	14	19	141
Phosphosilicate Glass	1.5	8	23	97
Phosphosilicate Glass	1.5	14	19	106
Aluminum on Silicon	1.5	8	23	31
Aluminum on Silicon	1.5	14	19	109
Silicon	2.5	8	37	−22
Silicon	2.5	14	30	164
Phosphosilicate Glass	2.5	8	37	27
Phosphosilicate Glass	2.5	14	30	13

6.6.2. Sputter Deposited Silicon Carbide (SiC)

Silicon carbide can be deposited using sputtering at room temperature. RF magnetron sputtering of SiC has been reported using a SiC target [131] and DC sputtering using dual source targets of silicon and graphite [132]. Unlike PECVD deposited SiC thin-films, sputtered SiC layers do not contain hydrogen. The deposited layers of SiC are amorphous in microstructure and electrically insulating. Some of the reported results of sputter deposited SiC thin-film layers including the processing conditions and mechanical properties are shown in Table 15. As can be observed from these published reports, the residual stress in the deposited SiC thin-film material layers are shown to vary over a large range depending on the processing conditions, particularly the chamber pressure which appears to have the most influence.

Table 15. Residual Stresses of Sputter-Deposited Silicon Carbide Layers for Various Processing Conditions. All Depositions were performed at Room Temperature.

Substrate	Power (kW)	Pressure (mtorr)	Temperature (°C)	Residual Stress (MPa)	Refs.
Silicon, Silicon Dioxide	0.05 to 0.3	4 to 31.95	Room Temperature	100 to −1400	[131]
Silicon, Silicon Dioxide	0.2	2.25 to 7.5	Room Temperature	−61 to 210	[132]

6.6.3. Sputter Deposited Silicon Dioxide (SiO_2)

Silicon dioxide thin-film layers can also be deposited using sputtering. The big advantage of sputtering is that the deposition can be performed at temperatures as low as room temperature. The deposited layers are amorphous in microstructure and electrically insulating. Like other sputtered deposited material types, the residual stress in sputter deposited silicon dioxide can exhibit either tensile or compressive stresses depending on the processing conditions. This is demonstrated in Table 16, which shows the measured residual stresses values of sputter deposited layers as a function of processing conditions.

Table 16. Mechanical Material Properties of Sputter-Deposited Silicon Dioxide Material for Various Processing Conditions [133].

Substrate	Power (kW)	Pressure (mtorr)	Temperature (°C)	Residual Stress (MPa)
Silicon, Quartz	0.1 to 0.3	5 to 20	25 to 285	−90 to 3000

6.7. Atomic Layer Deposition (ALD)

The residual stresses in the ALD deposited layers varies significantly depending on the processing conditions. Table 17 lists some of the reported measurement data on residual stresses of ALD Al_2O_3, and Table 18 for ZnO.

Table 17. Residual Stresses of Al_2O_3 Deposited using ALD methods.

Temp. (°C)	Gases	Residual Stress (MPa)	Refs.
130	$Al(CH_3)_3$; H_2O; N_2	228	[134]
177	$Al(CH_3)_3$; H_2O; N_2	383 to 474	[135]

Table 18. Residual Stresses of ZnO Deposited using ALD methods.

Temperature (°C)	Gases	Young's Modulus (GPa)	Refs.
100	$Zn(CH_2CH_3)_2$ H_2O N_2	134	[136]
177	$Zn(CH_2CH_3)_2$ H_2O N_2	143	[136]

6.8. Electrochemical Deposition

The material properties of electroplated materials are very dependent on the processing conditions. For example, it has been reported that electroplated nickel exhibited a residual stress that varied from −110 to 150 MPa over a range of plating current densities from 0 to 30 mA/cm^2 at a temperature of 60 °C [137]. Therefore, by adjusting the process settings, it is possible to plate a nickel film with a moderate to zero state of stress. As shown in Table 19, for the materials listed the residual stresses are dependent on the process conditions used in the electroplating and mostly have modest to neutral stress levels.

Table 19. Material Properties for Various Electroplated Materials as a Function of Processing Conditions.

Temp. (°C)	Mat. Plated	Subs.	Plating Mixture	Plating Current Den * (mA/cm^2)	Plating Rate * (nm/min)	Residual Stress (MPa)	Refs.
20 to 80	Ni	5 nm Cr and 60 nm Cu on Si wafer	$Ni(SO_3NH_2)4H_2($ −300 g/L $NiCl_26H_2O$ −10 g/L H_3BO_3 −40 g/L	0 to 30	0 to 550	−110 to 150 at 60 °C	[137]
-	Cu	Cu seed layer on top of Ta or TiW layer on Si wafer	$CuSO_4 \cdot 5H_2O$ with proprietary additives	-	-	0 to 200 as deposited depending on thickness, increased by 200 MPa after anneal at 350 °C	[138]

Table 19. *Cont.*

Temp. (°C)	Mat. Plated	Subs.	Plating Mixture	Plating Current Den * (mA/cm^2)	Plating Rate * (nm/min)	Residual Stress (MPa)	Refs.
-	Cu	SiNx on Si wafer	-	-	-	100	[139]
-	Au	50 nm Ti and 100 nm Au on Si wafer	Sulfite gold chemistry	1 to 9 pulsed	20 to 185	−90 to 90	[140]
50	Au	Si wafer	Sulfite gold chemistry with agitation	1 to 5	-	−105 to −10	[141]

* A dash in the Table means that the data was not given in the source.

7. Methods to Manage or Mitigate the Effects of Residual Stresses

The presence of a residual stress in a thin-film material layer will have an impact on the behavior of MNS devices made using this material layer. However, the impact needs to be analyzed with respect to the specific device application requirements in each situation. The methods used for conducting this type of analysis will not be reviewed here, but may be found elsewhere [3]. In most MNS device designs, the state of stress in the thin-film material layers is usually desired to be near zero in value or have a small magnitude of tensile stress, and a near zero stress gradient. There are several broad categories of methods that can be used to manage or mitigate the effects of residual stresses in thin-film material layers. Additionally, there are methods to enhance the behavior of some materials using the tailoring of the residual stress in the material layer. These methods are reviewed in this section.

7.1. Select an Alternative Deposition Process Resulting in the Desired Residual Stress

The review of the residual stresses in the deposited thin-film material layers showed that LPCVD methods often have larger residual stresses compared to other deposition methods. Nevertheless, as shown in Table 2 and depending on the deposition temperature, LPCVD polysilicon can also exhibit low to moderate residual stress values [79]. Therefore, it would seem that selecting a deposition temperature for a LPCVD process where the residual stress is lower would provide a solution. However, there may be undesirable effects associated with using some LPCVD deposition temperatures that needs to be understood and carefully analyzed. Specifically, the MNS device designer must also consider the impact on the other material properties of the thin-film layer. For example, in most circumstances the polysilicon layer is desired to be electrically conductive. That means the material must be suitably doped and the microstructure appropriate for electrical conductivity and the desired levels of conductivity may not be possible with different LPCVD deposition temperatures.

PECVD processes provide comparatively wider latitude for manipulating the process parameters in order to get a desired residual stress value. PVD Sputtering provides some latitude in tailoring the residual stress state of the deposited layers as well. Therefore, it would appear that selecting an alternative deposition process, such as PECVD or PVD, wherein the residual stresses are typically lower in magnitude compared to LPCVD would be attractive. Again, the other material properties must also be considered on their impact to the device performance and on the process sequence. For example, PECVD deposited silicon layers are often amorphous and are not good electrical conductors; same with PVD deposited silicon layers [111–114,128,129,142]. Additionally, some materials deposited using methods such as PECVD have densities, thermal conductivity, electrical breakdown voltages, etc. divergent from that of the same material in bulk form. For example, many silicon dioxide material layers deposited using PECVD are far less dense than bulk SiO_2 and have lower electrical breakdown voltages [15].

7.2. Elevated Temperature Anneal

Another method to lower a high value of residual stress is to perform a post-deposition elevated temperature anneal of the thin-film material layer that allows the residual stress state to relax to a lower magnitude [5]. This method was one of the first to be used and can be very effective. For example, it has been shown that the residual stress in LPCVD polysilicon layers can be reduced from 500 MPa compressive to less than about 10 MPa compressive after an anneal at 1000 °C [122]. The annealing is typically performed in a non-oxidizing ambient such as nitrogen. These anneals can be performed in conventional furnaces or using rapid thermal anneal (RTA) systems with the latter being preferable since it will reduce the thermal impact on dopants and other materials that may be present on the substrate [143–145] However, a RTA may require higher temperatures in order to obtain the same value of lowered residual stress.

Importantly, the annealing temperature required to reduce the residual stress may depend on the deposition parameters. For example, the annealing of polysilicon thin-films at temperatures of 1000 °C or higher deposited at or near the amorphous to crystalline transition (i.e., deposition performed at around 570 °C) an anneal of 1000 °C will reduce the residual stress to near zero, where polysilicon films that are highly textured (i.e., deposition performed at around 625 °C), an anneal at 1100 °C will be required to reduce the residual stress to near zero [80,81]. High temperature anneals reduce the stress in polysilicon thin-films for undoped and doped layers as seen in Tables 20 and 21, respectively.

Anneals can be used on most thin-film material layer types that are deposited using nearly any deposition methods including: LPCVD; PECVD; and PVD. The temperatures of the anneals to obtain modified material properties will depend on the material type, deposition method, and time of the anneal [3]. For example, upon exposure to an elevated temperature, PECVD silicon dioxide will undergo a densification [15]. This will reduce the thickness of the material layer considerably and also result in out-gassing during densification. It will also increase the electrical breakdown voltage. More problematically, if another material layer is deposited over the PECVD silicon dioxide and an elevated temperature exposure is involved, the PECVD silicon dioxide will densify and out-gas and may possibly cause the overlying material layer to delaminate, bubble, or crack. The solution to this problem is to expose the PECVD silicon dioxide layer to an elevated temperature to densify the layer prior to the deposition of another material layer on top of the oxide. However, this will require a high temperature exposure that may impact other material layers present on the substrate and the device functionality.

Anneals can serve multiple purposes including: reducing the residual stress; densifying the material layer; electrically activating dopants; etc. However, the issue with annealing is that it requires the exposure of the substrate to a higher temperature than the deposition was performed. A major reason for selecting some deposition processes such as PECVD and PVD is that the deposition temperatures are relatively low thereby reducing the impact of a high temperature exposure to other material layers present on the substrate. Therefore, performing an anneal can negate the benefit of a low deposition process. One caveat to this is that for some material layer types, a rapid thermal anneal (RTA) whereby the temperature may be high, but is performed over such a short period of time that the impact on other material layers may be minimized is one technique where an anneal can be performed while still retaining the desirable material properties of the other materials [3,146]. Importantly, each material type and situation must be experimentally explored and analyzed in order to determine if an RTA anneal is suitable or not. The use of composite layers, such as nickel silicide, and annealing has also been reported for reducing stresses in polysilicon layers [146].

Table 20. Residual Stress in Undoped LPCVD Polysilicon Thin-Film Material Layers After High-Temperature Annealing.

Deposition Temperature (°C)	Silane Flow Rate (sccm)	Pressure (mtorr)	Thickness (microns)	Anneal Conditions	Residual Stress (MPa)	Refs.
565	-	-	1	1050 °C for 10 s using RTA	142	[147]
570	80	150	1.3	1200 °C for 6 h	17	[80]
570	100	300	2	1100 °C for 30 min	30	[81]
580	-	-	3.5	1000 °C for 1 h	12 ± 5	[148]
615	100	300	2	1100 °C for 30 min	−20	[146]
620	70	300	0.5	900 to 1150 °C for 1 to 10 s, RTA	−340 to 90	[144]
620	70	100	0.46	1100 °C for 2 h	Low stress	[84]
625	80	180	3	1200 °C for 6 h	−205	[80]
630	-	-	4	1000 °C for 90 min	42	[149]

Table 21. Residual Stress in Doped LPCVD Polysilicon Thin-Film Material Layers After High-Temperature Annealing.

Deposition Temperature (°C)	Silane Flow Rate (sccm)	Deposition Pressure (mtorr)	Thickness (microns)	Doping and Anneal Conditions	Resultant Residual Stress (MPa)	Refs.
560 to 610	100	375 to 800	2	PH_3, anneal 900 °C, 10 to 120 s	−195 to 310	[150]
580	-	-	2	P implant, anneal 950 °C, 1 to 10 h	40.3 to 83.9	[151]
580	-	350	2	P implant or diff, variable anneal	26 to 72	[152]
585	50	200	0.5	$POCl_3$ diff, 8 50 to 950 °C	−110	[141]
625	80	180	3	$POCl_3$ diff, anneal 1200 °C for 6 h	−98 to 11	[80]

Similarly, an anneal can also be performed to reduce the residual stresses in LPCVD deposited silicon dioxide layers. Some of the reported data on the effects of annealing are provided in Table 22. Other thin-film material layers also show reduction in the residual stress after performing an anneal. Further information about the effects of anneals on thin-film material layers can be found in the literature [3].

Table 22. Resultant residual stresses in annealed SiO_2 thin-films deposited using LPCVD.

Material Type	Source Gases	Deposition Temperature (°C)	Anneal Conditions	Residual Stress (MPa)	Refs.
PSG	$SiH_4/O_2/PH_3$	425	600 °C for 30 min	0	[85]
LTO	SiH_4/O_2	425	600 °C for 30 min	20	[85]
PSG	$TEOS/O_2/PH_3$	700	950 °C for 10 min	100	[85]

7.3. Modification of the Material Stoichiometry

A method to change the residual stress in thin-film material layers is to modify the chemical makeup or stoichiometry of the deposition process. The most prominent example of this is silicon nitride. The extremely high values of stoichiometric silicon nitride, Si_3N_4, prompted MNS researchers to develop alternative recipes for LPCVD SiN that lower the residual stresses in these types of material layers.

Specifically, it has been shown that by modifying the ratio of dichlorosilane to ammonia to about 6 parts dichlorosilane to 1 part ammonia can result in almost zero residual stress in SiN thin-film layers deposited using LPCVD [153–156]. The resulting material layers are silicon-rich compared to the stoichiometric SiN layers, with a chemical makeup of about $Si_1N_{1.1}$ [154], although other chemistries are possible depending on the process conditions. Studies of the ratio of silicon and nitrogen and the effect on the residual stress in the deposited thin-film layers have been reported wherein it was shown that increasing the silicon content results in a lowering of the residual stress [89,157,158]. A significant lowering of the residual stress allows SiN layers of 2 microns to be deposited without the cracking seen in the stoichiometric materials.

Additionally, additional silicon content in these layers results in films that have lower etch rates in HF. The ability to deposit nearly stress-free SiN layers has been an important advancement for MNS devices [89,157].

Table 23 shows some of the reported values of the mechanical properties of SiN films as a function of the processing conditions, including the ratio of dichlorosilane to ammonia. As can be seen, the residual stress can vary over a very large range depending on the exact process recipe used for the deposition. More importantly, this table illustrates the relationship between the ratio of dichlorosilane to ammonia and residual stress in the deposited layers.

Table 23. Residual Stress Values of as-deposited LPCVD SiN thin-films layers.

Temperature (°C)	Gases	Ratio	Pressure (mtorr)	Thickness (microns)	Young's Modulus (GPa)	Residual Stress (MPa)	Refs.
785	SiH_2Cl_2/NH_3	6	368	-	230	430	[88]
-	SiH_2Cl_2/NH_3	4	-	0.3	295	322	[158]
850	SiH_2Cl_2/NH_3	5	-	1	186	108	[159]
850	SiH_2Cl_2/NH_3	5.7	150	-	230	125	[89]
880	SiH_2Cl_2/NH_3	4	600	-	-	1 ± 10	[160]

In addition to modifying the residual stresses in deposited layers, the other material properties of silicon-rich SiN layers are also different from those of stoichiometric material layers. For example, the tensile strength of the silicon-rich material was found to be 5.5 GPa compared to 6.4 GPa for Si_3N_4 and a similar decrease in the fracture toughness has been reported [158,161]. Low-stress SiN thin-film layers were reported to have a measured mass density of 3.0 g/cm^3, a thermal conductivity of 3.2 $\times$ 10^{-2} W/cm °K, and a heat capacity of 0.7 J/g °K [156].

Another material type where the residual stress is a function of the stoichiometry is silicon oxy-nitride. For example, it was reported that varying the nitrogen flow rate into the reactor chamber during deposition from 60 standard cubic centimeters per minute (sccm) to 75 sccm resulted in the residual stress varying from nearly 800 MPa compressive at 60 sccm, to nearly zero residual stress at 69 sccm, to around 600 MPa tensile at 75 sccm, wherein the stoichiometry of the deposited layers in terms of atomic concentrations tended to increase in nitrogen content and lower oxygen content at higher N_2 flow rates using PECVD [162].

7.4. Stacking Material Layers

The state of stress in a material layer can be compensated by creating of stack of multiple material layers wherein the residual stress values of the different layers are different from one another and mostly cancel out the differing residual stresses to create a composite layer material system layer that exhibits a more desirable residual stress value.

For example, a tensile-stressed silicon nitride layer combined with a compressively stressed polysilicon layer on the top and bottom sides was reported to exhibit no buckling in free-standing beams and membranes [163]. In other work, pressure sensors were fabricated wherein boron-doped etch-stop membranes having a tensile residual stress were compensated with an overlay of CVD silicon dioxide having a compressive residual stress [164]. While the composite layers have near zero resultant residual stress, the unbalanced nature of this configuration does result in a stress gradient. Fabry–Perot optical cavity sensors were made using $Si_xN_y/SiO_2/Si_xN_y$ membranes wherein the Si_xN_y layers are under tensile residual stress and the silicon dioxide layer is under compressive residual stress [165]. The result is a structure having a slightly tensile residual stress and a more balanced stress state. The use of multiple layers of polysilicon having alternating states of residual stress, wherein one layer of polysilicon with a tensile residual stress would be alternated with another layer of polysilicon with a compressive residual stress was reported to result in a composite layered structure with near zero residual stress and stress gradient and could have the residual stress in the composite structure made to a range of values depending on the thicknesses of the relative layers [166]. It is conceivable using the approach of multiple stacked layers that almost any value of resultant residual stress can be obtained.

There are some disadvantages of this approach. First, depositing multiple layers increases the cost and complexity of the manufacturing of the MNS devices in proportion to the number of layers involved in the composite material layer. Second, the overall material properties of the composite material system layer may be different from that of the different layers involved. For example, using a non-electrically conducting thin-film material layer such as silicon nitride to compensate for the stress in an electrically conductive polysilicon thin-film material layer will result in a composite layer having a reduced electrical conductivity. Third, since the deposition of additional material layers may involve exposure to higher temperatures, this may alter the material properties of any pre-existing material layers on the substrate. Therefore, the MNS device designer must evaluate the impact of these other depositions on the resultant MNS device behavior and process sequence. Fourth, if the composite structure employs different material layers then the overall structure may suffer from stress imbalances created in the differing thermal expansion coefficients of the materials involved.

7.5. Ion-Assisted Deposition

As noted in the section on deposition methods above, specifically related to sputtering, the ions in the plasma can be induced to impact the thin-film layer during growth as a means to modify the material properties of the layer. This occurs since the electrons in the plasma are extremely mobile and easily escape thereby resulting in a positively charged plasma that then repels positive ions which can impact the growing thin-film layer.

Additionally, a bias voltage can be placed on the substrate to add further energies to these positive ions. This is often termed plasma ion-assisted deposition. However, as can be appreciated from the description of the mechanism of this phenomenon, there is limited flexibility with regard to how much the process can be modified using this approach, and therefore, limited flexibility on how much the thin-film material layer can be modified. Nevertheless, based on this concept, the use of a separate source of ions should be able to perform this function and provide a more independent mechanism for the control of the resultant layer material properties.

This technique is often referred to as "ion-beam assisted deposition (IBAD)" and is one of the more recent developments in modifying the material properties in deposited material layers as well as controlling the state of residual stress. IBAD is combined with some form of physical vapor deposition (PVD), such as evaporation or sputtering [167]. In IBAD, the material species are liberated from a target using a PVD method and are concurrently bombarded by a separate and independently generated flux of ions impacting the thin-film material layer as it is being deposited. As the atoms or molecules condense on the substrate surface forming a thin-film material layer, energetic ions typically in the range of a few tens to a few hundreds of eV are directed and impact the surface of the growing thin-film layer.

There are different designs of ion-beam assisted deposition sources whereby ions, typically positively charged ions, are produced, extracted and then accelerated to impinge onto the substrate, with the Kauffmann source configuration more widely used [168,169]. The Kauffmann ion source is a broad-beam source and uses an electrified extraction grid for pulling ions from a source plasma and this creates a relatively narrow distribution of ion energies for impact on the substrate surface. This allows higher levels of precision for process parameter adjustment and thereby more accurate control over the resultant material properties. Another advantage is that the Kauffmann source can be integrated into many PVD deposition systems relatively easily and at a modest cost. The ions extracted from the source are positive and so would rapidly charge any dielectric material preventing further bombardment. Sufficient electrons to neutralize this charge are therefore added to the extracted beams yielding what is usually termed a neutral ion beam.

There are several effects that ion beam assisted deposition as on thin-film layers during deposition. First, the layers are denser compared to other PVD methods without the use of IBAD. A denser film normally exhibits material properties more closely aligned with those of the equivalent bulk material. For example, a denser layer typically will have a refractive index close to that of the bulk material [170]. It is believed the reason for the increased density is based on ion impingement disrupting the columnar growth that would otherwise occur in thin-film growth. Columnar growth is due to the piling up of adatoms on top of formed nuclei resulting in the shadowing of in-between regions resulting in columnar crystals having empty voids between the columns. Ion bombardment reduces the void spacing by creating a tighter packing of the crystals. There is an optimal ion energy to obtain a maximum density. However, beyond that optimal energy the material layer can be altered by shortening of the bond distances between the atoms and increased number of atoms in interstitial locations. This can result in a compressive residual stress in the thin-film material layer. The increased density of IBAD PVD layers also reduces or completely eliminates the moisture sensitivity of the film [171]. This results in more stable material properties.

In short, the advantages of using IBAD in PVD deposition is that it provides additional independent processing parameters which can be adjusted to tailor the material properties of the deposited thin-film layers. Further, the ion bombardment improves the thin-film adhesion to the underlying substrate, allows control of the film layer morphology, density, crystallinity, chemical stoichiometry, and residual stress state [171].

7.6. Device Design Modifications

Another method for managing the residual stresses in thin-film layers used in a MNS device is to modify the design of the device to compensate with the value of the residual stress [3]. Of course, this can only be done if the value of the residual stress is already known.

As an example, the free-standing beam-type resonator device examined above in Section 3 was assumed to be made from a polysilicon thin-film material layer using a surface micromachining process wherein the ends were clamped. In that analysis it was assumed that the polysilicon thin-film layer had a tensile residual stress of varying values which caused the resonant frequency of the resonator to be changed to a higher value than if no residual stress was present. Therefore, if a certain residual stress in the polysilicon layer was unavoidable (perhaps due to other processing concerns), then a possible method to implement a resonator with the correct resonate frequency would be to modify the other parameters in the design to compensate for the presence of the residual stress. Examining Equation (1), it can be seen that this could easily be achieved by making the length of the resonator beam, L, longer than otherwise. Of course, each MNS device situation needs to be examined carefully in order to understand and determine how changing some of the dimensional parameters of the device affects the device's behavior and performance.

7.7. Functional Layer Behavior Using Residual Stress

The application of mechanical strains on materials impacts the physical properties and this effect can be used in piezoelectrics, electrostrictives, and ferroelectrics to couple strain to electrical polarization and magnetostrictive, magnetoelastic and ferromagnetic materials to couple strain to magnetization. The functonal properties of these materials are significantly impacted by the residual stress and strain. In general, thin-film layers that are piezoelectric, electrostrictive ferroelectric, magnetostrictive, magnetoelastic or ferromagnetic exhibit material properties that are considerably lower compared to these sames materials in bulk forms [2]. It has been reported in thin-film piezoelectric aluminum-nitride (AlN) deposited on a silicon dioxide layer on a silicon substrate that the film stress and the piezoelectric coupling factor are strigly interrelated and making it challenging to obtain both a low residual stress state and attractive piezoelectric properties [172]. Similar issues are reported with magnetic materials [2].

Recently, some researchers have reported modifying the residual stress in material layers to tune the properties of thin-film material layers. For example, strain tuning of the magnetic properties of a thin-film antiferromagnetic semiconductor CrSBr wherein a reversible strain-induced antiferromagnetic-to-ferromagnetic phase transition at zero magnetic field and strain control of the out-of-plane spin-canting process was reported [173]. It was shown that tuning of the in-plane lattice constant strongly modified the interlayer magnetic exchange interaction, which changes sign at the critical strain. This demonstrates that strain control of magnetism and other electronic states in low-dimensional materials and heterostructures can be achieved. In another study, thin-film layers of the hard magnetic material CoMnP were combined with Cu to form CoMnP/Cu multilayers reducing the residual stress by 23% compared to the same thickness of CoMnP while enhancing the maximum energy product by 430~690% [174].

In other work, interesting new properties of materials including negative thermal expansion coefficicient have been explained with potential application in controlling the thermal expansion of layers [175]. Specifically, the negative thermal expansion coefficient of CuO was found to be related to an electron-transfer-driven superexchange interaction using neutron scattering and principal strain axes analysis.

In other work, the magnetic and electric properties could be coupled by creating a composite configuration of a magnetoelastic and a piezoelectric materials wherein the strains are transferred between the materials [176]. In related work, a heterostructure consisting of magnetoelastic thin-film layer was grown on a piezoelectric substrate and an applied magnetic field induced dimensional changes in the magnetoelastic layer. This was explained as the magnetostriction modified the residual strain in the magnetoelastic film that was then transferred through to the piezoelectric substrate resulting in a polarization in the the piezoelectric substrate [177]. More advanced material layer deposition methods using molecular beam epitaxy (MBE) to deposit cobalt ferrite, magnetite, and others were developed to improve these types of material layers and enhance the effect [178,179]. One potential application of this effect is in multiferroic memories [180].

8. Conclusions

This review paper has covered the topic of residual stresses in deposited thin-film material layers extensively employed in the fabrication and manufacturing of MNS devices. The causes of these stresses were explained (to the extent they are known) and their impact on device behavior was illustrated in the example of a beam resonator made from a thin-film material layer. The deposition methods were reviewed and the parameters affecting the resultant residual stresses in the deposited layers were reviewed. A variety of methods and techniques that can be used to measure the residual stresses in material layers were described. This was followed by a review of the reported values of residual stresses in thin-film layers to illustrate how widely the values and signs can vary. Finally, methods to manage or mitigate the effects of the residual stress in thin-film material layers were explained.

Funding: No external funding was involved.

Institutional Review Board Statement: Not applicable.

Informed Consent Statement: Not applicable.

Data Availability Statement: Not applicable.

Conflicts of Interest: The author declares no conflict of interest.

Appendix A. Review of Thin-Film Deposition Methods

Appendix A.1. Thermal Oxidation

Thermal oxidation is one of the most commonly used methods in MNS fabrication. The most common processing conditions for oxidation are growth temperatures in the range from 700 °C to over 1150 °C performed at an ambient pressure of 1.0 Atm, using either oxygen (called dry oxidation) or stream (called wet oxidation) as the oxidizing agent.

An illustration of a typical oxidation system is shown in Figure A1 and consists of a fused quartz glass tube holding 50 to 100 wafers or more. The substrates are positioned in the tube and the tube is heated to process temperatures using resistive heaters. Process gases are introduced into the tube and controlled using in-line mass flow controllers.

The oxidation of silicon proceeds by a silicon dioxide layer growing on the surface of the silicon substrate with the interface advancing into the depth of the silicon material. The silicon dioxide formed on the surface has a thickness about 2.27 times that of the thickness of the consumed silicon material [15]. This volume expansion is a potential source of a residual stress in the grown SiO_2 layer and as noted in a later section, the residual stresses in thermal oxides tends to be compressive [5].

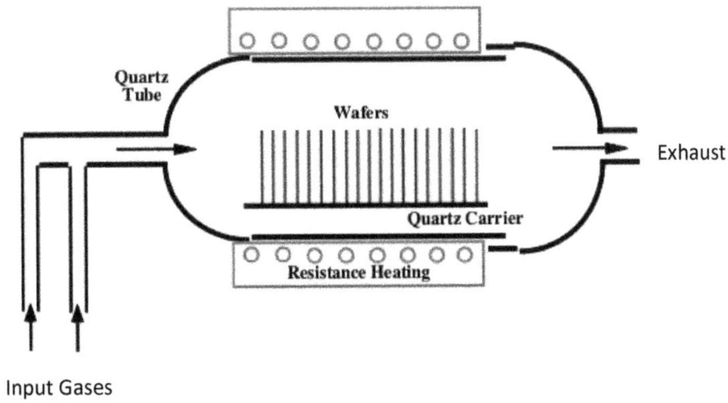

Figure A1. An oxidation system.

Appendix A.2. Chemical Vapor Deposition (CVD)

Chemical vapor deposition (CVD) is widely used in MNS manufacturing for the deposition of amorphous and polycrystalline thin-films, and under certain special conditions the deposition of single-crystal material layers [15].

A general model of CVD growth consists of a flux of reactants through a vapor boundary layer at the substrate surface in series with a second flux of the reactants that are involved in the chemical reactions which result in the thin-film material layer deposition [20]. These two fluxes are equal under steady-state conditions and the deposition rate is set by the lower flux. CVD processes exhibit growth rates dependent on temperature as illustrated in Figure A2. There are two major regions seen in this figure. The first, labeled the "mass transfer-controlled region", occurs at high process temperatures. In this region, growth rate is controlled by the arrival rate of chemical reactive species at the substrate surface. Operation in this region requires a uniform distribution of the process gases in order to get a uniform deposition. Therefore, systems operating in this region are usually single substrate process chambers and have high growth rates. At these temperatures, layers deposited in this region can exhibit high levels of crystallographic order; even single crystal material layers can be grown. Processes operating in this region are often called epitaxy [15,16].

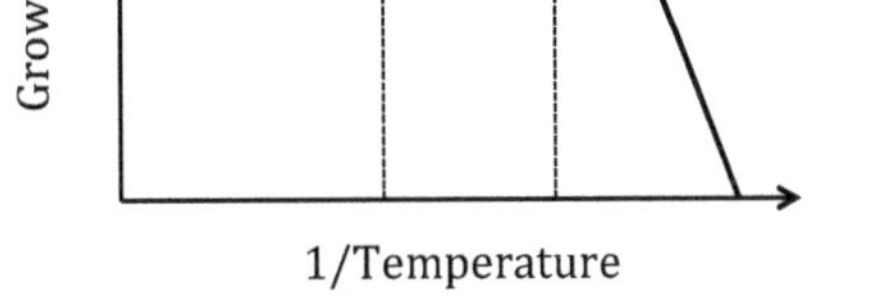

Figure A2. Relationship between the growth rate of the thin-film being deposited and the temperature (plotted as 1/temperature).

The second region is labeled "surface reaction controlled" and in this region the process is highly dependent on the temperature. Systems operating in this region should have excellent temperature control. These systems tend to process large batches of substrates so as to obtain large thermal masses and thereby good temperature control. In this region, the deposition rates are lower, but a large number of wafers can be processed simultaneously thereby offsetting the higher cost of a slower deposition rate. Layers deposited in this region tend to have less crystallographic order; they can be amorphous or polycrystalline [15,16].

There are a number of sub-categories of chemical vapor deposition including: atmospheric chemical vapor deposition (ACVD); low-pressure chemical vapor deposition (LPCVD); plasma-enhanced chemical vapor deposition (PECVD); and atomic layer deposition (ALD), which are reviewed below.

Appendix A.2.1. Atmospheric Chemical Vapor Deposition (ACVD)

Atmospheric chemical vapor deposition (ACVD) is performed at atmospheric pressure and the mass transfer-controlled region of Figure A2. ACVD is mostly used for the epitaxial deposition material layers such as silicon that is performed at temperature of 1000 °C or higher [21]. ACVD typically uses a single-wafer process tool configuration. The deposition rates of high-temperature ACVD can be relatively high, typically around 1 micron per minute or more. One major cause of residual stress for single-crystal depositions is any lattice mismatch that exists between the layer being deposited and the substrate and this usually occurs when the deposited layer is of a different material type from that of the substrate. Temperature is another process parameter having a major impact on the deposited material layer properties, particularly when the deposited material layer has a different TCE from that of the substrate.

Appendix A.2.2. Low-Pressure Chemical Vapor Deposition (LPCVD)

LPCVD depositions are usually performed in the surface reaction-controlled region of Figure A2. A typical deposition pressure ranges from about 10 mTorr to 100's mTorr and temperatures from about 400 °C to 950 °C. LPCVD is usually performed in a hot-walled reactor system composed of a fused silica tube (See Figure A3). Inlet process gas lines controlled by in-line mass flow controllers introduce gases into the tube. The tube is connected to a vacuum pump to obtain the process pressures and heating elements are positioned outside tube to obtain process temperatures.

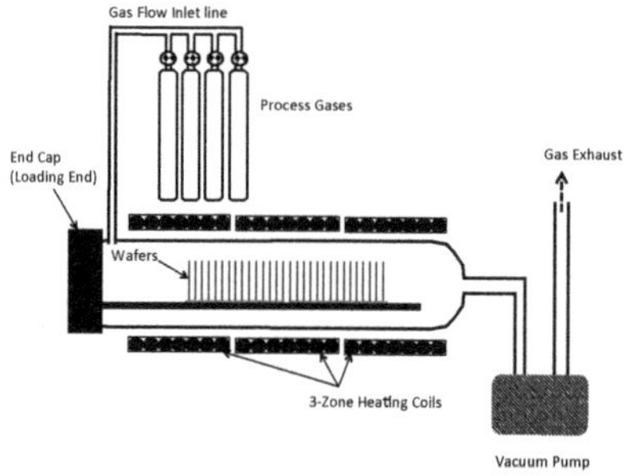

Figure A3. A low-pressure chemical vapor deposition (LPCVD) system.

Appendix A.2.3. Plasma-Enhanced Chemical Vapor Deposition (PECVD)

There are several different PECVD deposition reactor system configurations including: cold-walled parallel-plate; hot-walled parallel plate; and electron cyclotron resonance (ECR). Most PECVD deposition systems are the cold-walled parallel-plate reaction systems. Figure A4 is an illustration of the cold-walled parallel plate PECVD system. The reactor chamber pressure is lowered using a vacuum system. Process gases are introduced into the chamber using in-line mass transport controllers. Two electrodes, one above the wafer and the other under the wafer are connected to a RF-generator to create the plasma. The bottom electrode that the substrate is positioned onto is heated.

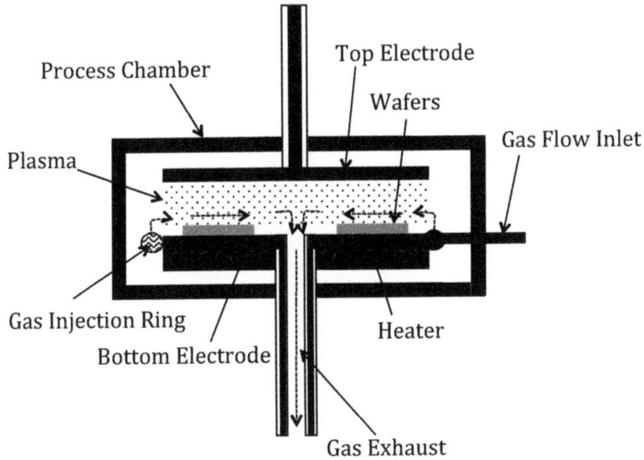

Figure A4. A cold-walled parallel plate plasma-enhanced chemical vapor deposition (PECVD) system.

Appendix A.2.4. Atomic Layer Deposition (ALD)

Atomic layer deposition (ALD) uses sequentially timed chemical processes to deposit material layers. In the first cycle, the first reactant gas is introduced and reacts with the substrate surface, followed by the next cycle where the second reactant gas is introduced and reacts with the reactants on the substrate surface from the first cycle [27,28]. This cycling of alternating reactive gases continues until the desired thickness of the layer is obtained. ALD deposition system consists of a process chamber, a heater substrate holder, a gas flow inlet for the introduction of the process gases, and an exhaust. The deposition pressures are typically in the range of mTorr and the deposition temperatures are typically in the range from 25 °C to over 400 °C [29,30]. ALD can also be done at lower temperatures using a plasma enhanced atomic layer deposition (PEALD) [31,32]. Among the material types that can be deposited using ALD include: SiO_2; Al_2O_3; ZrO_2; HfO_2; TiN; TaN; MgF_2; AlF_3; Pt; Ru; Ni; W; Ti; among others [33].

Appendix A.3. Physical Vapor Deposition (PVD)

Physical vapor deposition (PVD) use physical methods, such as heating or ion bombardment, to produce a vapor phase of the source material that condenses onto the substrate surface. One of the major advantages of PVD deposited materials are they can be deposited at relatively low temperatures in comparison to CVD processes. There are two main types of PVD, evaporation and sputtering.

Appendix A.3.1. Evaporation

Electron-beam bombardment of the target is the most common method of providing thermal energy in evaporation. The electrons in the beam are accelerated to a high kinetic energy (i.e., 5 to 30 keV) and steered to impinge on the target material using magnetic and electrical fields. Deposition rates of around 1 micron/min can be obtained depending on the e-beam energies and type of target material. Evaporation is performed at high vacuum levels in order for the electron beam not to be disrupted by gases in the chamber, typically a vacuum level of 10^{-7} torr or lower is used. Figure A5 shows an evaporative electron-beam physical vapor deposition (EBPVD) system. The system consists of a process chamber connected to a high-vacuum pumping system.

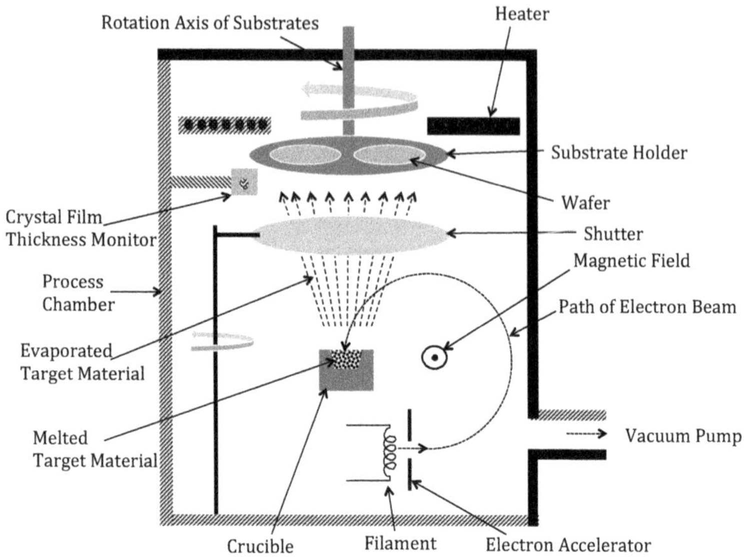

Figure A5. Evaporation tool configuration.

The substrates are usually rotated in order to improve the uniformity of the deposited layers. Heaters may be used to provide heating to the substrates during deposition. A crystal sensor for in situ monitoring the thickness of the deposited material layer is typically employed. Evaporation can be used to deposit multi-component thin-films by having two or more crucibles, each with a pure elemental target material and switching the electron-beam between the crucibles, or scanning each target with a dedicated electron beam. Additionally, reactive evaporation is performed by introducing controlled amounts of reactive gases into the process chamber during deposition.

Appendix A.3.2. Sputtering

The target may be electrically biased (with either DC or AC signals or both), and this can be used to provide additional energy to the ions impinging onto the target. A DC bias can also be placed on the substrate. If the target material is composed of an insulating material then a RF bias is used to reduce charge buildup on the target surface. The wafer (and target) can be heated (or cooled in some systems) during deposition. Sputtering is performed at a vacuum pressure in the mTorr range that translates into a mean-free path of around 1 cm. There are essentially three variations of the configuration of sputtering systems: DC bias sputtering; radio frequency (RF) bias sputtering; and magnetron sputtering.

Figure A6 is an illustration of a simple DC bias sputtering system configuration. The wafer is mounted onto a chuck electrode that is connected to a negative voltage potential acting as the anode. The material target is mounted onto another electrode separate from the first that is connected to a positive voltage potential acting as the cathode. An RF bias sputtering system (not shown) replaces the DC electrical connections to the anode and cathode.

Magnetron sputtering employs a powerful magnetic field (generated by using either a solenoid or magnetics) positioned behind the target in a manner that their magnetic field lines are parallel to the target surface. This results in significantly higher ion efficiency, sputter yield, deposition rates and uniformity than either DC or RF sputtering, at the expense of a more complicated design configuration.

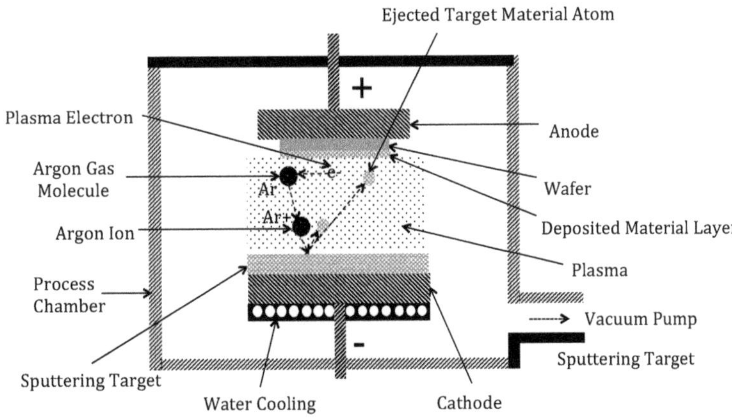

Figure A6. DC bias sputtering deposition system for physical vapor deposition.

Appendix A.4. Spin Casting

Spin casting is the deposition of a solution onto a wafer and then spinning it at a specific speed to obtain a uniform coating. Often the wafer after the layer has been spin coated is heated to drive off any solvents used to control the viscosity of the solution and harden the deposited layer. There are several materials that are spin casted.

Appendix A.4.1. SU-8

The lithographic process for SU-8 is very similar to a typical photoresist (e.g., spin coat, soft-bake, exposure, hard-bake; and development). It is also a thermoset polymer and a post-baking step is used to cross-link the polymers. Once cross-linked, SU-8 is an extremely robust material.

Appendix A.4.2. PDMS

Polydimethylsiloxane (PDMS) is a viscoelastic silicon-based organic polymer having a low density, is transparent over the visible spectrum, has a low mechanical stiffness, and is considered biocompatible. Moreover, since it has a low curing temperature, PDMS can be used with microsensors integrated with electronics without adverse effects on the electronics.

Appendix A.4.3. Polyimide

Polyimide is a polymer commonly used in the microelectronics industry for packaging applications [42,43]. It is an electrical insulating material resistant to heat. Polyimides exhibit either thermoset or thermoplastic behaviors. Some of the attractive material properties of polyimide for microsystems applications include: a low mechanical stiffness; biocompatibility, chemical and biological inertness, and low cytotoxicity. Other important characteristics of polyimides include: a high glass transition temperature, high thermal and chemical stability, low dielectric constant, high mechanical strength, low moisture absorption, and high solvent resistance. The processes for use of liquid forms of polyimide are very simple. The material is usually spun onto a wafer to obtain a uniform coating, followed by a cure at a temperature between 300 to 500 °C depending on the formulation. Polyimide exhibits shrinkage of between 40 to 50% during curing.

Appendix A.4.4. Sol-Gel PZT

Sol-gel PZT is a process whereby the constituents for a piezoelectric material, including lead, zirconate, and titanate (PZT) are in a polymer solution that can be spin casted onto substrates to form thin-film layers [44]. Typically, the sol-gel solution is spin casted in multiple layers, with soft-bakes between the spin casting of each layer, until the desired thickness is obtained. The spin-casted thin-film is then cured at a higher temperature, such as 650-C for 30 min or more.

Appendix A.5. Electrochemical Deposition

Electrochemical deposition methods are wet chemical processes, which involve the reduction of metal ions from an electrolyte solution resulting in the deposition of thin-film layer of metal atoms onto a surface [45–47]. There are two different types of electrochemical deposition processes: the first is electroplating where an electric current is passed between two electrodes placed in the electrolyte solution to supply the charges for the oxidation and reduction reactions; and, electro-less deposition wherein a reducing agent in the electrolyte solution provides the charges for these reactions. Electroless plating is rarely used due to challenges of reproducibility. A variety of metals can be electroplated including: Au; Ag; Cu; Cr; Pt; Ni; Zn; S; Cd; and Pb; as well as number of different alloys composed of metals that can be electrochemically deposited.

Figure A7 illustrates the setup for electroplating. Two electrodes placed into the electrolyte solution, one a cathode and the other an anode. The substrate on which the metal is to be plated from the solution is the cathode. The applied electrical current causes a reduction (addition of an electron) resulting in the deposition of the metal atoms on the surface. The anode is where an oxidation (removal of an electron) takes place. The current source is placed between the cathode and anode. The solution is stirred and temperature controlled, and the solution may be heated. The constant oxidation of the anode that is composed of the metal to be deposited maintains a steady replacement of metal ions for the electrolyte solution. Often wetting agents (i.e., surfactants), buffering agents and other agents are added to the electrolyte to enhance the plating process and/or the plated materials. Additionally, the applied currents can be DC or have a specific waveform that can enhance the plating process depending on the materials being plated.

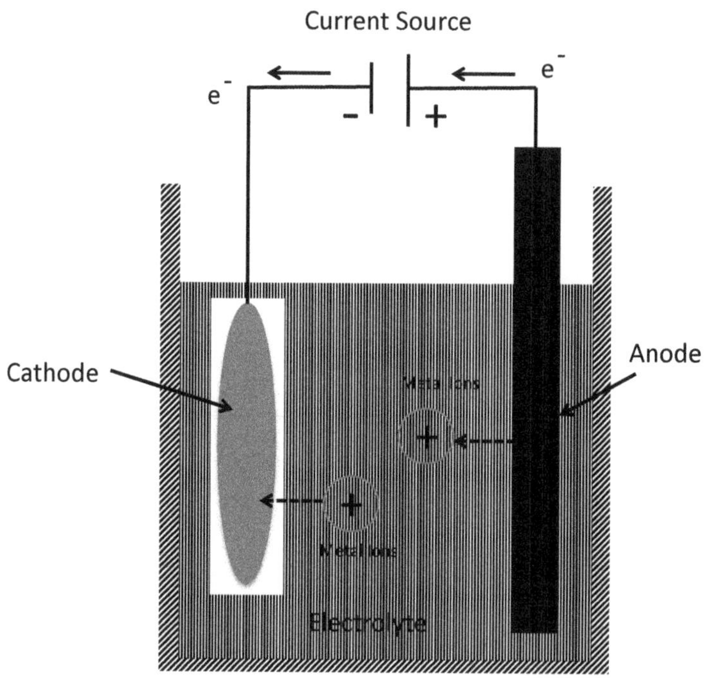

Figure A7. Setup for electroplating of metals onto substrates.

The material properties of electroplated layers are impacted by various process parameters including: chemistry solution used, temperature, stirring effectiveness, plating rate and electrical parameters used in plating including type of electrical waveforms used.

References

1. Senturia, S.D. *Microsystem Design*, 2nd ed.; Springer: New York, NY, USA, 2004.
2. Ghodssi, R.; Lin, P. (Eds.) *MEMS Materials and Processing Handbook*; Springer Press: New York, NY, USA, 2011.
3. Huff, M. *Process Variations Microsystem Manufacturing*; Springer/Nature Press: New York, NY, USA, 2020.
4. Fischetti, V.; Gamiz, F.; Hansch, W. On the enhanced electron mobility in strained-silicon inversion layers. *J. Appl. Phys.* **2002**, *92*, 7320. [CrossRef]
5. Ohring, M. *The Materials Science of Thin-Films*; Academic Press: New York, NY, USA, 1992.
6. Romig, A.D.; Dugger, M.T.; McWhorter, P.J. Material Issues in Microelectromechnical devices, science, engineering, manufacturability, and reliability. *Acta Mater.* **2003**, *51*, 5837. [CrossRef]
7. Mack, C. *Fundamental Principles of Optical Lithography: The Science of Microfabrication*, 1st ed.; Wiley: Hoboken, NJ, USA, 2008.
8. Smith, D.L. *Thin-Film Deposition: Principals and Practice*; McGraw-Hill: Boston, MA, USA, 1995.
9. Freund, L.B.; Suresh, S. *Thin-Film Materials: Stress, Defect Formation and Surface Evolution*; Cambridge University Press: Cambridge, UK, 2003.
10. Abadias, G.; Chason, E.; Keckes, J.; Sebastiani, M.; Thompson, G.B.; Barthel, E.; Doll, G.L.; Murray, C.E.; Stoessel, C.; Martinu, L. Review Article: Stress in thin-films and coatings: Current status, challenges, and prospects. *J. Vac. Sci. Technol. A Vac. Surf. Film.* **2018**, *36*, 020801. [CrossRef]
11. Huff, M. Silicon Micromachined Wafer Bonded Valves. Ph.D. Thesis, MIT, Cambridge, MA, USA, 1993.
12. Timoshenko, S. *Vibration Problems in Engineering*; Wolfenden Press: Singapore, 2007.
13. Material Property Database, MEMSNet, Polysilicon Material Properties. Available online: https://www.memsnet.org/material/polysiliconfilm/ (accessed on 2 September 2022).
14. Change, C.Y.; Sze, S.M. *ULSI Technology*; McGraw Hill: New York, NY, USA, 1996.
15. Wolf, S.; Tauber, R.N. *Silicon Processing for the VLSI Era, Volume 1-Process Technology*; Lattice Press: Sunset Beach, CA, USA, 1986.
16. Cambell, S.A. *The Science and Engineering of Microelectronic Fabrication*; Oxford Press: New York, NY, USA, 1996.
17. van Zant, P. *Microchip Fabrication*; McGraw Hill: New York, NY, USA, 2000.
18. Madou, M. *Fundamentals of Microfabrication*, 2nd ed.; CRC Press: Boca Raton, FL, USA, 2002.
19. Deal, B.E.; Grove, A.S. General Relationship for the Thermal Oxidation of Silicon. *J. Appl. Phys.* **1965**, *36*, 3770. [CrossRef]
20. Grove, A.S. Mass Transfer in Semiconductor Technology. *Ind. Eng. Chem.* **1966**, *58*, 48. [CrossRef]
21. Seshan, K. (Ed.) *Thin Film Deposition: Processes and Technologies*, 2nd ed.; Noyes Publications; William Andrew Publishing: Norwich, NY, USA, 2002.
22. Iscoff, R. Hotwall LPCVD Reactors: Considering the Choices. *Semicond. Int.* **1991**, 60.
23. Adams, A.C. *Dielectric and Polysilicon Film Deposition, VLSI Technology*; Sze, S.M., Ed.; McGraw Hill: New York, NY, USA, 1988.
24. Kamins, T.I. Structure and Properties of LPCVD Silicon Films. *J. Electrochem. Soc.* **1980**, *127*, 686. [CrossRef]
25. Compton, R.D. PECVD: A Versatile Technology. *Semicond. Int.* **1992**, 60.
26. Nguyen, S.; Burton, S.; Mackenzie, J.D. The Variation of Physical Properties of Plasma-Deposited Silicon Nitride and Oxynitride and Their Compositions. *J. Electrochem. Soc.* **1984**, *131*, 2348. [CrossRef]
27. Purunen, R.L. Surface chemistry of atomic layer deposition: A case study for the trimethlyaluminum/water process. *J. Appl. Phys.* **2005**, *97*, 12.
28. George, S.M. Atomic Layer Deposition: An Overview. *Chem. Rev.* **2010**, *110*, 111–131. [CrossRef] [PubMed]
29. Kim, H. Atomic layer deposition of metal and nitride thin films: Current research efforts and applications for semiconductor device processing. *J. Vac. Sci. Technol.* **2003**, *21*, 2231. [CrossRef]
30. Mackus, A.J.M.; Garcia-Alonso, D.; Knoops, H.C.; Bol, A.; Kessels, W.M.M. Room-Temperature Atomic Layer Deposition of Platinum. *Chem. Mater.* **2013**, *25*, 1769–1774.
31. Potts, S.E.; Kessels, W.M.M. Energy-enhanced atomic layer deposition for more process and precursor versatility. *Coord. Chem. Rev.* **2013**, *257*, 3254–3270. [CrossRef]
32. Langereis, E.; Knoops, H.C.M.; Mackus, A.J.M.; Roozeboom, F.; van de Sanden, M.C.M.; Kessels, W.M.M. Synthesis and characterization of low-resistivity TaNx films by remote plasma atomic layer deposition. *J. Appl. Phys.* **2007**, *102*, 83517. [CrossRef]
33. Parsons, G.N.; Elam, J.W.; George, S.M.; Haukka, S.; Jeon, H.; Kessels, W.M.M.; Leskelä, M.; Poodt, P.; Ritala, M. History of atomic layer deposition and its relationship with the American Vacuum Society. *J. Vac. Sci. Technol. A.* **2013**, *31*, 50818. [CrossRef]
34. Giang, R. *Vacuum Evaportion, Handbook of Thin Film Technology*; Maissel, L., Giang, R., Eds.; McGraw Hill: New York, NY, USA, 1970.
35. Hill, R.J. *Physical Vapor Deposition*; Temescal: Berkeley, CA, USA, 1976.
36. Werner, G.K.; Anderson, G.S. *The Nature of Physical Sputtering, Handbook of Thin Films*; Maissel, L.I., Glang, R., Eds.; McGraw Hill: New York, NY, USA, 1970.
37. Guerin, L.J.; Bossel, M.; Demierre, M.; Calmes, S.; Renaud, P. Simple and low-cost fabrication of embedded micro-channels by using a new thick-film photoplastic, Tech. In Proceedings of the Digest International Conference on Solid-State Sensors and Actuators, Chicago, IL, USA, 16–19 June 1997; 2001; p. 1419.

38. Ling, Z.G.; Lian, K.; Jian, L. Improved patterning quality of SU-8 microstructures by optimizing the exposure parameters. In Proceedings of the SPIE, International Society of Optical Engineering, Santa Clara, CA, USA, 23 June 2000; p. 1019.
39. Bhaget, A.A.S.; Jothimuthu, P.; Papautsky, L. Photodefinable polydimethylsiloxane (PDMS) for rapid lab-on-a-chip prototyping. *Lab Chip* **2007**, *7*, 1192. [CrossRef]
40. Cong, H.L.; Pan, T.R. Photopatternable conductive PDMS materials for microfabrication. *Adv. Funct. Mater.* **2008**, *18*, 1912. [CrossRef]
41. Tsougeni, K.; Tserepi, A.; Gogolides, E. Photosensitive poly(dimethylsiloxane) materials for microfluidic applications. *Microelectron. Eng.* **2007**, *84*, 1104. [CrossRef]
42. Wilson, A.M. Polyimide insulators for multilevel interconnections. *Thin Solid Films* **1981**, *83*, 145–163. [CrossRef]
43. Allen, M.G. Measurement of Adhesion and Mechanical Properties of Thin Films Using Microfabricated Structures. Ph.D. Thesis, MIT, Cambridge, MA, USA, May 1989.
44. Polcawich, R.G.; Pulskamp, J.S. Chapter 5—Additive Processes for Piezoelectric Materials: Piezoelectric MEMS. In *MEMS Materials and Processes Handbook*; Ghodssi, R., Lin, P., Eds.; Springer Press: New York, NY, USA, 2011.
45. Schlesinger, M.; Paunovic, M. *Modern Electroplating*; Wiley: New York, NY, USA, 2000.
46. Crow, D.R. *Principles and Applications of Electrochemistry*; Stanley Thornes Publishers Ltd.: Cheltenham, UK, 1998.
47. Zangari, G. Electro-deposition of Alloys and Compounds in Era of Microelectronics and Energy Conversion. *Coatings* **2015**, *5*, 195–218. [CrossRef]
48. Brown, M.A.; Rosakis, A.J.; Feng, X.; Huang, Y.; Ustundag, E. Thin film/substrate systems featuring arbitrary film thickness and misfit strain distributions. *Int. J. Solids Struct.* **2007**, *44*, 1755. [CrossRef]
49. Available online: http://www.tohotechnology.com/flx-series.php#Specification (accessed on 17 September 2022).
50. Chu, W.H.; Mehregany, M.; Ning, X.; Pirouz, P. Measurement of residual stress-induced bending moment of p+ silicon films. *Mat. Res. Soc. Symp.* **1992**, *239*, 160. [CrossRef]
51. Guckel, H.; Randazzo, T.; Burns, W.D. A simple technique for the determination of mechanical strain in thin films with application to polysilicon. *J. Appl. Phys.* **1985**, *57*, 1671. [CrossRef]
52. Timoshenko, S.P.; Goodier, J.N. *Theory of Elasticity*, 3rd ed.; McGraw-Hill: New York, NY, USA, 1934.
53. Chen, K.S. *Techniques for Residual Stress Measurement for MEMS and Their Applications*; MEMS/NEMS; Leondis, C.T., Ed.; Springer: Berlin/Heidelberg, Germany, 2006.
54. Van Drieenhiuzen, B.P.; Goosen, J.F.L.; French, P.J.; Wolffenbuttel, R.F. Comparison of techniques for measuring both compressive and tensile stress in thin films. *Sens. Actuators A* **1993**, *756*, 37–38. [CrossRef]
55. Ashurst, W.R.; Carraro, C.; Maboudian, R.; Frey, W. Wafer level anti-stiction coatings with superior thermal stability. *Sens. Actuators A* **2003**, *104*, 213. [CrossRef]
56. Guckel, H.; Burns, D.W.; Visser, C.C.G.; Tilmans, H.A.C.; Deroo, D. Fine-grained polysilicon films with built-in tensile strain. *IEEE Trans. Electron Devices* **1988**, *35*, 800. [CrossRef]
57. Sharma, A.; Bansal, D.; Kaur, M.; Kumar, P.; Kumar, D.; Rangra, K.J. Fabrication and analysis of MEMS test structures for residual stress measurement. *NSTI-Nanotech* **2011**, *2*, 344.
58. Allen, M.G.; Mehregany, M.; Howe, R.T.; Senturia, S.D. Microfabricated structures for the in situ measurement of residual stress, Young's modulus and ultimate strain of thin films. *Appl. Phys. Lett.* **1987**, *51*, 241. [CrossRef]
59. Osterberg, P.M.; Senturia, S.D. M-Test: A test chip for MEMS material property measurement using electrostatically actuated test structures. *J. Microelectromech. Syst.* **2002**, *6*, 107. [CrossRef]
60. Lin, L.A.P.; Pisano, R.T.; Howe, A. Micro Strain Gauge with Mechanical Amplifier. *IEEE JMEMS* **1997**, *6*, 313–321.
61. Srikar, V.T.; Spearing, S.M. A critical review of microscale mechanical testing methods used in the design of microelectromechanical systems. *Exp. Mech.* **2003**, *43*, 238. [CrossRef]
62. Klug, H.P.; Alexander, L.E. *X-ray Diffraction Procedures for Polycrystalline and Amorphous Materials*; Wiley and Sons: Hoboken, NJ, USA, 1974.
63. Ress, H.; Bruegemann, L. Good Diffraction Practice, Webinar, Bruker. 2010. Available online: https://www.bruker.com/fileadmin/user_upload/8-PDF-Docs/X-rayDiffraction_ElementalAnalysis/XRD/Webinars/Bruker_AXS_Good_Diffraction_Practice_III_-_Powder_XRD.pdf (accessed on 26 September 2022).
64. Keckes, J.; Bartosik, M.; Daniel, R.; Mitterer, C.; Maier, G.; Ecker, W.; Vila-Comamala, J.; David, C.; Schoeder, S.; Burghammer, M. X-ray nanodiffraction reveals strain and microstructure evolution in nanocrystalline thin films. *Scr. Mater.* **2012**, *67*, 748. [CrossRef]
65. Diebold, A.C. *Handbook of Silicon Semiconductor Metrology*; Marcel Dekker: New York, NY, USA, 2001.
66. Schroder, D.K. *Semiconductor Material and Device Characterization*; Wiley: New York, NY, USA, 1998.
67. Stefenelli, M.; Todt, T.; Riedl, A.; Ecker, W.; Muller, T.; Daniel, R.; Burghammer, M.; Keckes, J. X-ray analysis of residual stress gradients in TiN coatings by a Laplace space approach and cross-sectional nanodiffraction, a critical comparison. *J. Appl. Crystallogr.* **2013**, *46*, 1378–1385. [CrossRef]
68. Zeilinger, A.; Todt, T.; Krywka, C.; Muller, M.; Ecker, W.; Sartory, B.; Meindlhumer, M.; Stefenelli, M.; Daniel, R.; Mitterer, C.; et al. In-situ Observation of Cross-Sectional Microstructural Changes and Stress Distributions in Fracturing TiN Thin Film during Nanoindentication. *Sci. Rep.* **2016**, *6*, 22670. [CrossRef]
69. Gardiner, D.J. *Practical Raman Spectroscopy*; Springer: Berlin/Heidelberg, Germany, 1989.

70. Nakashima, S.; Inoue, Y.; Mitsuishi, A. The evaluation of the crystallinity of laser annealed polycrystalline silicon-on-insulator structures by Raman-microprobe polarization measurements. *J. Appl. Phys.* **1984**, *56*, 2989. [CrossRef]
71. Nix, W.D. Mechanical Properties of Thin Films. *Metallu. Trans. A* **1989**, *20A*, 2217. [CrossRef]
72. Suresh, S.; Gianakopoulos, A.E. A new method for estimating residual stresses by instrumented sharp indentation. *Acta Mater.* **1998**, *46*, 5755. [CrossRef]
73. Greco, A.; Sgambitterra, E.; Furgiuele, F. A new methodology for measuring residual stress using a modified Berkovich nano-indenter. *Int. J. Mech. Sci.* **2021**, *207*, 106662. [CrossRef]
74. Zeiss Orion Helium Ion Microscope Technical Data. Available online: https://asset-downloads.zeiss.com/catalogs/download/mic/102aa773-e50d-45eb-b8eb-3cf6d4542d24/EN_product-info_ORION-NanoFab_rel-2.0.pdf (accessed on 14 November 2022).
75. Korsunsky, A.M.; Sebastini, M.; Bemporad, E. Residual stress evaluation at the micrometer scale: Analysis of thin coatings by FIB milling and digital image correction. *Surf. Coat.* **2010**, *205*, 2393. [CrossRef]
76. Leconte, J.; Iker, F.; Jorez, S.; Andre, N.; Proost, J.; Pardoen, T.; Flandre, D.; Raskin, J.P. Thin films stress extraction using micromachined structures and wafer curvature measurements. *Microelectron. Eng.* **2004**, *76*, 219–226. [CrossRef]
77. Yang, J.; Gaspar, J.; Paul, O. Fracture properties of LPCVD silicon nitride and thermally grown silicon oxide thin films from the load-deflection of long Si3N4 and SiO2/Si3N4 diaphragms. *J. Microelectromech. Syst.* **2008**, *17*, 1120–1134. [CrossRef]
78. Kamins, T. *Polycrystalline Silicon for Integrated Circuits and Displays*, 2nd ed.; Kluwer: Boston, MA, USA, 1998.
79. Chen, L.; Miao, J.; Guo, L.; Lin, R. Control of stress in highly doped polysilicon multi-layer diaphragm structure. *Surf. Coat. Technol.* **2001**, *141*, 96–102. [CrossRef]
80. Gianchandani, Y.B.; Shinn, M.; Najafi, K. Impact of high-thermal budget anneals on polysilicon as a micromechanical material. *J. Microelectromech. Syst.* **1998**, *7*, 102–105. [CrossRef]
81. Yang, J.; Kahn, H.; He, A.Q.; Phillips, S.M.; Heuer, A.H. A new technique for producing large-area as-deposited zero-stress LPCVD polysilicon films: The multipoly process. *J. Microelectromech. Syst.* **2000**, *9*, 485–494. [CrossRef]
82. Dougherty, G.M.; Pisano, A.P.; Sands, T.D. Processing and morphology of permeable polycrystalline silicon thin films. *J. Mater. Res.* **2002**, *17*, 2235–2242. [CrossRef]
83. Dougherty, G.M.; Sands, T.D.; Pisano, A.P. Microfabrication using one-step LPCVD porous polysilicon films. *J. Microelectromech. Syst.* **2003**, *12*, 418–424. [CrossRef]
84. Maier-Schneider, D.; Maibach, J.; Obermeier, E.; Schneider, D. Variations in Young's modulus and intrinsic stress of LPCVD-polysilicon due to high-temperature annealing. *J. Micromech. Microeng.* **1995**, *5*, 121–124. [CrossRef]
85. Shioya, Y.; Mamoru, M. Comparison of phosphosilicate glass films deposited by three different chemical vapor deposition methods. *J. Electrochem. Soc.* **1986**, *133*, 1943–1950. [CrossRef]
86. Levin, R.M.; Adams, A.C. Low-pressure deposition of phosphosilicate glass films. *J. Electrochem. Soc.* **1982**, *129*, 1588–1592. [CrossRef]
87. Yang, J.; Paul, O. Fracture properties of LPCVD silicon nitride thin films from the load deflection of long membranes. *Sens. Act. A* **2002**, *97–98*, 520–526. [CrossRef]
88. Hong, S.; Weihs, T.P.; Bravman, J.C.; Nix, W.D. Measuring stiffness and residual stresses of silicon nitride thin films. *J. Electron. Mater.* **1990**, *19*, 903–909. [CrossRef]
89. French, P.J.; Sarro, P.M.; Mallee, R.; Fakkeldij, E.J.M.; Wolffenbuttel, R.F. Optimization of a low-stress silicon nitride process for surface micromachining applications. *Sens. Act. A* **1997**, *58*, 149–157. [CrossRef]
90. Sedky, S.; Witvrouw, A.; Baert, K. Poly SiGe, a promising material for MEMS monolithic integration with the driving electronics. *Sens. Act.* **2002**, *A97–A98*, 503–511. [CrossRef]
91. Witvrouw, A.; Tilmans, H.A.C.; De Wolf, I. Materials issues in the processing, the operation and the reliability of MEMS. *Microelectron. Eng.* **2004**, *76*, 245–257. [CrossRef]
92. Franke, A.E.; Heck, J.M.; King, T.J.; Howe, R.T. Polycrystalline silicon-germanium films for integrated microsystems. *J. Microelectromech. Syst.* **2003**, *12*, 160–171. [CrossRef]
93. King, T.-J.; McVittie, J.P.; Saraswat, K.C. Electrical properties of heavily doped polycrystalline silicon-germanium films. *IEEE Trans. Electron Dev.* **1994**, *41*, 228–232. [CrossRef]
94. Jeon, Y.C.; King, T.J.; Howe, R.T. Properties of phosphorus-doped poly-SiGe films for microelectromechanical system applications. *J. Electrochem. Soc.* **2003**, *150*, H1. [CrossRef]
95. Fu, X.A.; Jezeski, R.; Zorman, C.A.; Mehregany, M. Use of deposition pressure to control the residual stress in polycrystalline SiC films. *Appl. Phys. Lett.* **2004**, *84*, 341–343. [CrossRef]
96. Trevino, J.; Fu, X.; Mehregany, M.; Zorman, C. Low-stress, heavily-doped polycrystalline silicon carbide for MEMS applications. In Proceedings of the 18th International Conference on Micro Electro Mechanical Systems, Miami, FL, USA, 30 January–3 February 2005; pp. 451–454.
97. Fu, X.; Dunning, J.L.; Zorman, C.A.; Mehregany, M. Measurement of residual stress and elastic modulus of polycrystalline 3C-SiC films deposited by low-pressure chemical vapor deposition. *Thin Solid Films* **2005**, *492*, 195–202. [CrossRef]
98. Nagappa, S.; Zupan, M.; Zorman, C.A. Mechanical characterization of chemical-vapor deposited polycrystalline 3C silicon carbide thin films. *Scr. Mater.* **2008**, *59*, 995–998. [CrossRef]
99. Yamaguchi, Y.; Nagasawa, H.; Shoki, T.; Armaka, N.; Mitsui, H. Properties of heteroepitaxial 3C-SiC films grown by LPCVD. *Sens. Act.* **1996**, *A54*, 695–699. [CrossRef]

100. Hurtos, E.; Rodriguez-Viejo, J. Residual stress and texture in poly-SiC films grown by low pressure organometallic chemical-vapor deposition. *J. Appl. Phys.* **2000**, *87*, 1748–1758. [CrossRef]
101. Murooka, K.; Higashikawa, I.; Gomei, Y. Improvement of the Young's modulus of SiC film by low pressure chemical vapor deposition with B2H6 gas. *Appl. Phys. Lett.* **1996**, *69*, 37–39. [CrossRef]
102. Zhou, W.; Yang, J.; Sun, G.; Liu, X.; Yang, F.; Li, J. Fracture properties of silicon carbide thin films characterized by bulge test of long membranes. In Proceedings of the 3rd International Conference on Nano/Micro Engineered and Molecular Systems, Sanya, China, 6–9 January 2008; pp. 557–560.
103. Tarrif, A.; Daleiden, J.; Imer, S.; Prasai, D.; Hillmer, H. Stress investigation of PECVD dielectric layers for advanced optical MEMS. *J. Micromech. Microeng.* **2004**, *14*, 372–380. [CrossRef]
104. Zhang, X.; Chen, K.-S.; Ghodssi, R.; Ayon, A.A.; Spearing, S.M. Residual stress and fracture in thick tetraethylorthosilicate (TEOS) and silane-based PECVD oxide films. *Sens. Act. A* **2001**, *91*, 373–380. [CrossRef]
105. Walmsley, B.A.; Liu, Y.L.; Hu, X.Z.; Bush, M.B.; Dell, J.M.; Faraone, L. Poisson's ratio of low-temperature PECVD silicon nitride thin films. *J. Microelectromech. Syst.* **2007**, *16*, 622–627. [CrossRef]
106. Martyniuk, M.; Antoszewski, J.; Musca, C.A.; Dell, J.M.; Faraone, L. Dielectric thin films for MEMS-based optical sensors. *Microelectron. Reliab.* **2007**, *47*, 733–738. [CrossRef]
107. Huang, H.; Winchester, K.J.; Suvorova, A.; Lawne, B.R.; Liu, Y.; Hu, X.Z.; Dell, J.M.; Faraone, L. Effect of deposition conditions on mechanical properties of low-temperature PECVD silicon nitride films. *Mater. Sci. Eng. A* **2006**, *435–436*, 453–459. [CrossRef]
108. Zhou, W.; Yang, J.; Li, Y.; Ji, A.; Yang, F.; Yu, Y. Bulge testing and fracture properties of plasma-enhanced chemical vapor deposited silicon nitride thin films. *Thin Solid Films* **2009**, *517*, 1989–1994. [CrossRef]
109. Tabata, O.; Kawahata, K.; Sugiyama, S.; Igarashi, I. Mechanical property measurements of thin films using load-deflection of composite rectangular membrane. In Proceedings of the Micro Electro Mechanical Systems, Salt Lake City, UT, USA, 20–22 February 1989; pp. 152–156.
110. Gaspar, J.; Adrega, T.; Chu, V.; Conde, J.P. Thin-film paddle microresonators with high quality factors fabricated at temperatures below 110 °C. In Proceedings of the 18th International Conference on Micro Electro Mechanical Systems, Miami, FL, USA, 30 January–3 February 2005; pp. 125–128.
111. Chang, S.; Sivoththaman, S. Development of a low temperature MEMS process with a PECVD amorphous silicon structural layer. *J. Micromech. Microeng.* **2006**, *16*, 1307–1313. [CrossRef]
112. Alpuim, P.; Chu, V.; Conde, J.P. Amorphous and microcrystalline silicon films grown at low temperatures by radio-frequency and hot-wire chemical vapor deposition. *J. Appl. Phys.* **1999**, *86*, 3812–3821. [CrossRef]
113. Chung, C.-K.; Tsai, M.-Q.; Tsai, P.-H.; Lee, C. Fabrication and characterization of amorphous Si films by PECVD for MEMS. *J. Micromech. Microeng.* **2005**, *15*, 136–142. [CrossRef]
114. Chang, S.; Eaton, W.; Fulmer, J.; Gonzalez, C.; Underwood, B. Micromechanical structures in amorphous silicon. In Proceedings of the Technical Digest, International Conference on Solid State Sensors and Actuators, San Francisco, CA, USA, 24–27 June 1991; pp. 751–754.
115. Gromova, M.; Baert, K.; van Hoof, C.; Mehta, A.; Witvrouw, A. The novel use of low temperature hydrogenated microcrystalline silicon germanium (μcSiGe:H) for MEMS applications. *Microelectron. Eng.* **2004**, *76*, 266–271. [CrossRef]
116. Kannan, S.; Taylor, C.; Allred, D. PECVD growth of Six:Ge1-x films for high speed devices and MEMS. *J. Non-Cryst. Solids* **2006**, *352*, 1272–1274. [CrossRef]
117. Rusu, C.; Sedky, S.; Parmentier, B.; Verbist, A.; Richard, O.; Brijs, B.; Geenen, L.; Witvrouw, A.; Larmer, F.; Fischer, F.; et al. New low-stress PECVD poly-SiGe Layers for MEMS. *J. Microelectromech. Syst.* **2003**, *12*, 816–825. [CrossRef]
118. Schmid, U.; Eickhoff, M.; Richter, C.; Krötz, G.; Schmitt-Landsiedel, D. Etching characteristics and mechanical properties of a-SiC:H thin films. *Sens. Act.* **2001**, *A94*, 87–94. [CrossRef]
119. Rehder, G.; Carreño, M.N.P. PECVD a-SiC:H Young's modulus obtained by MEMS resonant frequency. *J. Non-Cryst. Solids* **2008**, *354*, 2359–2364. [CrossRef]
120. Flannery, A.F.; Mourlas, N.; Storment, C.W.; Tsai, S.; Tan, S.H.; Heck, J.; Monk, D.; Kim, T.; Gogoi, B.; Kovacs, G.T.A. PECVD silicon carbide as a mechanically resistant material for micromachined transducers. *Sens. Act.* **1998**, *A70*, 48–55. [CrossRef]
121. Du, J.; Zorman, C.A. Low temperature a-SiC/Si direct bonding technology for MEMS/NEMS. In Proceedings of the Technical Digest of the 14th International Conference on Solid State Sensors, Actuators and Microsystems, Lyon, France, 10–14 June 2007; pp. 2075–2078.
122. Gennissen, P.; Bartek, M.; French, P.J.; Sarro, P.M. Bipolar-compatible epitaxial poly for smart sensors: Stress minimization and applications. *Sens. Act.* **1997**, *A62*, 636–645. [CrossRef]
123. Lange, P.; Kirsten, M.; Riethmuller, W.; Wenk, B.; Zwicker, G.; Morante, J.R.; Schweitz, J.A. Thick polycrystalline silicon for surface-micromechanical applications: Deposition, structuring, and mechanical characterization. *Sens. Act.* **1996**, *A54*, 674–678. [CrossRef]
124. Greek, S.; Ericson, F.; Hansson, S.; Furtsch, M.; Rump, A. Mechanical characterization of thick polysilicon films: Young's modulus and fracture strength evaluated with microstructures. *J. Micromech. Microeng.* **1999**, *9*, 245–251. [CrossRef]
125. Kirsten, M.; Wenk, B.; Ericson, F.; Schweitz, J.A.; Reithmuller, W.; Lang, P. Deposition of thick doped polysilicon films with low stress in an epitaxial reactor for surface micromachining applications. *Thin Solid Films* **1995**, *259*, 181–187. [CrossRef]

126. De Sagazan, O.; Denoual, M.; Guil, P.; Gaudin, D.; Bonnaud, O. Microelectromechanical systems fast fabrication by selective thick polysilicon growth in epitaxial reactor. *Microsyst. Technol.* **2006**, *12*, 953–958. [CrossRef]
127. Hoffman, D.W.; Thornton, J.A. Internal stresses in Cr, Mo, Ta, and Pt films deposited by sputtering from a planar magnetron source. *J. Vac. Sci. Technol.* **1982**, *20*, 355. [CrossRef]
128. Abe, T.; Reed, M.L. Low strain sputtered polysilicon for micromechanical structures. In Proceedings of the 9th International Workshop on Micro Electro Mechanical Systems, San Diego, CA, USA, 11–15 February 1996; pp. 258–262.
129. Pal, P.; Chandra, S. RF sputtered silicon for MEMS. *J. Micromech. Microeng.* **2005**, *15*, 1536–1546. [CrossRef]
130. Honer, K.A.; Kovacs, G.T.A. Integration of sputtered silicon microstructures with prefabricated CMOS circuitry. *Sens. Act. A* **2001**, *91*, 392–403. [CrossRef]
131. Ledermann, N.; Baborowski, J.; Muralt, J.P.; Xantopoulos, N.; Tellenbach, J.M. Sputtered silicon carbide thin films as protective coating for MEMS applications. *Surf. Coat. Technol.* **2000**, *125*, 246–250. [CrossRef]
132. Inoue, S.; Namazu, T.; Tawa, H.; Niibe, M.; Koterazawa, K. Stress control of a-SiC films deposited by dual source dc magnetron sputtering. *Vacuum* **2006**, *80*, 744–747. [CrossRef]
133. Bhatt, V.; Chandra, S. Silicon dioxide films by RF sputtering for microelectronic and MEMS. *J. Micromech. Microeng.* **2007**, *17*, 1066–1077. [CrossRef]
134. Chang, Y.-J.; Gray, J.M.; Imtiaz, A.; Seghete, D.S.; Wallis, T.M.; George, S.M.; Kabos, P.; Rogers, C.T.; Bright, V.M. Micromachined resonators of high Q-factor based on atomic layer deposited alumina. *Sens. Act. A* **2009**, *154*, 229–237. [CrossRef]
135. Tripp, M.K.; Stampfer, C.; Miller, D.C.; Helbling, T.; Herrmann, C.F.; Hierold, C.; Gall, K.; George, S.M.; Bright, V.M. The mechanical properties of atomic layer deposited alumina for use in micro- and nano-electromechanical systems. *Sens. Act. A* **2006**, *130–131*, 419–429. [CrossRef]
136. Herrmann, C.F.; DelRio, F.W.; Miller, D.C.; George, S.M.; Bright, V.M.; Ebel, J.L.; Strawser, R.E.; Cortez, R.; Leedy, K.D. Alternative dielectric films for rf MEMS capacitive switches deposited using atomic layer deposited Al2O3/ZnO alloys. *Sens. Act. A* **2007**, *135*, 262–272. [CrossRef]
137. Luo, J.K.; Pritschow, M.; Flewitt, A.J.; Spearing, S.M.; Fleck, N.A.; Milne, W.I. Effects of Process Conditions on Properties of Electroplated Ni Thin Films for Microsystem Applications. *J. Electrochem. Soc.* **2006**, *153*, D155–D161. [CrossRef]
138. Volinsky, A.A.; Hauschildt, M.; Vella, J.B.; Edwards, N.V.; Gregory, R.; Gerberich, W.W. Residual Stress and Microstructure of Electroplated Cu Films on Different Barrier Films. *Mater. Res. Soc. Symp. Proc. Mater. Res. Soc.* **2002**, *695*, L1.11.1.
139. Xiang, Y.; Chen, X.; Vlassak, J.J. The Mechanical Properties of Electroplated Cu Thin Films Measured by Means of the Bulge Test Technique. *Mater. Res. Soc. Symp. Proc. Mater. Res. Soc.* **2002**, *695*, L4.9.1. [CrossRef]
140. Lawyer, P.H.; Fields, C.H. *Film Stress versus Plating Rate for Pulse-Plated Gold*; HRL Laboratories Report; HRL Laboratories: Malibu, CA, USA, 2001.
141. Elbrecht, L.; Catanescu, R.; Zacheja, J.; Binder, J. Highly phosphorus-doped polysilicon films with low tensile stress for surface micromachining using POCl3 diffusion doping. *Sens. Act.* **1997**, *A61*, 374–378. [CrossRef]
142. Brodsky, A.H.; Gambino, R.J. Electrical conduction in evaporated amorphous silicon films. *J. Non-Cryst. Solides* **1972**, *8*, 739–744. [CrossRef]
143. Howe, R.T.; Muller, R.S. Stress in polysilicon and amorphous silicon thin films. *J. Appl. Phys.* **1983**, *54*, 4674–4675. [CrossRef]
144. Zhang, X.; Zhang, T.Y.; Wong, M.; Zohar, Y. Rapid thermal annealing of polysilicon thin films. *J. Microelectromech. Syst.* **1998**, *7*, 356–364. [CrossRef]
145. Semmache, B.; Kleimann, P.; LeBerre, M.; Lemiti, M.; Barbier, D.; Pinard, P. Rapid thermal processing of piezoresistive polycrystalline silicon films: An innovative technology for low cost pressure sensor fabrication. *Sens. Act. A* **1995**, *46*, 76–81. [CrossRef]
146. Kang, T.J.; Lee, H.Y.; Kim, Y.H. Reduction in sheet resistance and low-thermal budget relaxation of stress gradients in polysilicon microcantilever beams using nickel-silicides. *IEEE JMEMS* **2007**, *16*, 279–288. [CrossRef]
147. Biebl, M.; von Philipsborn, H. Fracture strength of doped and undoped polysilicon. In Proceedings of the Technical Digest 8th International Conference on Solid-State Sensors and Actuators, Eurosensors IX, Stockholm, Sweden, 25–29 June 1995; pp. 72–75.
148. Kahn, H.; Tayebi, N.; Ballarini, R.; Mullen, R.L.; Heuer, A.H. Fracture toughness of polysilicon MEMS devices. *Sens. Act. A* **2000**, *82*, 274–280. [CrossRef]
149. Walker, J.A.; Gabriel, K.J.; Mehregany, M. Mechanical integrity of polysilicon films exposed to hydrofluoric acid solutions. *J. Electron. Mater.* **1991**, *20*, 665–670. [CrossRef]
150. Biebl, M.; Mulhern, G.T.; Howe, R.T. In situ phosphorus-doped polysilicon for integrated MEMS. In Proceedings of the Technical Digest 8th International Conference on Solid-State Sensors and Actuators, Eurosensors IX, Stockholm, Sweden, 25–29 June 1995; pp. 198–201.
151. Ericson, F.; Greek, S.; Soderkvist, J.; Schweitz, J. High sensitive internal film stress measurement by an improved micromachined indicator structure. In Proceedings of the Technical Digest 8th International Conference on Solid-State Sensors and Actuators, Eurosensors IX, Stockholm, Sweden, 25–29 June 1995; pp. 84–87.
152. Benitez, M.A.; Fonseca, L.; Esteve, J.; Benrakkad, M.S.; Morante, J.R.; Samitier, J.; Schweitz, J.A. Stress-profile characterization and test-structure analysis of single and double ion-implanted LPCVD polycrystalline silicon. *Sens. Act. A* **1996**, *54*, 718–723. [CrossRef]
153. Sekimoto, M.; Yoshihara, H.; Ohkubo, T. Silicon nitride single-layer x-ray mask. *J. Vac. Sci. Technol.* **1982**, *21*, 1017–1021. [CrossRef]

154. Gardeniers, J.G.E.; Tilmans, H.A.C.; Visser, C.C.G. LPCVD silicon-rich silicon nitride films for applications in micromechanics studied with statistical experimental design. *J. Vac. Sci. Technol. A* **1996**, *14*, 2879–2892. [CrossRef]
155. Temple-Boyer, P.; Rossi, C.; Saint-Etienne, E.; Scheid, E. Residual stress in low pressure chemical vapor deposition SiNx films deposited from silane and ammonia. *J. Vac. Sci. Technol. A* **1998**, *16*, 2003–2007. [CrossRef]
156. Mastrangelo, C.; Tai, Y.-C.; Muller, R. Thermophysical properties of low-residual stress, silicon-rich, LPCVD silicon nitride films. *Sens. Act.* **1990**, *A21–A23*, 856–860. [CrossRef]
157. Tsai, P.P.; Chen, I.-C.; Ho, C.J. Ultralow power carbon monoxide microsensor by micromachining techniques. *Sens Act.* **2001**, *B76*, 380–387. [CrossRef]
158. Kaushik, A.; Kahn, H.; Heuer, A.H. Wafer-level mechanical characterization of silicon nitride MEMS. *J. Microelectromech. Syst.* **2005**, *14*, 359–367. [CrossRef]
159. Bromley, E.I.; Randall, J.N.; Flanders, D.C.; Mountain, R.W. A technique for the determination of stress in thin films. *J. Vac. Sci. Technol. B* **1983**, *1*, 1364–1366. [CrossRef]
160. Olson, J.M. Analysis of LPCVD process conditions for the deposition of low stress silicon nitride. Part 1: Preliminary LPCVD experiments. *Mater. Sci. Semicond. Process.* **2002**, *5*, 51–60. [CrossRef]
161. Fan, L.S.; Howe, R.T.; Muller, R.S. Fracture toughness of brittle thin films. *Sens. Act.* **1990**, *A21–A23*, 872–874. [CrossRef]
162. Criado, D.; Alayo, M.I.; Fantini, M.C.A.; Pereyra, I. Study of the Mechanical and structural properties of silicon oxynitride films for optical applications. *J. Non-Cryst. Solids* **2006**, *352*, 2319. [CrossRef]
163. Farooqui, M.M.; Evans, A.G.R. Polysilicon microstructures. In Proceedings of the IEEE Micro Electro Mechanical Systems Workshop MEMS 91, Nara, Japan, 30 January–2 February 1991; pp. 187–191.
164. Cho, S.T.; Najafi, K.; Lowman, C.L.; Wise, K.D. An ultrasensitive silicon pressure-based flowmeter. In Proceedings of the. Intenational Electron Devices Meeting, Washington, DC, USA, 3–6 December 1989; pp. 499–502.
165. Han, J.; Kim, J.; Kim, T.-S.; Kim, J.-S. Performance of Fabry–Perot microcavity structures with corrugated diaphragms. *Sens. Actuators A* **2000**, *79*, 162–172. [CrossRef]
166. Heuer, A.H.; Kahn, H.; Yang, J. Low Stress Polysilicon Film and Method for Producing Same. U.S. Patent 626808B1, 31 July 2001.
167. Guglya, A.; Lyubchenko, E. Ion-beam-assisted deposition of thin films. In *Emerging Applications of Nanoparticles and Architecture Nanostructures*; Chapter 4; Elsevier: Amsterdam, The Netherlands, 2018.
168. Macleod, H.A. Recent developments in deposition techniques for optical thin films and coatings. In *Optical Thin Films and Coatings*, 2nd ed.; Elsevier: Amsterdam, The Netherlands, 2018.
169. Gago, R.; Jimenez, I.; Albella, J.M. Thin film Growth by Ion-Beam-Assisted Deposition Techniques. In *Materials Surface Processing by Directed Energy Techniques*; Elsevier: Amsterdam, The Netherlands, 2006.
170. Martin, P.J.; Macleod, H.A.; Netterfield, R.P.; Pacey, C.G.; Sainty, W.G. Ion-beam-assisted deposition of thin films. *Appl. Opt.* **1983**, *22*, 178–184. [CrossRef] [PubMed]
171. Martin-Palma, R.J.; Lakhtakia, A. Vapor-Deposition Techniques. In *Engineering Biomimicry*; Elsevier: Amsterdam, The Netherlands, 2013.
172. Iborra, E.; Olivares, J.; Clement, M.; Vergara, L.; Sanz-Hervas, A.; San-Grador, J. Piezoelectric properties and residual stress of sputtered AlN thin films for MEMS applications. *Sens. Acts. A* **2004**, *115*, 501–507. [CrossRef]
173. Cenker, J.; Sivakumar, S.; Xie, K.; Miller, A.; Thijssen, P.; Liu, A.; Dismukes, A.; Fonseca, J.; Anderson, E.; Xhu, X.; et al. Reversible strain-induced magnetic phase transition in a van der Waals magnet. *Nat. Nanotechnol.* **2022**, *17*, 256–261. [CrossRef]
174. Chen, Y.-S.; Lin, C.-C.; Chin, T.-S.; Change, J.-Y.; Sung, C.-K. Residual stress tuned magnetic properties of thick CoMnP/Cu multiplayers. *AIP Adv.* **2022**, *12*, 0350022.
175. Delaney, K.T.; Mostovoy, M.; Nicola, N.A. Superexchange-Driven Magnetoelectricity in Magnetic Vortices. *Phys. Rev. Lett.* **2009**, *102*, 157203. [CrossRef]
176. Newacheck, S.; Webster, T.; Youssef, G. The effect of multidirectional bias magnetic fields on the converse magnetoelectric response of multiferroic concentric composite ring. *Appl. Phys. Lett.* **2018**, *113*, 172902. [CrossRef]
177. Srinivasan, G. Magnetoelectric effects in bilayers and multilayers of magnetostrictive and piezoelectric perovskite oxides. *Phys. Rev. B* **2002**, *65*, 134402. [CrossRef]
178. Xie, S.; Cheng, J.; Wessels, B.W.; Dravid, V.P. Interfacial structure and chemistry of epitaxial $CoFe_2O_4$ thin films on $SrTiO_3$ and MgO substrates. *Appl. Phys. Lett.* **2008**, *93*, 181901–181903. [CrossRef]
179. Bibes, M.; Barthélémy, A. Multiferroics: Towards a magnetoelectric memory. *Nat. Mater.* **2008**, *7*, 425–426. [CrossRef] [PubMed]
180. Scott, J.F. Data storage: Multiferroic memories. *Nat. Mater.* **2007**, *6*, 256–257. [CrossRef] [PubMed]

micromachines

Article

Optimization of Quality, Reliability, and Warranty Policies for Micromachines under Wear Degradation

Alexandra D. Tseni [1], Panagiotis Sotiropoulos [2] and Stelios K. Georgantzinos [2,3,*]

1 General Department, National and Kapodistrian University of Athens, 34400 Psachna, Greece
2 School of Science and Technology, Hellenic Open University, 26222 Patras, Greece
3 Laboratory for Advanced Materials, Structures and Digitalization, Department of Aerospace Science and Technology, National and Kapodistrian University of Athens, 34400 Psachna, Greece
* Correspondence: sgeor@uoa.com

Abstract: This work presents an optimization technique to determine the inspection, warranty period, and preventive maintenance policies for micromachines suffering from degradation. Specifically, wear degradation is considered, which is a common failure process for many Micro-Electro-Mechanical Systems (MEMS). The proposed mathematical model examines the impact of quality control on reliability and the duration of the warranty period given by the manufacturer or the supplier to the customer. Each of the above processes creates implementation costs. All the individual costs are integrated into a single measure, which is used to build the model and derive the optimal parameters of the quality and maintenance policies. The implementation of various levels of the quality, warranty, and maintenance policies are compared with their optimum level options to highlight their contribution to the assurance and improving product quality. To the authors' best knowledge, the introduction of a warranty period is implemented for the first time in the open literature concerning this type of optimization model for MEMs and surely can bring additional advantages to their quality promotion strategy. The proposed optimization tool provides a comprehensive simultaneous answer to the optimal selection of all the values of the design variables determining the overall maintenance and quality management approach.

Keywords: MEMs; burn-in; quality; reliability; warranty; maintenance; optimization; degradation

Citation: Tseni, A.D.; Sotiropoulos, P.; Georgantzinos, S.K. Optimization of Quality, Reliability, and Warranty Policies for Micromachines under Wear Degradation. *Micromachines* **2022**, *13*, 1899. https://doi.org/10.3390/mi13111899

Academic Editor: Ion Stiharu

Received: 8 October 2022
Accepted: 31 October 2022
Published: 2 November 2022

1. Introduction

MEMS technology exhibits excessive potential for many critical applications in aerospace, automotive, medical, nuclear, and other areas. With more extensive commercialization of MEMs, many challenging manufacturing questions are into consideration including precise dimensional control and inspection, reliability testing and modeling, avoiding stiction, and maintenance strategies [1]. These productivity, quality, and reliability questions are critical issues that influence the path of MEMS to the conventional market. Therefore, MEMS manufacturers need effective tools for optimal operational decisions. These tools can be derived from the use of equivalent tools and methods developed in a traditional industry.

Preventive maintenance refers to work carried out to maintain the equipment at the desired level of operation and to avoid failures leading to production stops. In the context of preventive maintenance includes carrying out checks and inspections of equipment and the replacement of defective units. As part of preventive maintenance, the policy of periodic maintenance is applied replacement of a unit, whereby a unit is replaced at every specific operating interval [2]. The time-based replacement policy refers to the replacement of a unit at a predefined time interval, which has been derived from the manufacturer's guidelines or existing experience. However, in many cases, the unit fails in time less than the replacement time. This case has been analyzed in [3], in which failures are divided into type I (less significant) and type II (catastrophic) failures. Type I failures require minimal replacement of a component or part while Type II failures require replacement of the entire system.

The failure-based burn-in policy is applied before the product is on the market for a short period. The tests aim to identify products that show some early failure (infant mortality) so that they can be avoided from being introduced to the market and consequently to customers. After the burn-in procedure has been applied for the specified period, the products that do not show any failure are immediately put on the market for use, while those whose component fails are replaced and then the product is put on the market [4]. On the other hand, the burn-in policy based on degradation is applied after the product has been produced and before its release and aims to identify those products in which the investigated feature exceeds a certain cut-off level without necessarily ceasing to function. The cut-off level is often lower than the failure level [5]. In these tests, products are tested under more intense conditions (elevated temperature, higher mechanical stress) so that failure or degradation occurs, which under normal conditions would occur at a much later stage of product operation. Hu et al. [6] have developed a partially observed Markov decision process to minimize the expected total burn-in cost of a product and derive some interesting structures of the optimal policy. Yu et al. [7] have developed an improved Wiener process incorporating nonlinear terms to build the degradation model of incipient fault based on the fault estimation results. For prognosis, the fast krill herd algorithm has been proposed to estimate unknown degradation model coefficients. Shafiee et al. [8] develop a mathematical model from which derives the optimal choices of maintenance policies, and the optimal time of burn-in for which the total cost is minimized.

Lifetime is defined as the time at which degradation exceeds a predefined threshold [9]. Degradation, which is caused by endogenous and exogenous factors such as environment and operating conditions, can gradually lead to failure or a product operating at an undesirable level. For this reason, it is important to model it so that predicts and avoids possible future failures. The most widely used models, which have been extensively analyzed in the literature, are the Markov models, the Gamma approach, and the Wiener process [10–13]. Of these, the Markov models are suitable for systems with discrete states while the Gamma and Wiener approximation for continuous degradation systems. Shahraki et al. [14] provide a summary of both models, highlighting the advantages of applying each of them and the limitations in their application while providing food for thought for future analysis. Li et al. [15] use the Markov model to simulate the degradation of wind turbines and derive the reliability curve. Cholette et al. [16] use the Gamma approach to derive the degradation model of a heat exchanger while from the resulting model a maintenance policy is proposed based on the condition of the heat exchanger. Zhang et al. [17] use the Wiener approach to estimate the remaining useful life and comment on its usefulness in various applications.

Age-based preventive replacement policy has been extensively studied in the literature. Jiang et al. [18] investigated an optimal age-based replacement policy, where minimal repair, corrective replacement, and imperfect repair can be carried out upon an unexpected failure with different maintenance effects. An et al. [19] studied the joint optimization of preventive maintenance and flexible job-shop rescheduling with processing speed selection, and the dynamic arrival of the new machine is considered to enhance productivity. Dong et al. [20] investigated novel reliability models and schedules optimal preventive maintenance policies, in which the closed-form reliability quantities are derived analytically and the optimum preventive replacement interval is demonstrated theoretically. Furthermore, they developed generalized reliability models [21] for multi-component systems, where each component is subject to two dependent competing failure processes, i.e., a soft failure process caused jointly by internal performance degradation and incremental damage due to effective external shock sets, and a hard failure process caused by the same random shocks. Moreover, a system may fail due to the aging of its components, or it may fail due to fatal shocks arriving from external sources. Under this mechanism of system failure, Hashemi et al. [22] proposed optimal age-based and block preventive maintenance models by considering the costs of preventive maintenance, corrective maintenance, and minimal repairs providing some formulas for the average long-run cost rate of the proposed strategies.

The issue of property degradation of mechanical systems has been extensively studied and various models have been developed to describe it. The modeling of degradation has contributed to the prediction of material behavior as well as potential failure. In addition, there are many studies related to maintenance policies for mechanical systems. In this work, however, is developed a mathematical model for finding the optimal options of various quality policies such as e.g., the maintenance interval, the burn-in method, the replacement time, and the warranty time of a key component of micromachines, which is the pin joint. The pin joint of microengines wears out the more rotations the more revolutions it performs. Thus, the optimal choices of quality policies are examined which mentioned above are considered to minimize the overall costs.

2. Optimization Model

Devices at the microscale can be classified into four classes [23] according to their operational interactions. Class I devices have no free moving parts, like accelerometers, pressure sensors, or strain gauges; Class II devices have moving parts with no friction or contact surfaces, such as gyros, resonators, and filters; Class III devices have moving parts with contact surfaces such as relay and valve pump; Class IV equipment has moving parts with friction and contact surfaces such as shutters, scanners, optical switches. The first three classes can achieve high reliability if adequately manufactured and packaged. For class IV devices, where frictional surfaces cannot be avoided, failure analysis and reliability assessment must be performed to drive the robust commercialization of micromachines [24].

The failure modes in a micromachine can be wear, friction, fracture, contamination, stiction, etc. The micromachine used in this study is the electrostatically driven microactuator (microengine) developed at Sandia National Laboratories [25] (National Technology and Engineering Solutions of Sandia, LLC., Albuquerque, New Mexico, USA). The micromachine consists of orthogonal linear comb drive actuators that are mechanically connected to a rotating gear as seen in Figure 1. By applying voltages, the linear displacement of the comb drives is transformed into circular motion. The linkage arms are connected to the gear via a pin joint. The gear rotates about a hub, which is anchored to the substrate [25]. The dominant failure mechanism was mainly identified as visible wear on the friction surface, often resulting in a seized micromotor or a micromotor with a damaged shaft seal [25,26]. Wear can be defined as the removal of material from a solid surface by mechanical action. Abrasion degradation is a very complex phenomenon, involving both the mechanical and chemical properties of the objects in contact, as well as the pressure and surface speed with which the objects are in contact.

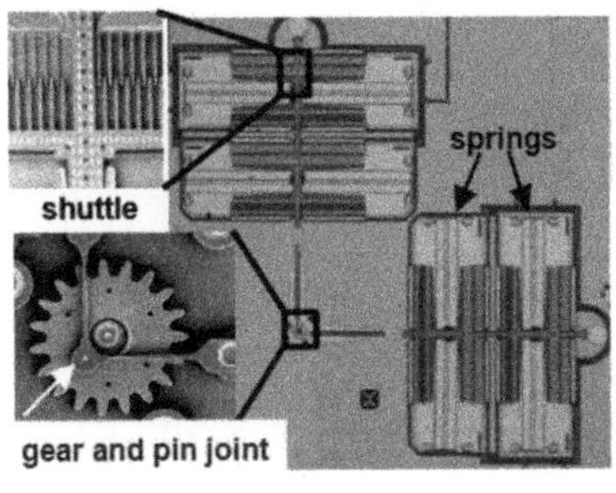

Figure 1. SEM image of a microengine [25,26].

The pin joint is a critical area for micromotors in which either less significant soft failures due to wear occur or where severe failures (hard failures) due to material breakage [27]. In the first case, the wear of the pin joint material exceeds a critical value but continues to operate while in severe failures the system stops operating. In addition, environmental conditions affect the degree of wear of the components as it has been shown that with a reduction in the humidity of the operating environment, there is an increase in the rate of wear [25,26].

Rubbing creates friction and often leads to the creation of abrasive materials or debris. The configuration of this material can lead to several different failure mechanisms. This is due to equipment-associated particle contamination, third-body abrasive particles that alter the motion tolerance, particle contamination that prevents or interferes with the movement and adhesion of surfaces, or frictional contact [28]. The wear mechanism may depend on the temperature achieved during the friction process. The material can be rubbed on the contact surface; surface materials can oxidize—Then fade, etc. Many parameters must be examined to determine the root cause of the wear, which makes analysis simple but time consuming.

To simultaneously improve quality, reliability and the warranty plan over the lifetime of micromachines, a systematic inspection, preventive replacement procedure, and warranty model have been developed, as described in Figure 2.

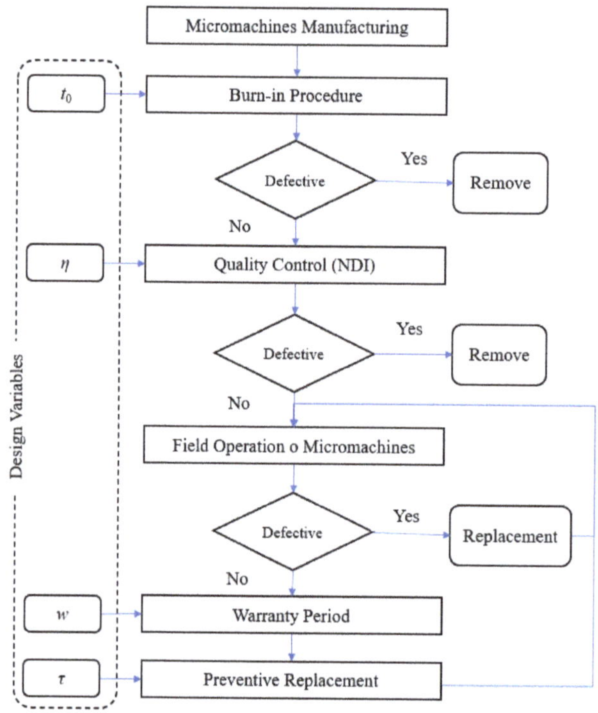

Figure 2. Burn-in, inspection, warranty, and preventive replacement procedures for micromachines.

A post-manufacturing burn-in process for micromachines is used to identify and remove defective and early failed components. Burn-in is a crucial method for achieving reliable parts and systems, but it also exposes them to stress. For burn-in components, non-destructive inspection is applied to separate the fraction of units whose wear surpasses the specific specification threshold. The screened units, with a high level of quality, are then released for field operation until they reach the time of periodic replacement when the cost of a forthcoming failure makes it economically advantageous to replace them with new ones. The preventive replacement procedure is used to avoid failure due to the wear of standard operating units. The introduction of a warranty period can bring additional advantages to the quality promotion strategy and could therefore be adopted.

For a simultaneous optimal selection of the values of the design variables determining the overall maintenance and quality management strategy, a comprehensive methodology is developed in the following sections.

2.1. Degradation Model

To derive the model describing the degradation of the characteristic, it is first necessary to determine the factors which influence the way affects it. The wear of the pin joint due to the rotation of the gear during the operation of a micromachine increases with the number of revolutions it performs. Therefore, the key factor in the model is the operating time, t [19]. The wear of the gear depends also on three other factors. These factors are the radius, r, of the pin, the coefficient, c, relating to the wear and hardness of the material, and the force, F, developed between the pin and the gear. This leads to a linear model in which all factors influence proportionally the wear of the pin, X, and the relationship is

$$X(t, r, c, F) = 2\pi rcFt. \tag{1}$$

For various microengines, some of the factors are constant, while others are considered random. The radius of the pin varies for each and is considered a random factor with a mean value of μ_r and a standard deviation of σ_r. The force depends on the rotation frequency of the gear and consequently on the input voltage of the drive. It is therefore a random factor with mean value μ_F and standard deviation σ_F. Thus, at any time, for any value of radius and force, the wear $X(t)$ is considered to follow the normal distribution with a mean value

$$\mu_t = 2\pi c\mu_r\mu_F t \tag{2}$$

and standard deviation

$$\sigma_t = 2\pi ct\sqrt{\sigma_r^2\sigma_F^2 + \sigma_r^2\mu_F^2 + \mu_r^2\sigma_F^2} \tag{3}$$

Here, we notice that more complex models considering simultaneously other failure modes could be developed. Such models may be useful for cases where the dominant failure modes have been demonstrated to be more than one.

2.2. Effect of Quality Control on Reliability

After the production stage of the microengines, the burn-in technique is applied to identify the defective units. Then, the non-destructive testing of all the units produced follows to remove those whose wear has exceeded the failure threshold H. To demonstrate the importance of quality control, it is necessary to calculate the reliability when 100% inspection of the units is applied and compare it with the reliability when none is applied.

When non-destructive testing is not applied after the burn-in process, the defective units are not detected and the reliability at any time, t, will be equal to the probability that the pin joint wear does not exceed the threshold failure threshold H

$$R(t) = P(X(t) < H) = \int_0^H f_{x(t)}(x,t)dx = \Phi\left(\frac{H-\mu_t}{\sigma_t}\right). \tag{4}$$

When post-burn-in testing is applied then only those units whose wear is less than the failure threshold are marketed. Thus, reliability will be a bound probability and will be equal to the probability that the wear $X(t)$ at any time does not exceed the failure threshold, H, given that it has successfully passed the burn-in procedure for a time t_0. Mathematically, the reliability function will be given by

$$R(t) = \frac{P(X(t) < H)}{P(X(t_0) < H)}, \tag{5}$$

where $t_0 < t < \tau$, and τ is the replacement time.

Comparing the previous equations, we observe that the reliability with the application of non-destructive testing is equal to the reliability without testing divided by a number less than unity (the probability that the post-burn-in decay is less than the failure threshold). Therefore, the application of the non-destructive control increases the reliability value.

2.3. *Quality and Reliability*

Joint optimization of quality and reliability requires the definition of the measure to be used for this purpose and the factors to be used in this model. The factors considered in finding the optimal model are the time of the burn in process, b, the value of the wear, H', below which the system is considered to fail and the replacement time, τ. The measure that can be used consists of the sum of three different costs divided by the expected time of use. These costs are the quality cost, the expected failure cost, and the replacement costs.

2.3.1. Quality Cost

The unit cost of quality is given by

$$QC(\eta) = C_Q(\eta) + C_S(\eta) + C_I, \tag{6}$$

where $C_Q(\eta)$ is the loss of quality, $C_S(\eta)$ is the cost of the rejected units, and C_I is the cost of the inspection of each unit, which is usually fixed. The loss of quality can be calculated from Taguchi's loss function, in which there are three different types depending on the nature of the feature under consideration. These types are the smaller the better (S-type), the larger the better (L-type), and the better the target value (T-type). In the case of the pin joint the ideal is for the wear value to be as low as possible. Therefore, the relationship that characterizes the S-Type case is chosen which is

$$L(X(t_0)) = kX(t_0)^2, \tag{7}$$

where k is the coefficient used to convert the deviations into economic values. The quality loss can be calculated using the expected value of $L(X(t_0))$ and will be given by

$$\begin{aligned} C_Q(\eta) &= \int_0^{USL} L(x,t_0) f_{X(t_0)}(x,t_0)dx = \int_0^{\eta} kx^2 f_{X(t_0)}(x,t_0)dx \\ &= -k\sigma_0[\mu_0+\eta]\Phi\left(\tfrac{\eta-\mu_0}{\sigma_0}\right) + \sigma_0\mu_0 k\Phi\left(\tfrac{-\mu_0}{\sigma_0}\right) - k[\sigma_0^2+\sigma_0^2]\Phi\left(\tfrac{-\mu_0}{\sigma_0}\right) + k[\sigma_0^2+\sigma_0^2]\Phi\left(\tfrac{\eta-\mu_0}{\sigma_0}\right), \end{aligned} \tag{8}$$

where $f_{X(t_0)}$ is the probability density function at the end of the burn-in application.

If a unit exceeds the failure limit, it shall either be repaired or rejected. If $q(\eta)$ is the percentage of units that successfully pass the burn-in stage, then this percentage will be given by

$$q(\eta) = \int_0^{\eta} f_{x(t_0)}(x,t_0)dx = \Phi\left(\frac{\eta-\mu_0}{\sigma_0}\right) - \Phi\left(\frac{-\mu_0}{\sigma_0}\right). \tag{9}$$

Moreover, if the scrap/reworked cost per unit is s then the scrapped portion is (1—$q(\eta)$) and the total expected scrapped cost will be

$$C_S(\eta) = (1-q)s = \left(1 - \Phi\left(\frac{\eta-\mu_0}{\sigma_0}\right) + \Phi\left(\frac{-\mu_0}{\sigma_0}\right)\right)s. \tag{10}$$

Inspection costs, C_I, are fixed and independent of the failure threshold η.

2.3.2. Failure Cost—Preventive Replacement Cost

The failure cost of each unit f_C is fixed and s-independent of the time of failure and can be estimated by a one-year warranty cost or a one-time repair cost. Thus, the total failure cost depends on the reliability value at the time of replacement and is given by the relation

$$FC(\tau) = f_C(1 - R(\tau|t_0)). \tag{11}$$

In the case where we have no failure before the replacement time, the replacement based on policy will be done at that time at a replacement cost, *RC*. However, in case we have a failure before the scheduled time of replacement, then an additional replacement cost, *RC*, arises. In the more general case, the average expected failure cost will be given by the sum of the failure cost and the replacement cost. Otherwise, if it has not failed by τ, it should be replaced based on economic considerations, and the cost is *RC*. Thus, the expected total failure plus replacement cost at τ is $FC(\tau)$ + *RC*.

2.3.3. Expected Time of Use

The expected time of use is a function of the burn-in time, t_0, and the replacement time, τ, and will be given by the following equation

$$E[U/t_0,\tau] = \int_0^{\tau - t_0} R(t + t_0|t_0)dt = \frac{1}{R(t_0)}\left(\tau R(\tau) - t_0 R(t_0) + \int_{t_0}^{\tau} t f_T(t)dt\right), \tag{12}$$

where $f_T(t)$ is the probability density function of the failure time for a Bernstein distribution with two parameters and will be given by the equation

$$f_T(t) = -\frac{dR(t)}{dt} = \frac{H}{\sqrt{2\pi}b}e^{\frac{(H-at)^2}{2bt^2}}, \tag{13}$$

where

$$a = 2\pi c\mu_r\mu_F, \tag{14}$$

and

$$b = t\sqrt{(2\pi c)^2(\sigma_r^2\sigma_F^2 + \sigma_r^2\mu_F^2 + \mu_r^2\sigma_F^2)}. \tag{15}$$

This is the pdf for a two-parameter Bernstein distribution.

2.4. *Effect of the Warranty*

A warranty is a contractual obligation of a manufacturer for the sale of its products. Manufacturers essentially declare that repair any damage that occurs for the duration of the warranty. But this creates an additional cost.

A product that has successfully passed the burn-in process is promoted to the market at the time, t_0. Assuming that the micromachines are given a warranty period, w. If the product fails during its use at time t_w (where $t_w < w$) then under the supplier's contract with the customer, the supplier must replace it, and this produces an additional cost $c_2(t_w)$.

The expected cost for a product that comes onto the market after a burn-in period, t_0, and has a warranty period, w, it is given by the relation

$$E[c_w(b)] = \int_0^{w} c_2(t_w)r_b(t_w)dt_w, \tag{16}$$

where $r_b(t_w)$ is the reliability function after application of the burn-in procedure. However, if the failures occurring after the application of the burn-in process are repaired, the reliability of the product will be

$$r_b(t_w) = r(t_0 + t_w), \tag{17}$$

and the expected warranty cost will be

$$E[c_w(b)] = \int_{t_0}^{w+t_0} c_2(t - t_0)r(t)dt. \quad (18)$$

2.5. Total Cost and Optimization

The expected total system cost per unit expected usage time including warranty is described as

$$TC(\eta, \tau, t_0, w) = \frac{QC(\eta) + FC(\tau) + RC + E[c_w(b)]}{E[U/t_0, \tau]}. \quad (19)$$

In fact, the upper bound of the replacement interval is usually specified and is denoted as B_τ. To minimize this cost, the optimal options of burn-in time, t_0, the replacement time, τ, the failure threshold, η, and the warranty interval, w, are sought. The constrained optimization model is expressed as

$$(\eta^*, \tau^*, t_0{}^*, w^*) = \text{argmin}\left\{TC(\eta, \tau, t_0, w) = \frac{QC(\eta) + FC(\tau) + RC + E[c_w(b)]}{E[U/t_0, \tau]}\right\}, \quad (20)$$

subjected to

$$\eta > \mu_{0,}, \; t_0 \leq \tau \leq B_\tau. \quad (21)$$

The Sequential Quadratic Programming (SPQ) method [29] is employed to solve the optimization problem. This method is a technique used to find the optimal solution to non-linear problems consisting of several subproblems. The algorithm starts from an initial hypothesis and stops after a series of iterations when the criteria set are satisfied. Calculates the optimal solution of the subproblems for which a quadratic model of the function of the original problem.

3. Results and Discussion

This section presents the implementation of the proposed methods using numerical examples. Firstly, selecting suitable data, the levels of the decision variables that minimize the total cost are calculated, and then the effect of the decision variables on the total cost is examined.

3.1. Numerical Data

To apply the model and extract the desired results and conclusions, the numerical data are presented in Table 1. It has been assumed that the surface wear of the microengines follows the normal distribution. The source of the data is given by Tanner et al. [30] considering the coefficient c in Equation (1), the radius of the pin joint r, the nominal value of the force applied between rubbing surfaces F, the quality loss factor, the failure per unit cost f_c, the replacement cost RC, and the rejection cost. Here, the replacement cost within warranty c_2 is also considered.

Table 1. Numerical data applied to the proposed models.

c (μm²/N)	r (μm)	F (N)	s.d. of r (μm)	s.d. of F (N)	Quality Loss Factor	f_C (€/Unit)	RC (€/Unit)	Rejection Cost (€/unit)	c_2 (€/Unit)
0.0003	1.5	3×10^{-6}	0.075	1.5×10^{-7}	10^{10}	1000	50	50	70

First, the values of the decision variables from which the model starts are given to calculate the cost values until the optimal solution are reached. The initial values selected are burn-in time t_0 = 200 revolutions, failure threshold η = 0.001 μm³, replacement time τ = 25,000 revolutions, and warranty period w = 10,000 revolutions. The values of the variables that minimize the total cost, which are derived from the application of the proposed model, are burn-in time $t_0* = 236$ revolutions, failure threshold $\eta^* = 5.81 \times 10^{-3}\ \mu m^3$, replacement time $\tau^* = 58{,}945$ revolutions, warranty period $w^* = 11{,}340$ revolutions. The total cost for the optimal solution is $TC = 0.0404$ €/unit. It is noticed that the above values obtained from the solution of the constrained optimization model expressed in Equations (20) and (21) applying the SQP method. Here, we notice that this test case and its results are directly comparable only with the ones presented in [1]. If we apply the values of the example in [1] and we simplify the model neglecting the warranty policies to be the same as the corresponding one in [1], the obtained results are identical.

3.2. Comparison of Quality Policies

In order to examine the effect of quality control after the application of the burn-in process, initially in this section the different values of the total cost without quality control, i.e., without a failure limit (η) of the defective units, will be calculated.

Firstly, the numerical model is applied, after having reset to zero the rejection threshold h, changing each time the value of the time interval of the burn-in application. The constant values selected are a replacement time $\tau = 50{,}000$ revolutions, and a warranty period $w = 25{,}000$ revolutions. The burn-in time t_0 varies from 0 to 5000 revolutions, while the failure threshold is $\eta = 0\ \mu m^3$, due to the absence of quality control. The results are summarized in Figure 3a. We note that the application of a longer time interval of the burn-in process reasonably increases the total costs. In addition, we find a first insight into the values of the total costs incurred in the absence of quality control.

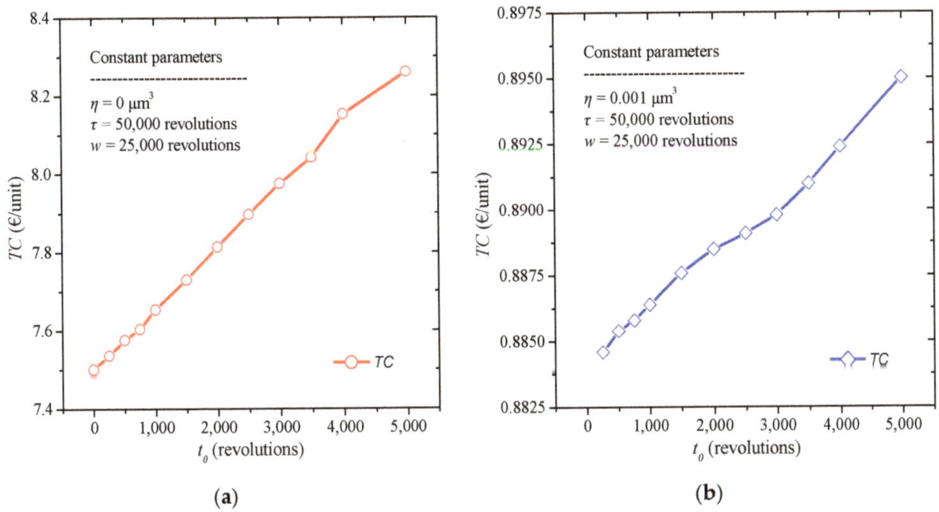

Figure 3. Total cost diagram for various values of burn-in time (**a**) without, and (**b**) with the presence of quality control.

To draw conclusions on the impact of quality control on the total cost, the proposed model is applied by considering the quality control expressed through the failure threshold. A failure threshold $\eta = 0.001\ \mu m^3$ is chosen to calculate the total cost. The constant values selected are a replacement time $\tau = 50{,}000$ revolutions, and a warranty period $w = 25{,}000$ revolutions. The burn-in time t_0 varies from 0 to 5000 revolutions, while the failure threshold is $\eta = 0.001\ \mu m^3$, due to the quality control process. The results are summarized in Figure 3b. From the above results we notice that the total cost increases with the increase of the burn-in time. However, we observe that while doubling the time of the burn-in process there is no meaningful change in the value of total cost. Specifically, by increasing by 1000 revolutions the burn-in time the total cost increases less than 5% in all cases. Moreover, we observe that, in this case, the absence of quality control can produce ten times (1000%) greater total cost than the one with quality control in most cases.

For a more detailed analysis of the results and to allow comparison of the results, the box plots for total costs as a function of the change in burn-in time when quality control is or is not applied are presented in Figure 4. We observe a sharp decrease in the cost values with application of quality control. Furthermore, the results obtained without quality control show a considerably larger dispersion.

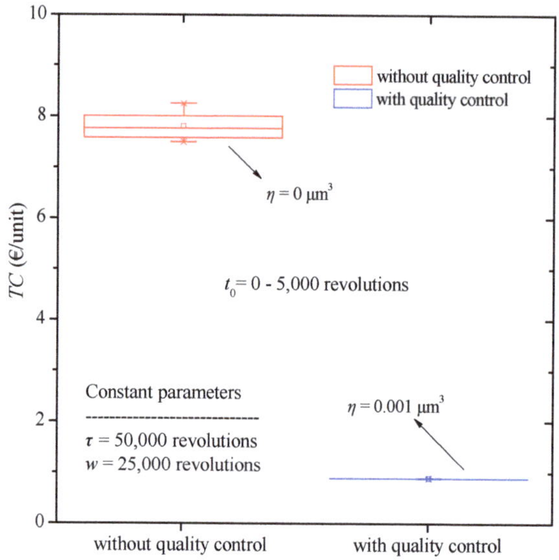

Figure 4. Total cost boxplot diagram for various values of burn-in time with and without quality control.

The same procedure is then repeated keeping all variables constant except for the replacement time interval, τ. The constant values selected are a burn-in time t_0 = 1000 revolutions, and a warranty period w = 25,000 revolutions. The replacement time, τ, varies from 25,000 to 150,000 revolutions, while the failure threshold is η = 0 μm^3, due to the absence of quality control. The results are summarized in Figure 5a. It is obvious that increasing the replacement interval leads the reduction of overall costs when no quality control is applied.

A similar procedure is repeated keeping all the factors of the model constant, each time changing the replacement time applying quality control. The constant values selected are a burn-in time $t_0 = 1000$ revolutions, a warranty period w = 25,000 revolutions, and a failure threshold η = 0.001 μm^3. The replacement time varies again from 25,000 to 150,000 revolutions. The results are illustrated in Figure 5b. We can observe that with the increase of the replacement time the total cost as the replacement cost decreases. Furthermore, increasing the replacement time, it contributes significantly to reducing costs. Near at the optimal value of 58,945 rotations the total cost is about 0.85 €/item, while when the value is under sampled the total cost increases and approaches 7.5 €/item. This is because the sooner a component is replaced the less likely it will fail. These results suggest that increasing the preventive replacement time has a significant effect on the reduction of the total cost. The rate of reduction decreases with increasing replacement time.

The boxplot diagram in Figure 6 is then assembled from the results of previous analysis for the various values of the replacement time. In this case as well, we observe a reduction in total cost when quality control is applied as well as a smaller dispersion of the results appears. Moreover, we observe that, in this case, the absence of quality control can produce 100 times greater total cost than the one with quality control in some cases.

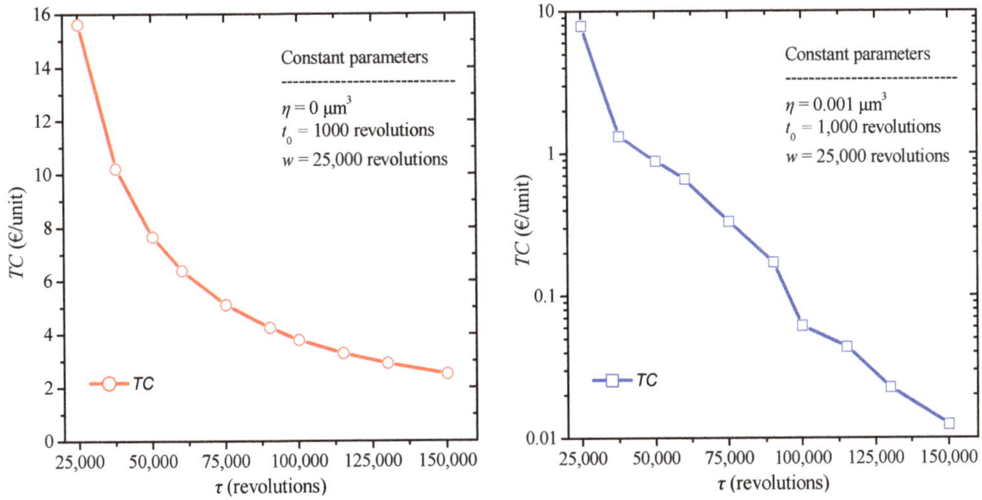

(a) (b)

Figure 5. Total cost diagram for various values of replacement time (**a**) without, and (**b**) with the presence of quality control.

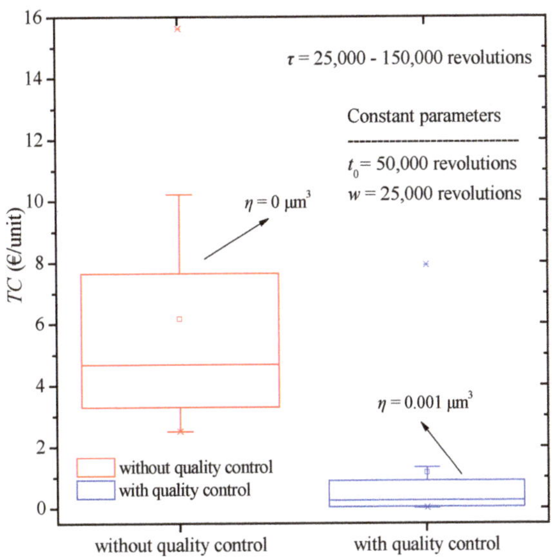

Figure 6. Total cost boxplot diagram for various values of burn-in time with and without quality control.

Furthermore, the application of the model without quality control is implemented by varying this time the length of the warranty period provided by the manufacturer. The constant values selected are a burn-in time t_0 = 1000 revolutions, and a replacement time τ = 50,000 revolutions. The warranty period, w, varies from 10,000 to 100,000 revolutions, while the failure threshold is η = 0 μm^3, due to the absence of quality control. The results obtained are shown in Figure 7a. It can be observed that the total cost increases with the increase in the total warranty period provided by the manufacturer.

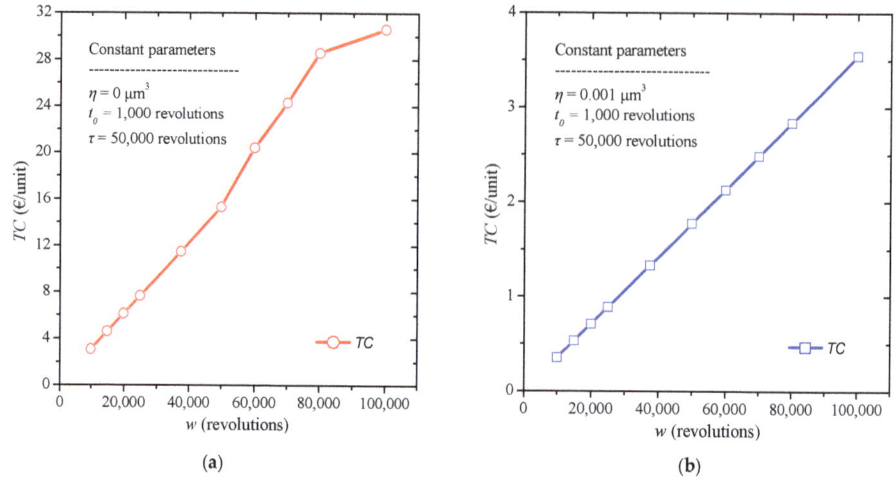

Figure 7. Total cost diagram for various values of warranty period provided by the manufacturer (**a**) without, and (**b**) with the presence of quality control.

The warranty period is also examined for its influence on total costs when quality control is applied. The constant values selected are a replacement time τ = 50,000 revolutions, and a burn-in period of t_0 = 1000 revolutions. The warranty time w varies from 10,000 to 100,000 revolutions, while the failure threshold is η = 0.001 μm^3, due to the quality control process. The results are summarized in Figure 7b. From the presented results we additionally conclude that with an increase in the warranty period the total cost increases as more failures occur during this period which must be replaced by the manufacturer.

To further compare the effect of period of the warranty on the total cost, a corresponding boxplot is presented in Figure 8. And in this case, we come to the same conclusion, i.e., the noticeable reduction in total costs and in the less dispersion of results by applying quality control.

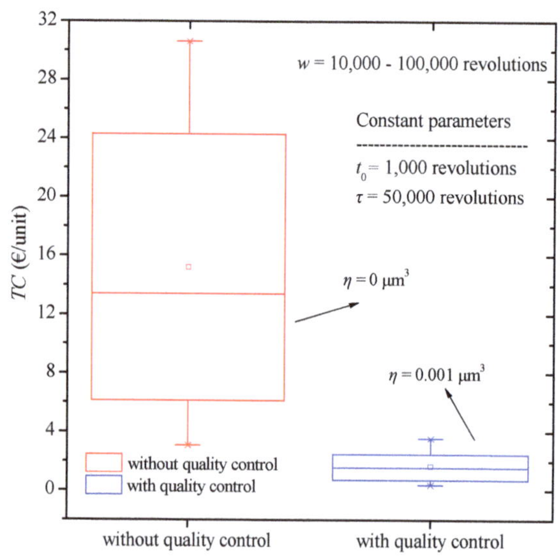

Figure 8. Total cost boxplot diagram for various values of warranty period with and without quality control.

The effect of the quality control threshold, i.e., the different values of the failure threshold, on the total cost is finally examined. The constant values selected are a replacement time τ = 50,000 revolutions, a burn-in time $t_0 = 1000$ revolutions, and a warranty period w = 25,000 revolutions. The failure threshold varies from 0.0005 to 0.0100 μm^3. The results are depicted in Figure 9.

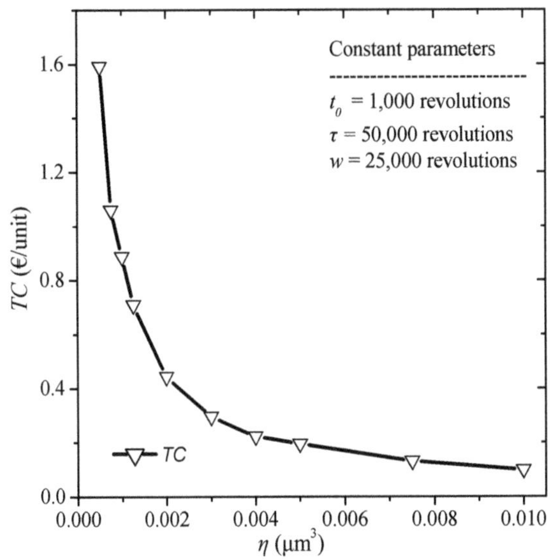

Figure 9. Total cost diagram for various values of failure threshold.

From the above results we observe that as the failure threshold decreases the total cost increases, which is since the lower the failure threshold, the more units are rejected after the burn-in stage and are considered to fail.

An important observation concerns the dispersion of results. We observe that when there is no quality control the results show a large dispersion from the mean value and a large range from the minimum and maximum values, while when there is a quality control for all three decision variables it is observed that the results are clustered around the mean value. We also observe that increasing the time of application of the burn-in procedure only slightly increases the total cost with values close to 0.885 €/unit, since its application for a longer period contributes to the detection of more defective units which are not passed on to the consumer.

4. Conclusions

In this paper, a mathematical model with economic cost factors was developed based on the degradation of an important component of micromotors, the pin. This model could be applied to other components of micromotors, whether electrical or mechanical. The wear of micromotors increases with the number of revolutions of the pin and depends on three other factors, namely the radius of the pin, the hardness coefficient of the material and the force developed between the gear and the pin. Based on the calculation of the degradation relationship, which is exhibited by the micromotors and the rotation pin which is the contact point between the actuator and the gear, the various costs were calculated. Quality costs, failure costs, replacement costs and warranty costs were considered. Deriving the total cost expression, the values of the decision parameters that minimize the total cost were calculated using computational techniques.

Using a parametric analysis to study the sensitivity of the solution on the decision variables, the following conclusions are arisen:

- the presence of quality control significantly reduces the total costs for all the decision variables. Concerning the numerical example, the total cost is reduced more than ten times (1000%) when quality control is applied in most cases.
- a large dispersion from the mean with a large range from the minimum and maximum values exists without a quality control, while when a quality control is implemented, the results are clustered around the mean for all the decision variables.
- the increase in the burn-in time slightly increases the overall cost, while its application for longer time contributes to the detection of more defective units which are not passed on to the consumer. Concerning the numerical example, increasing by 1000 revolutions the burn-in time the total cost increases by less than 5% in all cases.
- the increase in the replacement interval is observed to contribute significantly to the reduction of total cost since the sooner a component is replaced, the less likely it is to fail.
- the increase in the warranty period provided by the manufacturer increases the total cost significantly.

Author Contributions: Conceptualization, S.K.G.; methodology, A.D.T., P.S. and S.K.G.; software, S.K.G.; validation, A.D.T., P.S. and S.K.G.; formal analysis, S.K.G.; investigation, A.D.T., P.S. and S.K.G.; resources, A.D.T.; data curation, P.S. and S.K.G.; writing—original draft preparation, A.D.T. and P.S.; writing—review and editing, A.D.T. and S.K.G.; visualization, A.D.T.; supervision, S.K.G.; project administration, S.K.G. All authors have read and agreed to the published version of the manuscript.

Funding: This research received no external funding.

Conflicts of Interest: The authors declare no conflict of interest.

References

1. Peng, H.; Feng, Q.; Coit, D.W. Simultaneous quality and reliability optimization for microengines subject to degradation. *IEEE Trans. Reliab.* **2009**, *58*, 98–105. [CrossRef]
2. Chien, Y.H.; Chen, M. An optimal age for preventive replacement under fully renewing combination free replacement and pro rata warranty. *Commun. Stat.-Theory Methods* **2010**, *39*, 2422–2439. [CrossRef]
3. Chien, Y.H.; Sheu, S.H. Extended optimal age-replacement policy with minimal repair of a system subject to shocks. *Eur. J. Oper. Res.* **2006**, *174*, 169–181. [CrossRef]
4. Cha, J.H. On a better burn-in procedure. *J. Appl. Probab.* **2000**, *37*, 1099–1103. [CrossRef]
5. Ye, Z.S.; Xie, M.; Tang, L.C.; Shen, Y. Degradation-based burn-in planning under competing risks. *Technometrics* **2012**, *54*, 159–168. [CrossRef]
6. Hu, J.; Sun, Q.; Ye, Z.S.; Ling, X. Sequential degradation-based burn-in test with multiple periodic inspections. *Front. Eng. Manag.* **2021**, *8*, 519–530. [CrossRef]
7. Yu, M.; Lu, H.; Wang, H.; Xiao, C.; Lan, D.; Chen, J. Computational Intelligence-Based Prognosis for Hybrid Mechatronic System Using Improved Wiener Process. *Actuators* **2021**, *10*, 213. [CrossRef]
8. Shafiee, M.; Finkelstein, M.; Zuo, M.J. Optimal burn-in and preventive maintenance warranty strategies with time-dependent maintenance costs. *IIE Trans.* **2013**, *45*, 1024–1033. [CrossRef]
9. Belhaj Salem, M.; Fouladirad, M.; Deloux, E. Prognostic and Classification of Dynamic Degradation in a Mechanical System Using Variance Gamma Process. *Mathematics* **2021**, *9*, 254. [CrossRef]
10. Bogdanoff, J.L.; Kozin, F. *Probabilistic Models of Cumulative Damage*; Wiley-Interscience: New York, NY, USA, 1985; 350p.
11. Van Noortwijk, J.M. A survey of the application of gamma processes in maintenance. *Reliab. Eng. Syst. Saf.* **2009**, *94*, 2–21. [CrossRef]
12. Frangopol, D.M.; Kallen, M.J.; Noortwijk, J.M.V. Probabilistic models for life-cycle performance of deteriorating structures: Review and future directions. *Prog. Struct. Eng. Mater.* **2004**, *6*, 197–212. [CrossRef]
13. Abdel-Hameed, M. Degradation processes: An overview. In *Advances in Degradation Modeling*; Springer: Berlin/Heidelberg, Germany, 2010; pp. 17–25.
14. Shahraki, A.F.; Yadav, O.P.; Liao, H. A review on degradation modelling and its engineering applications. *Int. J. Perform. Eng.* **2017**, *13*, 299. [CrossRef]
15. Li, J.; Zhang, X.; Zhou, X.; Lu, L. Reliability assessment of wind turbine bearing based on the degradation-Hidden-Markov model. *Renew. Energy* **2019**, *132*, 1076–1087. [CrossRef]
16. Cholette, M.E.; Yu, H.; Borghesani, P.; Ma, L.; Kent, G. Degradation modeling and condition-based maintenance of boiler heat exchangers using gamma processes. *Reliab. Eng. Syst.* **2019**, *183*, 184–196.

17. Zhang, Z.; Si, X.; Hu, C.; Lei, Y. Degradation data analysis and remaining useful life estimation: A review on Wiener-process-based methods. *Eur. J. Oper. Res.* **2018**, *271*, 775–796. [CrossRef]
18. Jiang, L.; Li, B.; Hu, J. Preventive Replacement Policy of a System Considering Multiple Maintenance Actions upon a Failure. *Qual. Reliab. Eng. Int.* 2022; *early view*. [CrossRef]
19. An, Y.; Chen, X.; Hu, J.; Zhang, L.; Li, Y.; Jiang, J. Joint Optimization of Preventive Maintenance and Production Rescheduling with New Machine Insertion and Processing Speed Selection. *Reliab. Eng. Syst. Saf.* **2022**, *220*, 108269. [CrossRef]
20. Dong, W.; Liu, S.; Cao, Y.; Javed, S.A.; Du, Y. Reliability Modeling and Optimal Random Preventive Maintenance Policy for Parallel Systems with Damage Self-Healing. *Comput. Ind. Eng.* **2020**, *142*, 106359. [CrossRef]
21. Dong, W.; Liu, S.; Bae, S.J.; Cao, Y. Reliability Modelling for Multi-Component Systems Subject to Stochastic Deterioration and Generalized Cumulative Shock Damages. *Reliab. Eng. Syst. Saf.* **2021**, *205*, 107260. [CrossRef]
22. Hashemi, M.; Tavangar, M.; Asadi, M. Optimal Preventive Maintenance for Coherent Systems Whose Failure Occurs Due to Aging or External Shocks. *Comput. Ind. Eng.* **2022**, *163*, 107829. [CrossRef]
23. Walraven, J.A.; Headley, T.J.; Campbell, A.B.; Tanner, D.M. Failure analysis of worn surface-micromachined microengines. In *MEMS Reliability for Critical and Space Applications*; SPIE: Bellingham, WA, USA, 1999; Volume 3880, pp. 30–39. [CrossRef]
24. Tavrow, L.S.; Bart, S.F.; Lang, J.H. Operational characteristics of microfabricated electric motors. *Sens. Actuator A Phys.* **1992**, *35*, 33–44. [CrossRef]
25. Tanner, D.M.; Dugger, M.T. Wear mechanisms in a reliability methodology. In *Reliability, Testing, and Characterization of MEMS/MOEMS II, Proceedings of the Micromachining and Microfabrication, San Jose, CA, USA, 25–31 January 2003*; SPIE: Bellingham, WA, USA, 2003.
26. Wen, Y.; Liu, B.; Shi, H.; Kang, S.; Feng, Y. Reliability Evaluation and Optimization of a System with Mixed Run Shock. *Axioms* **2022**, *11*, 366. [CrossRef]
27. Gao, H.; Cui, L. Reliability analysis for a degradation system subject to dependent soft and hard failure processes. In Proceedings of the 2017 IEEE International Conference on Software Quality, Reliability and Security Companion, Prague, Czech Republic, 25–29 July 2017.
28. Walraven, J.A. Failure Mechanisms in Mems. In Proceedings of the International Test Conference, Charlotte, NC, USA, 30 September–2 October 2003; Volume 1, pp. 828–833.
29. Nocedal, J.; Wright, S. *Numerical Optimization*; Springer Science & Business Media: Berlin/Heidelberg, Germany, 2006.
30. Tanner, D.M.; Miller, W.M.; Peterson, K.A.; Dugger, M.T.; Eaton, W.P.; Irwin, L.W.; Senft, D.C.; Smith, N.F.; Tangyunyong, P.; Miller, S.L. Frequency dependence of the lifetime of a surface micromachined microengine driving a load. *Microelectron. Reliab.* **1999**, *39*, 401–414. [CrossRef]

micromachines

Article

An Annular Fresnel Zone Plate without Central Spots Fabricated by Femtosecond Laser Direct Writing

Xiaoyan Sun, Fang Zhou and Lian Duan *

State Key Laboratory of High Performance Complex Manufacturing, Central South University, Changsha 410083, China
* Correspondence: lianduan@csu.edu.cn

Abstract: In recent years, micro-annular beams have been widely used, which has expanded the possibilities for laser processing. However, the current method of generating an annular beam still has shortcomings, such as spot energy at the center of the produced beam. In this study, a Fresnel zone plate with an annular structure was machined using a femtosecond laser. After focusing, an annular laser beam without a spot in the center was obtained, and the radius and focal length of the annular beam could be easily adjusted. In addition, two annular Fresnel zone plates were concentrically connected to obtain a concentric double-ring beam in the same focal plane. The simulation and experimental results were consistent, providing effective potential for applications related to nontraditionally shaped laser beams.

Keywords: Fresnel zone plate; annular structure; femtosecond laser machining; concentric double-ring beams

Citation: Sun, X.; Zhou, F.; Duan, L. An Annular Fresnel Zone Plate without Central Spots Fabricated by Femtosecond Laser Direct Writing. *Micromachines* **2022**, *13*, 1285. https://doi.org/10.3390/mi13081285

Academic Editor: Stelios K. Georgantzinos

Received: 21 July 2022
Accepted: 8 August 2022
Published: 10 August 2022

1. Introduction

In laser processing, the typical laser parameters include power, spot size, wavelength, pulse duration, and repetition frequency. These parameters are necessary to determine the optimal machining process, and their effects on structure and performance have been extensively studied. The irradiance distribution when the laser reaches the material is also important in material processing. The most common is the Gaussian distribution, which can improve the energy utilization rate to the maximum efficiency. However, in some processes, energy distributions in nontraditional shapes are required. For example, when the diameter of the drilling hole is larger than that of the laser beam [1], the annular distribution of the laser beam can be fired continuously on the material's surface in the form of a pulse. This method, known as optical perforation, has been successfully used to drill submillimeter holes through stainless steel [2]. In recent years, annular beams have been increasingly used, thereby expanding the possibilities of laser processing [3]. When an annular beam is used together with two-photon polymerization, point-by-point scanning can be optimized to multipoint parallel scanning, and a circular structure can be generated in one process, significantly improving the processing efficiency [4,5]. New imaging techniques can be obtained in stimulated emission depletion microscopes when a circular beam is combined with a Gaussian beam [6,7]. Annular beams can also process microtubules, which have unique applications in biological sensing [8], cell manipulation [9], targeted drug delivery [10], and other fields, attracting the attention of researchers.

Many researchers have studied annular beams from different aspects [11–13]. Marti et al. [3] studied the properties, generation methods, and emerging applications of laser beams with nontraditional shapes, such as Bessel beams, annular beams, and vortex beams. They used an annular beam and optical polymerization to generate three-dimensional stacks of circular structures and annular patterns. Arash et al. [14] turned an incident plane wave into an annular beam at its focal plane using a Fresnel zone plate (FZP) with phase shifting radially outward. This method generated an annular beam with high efficiency and good flexibility, but it produced an apparent bright spot at the center of the annular beam. Zhang et al. [15] loaded the phase function of an FZP on a space light modulator to generate an annular beam with variable diameter and realized the rapid processing of tubular structures. They added a high-pass filter to the optical path to filter the light at the center of the annular beam. Fatemeh et al. [16] carried out phase superposition of two radial phase-shifted FZPs to obtain a long-depth bifocal diffraction lens with two focal planes, and a plane with a double-ring beam was obtained between the two focal planes. However, the plane of the double-ring beam was not the focal plane; therefore, energy could not be concentrated when used in laser processing.

In this study, a laser beam with an annular distribution was obtained by focusing the annular Fresnel zone plate (AFZP), and a method of eliminating the center energy of the annular beam was obtained through research. Two AFZPs were concentrically connected, and a focused double-ring beam was obtained on the focal plane. In this study, a simulation calculation was carried out, and an experimental study was carried out by femtosecond laser machining of the AFZP. The results of these two experiments were verified and found to be consistent.

2. Theoretical and Experimental Methods

2.1. Theoretical Approach

The energy of an annular laser beam is concentrated in the annular region. The different energy distributions in the annular region can be divided into various annular beams, such as the Bezier beam, the optical vortex beam, and the Gaussian distribution of the annular beam. The light field of the Bezier beam is a series of concentric rings with high energy in the center and low energy outside. Therefore, it is necessary to accurately control the processing energy when using a Bezier beam to make the energy density of the inner ring greater than the ablation threshold and the energy density of the outer ring lower than the threshold; otherwise, a concentric annular structure will appear in the processing [17]. The vortex is twisted in a spiral shape around an axis where the light waves cancel each other, making it a circular beam with a radial cross-section. When the topological charge of the vortex light is an integer, the energy distribution of the annular light is uniform in the cross-section. However, when the topological charge is a non-integer, the distribution of the annular light field becomes uneven, which limits the application of vortex light [18].

Researchers have studied methods for generating circular beams. First, an annular beam is produced using an aperture [3]. A Bessel beam can be obtained by irradiating an annular slit on the rear focal plane of the convergent lens with plane waves. However, its efficiency is low because the aperture blocks most of the incoming energy. Annular beams can also be formed using a cone lens [19]. The workpiece is placed sufficiently far away from the cone lens to encounter the far-field profile of the cone lens, which corresponds to a circular beam. However, this method cannot effectively control the size and shape of circular beams. The spatial light modulator (SLM) can change the phase of the incident laser [20–22] and generate an arbitrary beam profile and hologram. Therefore, the spatial light modulator can quickly generate an annular beam. However, SLM often has a low damage threshold, which limits micromachining in high-power lasers. In addition, FZP processing on the end face of a multimode fiber is an effective method for generating annular beams [23]. The FZP of the fiber end is conducive to integration into small systems and intrinsically compatible with rapidly developing fiber lasers [24–26]. However, optical-fiber FZPs are unsuitable for non-optical-fiber systems because of their narrow application range.

To overcome the shortcomings of Bessel and vortex lights, we used an AFZP to generate a Gaussian circular beam. According to the study of linear FZPs, when the FZP extends indefinitely along the X- or Y-direction, a grating with varying pitch can be obtained [27]. After the laser passes through the grating, the diffraction pattern in the focal plane is a diffraction line in the same direction as the FZP. A linear FZP can be considered an annular FZP with an infinite radius. When the radius of the annular FZP is reduced to a finite size, an AFZP can be obtained. As shown in Figure 1, the inner radius of the circular

FZP is r_0, and the outer diameter is r. When the number of circular FZP is n, $r = \sqrt{n}r_0$. Then, radius R of the AFZP is the distance from the center of the AFZP to the center of the circular FZP, the inner diameter is $R_i = R - r$, and the outer diameter is $R_o = R + r$. According to linear FZP theory, when a laser passes through an AFZP, it will exhibit an annular beam with radius R on the focal plane, and the energy on the focusing ring has a Gaussian distribution. The distance from the focusing ring to the AFZP is $f = r_0^2/\lambda$, where λ is the wavelength of the incident laser.

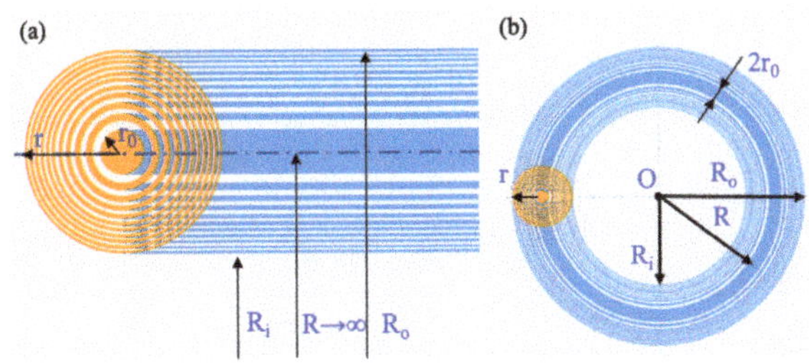

Figure 1. Schematic diagram of Fresnel zone plate (FZP) structure: (**a**) linear FZP; (**b**) AFZP.

A major problem in previous studies on the annular beam is a bright spot in the center of the generated annular beam [14,15]. In ablative machining or material modification, such an annular beam inevitably has a negative effect on the machining morphology, and the center energy should be eliminated. Because the central bright spot is generated during the circular bending of the linear FZP, appropriate structure parameters should be set when designing the AFZP.

Since the AFZP is composed of a transparent and opaque zone, and the area of the opaque zone is larger than that of the transparent zone, when incident light irradiates the AFZP, more than half of the laser light cannot pass through. There is no barrier at the center of the AFZP, and the light beam can pass directly through it. Therefore, incident light passing through the AFZP can be approximated as passing through a circular hole of radius. According to diffraction theory, after the laser passes through the circular hole, it successfully passes through three Z-axis zones: the projection area, Fresnel diffraction area, and Fraunhofer diffraction area (as shown in Figure 2). In the projection area of the circular hole, the laser penetrates the circular hole along the Z-axis with little energy. In the Fresnel diffraction area, the central spot on the Z-axis may be dark or bright, with alternating light and dark spots. In the Fraunhofer diffraction area, a bright spot is always present along the Z-axis. The energy at the center of the annular beam is the central bright spot caused by circular-hole Fresnel diffraction or Fraunhofer diffraction.

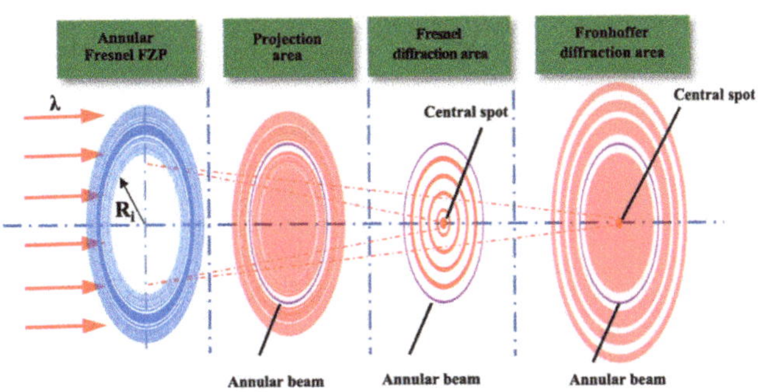

Figure 2. After the laser passes through the circular hole of the AFZP, it successively passes through the projection area, Fresnel diffraction area, and Fraunhofer diffraction area.

The approximate range of the Fresnel diffraction area and the Fraunhofer diffraction area produced on the Z-axis can be deduced according to the Fresnel diffraction formula. When the radius of the circular hole is R_i, the impulse response function of the light wave field is

$$h(x - x_0, y - y_0) = \frac{e^{jkR_i}}{j\lambda R_i}. \quad (1)$$

In rectangular coordinates,

$$R_i = \sqrt{z^2 + (x - x_0)^2 + (y - y_0)^2}, \quad (2)$$

where z is the distance between the hole and viewing screen. When z is large (paraxial condition), the first two terms of the Taylor expansion can be considered.

$$R_i \approx z\left[1 + \frac{1}{2}\left(\frac{x - x_0}{z}\right)^2 + \frac{1}{2}\left(\frac{y - y_0}{z}\right)^2\right] - \frac{\left[(x - x_0)^2 + (y - y_0)^2\right]^2}{8z^3} + \cdots \quad (3)$$

The essence of the Fresnel approximation is to replace the Huygens wavelet of the sphere with a quadrilateral. Because the phase change caused by the third term of the Taylor expansion is far less than $\pi/2$, the Fresnel diffraction area can be deduced as

$$z^3 \geq \frac{\left[(x - x_0)^2 + (y - y_0)^2\right]^2_{max}}{2\lambda} = \frac{\left(2(2R_i)^2\right)^2}{2\lambda} = 32\frac{{R_i}^4}{\lambda}. \quad (4)$$

Furthermore, Equation (3) can be written as

$$R_i \approx z\left[1 + \frac{1}{2}\left(\frac{x - x_0}{z}\right)^2 + \frac{1}{2}\left(\frac{y - y_0}{z}\right)^2\right] = z + \frac{x^2 + y^2}{2z} - \frac{xx_0 + yy_0}{z} + \frac{x_0^2 + y_0^2}{2z}. \quad (5)$$

In this formula, the phase change caused by the last term is less than $\pi/2$; that is, the phase change caused by different hole positions can be ignored. The Fraunhofer diffraction condition is satisfied when $k\frac{(x_0^2+y_0^2)_{max}}{2z} \leq \frac{\pi}{2}$. Because $(x_0^2 + y_0^2)_{max} = 2{R_i}^2$, the following can be obtained:

$$z \geq 4\frac{{R_i}^2}{\lambda}. \quad (6)$$

The AFZP is composed of many concentric zones; thus, the energy of its Z-axis plane is due to the combined superposition of the diffraction of all the circular holes. According to Equations (4) and (6), the positions of the projection area, Fresnel diffraction area, and Fraunhofer diffraction area on the Z-axis are positively correlated with the radius of the circular hole.

2.2. Experimental Methods

The femtosecond laser has an ultrashort pulse and ultrahigh energy peak [28–30], which can realize non-hot-melt cold processing and high-precision processing, and which has a wide range of applications in micro and nano processing [31]. In the experiment, a Yb:KGW femtosecond laser (Pharos, Light Conversion, Vilnius, Lithuania)—with a center wavelength of 1030 nm, pulse duration of 216 fs, and repetition rate of 10 kHz—was used to generate Gaussian beams. The optical path diagram of the femtosecond laser processing system is shown in Figure 3. When exiting from the femtosecond laser provider, it passes through the shutter, attenuator, mirrors, power meter, and apertures, before finally entering the objective lens. The shutter controls the laser switch, the attenuator regulates the light intensity, and two apertures are used to collimate the beam entering the objective [32]. The lens is a 20× objective lens (Nikon, Tokyo, Japan, NA = 0.4). After focusing the laser beam on the vertical incident mode, it is ablated and processed on the surface of the sample. The sample used was fused silica (Gulo company, Luoyang, China, K9). The light transmittance was higher than 90% with high precision, high parallels, and high finish, with a size of 20 × 20 × 0.4 mm. The samples were mounted on a three-dimensional mobile platform (Suruga Seiki Corporation, Tokyo, Japan) with computer-controlled *XYZ* motion in three directions with a movement accuracy of 20 nm.

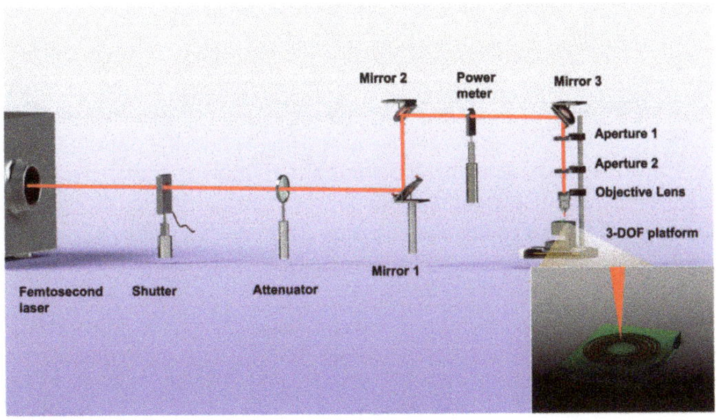

Figure 3. Schematic diagram of femtosecond laser processing system.

3. Results and Discussion

3.1. Study of Single-Ring Annular Fresnel Zone Plate (AFZP)

To study the process of the AFZP generating an annular beam, the AFZP model was established in ZEMAX software. The designed AFZP radius R was 0.6 mm, the inner zone radius r_0 of the circular FZP was 80 μm, and the zone number of circular FZP n was 29. Therefore, the inner diameter R_i of the AFZP was 0.169 mm, and the outer diameter R_0 was 1.031 mm. The incident light was a flat-topped beam with a wavelength of 800 nm, which was obtained in a focal plane 8 mm from the AFZP, as shown in Figure 4. The radius of the annular beam was 0.6 mm, which is the same as that of the AFZP. The beam on the ring had high energy in the center and low energy on the sides. In addition to the annular beam, a bright spot was observed at the center. It can be seen more clearly in Figure 4b that the energy of the central bright spot was much greater than that of the annular area.

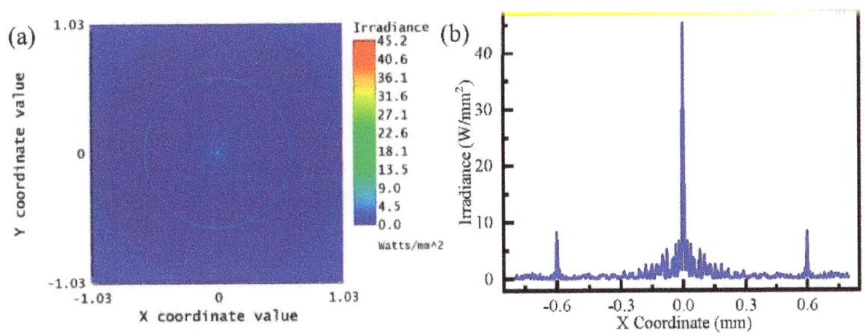

Figure 4. Simulation of annular Fresnel zone plate (AFZP) with R as 0.6 mm and r_0 as 80 μm: (**a**) annular beam and central spot in focal plane; (**b**) energy distribution in the focal plane.

To eliminate the bright spot in the center, appropriate structural parameters need to be selected. For the AFZP with constant r, when radius R increases, the radius of each concentric zone increases, and the position of the diffraction area on the Z-axis also extends backward. The software modeled AFZPs with different radii to study the influence of the radius R on the central energy of the annular beam.

When the incident laser power was 1 W/mm^2, the energy in the annular area and center area in the focal plane was as shown in Figure 5. The energy in the annular area remained constant at approximately 10 W/mm^2, and the energy in the center fluctuated. When R was 0.43 mm-i.e., $R = r$-the energy of the central bright spot was seven times that of the annular area. When the radius R was 0.8 mm, the central bright spot energy reached a maximum of 58.22 W/mm^2. However, when R was 1.0 mm, the energy at the center decreased sharply to as low as 0.01 W/mm^2. These energy fluctuations were caused by Fresnel diffraction, resulting in alternating light and dark in the center. When R expanded to more than 1.6 mm, the central energy remained at a persistently low level of about 0.1 W/mm^2. According to Equation (4), with the increase in radius R of the AFZP, the position of the Fresnel diffraction area of the circular hole is gradually pushed back on the Z-axis, such that it occurs behind the focal plane, causing the focusing ring to appear in the circular hole projection area with no energy in the center. Figure 6 shows the annular beam with a radius R of 1.6 mm, where the focusing ring is clear and there is no bright spot in the center.

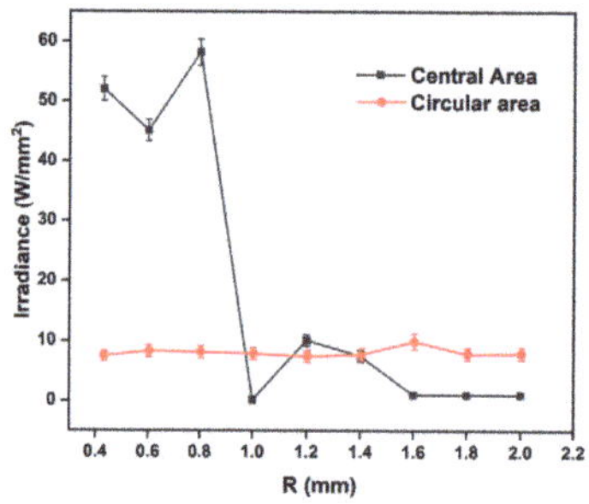

Figure 5. Simulation of the energy on the annular and central areas of the focal plane of AFZPs with different radii R.

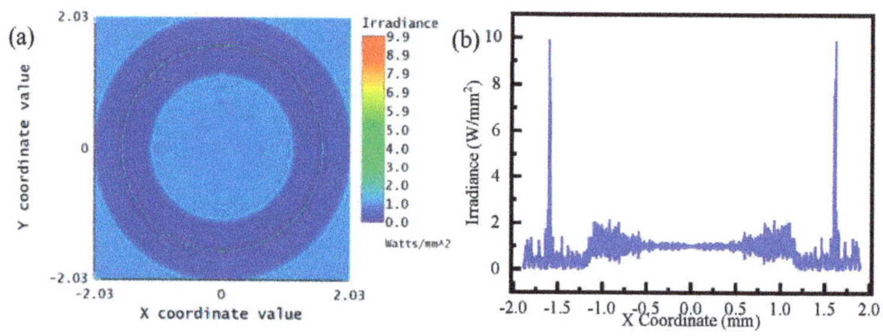

Figure 6. Simulation of AFZP with R as 1.6 mm and r_0 as 80 μm: (**a**) annular beam in the focal plane; (**b**) energy distribution in the focal plane.

In laser processing, it is usually necessary to obtain an annular beam of a specified radius R. Therefore, for an annular beam with an arbitrary radius R, it is crucial to eliminate the central bright spot. As can be seen from the above analysis, the central bright spot can be avoided by focusing the AFZP on the projection area of the circular hole. The position of the focusing ring on the Z-axis can be realized by adjusting the radius of the circular FZP r_0.

Figure 7 shows the focal plane energy distribution of different AFZPs obtained by adjusting r_0 when the radius R was 0.6 mm. As can be seen from the figure, although r_0 was different, the energy on the focusing ring did not change significantly, and it was stable at approximately 8 W/mm^2. Simultaneously, there was a significant change in energy at the center of the Z-axis. When the radius of the AFZP r_0 was 20–30 μm, and the focal plane was within 1.2 mm of the Z-axis, there was no energy in the center of the annular beam. This shows that the focal plane was in the circular hole projection area, and there was no circular hole Fresnel diffraction. When r_0 was 35–80 μm, the focal plane appeared 1.2 mm behind the Z-axis, there was energy in the center, and the focal plane was in the Fresnel diffraction area of the circular hole. In summary, a smaller r_0 resulted in a smaller focal length of the focusing ring and a closer distance between the annular beam and the FZP. However, when r_0 increased, the focal plane of the annular beam appeared in the Fresnel diffraction area or even in the Fraunhofer diffraction area, which resulted in the enhancement of interference at the center and the appearance of a central bright spot. Therefore, a smaller r_0 and larger R should be selected to eliminate the bright spot in the center of the circular beam.

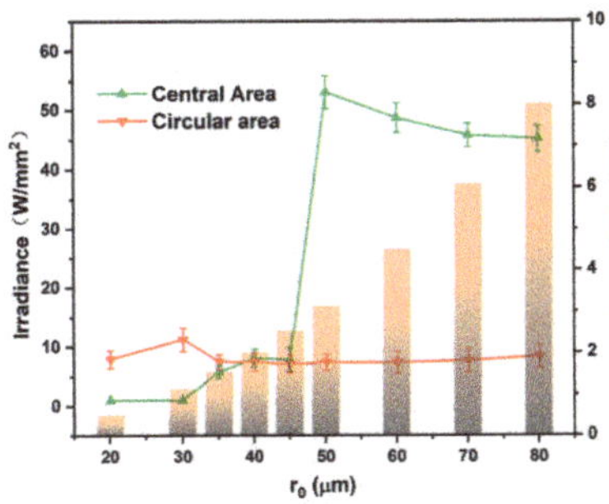

Figure 7. Simulation of the energy on the annular and central areas of the focal plane of AFZPs with different radii r_0.

The focusing process of an AFZP with r_0 80 μm and R 1.60 mm was further studied. Distances of 7.1–8.7 mm before and after the focal plane were selected along the Z-axis. The sections presented annular beams without bright spots in the center, but the radii of the annular beams were different. From 7.2 mm to 8.4 mm along the Z-axis, the radius increased from 1.45 mm to 1.70 mm. In the 8 mm focal plane of the Z-axis, the radius of the annular beam was slightly larger than that of the AFZP, which was 1.61 mm. For a linear FZP whose radius tended to be infinite, the diffraction lines were at the same position before and after the focal plane. However, for an AFZP with a finite radius, the position of the annular beam changed along the Z-axis, which was caused by the different energies reaching the focus through the inner and outer zones. As shown in Figure 8, the laser energy brought in by the outer zones was higher than that brought in by the inner zones; thus, the annular beam before and after the focal plane appeared to be scattered [15]. Therefore, the radius of the annular beam could be adjusted by adjusting the position of the Z-axis, and the highest energy of the annular beams remained in the focal plane.

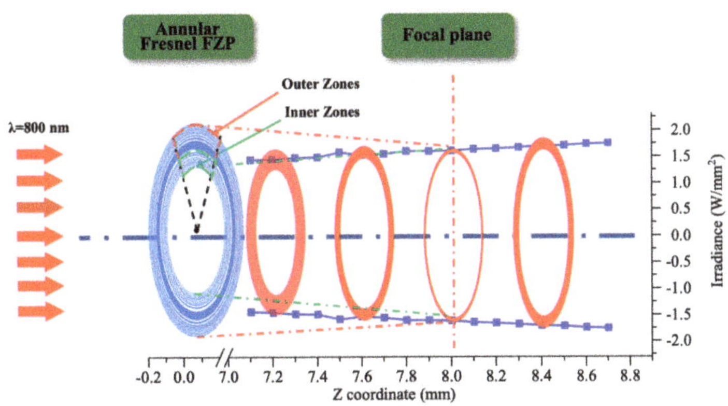

Figure 8. Focusing process of the annular beam before and after focal plane.

The femtosecond laser was applied to micromachine fused silica to verify the theory that the AFZP generates an annular beam. The design of the processing path of AFZP was carried out on the original software. Since the radius and width of each zone were not consistent, the processing path of each zone needed to be designed separately. The zone width was wider than the width of single microgroove processing; thus, it was necessary to carry out ring processing many times. The laser power and speed were kept constant during the processing, and the machining radius was successively increased until the whole zone was prepared. The groove width of single microgroove machining was affected by moving speed and laser power. After careful consideration, the parameters of the femtosecond laser were adjusted to 10 mW and 1 μJ of energy, and the focus size was 3.14 μm. The platform was translated perpendicular to the laser beam at a speed of 1000 μm/s, and the width of the single microgroove machining groove was 8 μm. An AFZP with R = 0.6 mm and r_0 = 80 μm was fabricated. As shown in Figure 9a, each zone of the FZP was clear and concentric. From a radial point of view, the radii of the different FZPs were consistent with the parameters of the linear FZPs.

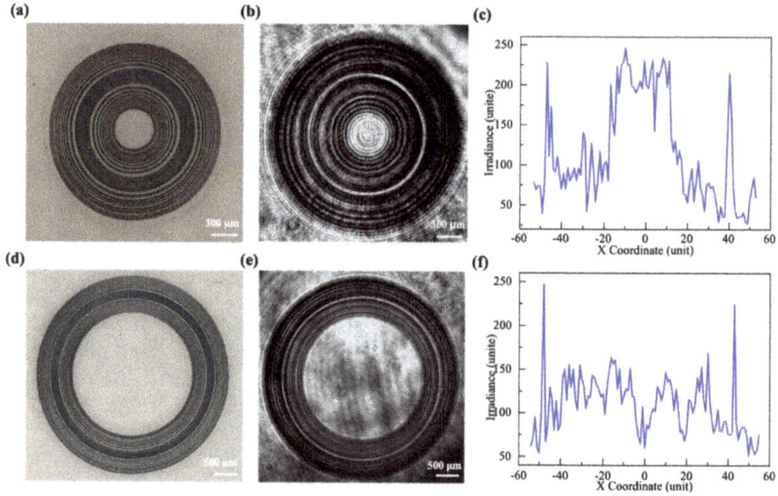

Figure 9. AFZP with R of 0.6 mm with central spots: (**a**) the AFZP prepared by a femtosecond laser on a fused silica sample; (**b**) the focusing ring behind the AFZP photographed by a CCD; (**c**) the energy distribution diagram of the focal plane, for the AFZP with R = 1.6 mm without central spots; (**d**) the AFZP prepared by a femtosecond laser on a fused silica sample; (**e**) the focusing ring behind the AFZP photographed by a CCD; (**f**) the energy distribution diagram of the focal plane.

The processed sample was placed in the diffraction test light path and irradiated by plane light with a wavelength of 800 nm, and the focal plane pattern was received 8 mm behind the FZP. A clear annular beam is shown in Figure 9b. The radius of the annular beam was equal to the radius R of the AFZP. At the same time, a bright spot can be seen at the center of the circular beam. Figure 9e shows the diffraction pattern when R increased to 1.6 mm. It shows that the center of the annular beam only had the laser passing through the circular hole. These lasers in the center were not focused and had very little energy. Since the theoretical maximum diffraction efficiency of the amplitude-type FZP was only 10.13%, the energy of the beam diffracted on the ring did not significantly exceed that of the center. Therefore, the effect of the energy in the center was negligible when processing with an annular beam.

3.2. Study of Double-Ring AFZP

Two annular beams or even multiple annular beams can be obtained by concentrically connecting multiple AFZPs. Figure 10 shows a schematic of the double-ring AFZP. The radius of the inner AFZP is R_1, the radius of the outer AFZP is R_2, and the spacing between the two AFZPs is R_d. Because $R_2 > R_1$, according to Equation (4), the Fresnel diffraction area of the outer AFZP on the Z-axis is outside the diffraction zone of the inner AFZP. If the annular beam of the inner AFZP has no central spot, then the outer AFZP would also have no central spot.

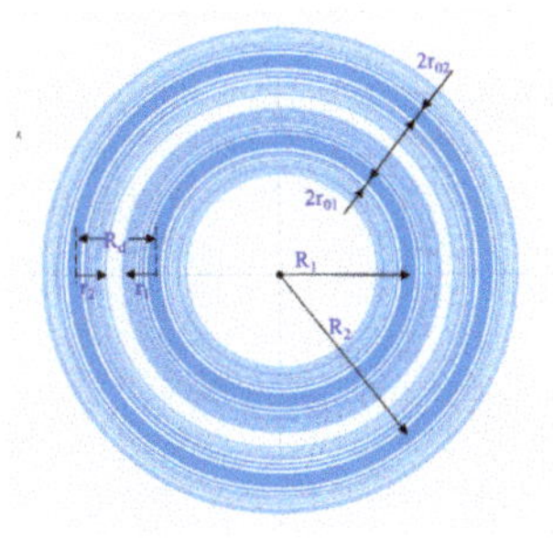

Figure 10. Schematic diagram of double-ring AFZP.

A double-ring AFZP with r_0 = 80 μm, R_1 = 1.6 mm, and R_2 = 2.4 mm was simulated. As shown in Figure 11a, two clear concentric rings appeared in the focal plane with no spot at the center, and the spacing R_d between the two rings was 0.8 mm. For multi-annular beams, the consistency of the energy on different rings is very crucial for machining. As shown in Figure 11c, when R_d = 0.8 mm, the peak value of the outer AFZP was 9.8 W/mm^2, and that of the inner AFZP was 5.3 W/mm^2, with a significant difference between them. Figure 11b shows that the outer AFZP was moved inward to remove the spacing between the inner and outer FZPs, and R_d was reduced to 0.73 mm. The energy of the inner and outer rings was approximately 4 W/mm^2, and the energy consistency of the two rings was greatly improved.

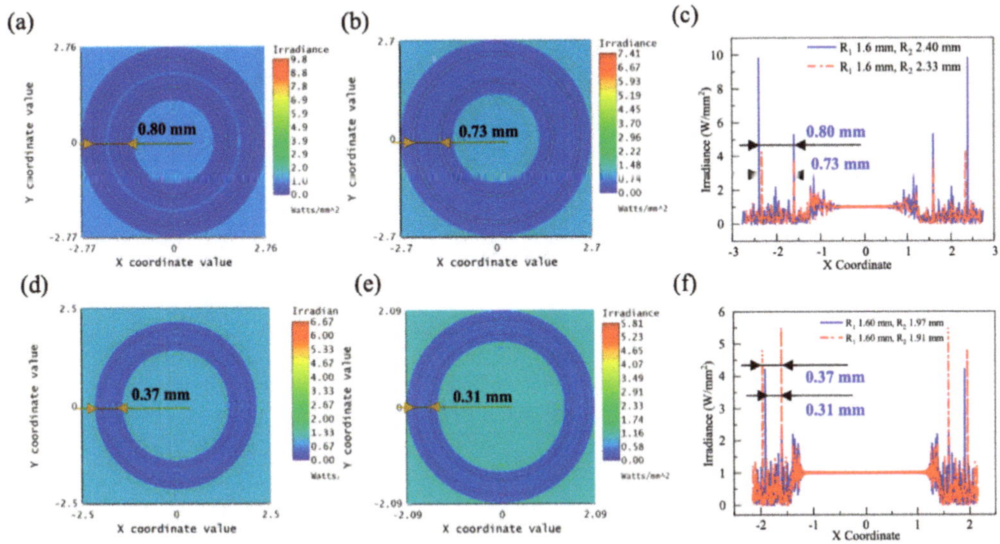

Figure 11. Simulation pattern of double-ring AFZP with R_1 of 1.6 mm: (**a**) R_2 = 2.40 mm, R_d = 0.80 mm; (**b**) R_2 = 2.33 mm, R_d = 0.73 mm; (**c**) comparison of energy distribution in a/b diagram; (**d**) R_2 = 1.97 mm, R_d = 0.37 mm; (**e**) deletion of 12 zones, with R_2 = 1.91 mm, R_d = 0.31 mm; (**f**) comparison diagram of energy distribution of d/e diagram.

In practical applications, it is important to obtain a double-ring beam with a specified radius and ring spacing. The spacing R_d can be adjusted by adjusting R_1 and R_2; however, when the required spacing R_d is less than the sum of the widths of the inner and outer AFZPs, i.e., $R_d < r_1 + r_2$, the required spacing R_d cannot be obtained by adjusting R_1 and R_2. At this point, there are two ways to reduce the spacing R_d, one of which is to reduce r_{01} and r_{02}. When r_{01} and r_{02} decrease, r_1 and r_2 decrease proportionally, and the spacing R_d is significantly reduced. Because multiple focusing rings should be in the same focal plane, r_{01} and r_{02} must be the same when adjusting these parameters. As shown in Figure 11c, when r_0 was reduced from 80 to 40 μm, R_d decreased significantly from 0.73 mm to 0.37 mm. The energy values of the inner and outer rings were similar, both above 4.50 W/mm^2. In addition, R_d can be reduced further by decreasing the zone number n. A smaller zone number n results in smaller values of r_1 and r_2. As shown in Figure 11d, to shorten the spacing R_d, six zones on one side of the AFZP were deleted. After deleting 12 zones of the inner and outer AFZPs, the spacing R_d was reduced from 0.37 mm to 0.31 mm. However, the energy on the outer ring decreased to 4.21 W/mm^2, and the energy on the inner ring decreased to 2.27 W/mm^2. Therefore, when adjusting the double-ring spacing R_d, the method of adjusting the inner ring radius r_{01} and r_{02} should be the first choice, which cannot only quickly and substantially adjust R_d, but also maintain the energy on the ring.

Figure 12a shows the double-ring AFZP fabricated using a femtosecond laser, with r_0 = 80 μm, R_1 = 1.60 mm, and R_2 = 2.33 mm. Figure 12b shows the focusing pattern 8 mm behind the AFZP. Two clear focusing rings can be observed. The two focusing rings had high concentricity and appeared simultaneously in the same focal plane, and there was no bright spot in the center. More focusing rings can be obtained in the focal plane if the same r_0 AFZP is sheathed on the outer side of the double-ring AFZP. Moreover, the distance between different AFZPs R_d can be easily adjusted using the forementioned method.

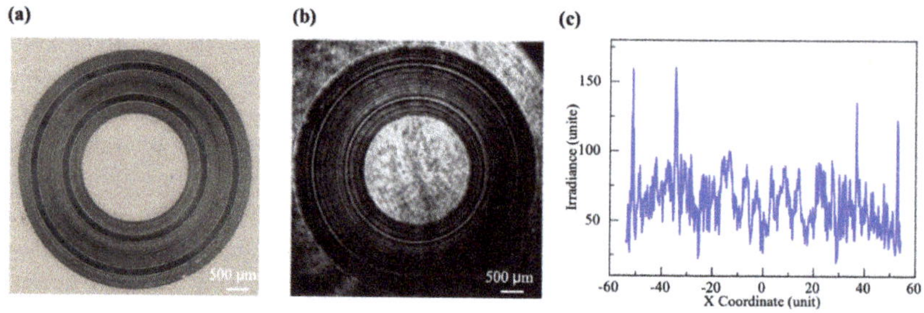

Figure 12. (**a**) The double-ring AFZP with R_1 = 1.60 mm and R_2 = 2.33 mm was fabricated using a femtosecond laser on a fused silica sample; (**b**) focusing double-ring behind the AFZP captured by a CCD; (**c**) energy distribution diagram of the focal plane.

4. Conclusions

The AFZP designed in this study not only eliminates the central energy of the single annular beam, but it also obtains a double-ring beam with a controllable radius in the same focal plane. In addition, the AFZP can be easily integrated into the radial direction to obtain a multi-ring beam, and it can also be arrayed in a plane to obtain an annular beam array. Further research on annular beams will provide a more productive potential for applications related to nontraditionally shaped laser beams.

Author Contributions: Conceptualization, L.D. and X.S.; Investigation, F.Z. and X.S.; Writing—original draft, X.S.; Writing—review and editing, X.S. and F.Z. All authors have read and agreed to the published version of the manuscript.

Funding: This work was supported by the National Natural Science Foundation of China 51875584, 51875585, and 51935013.

Conflicts of Interest: The authors declare no conflict of interest.

References

1. Zeng, D.; Latham, W.P.; Kar, A. Temperature distributions due to annular laser beam heating. *J. Laser Appl.* **2005**, *17*, 256–262. [CrossRef]
2. Zeng, D.; Latham, W.P.; Kar, A. Optical trepanning with a refractive axicon lens system. In *Laser Beam Shaping VII*; SPIE: Bellingham, WA, USA, 2006; Volume 6290.
3. Marti, D.; Craig, B.A. Bessel and annular beams for materials processing. *Laser Photonics Rev.* **2012**, *6*, 607–621.
4. Winfield, R.J.; Bhuian, B.; O'Brien, S.; Crean, G.M. Refractive femtosecond laser beam shaping for two-photon polymerization. *Appl. Phys. Lett.* **2007**, *90*, 111115. [CrossRef]
5. Bhuian, B.; Winfield, R.J.; O'Brien, S.; Crean, G.M. Pattern generation using axicon lens beam shaping in two-photon polymerization. *Appl. Surf. Sci.* **2007**, *254*, 841–844. [CrossRef]
6. Nathan, B. Cell imaging: New ways to see a smaller world. *Nature* **2008**, *456*, 825–830.
7. Hell, S.W. Microscopy and its focal switch. *Nat. Methods* **2009**, *6*, 24–32. [CrossRef]
8. Smith, E.J.; Schulze, S.; Kiravittaya, S.; Mei, Y.; Sanchez, S.; Schmidt, O.G. Lab-in-a-tube: Detection of individual mouse cells for analysis inflexible in flexible split-wall microtube resonator sensors. *Nano Lett.* **2010**, *11*, 4037–4042. [CrossRef]
9. Gaoshan, H.; Yongfeng, M.; Thurmer, D.J.; Coric, E.; Schmidt, O.G. Rolled-up transparent microtubes as two-dimensionally confined culture scaffolds of individual yeast cells. *Lab A Chip* **2009**, *9*, 263–268.
10. Takei, K.; Kawashima, T.; Kawano, T.; Kaneko, H.; Sawada, K.; Ishida, M. Out-of-plane microtube arrays for drug delivery—liquid flow properties and an application to the nerve block test. *Biomed. Microdevices* **2009**, *11*, 539–545. [CrossRef]
11. Hsu, K.-Y.; Tung, Y.-C.; Chung, M.-H.; Lee, C.-K. Design and Fabrication of Sub-wavelength Annular Apertures for Femtosecond Laser Machining. In *Laser-Based Micro- and Nanoprocessing IX*; SPIE: Bellingham, WA, USA, 2015.
12. Liu, W.; Yu, Z.; Lu, D.; Zhang, N. Multi-wavelength annular optical pulses generated by double inter-ferent femtosecond Bessel laser beams in silica glass. *Opt. Lasers Eng.* **2021**, *136*, 106330. [CrossRef]
13. Tung, Y.-C.; Chung, M.-H.; Sung, I.-H.; Lee, C.-K. Design and Fabrication of Sub-wavelength Annular Apertures on Fiber Tip for Femtosecond Laser Machining. In *Micromachining and Microfabrication Process Technology XIX*; SPIE: Bellingham, WA, USA, 2014.
14. Arash, S.; Bahar, M. Generation of annular beam by a novel class of Fresnel zone plate. *Appl. Opt.* **2014**, *53*, 5995–6000.
15. Zhang, C.; Hu, Y.; Li, J.; Li, G.; Chu, J.; Huang, W. A rapid two-photon fabrication of tube array using an annular Fresnel lens. *Opt. Express* **2014**, *22*, 3983–3990. [CrossRef]
16. Fatemeh, S.; Arash, S. Coherence combination to create a long depth bifocal diffractive lens capable of generating tailorable biannular and bottle beam. *Opt. Quantum Electron.* **2021**, *53*, 144.
17. Yu, Y.Y.; Chang, C.K.; Lai, M.W.; Huang, L.S.; Lee, C.K. Ablation of silicon by focusing a femtosecond laser through a subwavelength annular aperture structure. In *Laser Beam Shaping XI*; SPIE: Bellingham, WA, USA, 2010.
18. Bekshaev, A.; Bliokh, K.Y.; Soskin, M. Internal flows and energy circulation in light beams. *J. Opt.* **2011**, *13*, 053001. [CrossRef]
19. Alexeev, I.; Leitz, K.-H.; Otto, A.; Schmidt, M. Application of Bessel beams for ultrafast laser volume structuring of non transparent media. *Phys. Procedia* **2010**, *5*, 533–540. [CrossRef]
20. Maurer, C.; Jesacher, A.; Bernet, S.; Ritsch-Marte, M. What spatial light modulators can do for optical microscopy. *Laser Photonics Rev.* **2011**, *5*, 81–101. [CrossRef]
21. Beck, R.J.; Parry, J.P.; MacPherson, W.N.; Waddie, A.; Weston, N.J.; Shephard, J.D.; Hand, D.P. Application of cooled spatial light modulator for high power nanosecond laser micromachining. *Optics Express* **2010**, *18*, 17059–17065. [CrossRef]
22. Yang, L.; El-Tamer, A.; Hinze, U.; Jiawen, L.; Yanlei, H.; Wenhao, H.; Jiaru, C.; Chichkov, B.N. Two-photon polymerization of cylinder microstructures by femtosecond Bessel beams. *Appl. Phys. Lett.* **2014**, *105*, 041110. [CrossRef]
23. Fang, Z.; Xiaoyan, S.; Hongmin, Z.; Lian, D.; Youwang, H.; Ji'an, D.; Ming, L. Optimization of the focusing characteristics of Fresnel zone plates fabricated with a femtosecond laser. *J. Mod. Opt.* **2021**, *68*, 100–107.
24. Kim, J.; Ha, W.; Park, J.; Kim, J.K.; Sohn, I.B.; Shin, W.; Oh, K. Micro Fresnel Zone Plate Lens Inscribed on a Hard Polymer Clad Fiber Using Femtosecond Pulsed Laser. *IEEE Photonics Technol. Lett.* **2013**, *25*, 761–763. [CrossRef]
25. Zhang, Y.; Yuan, L.; Liu, Z.; Yang, J. Dual optical tweezers integrated in a four-core fiber: Design and Simulation. In *Fourth Asia Pacific Optical Sensors Conference*; SPIE: Bellingham, WA, USA, 2013.
26. Zhang, Y.; Liu, Z.; Yang, J.; Yuan, L. A non-contact single optical fiber multi-optical tweezers probe: Design and fabrication. *Opt. Commun.* **2012**, *285*, 4068–4071. [CrossRef]
27. Arash, S.; Shima, G. Generation of double line focus and 1D non-diffractive beams using phase shifted linear Fresnel zone plate. *Opt. Laser Technol.* **2015**, *69*, 65–70.
28. Bricchi, E.; Mills, J.D.; Kazansky, P.G.; Klappauf, B.G.; Baumberg, J.J. Birefringent Fresnel zone plates in silica fabricated by femtosecond laser machining. *Opt. Lett.* **2002**, *27*, 2200–2202. [CrossRef]
29. Yamada, K.; Watanabe, W.; Li, Y.; Itoh, K.; Nishii, J. Multilevel phase-type diffractive lenses in silica glass induced by filamentation of femtosecond laser pulses. *Opt. Lett.* **2004**, *29*, 1846–1848. [CrossRef]
30. Wataru, W.; Daisuke, K.; Kazuyoshi, I. Fabrication of Fresnel zone plate embedded in silica glass by femtosecond laser pulses. *Opt. Express* **2002**, *10*, 978–983.

31. Delullier, P.; Druart, G.; De La Barrière, F.; Calvez, L.; Lancry, M. Femtosecond Laser Direct Writing of Gradient Index Fresnel Lens in GeS2-Based Chalcogenide Glass for Imaging Applications. *Appl. Sci.* **2022**, *12*, 4490. [CrossRef]
32. Sun, X.; Zhou, F.; Dong, X.; Zhang, F.; Liang, C.; Duan, L.; Hu, Y.; Duan, J.A. Fabrication of GaAs micro-optical components using wet etching assisted femtosecond laser ablation. *J. Mod. Opt.* **2020**, *67*, 1516–1523. [CrossRef]

micromachines

Article

Design of a Cylindrical Compliant Linear Guide with Decoupling Parallelogram Mechanisms

Tinghao Liu and **Guangbo Hao** *

Electrical and Electronic Engineering, School of Engineering and Architecture, University College Cork, T12 K8AF Cork, Ireland
* Correspondence: g.hao@ucc.ie

Abstract: A conventional linear guiding mechanism refers to the slide rail guides composed of multiple assemble parts. These guiding mechanisms suffer from many adverse effects, including lubrication, wear and assembly issues. A novel compliant guiding mechanism is proposed in this paper to address these common problems, and this mechanism transfers or transforms motion, force and energy via the deformation of flexible members. This linear guide is designed in a cylindrical shape, and the centre platform moves along its axis (i.e., the motion direction). The proposed linear guide consists of several in-parallel curved compound double parallelogram mechanisms (CDPMs) connected by the same number of decoupling parallelogram mechanisms. Nonlinear finite element analysis (FEA) is used for stiffness analysis and shows that applying the decoupling mechanisms to the detached linear guide (the in-parallel curved CDPMs only) can dramatically improve the stiffness in undesired movement (bearing) directions while keeping its original stiffness along its axis. The nonlinear FEA can capture the stiffness variation by considering all the structural deformation. The issue of bearing-direction stiffness degradation of the detached linear guide is dealt with by applying decoupling mechanisms. The static experimental test is conducted on a 3D printed prototype and shows that the stiffness in the motion direction is nearly constant (linear). The results obtained from the experimental test show good agreement with those obtained from the nonlinear FEA with a maximum error of 9.76%.

Keywords: compliant mechanism; linear guide; decoupling mechanism; finite element analysis

Citation: Liu, T.; Hao, G. Design of a Cylindrical Compliant Linear Guide with Decoupling Parallelogram Mechanisms. *Micromachines* **2022**, *13*, 1275. https://doi.org/10.3390/mi13081275

Academic Editor: Stelios K. Georgantzinos

Received: 6 July 2022
Accepted: 5 August 2022
Published: 8 August 2022

1. Introduction

Linear guiding mechanisms have attracted significant attention from both manufacturing industries and researchers [1]. Traditional linear positioning is achieved by using slide rail guides. These linear guides are generally composed of multiple assemble parts, including railways and rods [2]. Although the conventional linear guides benefit from their relatively extensive range of motion, they suffer from many adverse aspects. For example, insufficient lubrication can cause a choppy linear motion or reduce the service life of slide rail guides. Wear and tolerance between assembled parts can result in low positioning accuracy and repeatability. The concept of the compliant mechanism has been introduced into the design of guiding mechanisms to address these common issues due to its merits in high performance and low-cost [3,4].

The compliant mechanism has the capability of transferring or transforming motion, force and energy via the deformation of flexible members [5]. The compliant mechanisms are widely used in the field of surgical robotics [6–9], positioning stages [10–13], energy harvesting [14–16], mechanical metamaterial [17–20] and micro-electromechanical system (MEMS) [20–22]. Previous researchers have proposed diverse linear guides based on compliant mechanisms. These compliant guides can be classified into planar and spatial linear guides [23].

The parallelogram mechanism is the most common compliant planar guide, and it can perform a single-axis translational movement and has a compact footprint [24]. Generally, the compliant parallelogram mechanisms are composed of two compliant sheets whose ends are fixed to the base and motion stage separately. However, the parallelogram mechanism suffers from parasitic error along the bearing directions.

The parasitic error of the parallelogram can be mitigated by nesting two same mechanisms symmetrically [25]. This mechanism is named as the compound parallelogram mechanism. Furthermore, a fully compliant planar linear guiding mechanism—named XBob—was presented in [26]. The XBob derived from the Roberts four-bar approximate straight line mechanism has the potential use in macro-scale devices or microsystems due to its flat design. In addition, Watt's mechanism is used as a compliant planar linear guide as well [27].

The diaphragm is one special type of spatial linear guiding mechanism since it is designed flat and can perform an out-of-plane linear movement [28]. Another type of spatial linear guide consisting of folded leaf springs was proposed in [23]. In this design, torsion reinforcement structures are added to increase the stiffness in the bearing directions. Additionally, a cylindrical compliant spring based on the curved parallelogram mechanisms was proposed in [29]. This cylindrical guide benefits from its long stroke, high positioning resolution and movement accuracy. This linear guide is named the detached linear guide in this paper since it comprises the in-parallel curved compound double parallelogram mechanisms (CDPMs) only.

Despite the above benefits existing in the detached linear guide, there are still some open issues, such as significantly reduced bearing-direction stiffness over primary motion. We have improved the detached linear guide by embedding a novel decoupling mechanism between each pair of CDPMs. The design of decoupling mechanisms was inspired by [28].

Applying the decoupling mechanism dramatically improved the bearing-direction stiffness without compromising the stiffness in the motion direction. Furthermore, the issue of bearing-direction stiffness degradation of the detached linear guide is addressed. The decoupling mechanism was developed based on a curved parallelogram mechanism, which deforms along the annular direction.

The methodology used in this paper includes designing a new compliant linear guide mechanism by adding decoupled parallelogram mechanisms [30], nonlinear finite element analysis (FEA) and validation, parametric optimization, prototyping and experimental testing. Nonlinear FEA is used to capture nonlinear characteristics of the design that could not be obtained using the available curved-sheet spatial model (i.e., the linear model as reported in [29,31]) and can also consider the deformation from all structural elements.

This work identifies several main contributions:

- The design of a compliant linear guide (i.e., a decoupled linear guide) that has an improved bearing-direction stiffness without sacrificing its linear stiffness along the motion direction, which is based on connecting decoupling mechanisms to the detached linear guide.
- The comparison of the static stiffness along each axis of decoupled linear guide and that of the detached linear guide via nonlinear FEA simulations.
- A static experimental test conducted on a 3D printed prototype to verify the linear stiffness along the motion direction.

This paper is organized as below. Section 2 proposed the structural design of the cylindrical compliant linear guide with decoupling mechanisms (decoupled linear guide). The deformations of the detached linear guide are shown in this section as well. Section 3 mainly involves the analysis and optimization strategy. A series of nonlinear FEA simulations are performed to discover the nonlinear stiffness before and after applying the decoupling mechanism to the detached linear guide. Then, the decoupled linear guide is optimized and aims to hold a lower primary stiffness.

After determining the geometrical parameters of the detached linear guide, a prototype is manufactured using additive manufacturing technology. An experiment is detailed in this section as well. Lastly, our conclusions are drawn in Section 6.

2. Structural Design of Decoupled Linear Guide

The components of the decoupled linear guide proposed in this paper mainly involve the compound parallelogram mechanism, the decoupling mechanism and the centre platform. These three parts form a linear guide together with rigid bodies. The main content of this section is to illustrate the structural design of the decoupled linear guide.

As shown in Figure 1a, the fundamental compound parallelogram consists of a centre shuttle and four straight compliant beams. This mechanism can be regarded as two identical parallelogram mechanisms in a mirror-symmetrical arrangement. The centre shuttle could move in the vertical direction due to the deformation of these beams. Previous guiding mechanisms are primarily designed based on it since they can provide a relatively linear path [32]. However, the stiffness in the desired motion direction is relatively large due to its overconstrained nature.

Hence, the compound double parallelogram mechanism was proposed to solve the problem of large primary stiffness [33]. The structure of the CDPM and its deformed configuration can be seen in Figure 1b,c. The upper and lower surface are fixed. The shuttle of CDPM moves in the vertical direction, which is similar to the motion of the compound parallelogram mechanism. All of the eight compliant beams in CDPM deform, contributing to the movement of the shuttle. The intermediate rigid body also moves vertically during the deformation process. The curved CDPM and its deformation under vertical load is shown in Figure 1d.

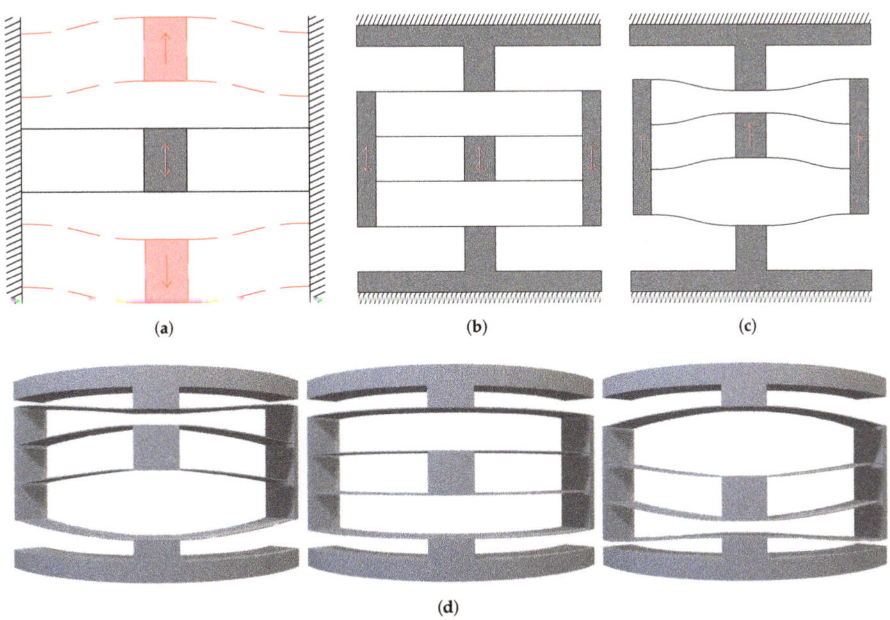

Figure 1. The parallelogram mechanisms. (**a**) The fundamental compound parallelogram mechanism and its deformation. (**b**) Planar CDPM. (**c**) Deformed planar CDPM. (**d**) The deformation process of curved CDPM.

Another vital component of the detached linear guide is the decoupling mechanism, which is the part that gives the guiding mechanism higher performance. The decoupling mechanism deforms along the annular direction. The decoupling mechanism is embedded between intermediate rigid bodies of each couple of CDPMs. After applying the decoupling mechanism, the CDPMs in the detached linear guide are connected to form an entire cylindrical structure. When the CDPMs deform, the parasitic motion of the intermediate rigid bodies will tend to cause the decoupling mechanisms to deform.

Figure 2a,b shows the deformation process of the decoupling mechanism in a 2D sketch. The Figure 2c,d illustrate the 3D model of decoupling before and after tension. The tension works along the annular direction, which is perpendicular to the radius direction. The decoupling mechanism keeps its radius constant when it deforms.

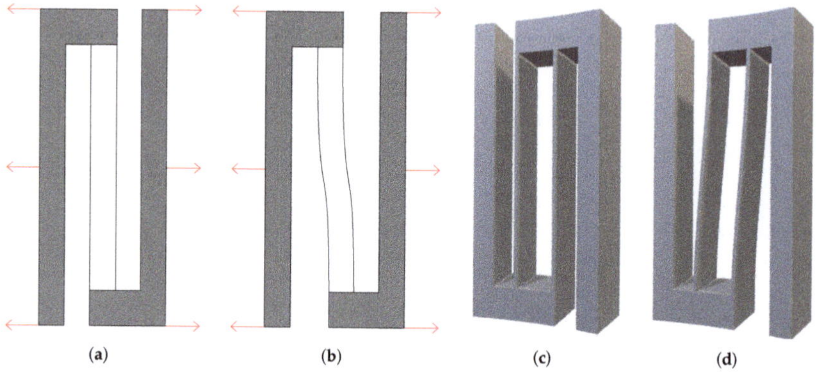

Figure 2. The decoupling mechanism. (**a**) 2D decoupling mechanism. (**b**) Deformed 2D decoupling mechanism. (**c**) 3D decoupling mechanism (load is added in the direction perpendicular to the radius direction). (**d**) Deformed 3D decoupling mechanism.

Figure 3a illustrates how to embedding the decoupling mechanism between CDPMs. Figure 3b shows the deformed mechanism where both the CDPMs and decoupling mechanism deform. The compliant beams of CDPM deform in the vertical direction, while the beams of the decoupling mechanism deform transversely.

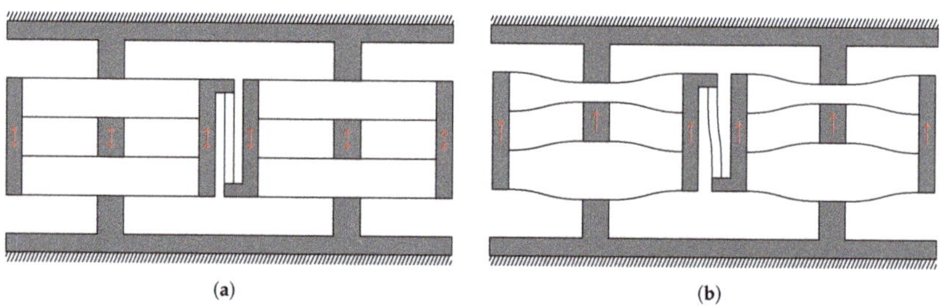

Figure 3. The in-parallel planar CDPMs connected by decoupling mechanism and its deformed shape.

Figure 4a is the front view of the cylindrical compliant guide with three CDPMs. In this case, three in-parallel curved CDPMs are connected by three decoupling mechanisms. A centre platform is fixed to the centre moving shuttle of the CDPM. The moving platform can perform reciprocating motion since this mechanism has only one degree of freedom (DoF). Figure 4b shows the deformation of the entire linear guiding mechanism. This paper assumes the upper and lower annular structures as rigid bodies. The compliant beams are fixed to them.

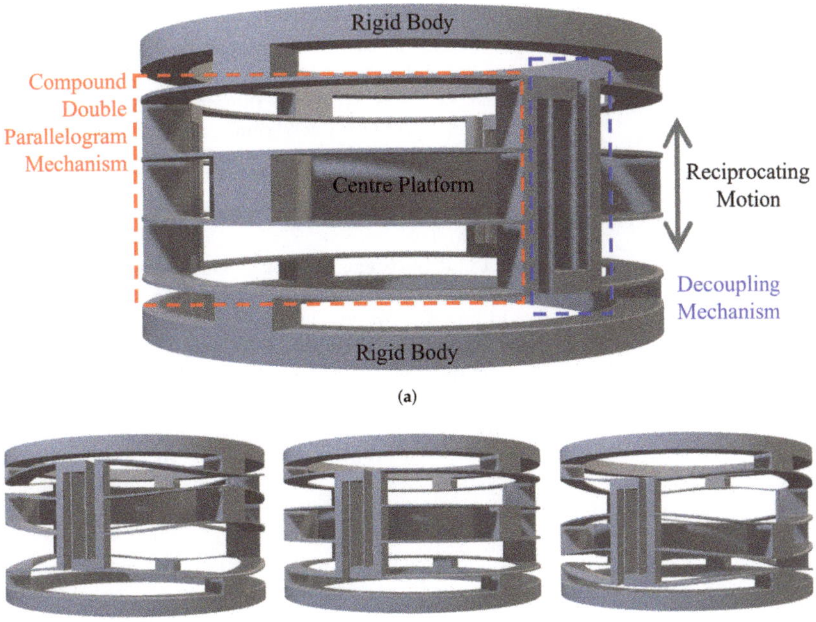

Figure 4. The 3D model of the decoupled linear guide. (**a**) The geometrical design and explanation. (**b**) The deformation process of the decoupled linear guide.

3. Analysis and Optimization

The main aim of this section is to compare the detached linear guide and decoupled linear guide in the aspect of static stiffness. The performed simulations are based on finite element analysis. The nonlinear simulations reflect the stiffness variation and force–displacement relationships with high accuracy. A global coordinate system is defined to explain the spatial position and arrangement of the model. As seen from Figure 5, the origin of the coordinate system is determined at the geometry centre of the cylinder. The primary motion direction of the decoupled linear guide is defined as the Z-axis. The X-axis and Y-axis are defined based on the right-hand rule.

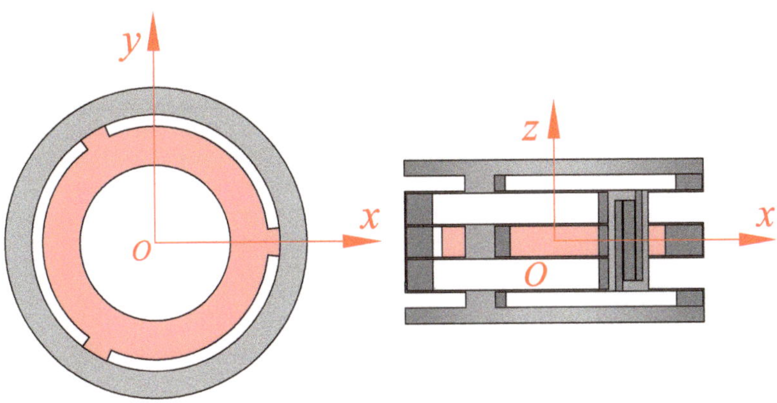

Figure 5. The defined coordinate system.

3.1. Nonlinear Stiffness Analysis

The issue of the stiffness degradation of detached linear guide is dealt with by applying the decoupling mechanisms. The stiffness degradation refers to the stiffness in bearing directions that decreases during the deformation process. For the linear guide mechanism, the degradation of stiffness would cause lower precision during the deformation. The centre platform would be easier to swivel or pitch during deformation. Hence, the stiffness degradation should be avoided or weakened.

In this design, we only consider the effects of the geometrical shape on stiffness. The material nonlinearity is not under consideration. The stiffness of the detached linear guide in the bearing directions decreases dramatically when the centre platform moves. The decoupled linear guide addressed this problem by adding the decoupling mechanisms. In addition, the stiffness variation cannot be captured from the linear FEA simulation or linear model. This is because the linear simulation can only produce a linear relationship between the load and displacement, which cannot clearly show the variation process of stiffness. Hence, a series of nonlinear FEA is involved in this part.

The nonlinear simulations further discover the force–displacement relationship and the stiffness-displacement relationship of the compliant linear guiding mechanism. The nonlinear FEA simulation is achieved by the nonlinear static analysis module of Strand7 (R2.4). The geometrical shape of the linear guide is meshed using a rectangular plate element with four endpoints. The maximum length of the mesh edge is fixed as 3% to gain acceptable resolution. The freedom condition of nonlinear FEA is the same as the linear analysis. The upper and bottom rigid bodies are fixed. The constructed points on the centre platform are assigned as one rigid part by using the function called "Auto Assign".

Then, a mandatory displacement along the motion direction is added to the rigid part. The displacement along the motion direction is set as increasing gradually. The modulus of the material is defined as 3150 MPa, which is obtained from the technical data sheet in [34]. Based on the nonlinear results, the reaction force of the centre platform reflects the magnitude of stiffness. When discovering the stiffness in bearing-directions, the small mandatory translational or rotational displacements are added into the system. The variation of reaction forces in these directions can reflect the stiffness variation.

Figure 6a is the nonlinear simulation result, which shows the relationship between the reaction force in the z-direction and displacement along the motion direction. This figure shows that the difference between decoupled and detached linear guides is tiny. Hence, we can draw a conclusion: adding the decoupling mechanism to the detached linear guide has a small effect on the primary stiffness. Figure 6b shows the relationship between the translational stiffness and displacement along the motion direction.

As can be seen, when the centre platform moves, the stiffness decreases dramatically. This phenomenon is exactly the stiffness degradation, which is mentioned above. After applying the decoupling mechanism to the detached linear guide, the stiffness K_{dz} will not change dramatically. The stiffness degradation of the rotational stiffness along x-axis K_{rx} is also addressed by adding the decoupling mechanism.

The difference is that the K_{rx} of the decoupled linear guide increases instead of decreases. This means that the centre platform of the decoupled linear guide has a higher resistance ability to the rotational external load during the deformation process than does the detached linear guide. In terms of the rotational stiffness about the motion direction K_{dz}, the stiffness degradation is addressed as well.

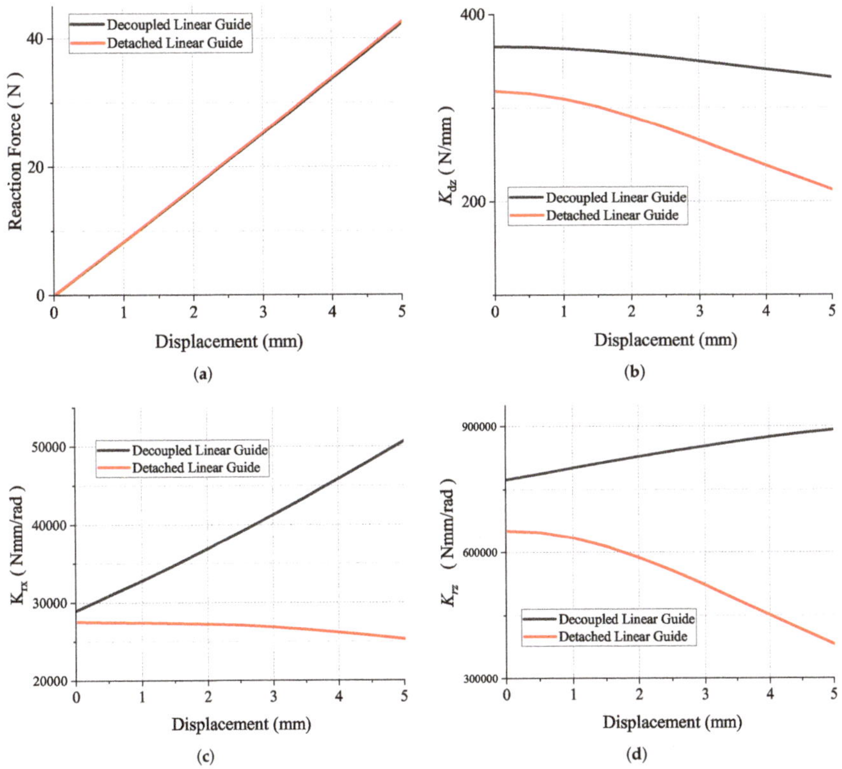

Figure 6. The comparison between the decoupled linear guide and the detached linear guide. The results are derived from the nonlinear FEA simulation. (**a**) The linear relationship between reaction force and displacement along motion direction. (**b**) The translational stiffness along Z-axis. (**c**) The rotational stiffness along X-axis. (**d**) The rotational stiffness along Z-axis.

3.2. Parametric Optimization

In this part, the main aim is to optimize the decoupled linear guide. The optimization aims to find a lower stiffness in the motion direction (K_{dz}). The optimization strategy is to change the geometry parameters to obtain a linear guide with lower stiffness. Figure 7 is the section view of the linear guide. The geometry parameters are defined and shown in this figure. The radius R represent the average radius of the annulus. The parameter w represents the width of the compliant beam. Angles θ_1, θ_2 and θ_3 are defined to illustrate the distribution of rigid block, compliant beam and decoupling mechanism.

The relationship between these angles can be defined as:

$$\theta_1 + 2\theta_2 + \theta_3 = \frac{2\pi}{n} \tag{1}$$

where n is the number of CDPMs in the detached linear guide. In this design, the number n is fixed as 3. Another parameter, "Percentage of the beam P_b", is introduced to this system for a more straightforward representation of the proportion of compliant beam. In this design, the θ_1 is fixed as 13° to maintain enough support between the centre platform and compliant linear guide. The Pb can be described as:

$$P_b = \frac{\theta_1 + 2\theta_2}{\theta_1 + 2\theta_2 + \theta_3} \tag{2}$$

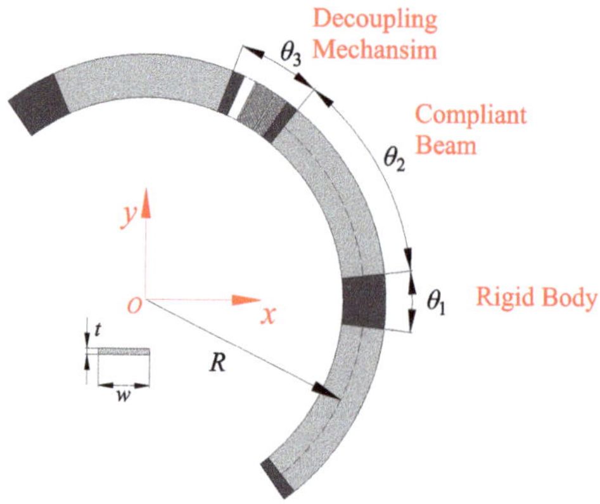

Figure 7. The definition of parameters from the section view of the decoupled linear guide.

The geometry radius R is fixed as 40 mm in this design. The larger the radius R is, the smaller the stiffness K_{dz} is. This compliant mechanism is designed in a compact configuration for trying to suit more application scenarios. Figure 8 shows the main stiffness K_{dz} with different geometrical parameters. The first parameter to be determined is the *Pb*, which determines the length of the beam and the proportion of the decoupling mechanisms. As can be seen from the results, the higher the *Pb* is, the lower the stiffness is.

The percentage of the beam is limited to a range from 0.70 to 0.90. This is because, if *Pb* is too small, the beam would be too short for large range displacement. Furthermore, if *Pb* is too large, the space for the decoupling mechanism would be insufficient to work properly. Hence, the percentage of beam *Pb* is determined as 0.90. The next parameter that needs to be determined is the thickness of the beam t. Determining the thickness of the beam highly depends on the manufacturing precision. As can be seen from the result, the stiffness k_{dz} increases with the increment of beam thickness. Hence, to obtain a lower stiffness guide, the beam thickness is chosen to be 0.5 mm. Similarly, the width of the compliant beam is chosen to be 7 mm.

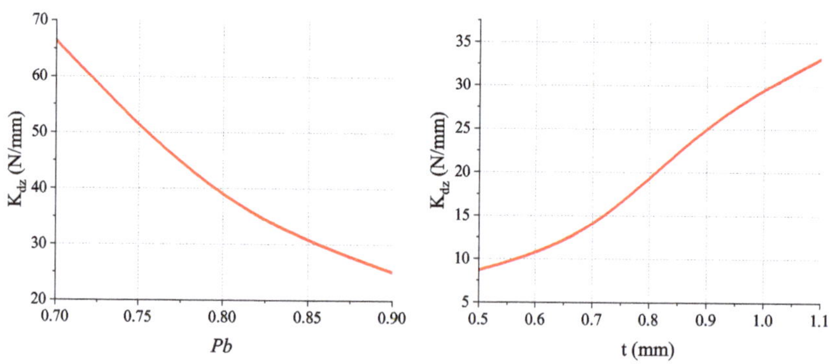

Figure 8. *Cont.*

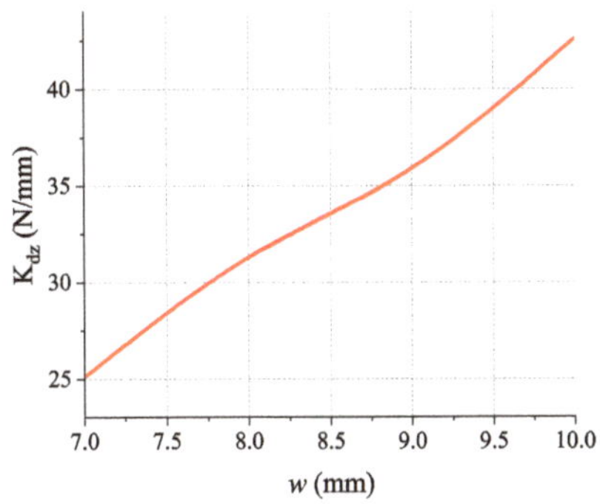

Figure 8. Stiffness under different geometrical parameters.

4. Prototype and Experiment

A prototype based on the optimized decoupled linear guide is manufactured using 3D printing technology to verify the FEA analysis. The 3D printer used for prototyping is Ultimaker S3. The slicing software is Ultimaker Cura. The primary material for printing the compliant guide is Polylactide (PLA), and the material used for support structure is Polyvinyl Alcohol (PVA). PVA is one type of water-soluble support material. By applying PVA as the support material, the risk of destroying the thin beams while removing the support structure has been reduced a lot. The print core used for building the primary structure is AA 0.4. The print core used for building the support structure is BB 0.4. The printed model is shown in Figure 9.

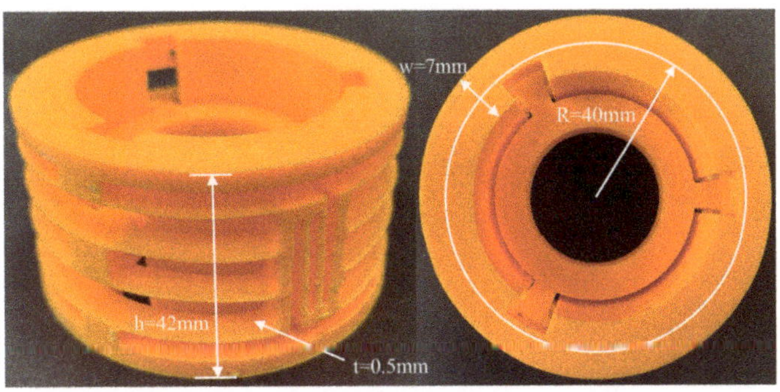

Figure 9. The side view and top view of the printed decoupled compliant guide.

The experiment was performed on the texture analyser (TA.Hd plus texture static test system). This system can measure the relationship between load and displacement. The test machine obtains the reaction forces from sensors. In addition, a probe was printed using the 3D printer as well. The probe added a uniform load to the centre platform. To gain an experimental result with higher resolution, the load cell of the texture analyser is selected as 5 kg. The resolution of the force sensor is 0.1 g.

In the experiment system, the input is chosen to be displacement along the motion direction. The output is the reaction force. To minimize the dynamic effect, the moving speed is set as 0.1 mm/s. The experimental procedure was repeated five times to gain an average value. The stiffness of the prototype in its motion direction is 1.10773 N/mm. The experiment setup can be seen in Figure 10.

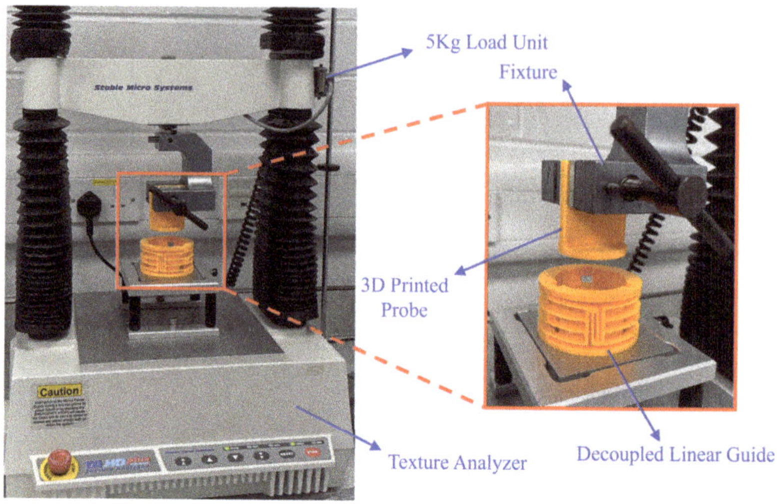

Figure 10. The experimental setup.

A comparison between the average experimental results and FEA results is shown in Figure 11. As can be seen, the experimental results are slightly smaller than the FEA results. Several aspects might cause the error. The first aspect is the limited precision of additive manufacturing. The additive manufactured model suffers from orthotropic properties. In addition, the error might be caused by plastic deformation during the repeat deforming process.

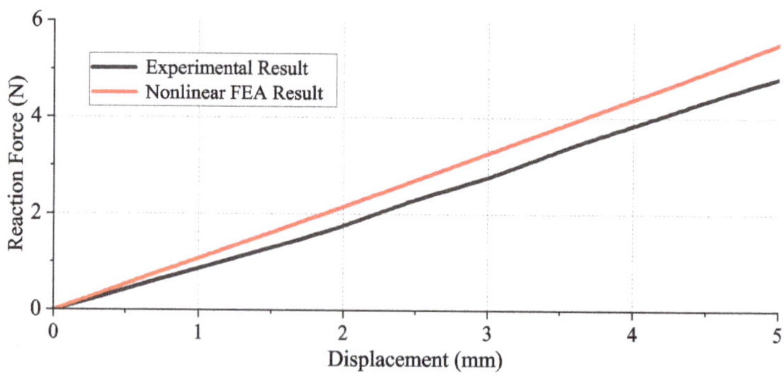

Figure 11. The comparison between the experimental results and nonlinear FEA result.

5. Discussion

The proposed decoupled linear guide offers a new approach for the design of guiding mechanisms.

- First, compared with the conventional slide rail guides [2], the cylindrical shape of the decoupled linear guide makes it more suitable to use than these conventional linear guides in certain scenarios, such as voice coil actuators and electromagnetic energy harvesting. The cylindrical shape ensures the target moving platform moves along its motion direction and avoids wear and friction. The decoupled linear guide has the benefits of lower maintenance costs, high positioning precision and repeatability.
- The decoupled linear guide has a smaller footprint than the spatial diaphragm mechanism [28]. This is because the compliant beams are arranged in a vertical direction instead of a distribution along the radial direction. Furthermore, the decoupled linear

guide has higher bearing-direction stiffness than the diaphragm mechanism, which contributes to higher motion precision under payload in bearing directions.

- Compared with the folded leaf springs with torsion reinforcement structures [23], the decoupled linear guide benefits from its simple structure in terms of manufacturing. The application of the decoupling mechanism to improve the bearing-direction stiffness is a different method compared to the use of torsion reinforcement structures for the same motivation.
- Compared to the detached linear guide [29], the primary advantage of the decoupled linear guide is the higher bearing-direction stiffness. They have the same footprint and symmetry.

In summary, the proposed decoupled linear guide has advantages over other linear guides, such as assembly issues, positioning precision and manufacturing complexity. A summarized table is shown in Table 1, which illustrates the comparison between different compliant spatial linear guides.

Table 1. The summarized comparison between types of compliant spatial linear guides.

	Linear Stiffness in Motion Direction	Alleviated Bearing-Direction Stiffness Reduction	Symmetrical Design for Eliminating Parasitic Motion	Compactness	Simple Manufacturing
Decoupled Linear Guide	✓	✓	✓	✓	✓
Spatial Diaphragm [28]	✓		✓		✓
Folded leaf springs [23]	✓	✓	✓		
Detached Linear Guide [29]	✓		✓	✓	✓

6. Conclusions

In this work, we presented a novel cylindrical compliant linear guide with decoupling mechanisms. The compliant-mechanism-based linear guide has the benefits of no friction, backlash, wear and high precision. In addition, a decoupling mechanism was designed and embedded in the detached linear guide. The nonlinear FEA showed that the decoupled linear guide had higher bearing-direction stiffness while keeping the same primary stiffness. Furthermore, the nonlinear FEA simulations proved that the new design addressed the problem of stiffness degradation as can be seen in most reported compliant linear guide. Finally, the static experimental test was performed on a 3D printed prototype and showed that the stiffness in the motion direction was linear.

The main outcomes of this paper are listed below:

- The decoupled linear guide had higher stiffness compared with the detached linear guide in all the bearing directions during deformation.
- As proved from nonlinear FEA, the stiffness degradation of the detached linear guide was addressed by adding the decoupling mechanisms.
- A prototype of the decoupled linear guide was manufactured where the stiffness in its motion direction was 1.0073 N/mm. The stiffness derived from the nonlinear FEA result was 1.1162 N/mm. The error between the FEA and experimental results was 9.76%.

The main limitation of this paper is the absence of a nonlinear mathematical model. The mechanism can be better explained by modelling the relationship between force and displacement, which forms part of our future work. Applications of the decoupled linear guide will also be investigated in some promising fields, such as voice coil actuators and electromagnetic energy harvesting.

Author Contributions: Conceptualization, G.H. and T.L.; methodology, G.H. and T.L.; software, T.L.; validation, T.L.; formal analysis, T.L.; investigation, T.L.; resources, G.H. and T.L.; data curation, T.L.; writing—original draft preparation, T.L.; writing—review and editing, G.H.; visualization, G.H. and T.L.; supervision, G.H.; project administration, G.H.; funding acquisition, G.H. All authors have read and agreed to the published version of the manuscript.

Funding: This work was funded by the EnABLES TA project (No. 122). EnABLES (http://www.enables-project.eu/) received funding from the EU Horizon 2020 research and innovation programme, under Grant Agreement No. 730957.

Institutional Review Board Statement: Not applicable.

Informed Consent Statement: Not applicable.

Acknowledgments: The authorswould like to acknowledge Shiyao Li and Jiaxiang Zhu for their help on the FEA simulations and modelling.

Conflicts of Interest: The authors declare no conflict of interest.

References

1. Li, R.; Yang, Z.; Cai, B.; Chen, G.; Wu, B.; Wei, Y. A compliant guiding mechanism utilizing orthogonally oriented flexures with enhanced stiffness in degrees-of-constraint. *Mech. Mach. Theory* **2022**, *167*, 104555. [CrossRef]
2. Sato, K.; Nakamoto, K.; Shimokohbe, A. Practical control of precision positioning mechanism with friction. *Precis. Eng.* **2004**, *28*, 426–434. [CrossRef]
3. Hao, G. Determinate design and analytical analysis of a class of symmetrical flexure guiding mechanisms for linear actuators. *J. Mech. Des.* **2017**, *139*, 012301. [CrossRef]
4. Howell, L.L. *Compliant Mechanisms;* John Wiley & Sons: New York, NY, USA, 2001.
5. Howell, L.L.; Magleby, S.P.; Olsen, B.M. *Handbook of Compliant Mechanisms;* John Wiley & Sons: Hoboken, NJ, USA, 2013.
6. Frecker, M.I.; Dziedzic, R.; Haluck, R. Design of multifunctional compliant mechanisms for minimally invasive surgery. *Minim. Invasive Ther. Allied Technol.* **2002**, *11*, 311–319. [CrossRef]
7. Yoneyama, T.; Watanabe, T.; Kagawa, H.; Hamada, J.; Hayashi, Y.; Nakada, M. Force detecting gripper and flexible micro manipulator for neurosurgery. In Proceedings of the 2011 Annual International Conference of the IEEE Engineering in Medicine and Biology Society, Boston, MA, USA, 30 August–3 September 2011; pp. 6695–6699.
8. Hu, Y.; Zhang, L.; Li, W.; Yang, G.Z. Design and fabrication of a 3-D printed metallic flexible joint for snake-like surgical robot. *IEEE Robot. Autom. Lett.* **2019**, *4*, 1557–1563. [CrossRef]
9. Cronin, J.A.; Frecker, M.I.; Mathew, A. Design of a compliant endoscopic suturing instrument. *J. Med. Devices* **2008**, *2*, 025002. [CrossRef]
10. George B, L.; Bharanidaran, R. Design of compliant gripper for surgical applications. *Aust. J. Mech. Eng.* **2022**, *20*, 256–262. [CrossRef]
11. Gu, X.; Li, C.; Xiao, X.; Lim, C.M.; Ren, H. A compliant transoral surgical robotic system based on a parallel flexible mechanism. *Ann. Biomed. Eng.* **2019**, *47*, 1329–1344. [CrossRef]
12. Hao, G.; Kong, X. A novel large-range XY compliant parallel manipulator with enhanced out-of-plane stiffness. *J. Mech. Des. Trans. ASME* **2012**, *134*, 061009. [CrossRef]
13. Wu, H.; Lai, L.; Zhang, L.; Zhu, L. A novel compliant XY micro-positioning stage using bridge-type displacement amplifier embedded with Scott-Russell mechanism. *Precis. Eng.* **2022**, *73*, 284–295. [CrossRef]
14. Liang, H.; Hao, G.; Olszewski, O.Z.; Pakrashi, V. Ultra-low wide bandwidth vibrational energy harvesting using a statically balanced compliant mechanism. *Int. J. Mech. Sci.* **2022**, *219*, 107130. [CrossRef]
15. Yamomo, G.; Hossain, N.; Towfighian, S.; Willing, R. Design and analysis of a compliant 3D printed energy harvester housing for knee implants. *Med. Eng. Phys.* **2021**, *88*, 59–68. [CrossRef] [PubMed]
16. Berdy, D.F.; Srisungsitthisunti, P.; Jung, B.; Xu, X.; Rhoads, J.F.; Peroulis, D. Low-frequency meandering piezoelectric vibration energy harvester. *IEEE Trans. Ultrason. Ferroelectr. Freq. Control* **2012**, *59*, 846–858. [CrossRef] [PubMed]
17. Jin, L.; Khajehtourian, R.; Mueller, J.; Rafsanjani, A.; Tournat, V.; Bertoldi, K.; Kochmann, D.M. Guided transition waves in multistable mechanical metamaterials. *Proc. Natl. Acad. Sci. USA* **2020**, *117*, 2319–2325. [CrossRef] [PubMed]
18. Wu, K.; Sigmund, O.; Du, J. Design of metamaterial mechanisms using robust topology optimization and variable linking scheme. *Struct. Multidiscip. Optim.* **2021**, *63*, 1975–1988. [CrossRef]
19. Lee, T.U.; Chen, Y.; Heitzmann, M.T.; Gattas, J.M. Compliant curved-crease origami-inspired metamaterials with a programmable force-displacement response. *Mater. Des.* **2021**, *207*, 109859. [CrossRef]
20. Masters, N.D.; Howell, L.L. A self-retracting fully compliant bistable micromechanism. *J. Microelectromech. Syst.* **2003**, *12*, 273–280. [CrossRef]
21. Wittwer, J.W.; Baker, M.S.; Howell, L.L. Simulation, measurement, and asymmetric buckling of thermal microactuators. *Sens. Actuators A Phys.* **2006**, *128*, 395–401. [CrossRef]
22. Sano, P.; Verotti, M.; Bosetti, P.; Belfiore, N.P. Kinematic synthesis of a d-drive mems device with rigid-body replacement method. *J. Mech. Des.* **2018**, *140*, 075001. [CrossRef]
23. Rommers, J.; Naves, M.; Brouwer, D.; Herder, J.L. A flexure-based linear guide with torsion reinforcement structures. *J. Mech. Robot.* **2022**, *14*, 031013. [CrossRef]
24. Hao, G.; He, X.; Awtar, S. Design and analytical model of a compact flexure mechanism for translational motion. *Mech. Mach. Theory* **2019**, *142*, 103593. [CrossRef]

25. Hao, G.; Li, H. Nonlinear analytical modeling and characteristic analysis of a class of compound multibeam parallelogram mechanisms. *J. Mech. Robot.* **2015**, *7*, 041016. [CrossRef]
26. Hubbard, N.B.; Wittwer, J.W.; Kennedy, J.A.; Wilcox, D.L.; Howell, L.L. A novel fully compliant planar linear-motion mechanism. In Proceedings of the International Design Engineering Technical Conferences and Computers and Information in Engineering Conference, Salt Lake City, UT, USA, 28 September–2 October 2004; Volume 46954, pp. 1–5.
27. Brouwer, D.M.; Otten, A.; Engelen, J.; Krijnen, B.; Soemers, H. Long-range elastic guidance mechanisms for electrostatic comb-drive actuators. In Proceedings of the EUSPEN International Conference, Delft, The Netherlands, 31 May–4 June 2010.
28. Awtar, S.; Slocum, A. Flexure systems based on a symmetric diaphragm flexure. In Proceedings of the ASPE 2005 Annual Meeting, Norfolk, VA, USA, 9–14 October 2005.
29. Yang, M.; Zhang, C.; Huang, X.; Chen, S.L.; Yang, G. A Long Stroke Nanopositioning Stage with Annular Flexure Guides. *IEEE/ASME Trans. Mechatron.* **2021**, *27*, 1570–1581. [CrossRef]
30. Awtar, S.; Mariappan, D.D. Experimental measurement of the bearing characteristics of straight-line flexure mechanisms. *Precis. Eng.* **2017**, *49*, 1–14. [CrossRef]
31. Wu, K.; Zheng, G.; Hao, G. Efficient spatial compliance analysis of general initially curved beams for mechanism synthesis and optimization. *Mech. Mach. Theory* **2021**, *162*, 104343. [CrossRef]
32. Hao, G.; Li, H. Extended static modeling and analysis of compliant compound parallelogram mechanisms considering the initial internal axial force. *J. Mech. Robot.* **2016**, *8*, 041008. [CrossRef]
33. Hao, G.; Li, H.; He, X.; Kong, X. Conceptual design of compliant translational joints for high-precision applications. *Front. Mech. Eng.* **2014**, *9*, 331–343. [CrossRef]
34. Ultimaker. *Ultimaker PLA Technical Data Sheet*; v2.00; Ultimaker: Utrecht, The Netherlands, 2022.

micromachines

MDPI

Article

Performance Evaluation of Different Coating Materials in Delamination for Micro-Milling Applications on High-Speed Steel Substrate

Sandeep Bhoi [1], Ashwani Kumar [2], Arbind Prasad [3], Chandan Swaroop Meena [4,*], Rudra Bubai Sarkar [5], Bidyanand Mahto [6] and Aritra Ghosh [7,*]

1 Department of Mechanical Engineering, Parala Maharaja Engineering College, Berhampur 761003, Odisha, India
2 Technical Education Department Uttar Pradesh, Kanpur 208024, Uttar Pradesh, India
3 Department of Mechanical Engineering, Katihar Engineering College (Under Department of Science & Technology, Government of Bihar), Katihar 854109, Bihar, India
4 CSIR-Central Building Research Institute, Roorkee 247667, Uttarakhand, India
5 Research and Development Division, Tata Steel Ltd., Burma Mines, Jamshedpur 831007, Jharkhand, India
6 Government Engineering College Vaishali (Under Department of Science & Technology, Government of Bihar), Vailshali 844115, Bihar, India
7 College of Engineering, Mathematics and Physical Sciences, Renewable Energy, University of Exeter, Penryn, Cornwall TR109FE, UK
* Correspondence: chandanswaroop2008@gmail.com (C.S.M.); a.ghosh@exeter.ac.uk (A.G.)

Citation: Bhoi, S.; Kumar, A.; Prasad, A.; Meena, C.S.; Sarkar, R.B.; Mahto, B.; Ghosh, A. Performance Evaluation of Different Coating Materials in Delamination for Micro-Milling Applications on High-Speed Steel Substrate. *Micromachines* **2022**, *13*, 1277. https://doi.org/10.3390/mi13081277

Academic Editor: Stelios K. Georgantzinos

Received: 18 July 2022
Accepted: 4 August 2022
Published: 8 August 2022

Publisher's Note: MDPI stays neutral with regard to jurisdictional claims in published maps and institutional affiliations.

Abstract: The objective of the present work is to carry out analytical and finite element analysis for commonly used coating materials for micro-milling applications on high-speed steel substrate and evaluate the effects of different parameters. Four different coating materials were selected for micro-milling applications: titanium nitride (TiN), diamond-like carbon (DLC), aluminium titanium nitride (AlTiN) and titanium silicon nitride (TiSiN). A 3D finite element model of coating and substrate assembly was developed in Abaqus to find the Hertzian normal stress when subjected to normal load of 4 N, applied with the help of a rigid ball. The radius of the rigid ball was 200 μm. For all the coating materials, the length was 3 mm, the width was 1 mm, and the thickness was 3 μm. For the high-speed steel substrate, the length was 3 mm, the width was 1 mm, and the thickness was 50 μm. Along the length and width, coating and substrate both were divided into 26 equal parts. The deformation behaviour of all the coating materials was considered as linear–elastic and that of the substrate was characterized as elastic–plastic. The maximum normal stress developed in the FEA model was 12,109 MPa. The variation of the FEA result from the analytical result (i.e., 12,435.97 MPa is 2.60%) which is acceptable. This confirms that the FEA model of coating–substrate assembly is acceptable. The results shows that the TiSiN coating shows least plastic equivalent strain in the substrate, which serves the purpose of protecting the substrate from plastic deformation and the TiSiN of 3 micron thickness is the most optimum coating thickness for micro-milling applications.

Keywords: micro-milling; wear; tool failure; coating; substrate; delamination

1. Introduction

Micro-manufacturing processes have been used extensively in the aerospace, biomedical and defense industries. Presently, photolithography-based manufacturing techniques are used for selective materials such as Ni, Cu, Si and polymers to produce high aspect-ratio components. Micro-machining processes like micro-milling are able to generate three-dimensional surfaces in ceramics, metals and polymers [1–6]. Micro-machining is becoming popular because of its ability to produce three-dimensional parts of different sizes varying from a few micrometers to a few millimeters across various materials [7]. The micro-milling, micro-EDM, micro-grinding and micro-grooving processes are part of the micro-machining process.

Micro-milling is a micro-cutting process which is used for the fabrication of micro-and meso-scale components and devices. It can also be said that it is a milling operation at the micro-scale. However, there are vital drawbacks of the micro-milling process, particularly when machining hard materials having a hardness of 7.25–8.43 GPa and sintered ceramics having a hardness of 12.75–14.71 GPa. These drawbacks are due to the small size of the cutting tools, low flexural stiffness and strength compared to conventional-scale tools due to low material removal rates, rapid tool wear/failure and poor part feature accuracy, especially when cutting hard materials occurs [8–10]. Figure 1 shows a two-flute micro-end mill cutting tool with important dimensions [11]. The low flexural stiffness of these tools results in catastrophic failure because of the bending stresses generated by the cutting forces.

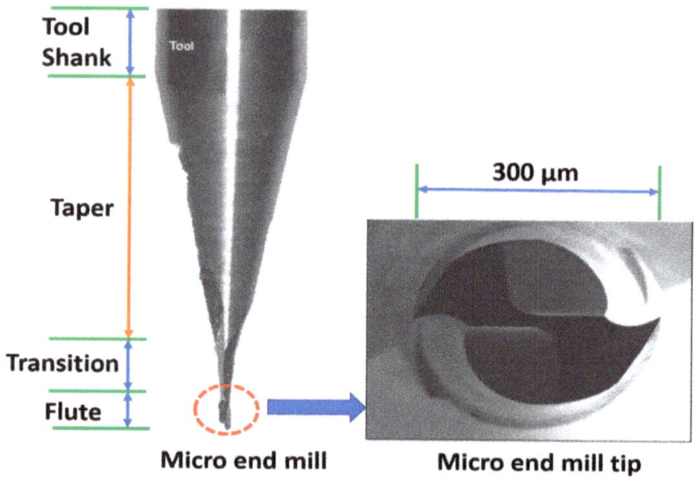

Figure 1. Micro-end mill and cross-section of two-flute micro-end mill [11].

To prevent tool failure, a few methods have been proposed in previous research work. The first method is using cutting fluids to dissipate heat and provide lubrication. This facilitates the minimization of friction at the tool–workpiece interface and, in turn, reduces tool wear. However, this method is not very successful in micro-milling machining. The main reason is that the application of the cutting fluid to the cutting zone and the tool–workpiece interface is very difficult because of high cutting speeds and the small size of the contact zone [12]. The second method is the use of coatings on the micro-tool to reduce wear. This enhances tool life [1]. However, very little has been reported in tool-wear studies of coated micro-tools. The aim of the present work is to study the wear of micro-machining. Therefore, later in the introduction section, a brief description of wear is presented. There are several types of wear phenomena occurring in the field of mechanics, such as adhesive, abrasive, fatigue, corrosive and fretting wear as well as erosive wear by solid particles, fluids and cavitation and electric arc–induced and delamination wear. Wear of non-lubricated metal pairs sliding in a dusty environment may be termed as dry sliding wear and abrasive wear. The classification of wear processes is based on the type of wearing contacts, such as single-phase and multiple-phase.

In micro-machining, smaller tools are used. Researchers have reported that low flexural stiffness and strength causes huge bending of the tool, hampering the cutting process and leading to tool failure [5,13,14]. This is avoided by minimizing the cutting forces below a certain critical value in order to ensure that the uncut chip thickness remains sufficiently small. During the machining of steel, titanium, nickel alloys, etc. the maximum permissible chip thickness is on the order of or less than the cutting-edge radius [14,15].

In the conventional milling process, the work pieces act as isotropic and homogeneous materials, whereas in micro-milling, the smaller grains in the work piece are comparable to the size of the tool. In view of the above, the micro-milling process is very complicated [14]. The chips, which adhere to the tool, block the path at the cutting zone, and this results in an increase in the cutting forces and leads to a catastrophic failure of the tool because of its low flexural stiffness. Moreover, the small size of the micro-milling cutters makes the tip very weak because of its low stiffness value. Diamond-coated cutters are promising because of their ability to improve tool stiffness and tool life [16].

In micro-machining, the uncut chip thickness (h) is less than the cutting-edge radius (re) due to the negative effective rake angle ($-\alpha$) influencing the ploughing effect in the work piece [5]. Therefore, the ratio of uncut chip thickness to cutting-edge radius is an important parameter in micro-machining [5,8]. A sharper cutting edge is required to remove the smallest amount of undeformed chip thickness [17].

The major limitations of the micro-milling process are unpredictable tool life and premature tool failure, deterioration of the cutting edge and tool wear leading to high friction generation [18–29]. It has been reported that coated micro-cutting tools have longer tool life and improved cutting performance [25]. Many researchers have used TiAlN-coated micro tools in cutting tests [19,20,23,24,28,30]. It has also been reported that CrTiAlN-coated micro-end mills provide great advantage in tool wear reduction and smooth surface finish [31,32]. The small size of micro-milling tools makes coating deposition difficult around the cutting edge. The desirable properties of the coatings for micro-machining tools are high hardness, toughness, chemical/erosive and abrasive wear resistance as well as dense and fine microstructure. Coating also provides smooth machined surfaces with a reduced coefficient of friction compared with uncoated tools [33].

TiN, TiCN, TiAlN and Al_2O_3 are coatings that have been frequently used in industry. Earlier studies reveal that an increase in tool life is due to an increase in hardness, greater bonding energy of the coating elements and lower friction coefficients. Due to oxidation resistance and wear resistance at higher temperatures, TiAlN coating improved cutting performance. These properties make TiAlN an appropriate coating for cutting abrasive work pieces at high speeds [34]. It has been found that the coating on a micro-milling cutting tool fails due to delamination, which was confirmed by the SEM images and EDS spectra of the worn tools [35]. Delamination wear occurs due to surface layer deformation, crack nucleation and propagation of cracks parallel to the surface. Cracks finally shear the surface, resulting in long and thin wear sheets.

Earlier it was reported in the literature through the four-point bending test that thick coatings, usually more than 2 μm, delaminate from the surface of the substrate because of their high summary toughness. Along the interface of the coating and substrate, there is a difference in material properties which facilitates delamination. Finally, the coating fails due to the buckling and spalling of the delaminated portion [36].

2. Research Objective

From the literature review, it was found that delamination in the coating of micro-milling cutting tools is confirmed only through SEM images and EDS spectra. However, coating material which is suitable for micro-milling applications and of appropriate thickness has not been reported in the literature. The delamination of coating from the substrate from the mechanics point of view has not been reported in the literature. Moreover, the factors on which delamination depend need to be examined. It was also found in the literature review that the coatings which are commonly used for micro-milling applications are titanium nitride (TiN), diamond-like carbon (DLC), aluminium titanium nitride (AlTiN) and titanium silicon nitride (TiSiN). It has been reported that the thickness of these coatings usually ranges from 2 to 4 microns [37–45]. Authors have also studied various metal deposition techniques in the past 5 years and studied their performance to reduce delamination [46–55].

The objective of the present work is to carry out finite element analysis of commonly used coatings for micro-milling applications on high-speed steel substrates and evaluate the following.

- Objective 1: to model delamination of the coating from the substrate for micro-milling applications and find out factors on which delamination depends.
- Objective 2: to evaluate the performance of different coating materials for delamination and report the best coating material for micro-milling applications and their corresponding thickness.

The above mentioned objectives can be achieved by carrying out finite element analysis with high speed steel as a substrate with different coating materials of different thicknesses. In the present study, three-point bending was examined to simulate the practical conditions of the micro-milling tools during machining. The FEA results were validated using analytical results.

3. CAD Model of Coating and Substrate Design to Study Delamination (Objective 1)

3.1. Designing of Coating and Substrate Assembly

A 3D finite element model of coating and substrate assembly was developed in Abaqus to find the Hertzian normal stress when subjected to normal load of 4N, applied with the help of a rigid ball, as shown in Figure 2. To validate the FEA results, it was compared with the analytical result of Hertzian normal stress. Earlier, the ball on a flat coating–substrate assembly was used to simulate the scratch test [37]. The dimensions of the FEA model are as follows: the radius of rigid ball is 200 μm. For the TiN coating, the length is 3 mm, the width is 1 mm, and the thickness is 3 μm. For the high speed steel substrate, the length is 3 mm, the width is 1 mm, and the thickness is 50 μm. Table 1 elaborates the material properties of the coating and substrate.

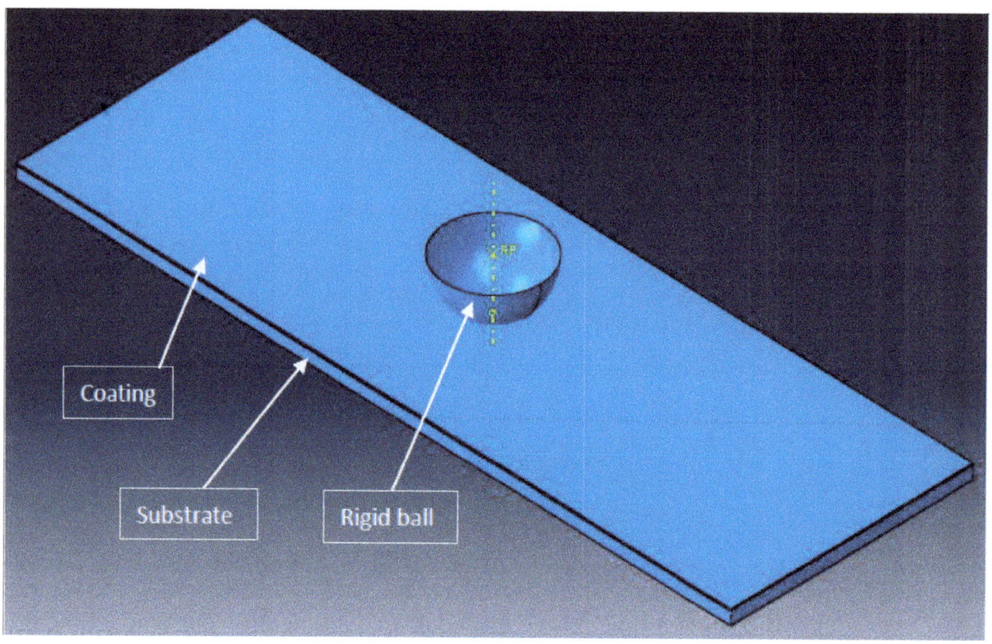

Figure 2. Abaqus model of coating and substrate assembly with rigid ball.

Table 1. Mechanical properties of coating and substrate.

Sl No.	Type	Material	Young's Modulus	Poisson's Ratio
1	Coating	TiN	300 GPa	0.22
2	Substrate	High-speed steel	200 GPa	0.29

3.2. Mechanical Properties of Different Coatings and the Substrate

The deformation behaviour of all the coating materials is considered as linear–elastic and that of the substrate is characterized as elastic–plastic, which was already mentioned earlier. The substrate material for all four coating materials was taken as high-speed steel (HSS). Material properties of different coating materials and substrate are given in Table 2 [37,43–45]. The following values were considered for the analysis.

Table 2. Mechanical properties of different coating materials and substrate.

S. No.	Type	Material	Young's Modulus (GPa)	Hardness (GPa)	Poisson's Ratio
1	Coating	TiN	300	27	0.22
2	Coating	DLC	70	10.5	0.22
3	Coating	AlTiN	560	35	0.22
4	Coating	TiSiN	510	56	0.20
5	Substrate	HSS	200	7.5	0.29

3.3. Assumptions in the Present Study

In the present study, the following assumptions are made based on which analysis was carried out.

- The deformation behaviour of the substrate was characterized as elastic–plastic with isotropic hardening.
- The deformation behaviour of the coating materials was modelled as linear–elastic.
- The hemispherical ball was modelled as analytical rigid.
- For the surface interaction between the ball and coating material, the outer surface of the hemispherical ball was considered as the master surface and the top surface of the coating material was considered as the slave surface. This was done because the hemispherical ball was modelled as analytical rigid [37].
- For the surface interaction between the coating material and substrate, the bottom surface of the coating was considered as the master surface and the top surface of the substrate was modelled as the slave surface as the coating material was harder than the substrate.
- The interaction property for the junction of the coating and substrate was modelled as a 'tie' such that there was no slip, separation and penetration.
- Brick elements were taken for both the coating and substrate, and the element type was taken as the quadratic with the hybrid formulation and reduced integration. The structured type of mesh control is used.

3.4. Dimensioning and Boundary Conditions

- Along the length and width, the coating and substrate both were divided into 26 equal parts. Along the thickness, the coating material was divided into two equal parts and the substrate was divided into five parts with single bias with the bias ratio as five. Along the length, the mesh size of the coating material and substrate was 115 μm; along the width, the mesh size of the coating material and substrate was 38 μm; along the thickness, the mesh size for the coating material was 1.5 μm; and for the substrate, the mesh size ranged from 4 μm to 19 μm. The interaction property for the junction of the coating and substrate was modelled as a 'tie' such that there was no slip, separation and penetration.

- The bottom of the substrate was given a rigid support and a normal load of 4 N was applied at the reference point of the rigid ball. The stress contour thus obtained is shown in Figure 3a,b.

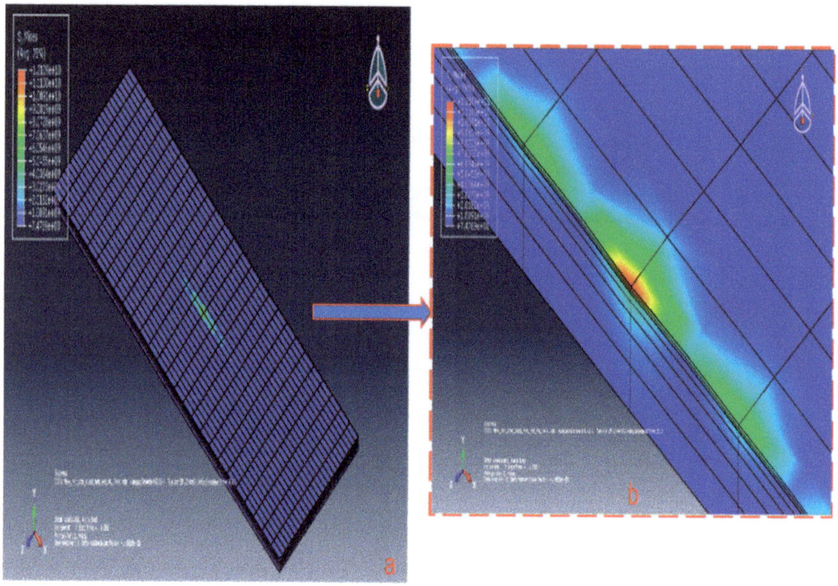

Figure 3. Stress contour for 4 N normal loads on coating–substrate assembly.

Now the above FEA model of coating-substrate assembly is treated as simply supported beam and is subjected to a normal load of 4 N centrally throughout the width of the coating-substrate assembly and analysed for three-point bending test. From previously reported results on micro cutting tool it was found experimentally that cutting force varies between 1.5 N to 2 N. Therefore, in view of the above, for the present analysis a load of 4 N is taken considering extreme conditions. Four different coating materials are used for analysis, which are commonly used for micro-milling applications. Dimensions of each coating materials and substrate used in the FEA model are tabulated in Table 3.

Table 3. Dimensions of different coating materials and substrate in FEA models.

Sl No.	Type	Material	Length	Width	Thickness
1	Coating	Titanium Nitride	3 mm	1 mm	2, 3 and 4 μm
2	Coating	Diamond-like Carbon	3 mm	1 mm	2, 3 and 4 μm
3	Coating	Aluminium Titanium Nitride	3 mm	1 mm	2, 3 and 4 μm
4	Coating	Titanium Silicon Nitride	3 mm	1 mm	2, 3 and 4 μm
5	Substrate	High-speed Steel	3 mm	1 mm	50 μm

3.5. Surface-Based Cohesive Behaviour

To simulate the delamination of the coating, the interface of different coatings and the substrate was modelled with cohesive surface in Abaqus. Surface-based cohesive behaviour was used to simulate the interface adhesion between the different coatings and substrate since the interface thickness was negligibly small. Cohesive behaviour is defined by means of the surface interaction property between the surfaces of the coating material and substrate material coming in contact with each other. The surface of the coating interacting with substrate was considered as the master surface since all the coating materials are harder than the substrate. The surface of the substrate interacting with coating was considered as the slave surface. Cohesive behaviour is defined by specifying the stiffness coefficients: Knn, Kss and Ktt for uncoupled traction–separation behaviour, where Knn represents the stiffness coefficient for the cohesive behaviour–enabled surface interaction in the normal direction and Kss and Ktt represent the stiffness coefficients for the cohesive behaviour–enabled surface interaction in the shear directions. However, it is advisable to keep the same stiffness value for the stiffness coefficients Knn, Kss and Ktt [47].

The stiffness value of the interface of the coating material and substrate is given by the relation given in Equation number (1) [48].

$$(E/H)_i^{1/2} = \frac{(E/H)_s^{1/2}}{1+(H_s/H_c)^{1/2}} + \frac{(E/H)_c^{1/2}}{1+(H_c/H_s)^{1/2}} \quad (1)$$

In equation number (1), 'i' represents the interface, 's' represents the substrate, 'c' represents the coating, E represents the modulus of elasticity and H represents the Vickers hardness. The hardness of the interface between the coating material and substrate is assumed as the average hardness of the coating material and substrate. The interface hardness, Hi, is given by Equation number (2).

$$H_i = \frac{H_S + H_C}{2} \quad (2)$$

The stiffness coefficients of the interface of the different coating–substrate assemblies Ei or Knn, Kss and Ktt are calculated by using Equation number (2) and are given in Table 4. These values are used for the analysis.

Table 4. Interface stiffness coefficients of different coating–substrate assemblies.

Sl No.	Coating–Substrate Assembly	Ec (GPa)	Hc (GPa)	Es (GPa)	Hs (GPa)	Hi = (Hs + Hc)/2 (GPa)	Ei or Knn, Kss, Ktt (GPa)
1	TiN-HSS	300	27	200	7.5	17.25	357
2	DLC-HSS	70	10.5	200	7.5	9	143
3	AlTiN-HSS	560	35	200	7.5	21.25	487
4	TiSiN-HSS	510	56	200	7.5	31.75	665

All the required inputs are fed into the FEA model of the different coating–substrate assemblies with three different coating thicknesses of 2, 3 and 4 μm with the corresponding mesh size of the coating material along the thickness as 2, 1.5 and 2 μm, respectively. All the models run for three-point bending load condition to evaluate desirable outputs such as the von Mises stress, the plastic equivalent strain and the deformation in the coating material and substrate for each case. Consolidated results of the desirable outputs of all FEA models are shown in the results section.

4. Analytical Calculation for Hertzian Normal Stress

When a spherically shaped summit of radius R is brought into contact with a flat surface with a load L, as shown in Figure 4, the surfaces deform to create the contact zone of radius a. According to Hertz's equations for the elastic deformation of a sphere on a flat surface, the radius of the contact zone is given by

$$a = \left(\frac{3RL}{4Ec}\right)^{1/3} \quad (3)$$

where Ec is the composite elastic modulus of the two contacting materials with the elastic modulus E_1 and E_2 and the Poisson's ratio ν_1 and ν_2, respectively. The value of Ec is given by the relation as given below.

$$\frac{1}{E_c} = \frac{1-\nu_1{}^2}{E_1} + \frac{1-\nu_2{}^2}{E_2} \quad (4)$$

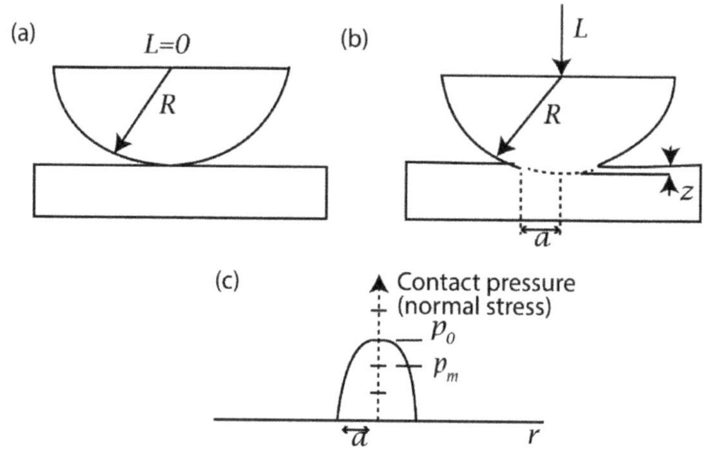

Figure 4. Sphere on flat-contact geometry with (**a**) no load and (**b**) load = L. (**c**) The distribution of normal stress across the contact zone of (**b**).

For this geometry, the real area of contact A is given by

$$A = \pi a^2 \quad (5)$$

$$A = \pi\left(\frac{3RL}{4Ec}\right)^{2/3} \quad (6)$$

The mean normal stress, pm, is given by

$$P_m = \frac{L}{A} \quad (7)$$

or,

$$P_m = \frac{1}{\pi}\left(\frac{4E_c}{3R}\right)^{2/3} L^{1/3} \quad (8)$$

The maximum normal stress, po, is given by

$$po = \frac{3P_m}{2} \quad (9)$$

When a rigid ball of radius R = 200 μm and elastic modulus $E_1 = \infty$ is pressed with a normal load of 4 N against a flat plate of TiN with elastic modulus E_2 = 300 GPa and Poisson's ratio ν = 0.22, then the maximum normal stress obtained, po, is given in Table 5.

Table 5. Value of composite elastic modulus and normal stresses.

S. N.	Particulars	Symbol	Equation Number Used	Value
1	Composite elastic modulus	Ec	3.2	315.26 GPa
2	Mean normal stress	pm	3.6	8290.64 MPa
3	Maximum normal stress	po	3.7	12,435.97 MPa

The above loading conditions of a load of 4 N through a rigid ball pressed against a flat plate of TiN is simulated in Abaqus 6.11–1. The maximum normal stress developed in FEA model was 12,109 MPa. The variation of the FEA result from the analytical result, i.e., 12,435.97 MPa, is 2.63%, which is acceptable. This confirms that the FEA model of the coating–substrate assembly is acceptable.

The plastic strain curve of the high-speed steel material used as a substrate is shown in Figure 5 [46]. With reference to Table 6, for compressive strain, v/s compressive stress values of the HSS five data point values are extracted for the plastic behaviour of high-speed steel corresponding to a 20 °C curve, as shown in Figure 5. This plastic behaviour data were added into the Abaqus 6.11–1 software as the plastic behaviour of the high-speed steel.

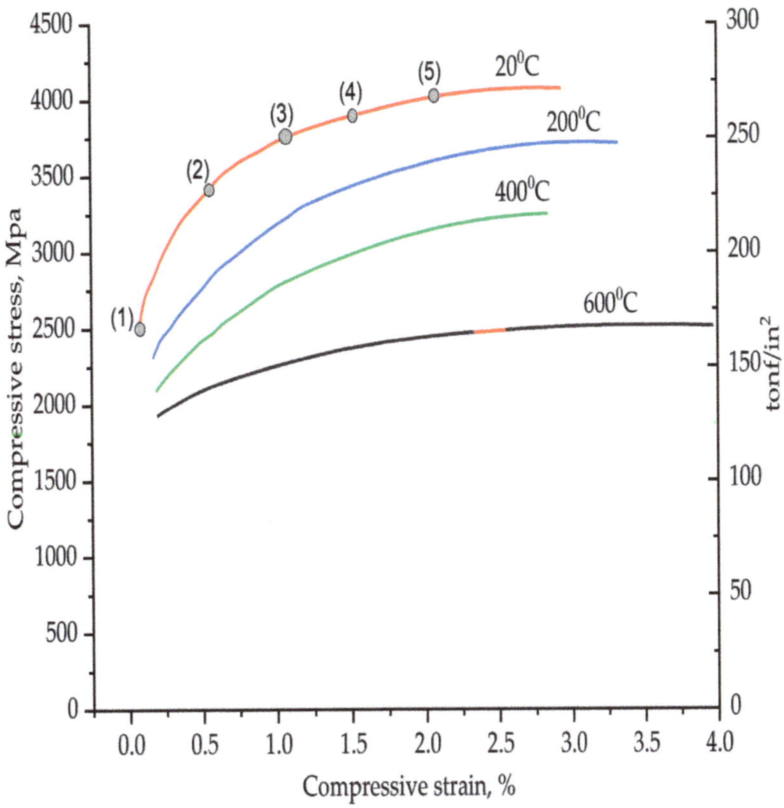

Figure 5. Compressive stress v/s compressive strain curve of high-speed steel [46].

Table 6. Plastic behaviour of HSS.

Data Point	Compressive Stress (MPa)	Compressive Strain
1	2450	0
2	3470	0.005
3	3750	0.01
4	3820	0.015
5	4000	0.02

5. Results and Discussion: Performance Evaluation of Different Coating Materials in Delamination (Objective 2)

All the FEA models of the different coating–substrate assemblies with three different coating thicknesses of 2, 3 and 4 μm were run for the three-point bending load condition to evaluate the von Mises stress, the plastic equivalent strain and the deformation in the coating materials and substrate. Contour plots of the desirable outputs are given in the Appendix A. The consolidated results of the desirable outputs of all FEA models are shown in this section. Figure 6 shows the maximum von Mises stress acting on the coating material at the junction of the coating and substrate, situated below the loading region, where AlTiN_3 represents the coating of aluminium titanium nitride material of 3 micron thickness.

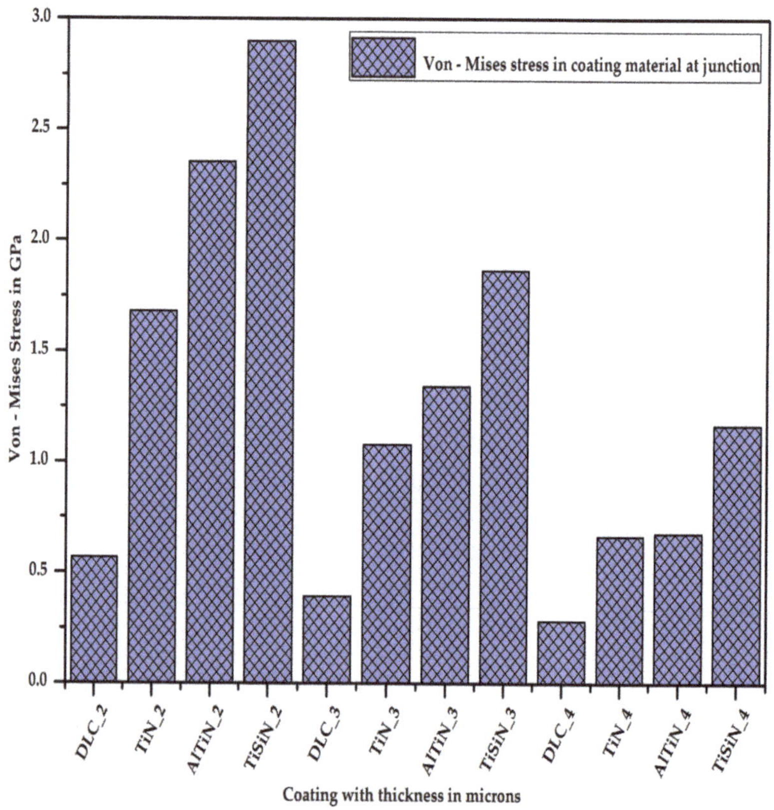

Figure 6. Von Mises stress in coating material at junction.

It is quite clear from Figure 6 that as the thickness of the coating material increases, the stress developed on the coating material decreases since the section modulus of the coating increases. Additionally, for a given thickness of coating material, the stress developed in the diamond-like carbon coating material is minimum (DLC_4) and that in the titanium silicon nitride coating material is maximum (TiSiN_4), which means the stress bearing capacity of the DLC coating is minimum and that of the TiSiN is maximum. Figure 7 shows the maximum von Mises stress acting on the substrate at the junction of the coating material and substrate, situated below the loading region with different coating materials of different thicknesses.

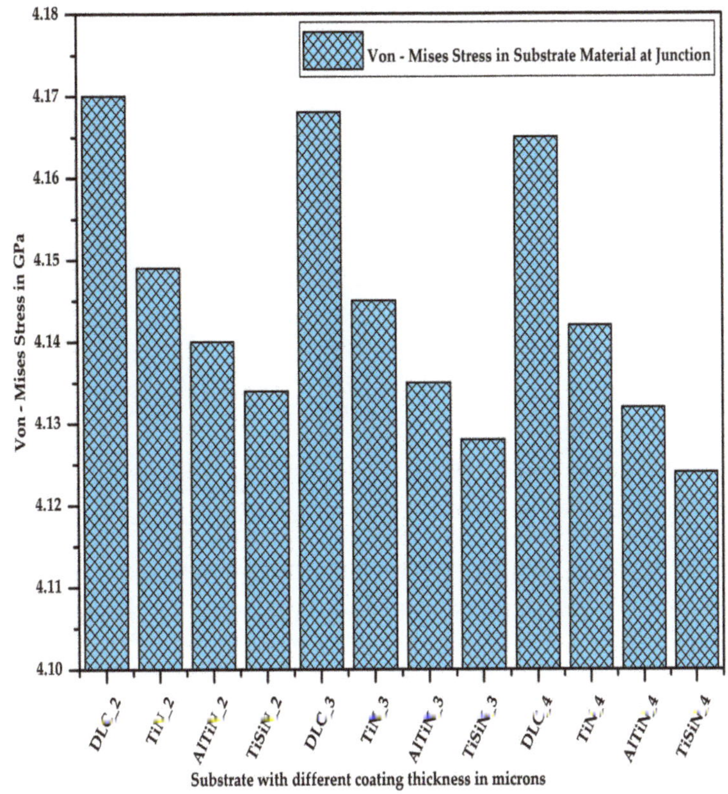

Figure 7. Von Mises stress in substrate material at junction.

It can be observed from Figure 7 that the stress developed on the substrate material decreases with the increase in the coating thickness due to the increase in the section modulus of the coating–substrate assembly. Additionally, for a given thickness of the coating material, the stress developed on the substrate is maximum in the case of the DLC coating and minimum in the case of the TiSiN coating. This shows that the TiSiN coating material bears most of the stress developed due to the application of the load preventing the substrate from experiencing high stress, unlike the other coating materials.

Figure 8 shows the differential stress at the junction of the coating and substrate, which is nothing but the difference of the stresses experienced by the coating material and substrate. It is clear from Figure 8 that for a given thickness of the coating material, the differential stress between the coating material and substrate at the junction is maximum for DLC and minimum for TiSiN. As the thickness of the coating material increases, the differential stress at the junction of the coating material and substrate increases.

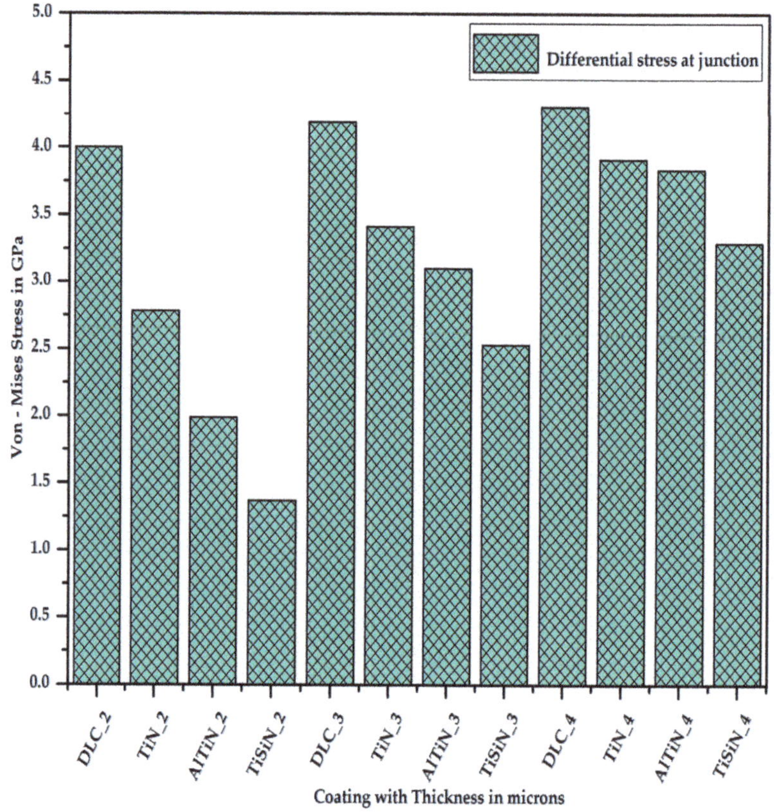

Figure 8. Differential stress at junction of coating material and substrate.

The differential stress at the junction of the substrate and coating material causes some plastic strain at the substrate surface due to the difference in the hardness of the coating material and substrate material. Figure 9 shows the plastic equivalent strain developed on the substrate with different coating materials of different thicknesses. It can be seen from Figure 9 that for a given coating material, as the thickness of the coating increases, the plastic equivalent strain in the substrate decreases as the stress developed in the substrate decreases. Additionally, for a given thickness of the coating, the plastic equivalent strain in the substrate is maximum for the DLC coating and minimum for the TiSiN coating.

Figure 10 shows the deformation in the substrate at one of the corners at the free edge having coatings of different materials with varying thicknesses. It is very clear from Figure 10 that as the thickness of the coating increases, the deformation in the substrate decreases. It can be also seen (Figure 13) that for a given thickness of coating, the deformation in the substrate is maximum for the DLC coating and minimum for the TiSiN coating.

Figure 11 shows the deformation in the different coating materials of varying thicknesses at the junction of the coating material and the substrate at one of the corners of the free edge. It can be observed from Figure 13 that for a given thickness of coating, the deformation is maximum for the DLC coating and minimum for the AlTiN coating. This is because the deformation is inversely proportional to the elastic modulus. As the thickness of coating material increases, its deformation decreases due to the increase in its section modulus.

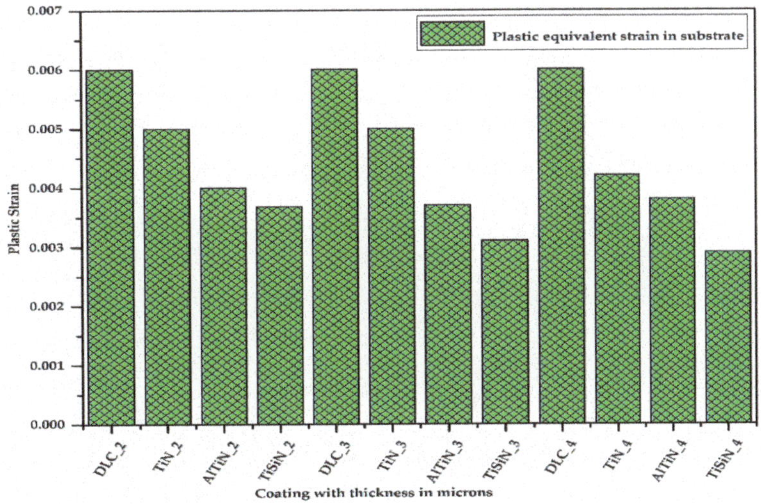

Figure 9. Plastic equivalent strain in substrate with different coating materials of varying thicknesses.

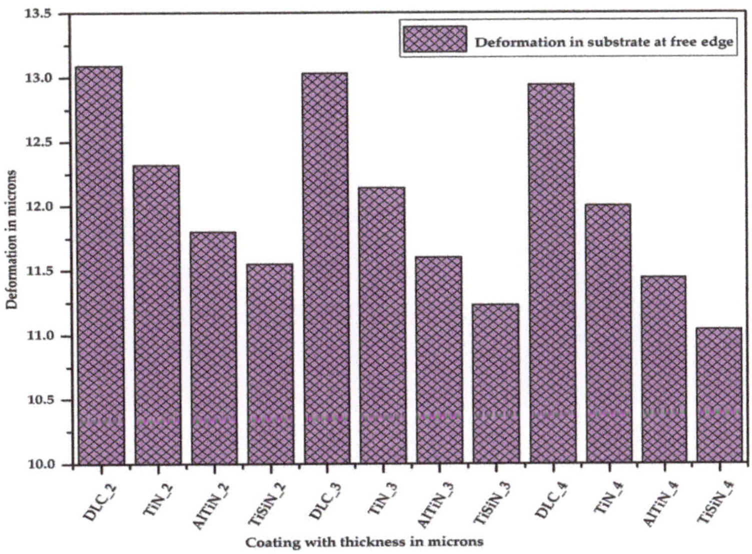

Figure 10. Deformation in substrate at free edge with different coating materials of varying thicknesses.

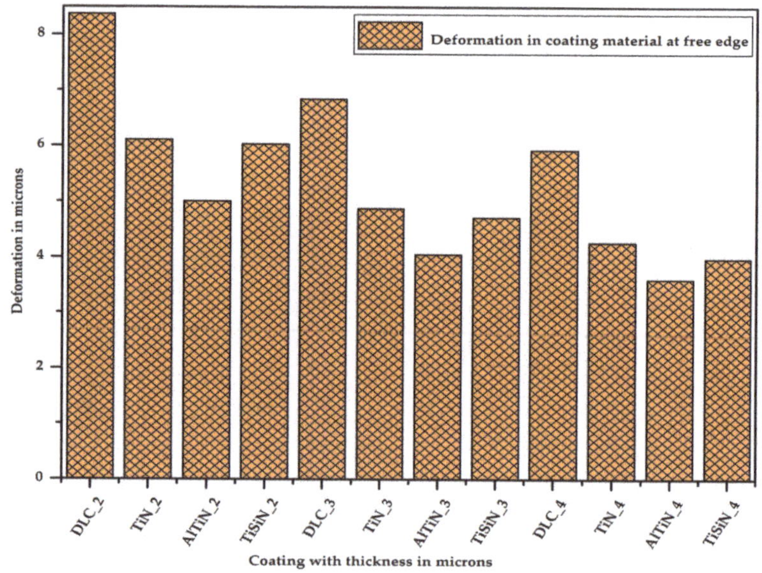

Figure 11. Deformation in different coating materials of varying thicknesses.

Figure 12 shows the difference in the deformation between the different coating materials of varying thicknesses and the substrate located at one of the corners at the free edge, thereby showing the extent of the delamination. It can be said from Figure 12 that for a given thickness of coating, the difference in the deformation between the coating material and substrate is the maximum for the AlTiN coating and minimum for the DLC coating. From Figure 13 it can be concluded that as the thickness of coating material increases, the difference in the deformation between the coating material and substrate, i.e., delamination, increases.

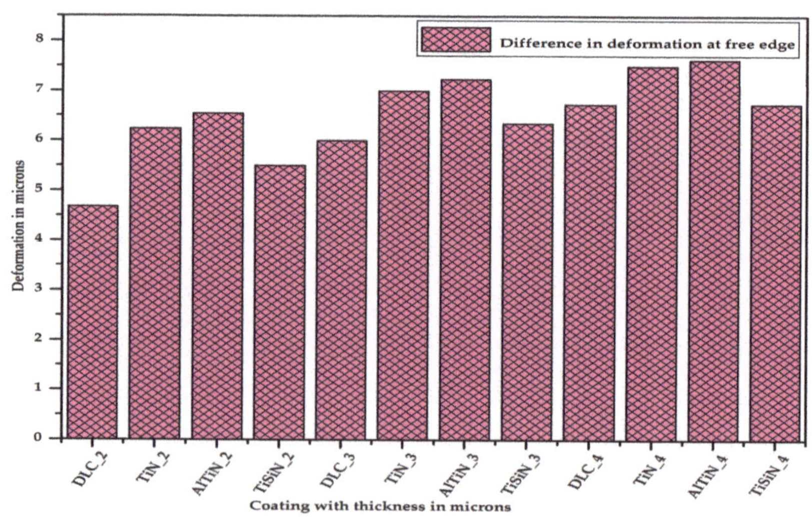

Figure 12. Difference in deformation between different coating materials of varying thicknesses and substrate.

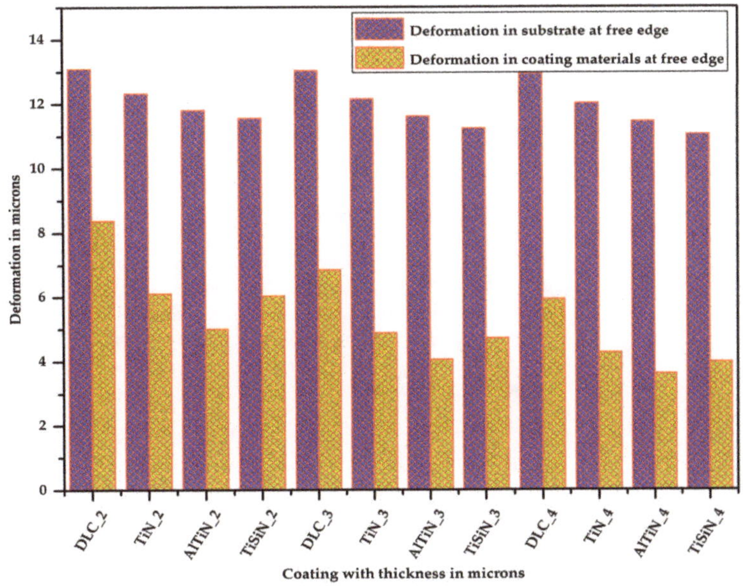

Figure 13. Deformation comparison between substrate and coating materials at free edge.

Because of the non-availability of FEA results in the literature, the results of the present study were not compared. However, it is clear from the result that aluminium titanium nitride coating material performs better than titanium nitride, which was reported earlier in the literature.

6. Conclusions

From the different consolidated results of the desired outputs, the following conclusions can be made:

1. Using a 3D modelling approach, the coating and substrate assembly was modelled and studied for delamination. Various factors were evaluated which impacted the delamination in micro-milling.
2. On the basis of the plastic equivalent strain occurring in the substrate material, the TiSiN and AlTiN coatings are the best since the plastic equivalent strain occurring in the substrate with the TiSiN and AlTiN coating materials is less (Appendix A). Between these two, the TiSiN coating shows the least plastic equivalent strain in the substrate, which serves the purpose of protecting the substrate from plastic deformation. The plastic equivalent strain in the substrate decreases with increase in the thickness of the coating material.
3. On the basis of the difference in deformation, i.e., delamination, the TiSiN coating is better than the AlTiN coating since delamination in the coating–substrate assembly with the TiSiN coating is less than that of the AlTiN coating. The delamination of the coating material from the substrate increases with increase in the thickness of the coating material.
4. By combining the plastic equivalent strain in the substrate and the delamination of the coating from the substrate, we can conclude that the TiSiN coating of 3 micron thickness is the most optimum coating thickness for micro-milling applications. These results fulfill the requirements of objectives 1 and 2.
5. The higher the interface stiffness coefficient of the coating–substrate assembly is, the less the delamination of the coating material from the substrate is.

6. The delamination depends on the Young's modulus and the hardness of both the coating material and substrate.

Suggestions for future work: In order to validate the FEA results obtained, experimental studies are essential. Therefore, these values will help in comparing the FEA results. An experimental setup is required to see the delamination process. Different coatings on the substrate are required to be generated to conduct the experiment. The mechanical properties of the coatings need to be evaluated. SEM images can be taken to evaluate the extent of the delamination occurring on different coating materials.

Author Contributions: S.B.: visualization, project administration; A.K. and C.S.M.: conceptualization, methodology, writing—original draft preparation; A.P. and C.S.M.: writing—review and editing, supervision, data curation; C.S.M.: conceptualization, methodology, writing—review and editing original draft, project administration; R.B.S.: investigation, writing—review and editing; B.M.: review and editing; A.G.: supervision, data curation. All authors have read and agreed to the published version of the manuscript.

Funding: This research received no external funding.

Conflicts of Interest: The authors declare no conflict of interest.

Appendix A

The contour plots of the desirable outputs of the FEA model with titanium silicon nitride coating of 3 μm thickness over high-speed steel substrate are as follows.

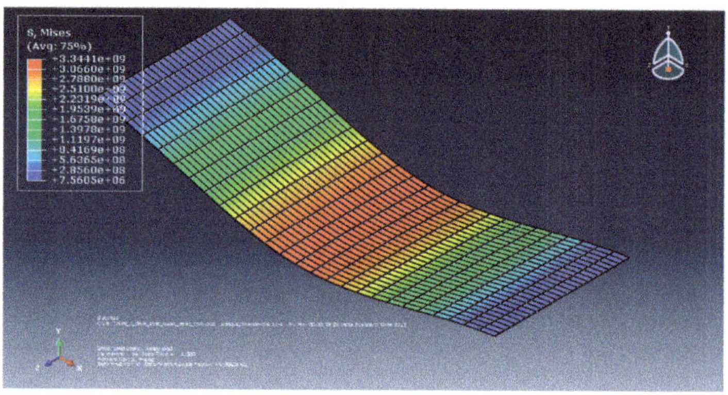

Figure A1. Von Mises stress contour plot of TiSiN coating with 3 μm thickness.

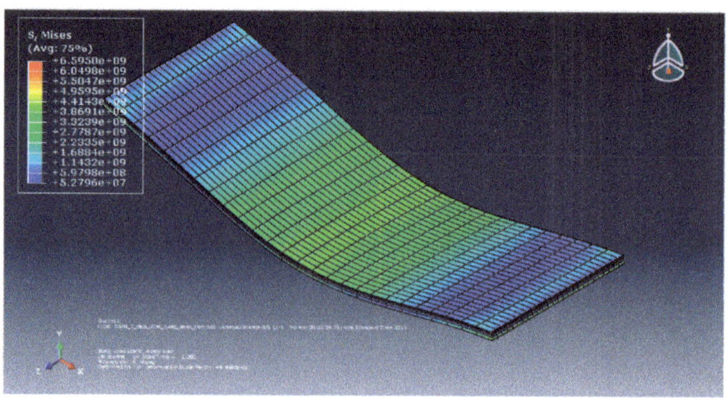

Figure A2. Von Mises stress contour plot of HSS substrate.

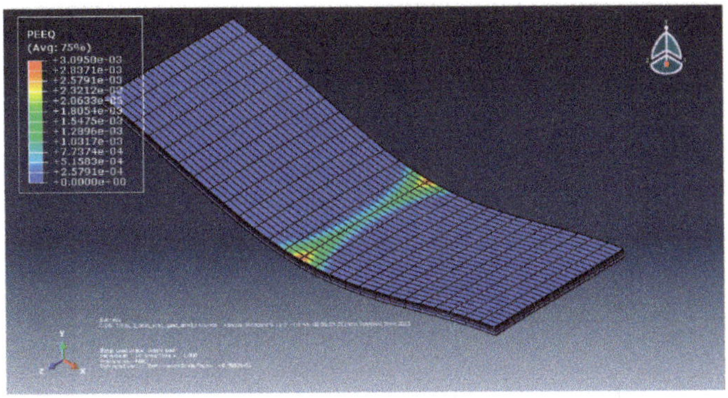

Figure A3. Plastic equivalent strain contour plot of HSS substrate with TiSiN coating thickness of 3 μm.

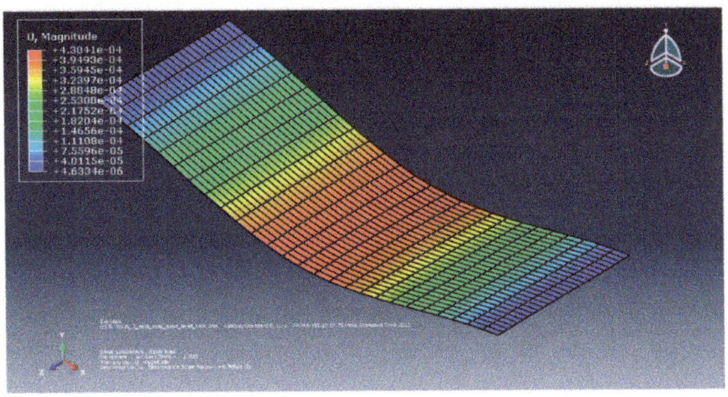

Figure A4. Deformation contour plot of TiSiN coating with 3 μm thickness.

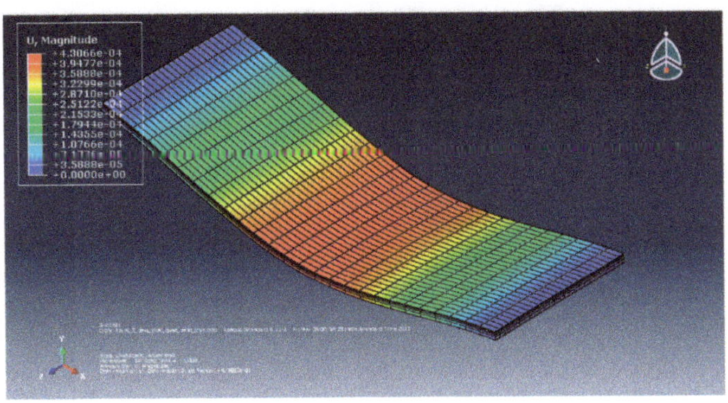

Figure A5. Deformation contour plot of HSS substrate with 3 μm coating thickness of TiSiN.

References

1. Aramcharoen, A.; Mativenga, P.T.; Yang, S.; Cooke, K.E.; Teer, D.G. Evaluation and selection of hard coatings for micro milling of hardened tool steel. *Int. J. Mach. Tools Manuf.* **2008**, *48*, 1578–1584. [CrossRef]
2. Dornfeld, D.; Min, S.; Takeuchi, Y. Recent advances in mechanical micromachining. *CIRP Ann.—Manuf. Technol.* **2006**, *55*, 745–768. [CrossRef]
3. Masuzawa, T. State of the art of micromachining. *CIRP Ann.—Manuf. Technol.* **2000**, *49*, 473–488. [CrossRef]
4. Weule, H.; Huntrup, V.; Tritschler, H. Micro-cutting of steel to meet new requirements in miniaturization. *CIRP Ann.—Manuf. Technol.* **2001**, *50*, 61–64. [CrossRef]

5. Liu, X.; de Vor, R.E.; Kapoor, S.G.; Ehmann, K.F. The mechanics of machining at the microscale: Assessment of the current state of the science. *J. Manuf. Sci. Eng.* **2004**, *126*, 666–678. [CrossRef]
6. Min, S.; Sangermann, H.; Mertens, C.; Dornfeld, D. A study on initial contact detection for precision micro-mold and surface generation of vertical side walls in micromachining. *CIRP Ann.—Manuf. Technol.* **2008**, *57*, 109–112. [CrossRef]
7. Masuzawa, T.; Toenshoff, H.K. Three-dimensional micromachining by machine tools. *CIRP Ann. —Manuf. Technol.* **1997**, *46*, 621–628. [CrossRef]
8. Bissacco, G.; Hansen, H.N.; de Chiffre, L. Micromilling of hardened tool steel for mould making applications. *J. Mater. Process. Technol.* **2005**, *167*, 201–207. [CrossRef]
9. Bissacco, G.; Hansen, H.N.; de Chiffre, L. Size effects on surface generation in micro milling of hardened tool steel. *CIRP Ann.—Manuf. Technol.* **2006**, *55*, 593–596. [CrossRef]
10. Li, P.; Oosterling, J.A.J.; Hoogstrate, A.M. Performance evaluation of micromilling of hardened tool steel. In Proceedings of the ICOMM, Greenville, NC, USA, 11–13 September 2007; pp. 219–224.
11. Aramcharoen, A.; Mativenga, P.T.; Yang, S. The effect of AlCrTiN coatings on product quality in micro-milling of 45 HRC hardened H13 die steel. In Proceedings of the 35th International Matador Conference, Taipei, Taiwan, 27 July 2007; pp. 203–206.
12. Maekawa, K.; Obikawa, T.; Yamane, Y.; Childs, T.H.C. *Metal Machining—Theory and Applications*; Butterworth-Heinemann: Oxford, UK, 2000.
13. Kobayashi, A. The features and application of UPC nano-micro forming tools. *Ind. Diam. Rev.* **2005**, *65*, 28–30.
14. Kim, C.J.; Bono, M.; Ni, J. *Experimental Analysis of Chip Formation in Micro-Milling*; Technical Papers Society of Manufacturing Engineers-All Series; NAMRIA: Manila, Philippines, 2002.
15. Jun, M.B.G.; DeVor, R.E.; Kapoor, S.G. Investigation of the dynamics of microend milling-Part II: Model validation and interpretation. *J. Manuf. Sci. Eng.* **2006**, *128*, 901–912. [CrossRef]
16. Torres, C.D.; Heaney, P.J.; Sumant, A.V.; Hamilton, M.A.; Carpick, R.W.; Pfefferkorn, F.E. Analyzing the performance of diamond-coated micro end mills. *Int. J. Mach. Tools Manuf.* **2009**, *49*, 599–612. [CrossRef]
17. Ozel, T.; Lui, X.; Dhanorker, A. Modelling and simulation of micro-milling process. In Proceedings of the 4th International Conference and Exhibition on Design and Production of Machines and Dies/Molds, Cesme, Turkey, 21–23 June 2007.
18. Tansel, I.; Rodriguez, O.; Trujillo, M.; Paz, E.; Li, W. Micro-end-milling-I. Wear and breakage. *Int. J. Mach. Tools Manuf.* **1998**, *38*, 1419–1436. [CrossRef]
19. Baharudin, B.T.H.T.; Dimou, N.; Hon, K.K.B. Tool wear behaviour of micro-tools in high speed CNC machining. In Proceedings of the 34th International MATADOR Conference, Manchester, UK, 7 July 2004; Springer: London, UK; pp. 111–118.
20. Uhlmann, E.; Schauer, K. Dynamic load and strain analysis for the optimization of micro end mills. *CIRP Ann.—Manuf. Technol.* **2005**, *54*, 75–78. [CrossRef]
21. Rahman, M.; Kumar, A.S.; Prakash, J.R.S. Micro milling of pure copper. *J. Mater. Process. Technol.* **2001**, *116*, 39–43. [CrossRef]
22. Bissacco, G.; Hansen, H.N.; Chiffre, L.D. Wear of micro end mills. In Proceedings of the 5th EUSPEN International Conference, Montpellier, France, 8–11 May 2005; pp. 561–564.
23. Zaman, M.T.; Kumar, A.S.; Rahman, M.; Sreeram, S. A three-dimension analytical cutting force model for micro end milling operation. *Int. J. Mach. Tools Manuf.* **2006**, *46*, 353–366. [CrossRef]
24. Uriarte, L.; Zatarian, M.; Albizuri, J.; Lacalle, L.N.L.d.; Lamikiz, A. Effect of the tool wear in micro-milling cutting forces. In Proceedings of the 2nd International Conference on High Performance Cutting, Tirupati, India, 4–5 December 2020.
25. Takacs, M.; Vero, B.; Meszaros, I. Micromilling of metallic materials. *J. Mater. Process. Technol.* **2003**, *138*, 152–155. [CrossRef]
26. Tansel, I.N.; Arkan, T.T.; Bao, W.Y.; Mahendrakar, N.; Shisler, B.; Smith, D.; McCool, M. Tool wear estimation in micro-machining, part II: Neural network-based periodic inspector for non-metals. *Int. J. Mach. Tools Manuf.* **2000**, *40*, 609–620. [CrossRef]
27. Schmidt, J.; Tritschler, H. Micro cutting of steel. *Microsyst. Technol.* **2004**, *10*, 167–174. [CrossRef]
28. Uhlmann, E.; Piltz, S.; Schauer, K. Micro milling of sintered tungsten-copper composite materials. *J. Mater. Process. Technol.* **2005**, *167*, 402–407. [CrossRef]
29. Vogler, M.P.; DeVor, R.E.; Kapoor, S.G. On the modeling and analysis of machining performance in micro-endmilling, Part I: Surface generation. *J. Manuf. Sci. Eng.* **2004**, *126*, 685–694. [CrossRef]
30. Weinert, K.; Petzoldt, V. Machining NiTi micro-parts by micro-milling. *Mater. Sci. Eng. A* **2008**, *481*, 672–675. [CrossRef]
31. Aramcharoen, A.; Mativenga, P.T.; Yang, S. The contribution of CrTiAlN coatings in micro milling of hardened die steel. In Proceedings of the 40th CIRP International Seminar on Manufacturing Systems, Liverpool, UK, 30 May–1 June 2007.
32. Bin Saedon, J. Micromilling of Hardened (62 HRC) AISI D2 Cold Work Tool Steel. Ph.D. Thesis, The University of Birmingham, Edgbaston, UK, October 2011.
33. Dobrzanski, L.A.; Golombek, K.; Kopac, J.; Sokovic, M. Effect of depositing the hard surface coatings on properties of the selected cemented carbides and tool cermets. *J. Mater. Process. Technol.* **2004**, *157*, 304–311. [CrossRef]
34. Talib, R.J.; Ariff, H.M.; Fazira, M.F. Machining performance and wear mechanism of TiAlN-coated insert. *Int. J. Mech. Mater. Eng. (IJMME)* **2011**, *6*, 414–418.
35. Kumar, M. Laser Assisted Micro Milling of Hard Materials. Ph.D. Thesis, Georgia Institute of Technology, Atlanta, GA, USA, 2011.
36. Scherzer, M.; Glaser, H. Mechanical Modelling of Failure Patterns in TiN-Coatings Loaded by Homogenous Stresses of a Bending Specimen. *Int. J. Fract.* **2002**, *114*, 291–306. [CrossRef]

37. Holmberg, K.; Laukkanen, A.; Ronkainen, H.; Wallin, K.; Varjus, S.; Koskinen, J. Tribological contact analysis of a rigid ball sliding on a hard coated surface. Part II: Material deformations, influence of coating thickness and Young's modulus. *Surf. Coat. Technol.* **2006**, *200*, 3810–3823. [CrossRef]
38. Holmberg, K.; Laukkanen, A.; Ronkainen, H.; Wallin, K.; Hogmark, S.; Jacobson, S.; Wiklund, U.; Souza, R.M.; Stahle, P. Residual stresses in TiN, DLC and MoS2 coated surfaces with regard to their tribological fracture behaviour. *Wear* **2009**, *267*, 2142–2156. [CrossRef]
39. Wu, T.; Cheng, K. Micro milling performance assessment of diamond-like carbon coatings on a micro-end mill. *Proc. Inst. Mech. Eng. Part J J. Eng. Tribol.* **2013**, *227*, 1038–1046. [CrossRef]
40. Ucun, I.; Aslantas, K.; Bedir, F. An experimental investigation of the effect of coating material on tool wear in micro milling of Inconel 718 super alloy. *Wear* **2013**, *300*, 8–19. [CrossRef]
41. Tao Wu, T. Tooling Performance in Micro Milling: Modelling, Simulation and Experimental Study. Ph.D. Thesis, Brunel University, Uxbridge, UK, December 2012.
42. Cselle, T.; Coddet, O.; Galamand, C.; Holubar, P.; Jilek, M.; Jilek, J.; Luemkemann, A.; Morstein, M. Triple coatings—New Generation of PVD-Coatings for Cutting Tools. *J. Mach. Manuf.* **2009**, *49*, 19–25.
43. Lille, H.; Koo, J.; Gregor, A.; Ryabchikov, A.; Sergejev, F.; Traksmaa, R.; Kulu, P. Comparison of Curvature and X-Ray Methods for Measuring of Residual Stresses in Hard PVD Coatings. *Mater. Sci. Forum* **2011**, *681*, 455–460. [CrossRef]
44. Ahmed, M.S.; Munroe, P.; Jiang, Z.T.; Rickard, W.; Xie, Z. Corrosion behaviour of nano-composite TiSiN coatings on steel substrates. *Corros. Sci.* **2011**, *53*, 3678–3687. [CrossRef]
45. Chang, Y.Y.; Yang, S.J.; Wang, D.Y. Structural and mechanical properties of AlTiN/CrN coatings synthesized by a cathodic-arc deposition process. *Surf. Coat. Technol.* **2006**, *201*, 4209–4214. [CrossRef]
46. Trent, E.M.; Paul, K. *Metal Cutting*, 4th ed.; Wright: Bristol, UK, 2000; p. 145.
47. *Surface Based Cohesive Behaviour 35.1.10*; 6.11 Documentation; Abaqus: Providence, RI, USA.
48. Lesage, J.; Chicot, D. Role of residual stresses on interface toughness of thermally sprayed coatings. *Thin Solid Films* **2002**, *415*, 143–150. [CrossRef]
49. Zhang, J.; Cho, Y.; Kim, J.; Malikov, A.K.; Kim, Y.H.; Yi, J.-H.; Li, W. Non-Destructive Evaluation of Coating Thickness Using Water Immersion Ultrasonic Testing. *Coatings* **2021**, *11*, 1421. [CrossRef]
50. Šporin, J.; Mrvar, P.; Janc, B.; Vukelić, Ž. Expression of the Self-Sharpening Mechanism of a Roller Cone Bit during Wear Due to the Influence of the Erosion Protection Carbide Coating. *Coatings* **2021**, *11*, 1308. [CrossRef]
51. Dettlaff, A.; Brodowski, M.; Kowalski, M.; Stranak, V.; Prysiazhnyi, V.; Klugmann-Radziemska, E.; Ryl, J.; Bogdanowicz, R. Highly oriented zirconium nitride and oxynitride coatings deposited via high-power impulse magnetron sputtering: Crystal-facet-driven corrosion behavior in domestic wastewater. *Adv. Eng. Mater.* **2021**, *23*, 2001349. [CrossRef]
52. Zheng, L.; Chen, W.; Huo, D. Investigation on the Tool Wear Suppression Mechanism in Non-Resonant Vibration-Assisted Micro Milling. *Micromachines* **2020**, *11*, 380. [CrossRef]
53. Kino, H.; Imada, T.; Ogawa, K.; Nakagawa, H.; Kojima, H. An Experimental Investigation on Micro End Milling with High-Speed Up Cut Milling for Hardened Die Steel. *Materials* **2020**, *13*, 4745. [CrossRef]
54. Grigoriev, S.N.; Migranov, M.S.; Melnik, Y.A.; Okunkova, A.A.; Fedorov, S.V.; Gurin, V.D.; Volosova, M.A. Application of Adaptive Materials and Coatings to Increase Cutting Tool Performance: Efficiency in the Case of Composite Powder High Speed Steel. *Coatings* **2021**, *11*, 855. [CrossRef]
55. Asghar, O.; Lou, L.Y.; Yasir, M.; Li, C.J.; Li, C.X. Enhanced Tribological Properties of LA43M Magnesium Alloy by Ni60 Coating via Ultra-High-Speed Laser Cladding. *Coatings* **2020**, *10*, 638. [CrossRef]

micromachines

Article

Implementation of Flip-Chip Microbump Bonding between InP and SiC Substrates for Millimeter-Wave Applications

Jongwon Lee [1,*,†], Jae Yong Lee [2,†], Jonghyun Song [1,3], Gapseop Sim [1], Hyoungho Ko [3,*] and Seong Ho Kong [2,*]

1 Nano Convergence Technology Division, National Nanofab Center, Daejeon 34141, Korea; jhsong@nnfc.re.kr (J.S.); gssim@nnfc.re.kr (G.S.)
2 School of Electronic and Electrical Engineering, Kyungpook National University, Daegu 41566, Korea; cheersssss@naver.com
3 Department of Electronics Engineering, Chungnam National University, Daejeon 34134, Korea
* Correspondence: temuchin80@nnfc.re.kr (J.L.); hhko@cnu.ac.kr (H.K.); shkong@knu.ac.kr (S.H.K.); Tel.: +82-042-366-1610 (J.L. & H.K. & S.H.K.)
† These authors contributed equally to this work.

Abstract: Flip-chip microbump (μ-bump) bonding technology between indium phosphide (InP) and silicon carbide (SiC) substrates for a millimeter-wave (mmW) wireless communication application is demonstrated. The proposed process of flip-chip μ-bump bonding to achieve high-yield performance utilizes a SiO_2-based dielectric passivation process, a sputtering-based pad metallization process, an electroplating (EP) bump process enabling a flat-top μ-bump shape, a dicing process without the peeling of the dielectric layer, and a SnAg-to-Au solder bonding process. By using the bonding process, 10 mm long InP-to-SiC coplanar waveguide (CPW) lines with 10 daisy chains interconnected with a hundred μ-bumps are fabricated. All twelve InP-to-SiC CPW lines placed on two samples, one of which has an area of approximately 11×10 mm^2, show uniform performance with insertion loss deviation within ±10% along with an average insertion loss of 0.25 dB/mm, while achieving return losses of more than 15 dB at a frequency of 30 GHz, which are comparable to insertion loss values of previously reported conventional CPW lines. In addition, an InP-to-SiC resonant tunneling diode device is fabricated for the first time and its DC and RF characteristics are investigated.

Keywords: InP; SiC; flip-chip bonding; millimeter wave; heterogeneous integration

Citation: Lee, J.; Lee, J.Y.; Song, J.; Sim, G.; Ko, H.; Kong, S.H. Implementation of Flip-Chip Microbump Bonding between InP and SiC Substrates for Millimeter-Wave Applications. *Micromachines* **2022**, *13*, 1072. https://doi.org/10.3390/mi13071072

Academic Editor: Stelios K. Georgantzinos

Received: 14 June 2022
Accepted: 2 July 2022
Published: 5 July 2022

1. Introduction

Because the InP substrate is lattice-matched with InGaAs materials featuring a high electron mobility of more than 8000 cm^2/vs, the InP-substrate-based low-noise amplifier (LNA) and power amplifier (PA) monolithic integrated circuits (MICs) using InGaAs high-electron-mobility transistor (HEMT) and heterojunction bipolar transistor (HBT) devices have operated at high operating frequencies above the millimeter-wave (mmW) regime [1,2]. Because the SiC substrate enables the growth of gallium nitride (GaN) materials exhibiting a high electric breakdown field of 3.3×10^6 V/cm, SiC-substrate-based mmW PA MICs using GaN HEMTs have shown high RF power performance [3]. By placing the SiC substrate as the first layer and the InP substrate as the second layer, an InP/GaN three-dimensional (3D) integration structure can be implemented, leading to high-performance mmW MICs and transceiver frontends [4,5]. The InP/GaN 3D structure can basically achieve a higher chip density and lower interconnection resistance compared to those of two-dimensional (2D) structures [4–6], and can enhance the current drivability of InGaAs HEMT and HBT devices owing to the excellent thermal conductivity of the SiC substrate (4.9 W/cmk) [7–10], resulting in an improved RF power and frequency performance of the mmW MICs or transceiver frontends.

An appropriate bonding method should be selected to realize the mmW InP/GaN 3D structure. Wire bonding, direct wafer bonding, and microbump (μ-bump) bonding for a heterogeneous integration have been utilized. Although wire bonding is easily accessible to researchers, it causes severe system performance degradation due to significant signal loss in the mmW band or higher, arising from the lengthening of interconnect lines [11]. Although direct wafer bonding makes it possible to achieve a high chip density in the MIC and transceiver owing to the utilization of a sequential fabrication process after the direct bonding followed by substrate removal, the substrate removal process for leaving active thin-film layers requires a high degree of fabrication proficiency [7–10,12]. On the other hand, μ-bump bonding consisting of electro-plated copper (Cu) pillar bumps and solder bonding [13,14] is a mature process technology that has been used in commercial 3D stacking memory products [14,15], and thereby it can be introduced to implement mmW 3D MICs and transceiver systems reproducibly. With the improvement of the alignment accuracy of bonding equipment, it is possible to densely form μ-bumps with a diameter (or width) of only a few to a few tens of micrometers [14,16]. Recently, heterogeneous μ-bump bonding technologies concerning various substrates, such as InP-to-SiC [5], InP-to-Si [17], and AlN/diamond-to-Si [18], have been reported for utilization in mmW wireless communication applications. However, [5] did not disclose any measurement results for bonded samples and [17] showed the RF measurement results for only a bonded HBT device, which are presumed to be implemented by a few to a few tens of μ-bump connections. The authors of [18] also measured only an interconnect line with four μ-bumps. To apply bump bonding technology to mmW ICs and transceiver systems, the process methodology and the implementation results of the μ-bump bonding technology to implement a much larger number of μ-bumps should be presented.

In this work, we report a flip-chip μ-bump bonding technology between InP and SiC substrates for mmW wireless communication applications. A process methodology for InP-to-SiC flip-chip μ-bump bonding with high-yield characteristics while being easily accessible to researchers is proposed. By utilizing the bonding process, 10 mm long InP-to-SiC coplanar waveguide (CPW) lines interconnected with a hundred μ-bumps were fabricated. The fabricated InP-to-SiC CPW lines showed uniformly good performance with an insertion loss deviation within 10% and an average insertion loss of 0.25 dB/mm at a frequency of 30 GHz, which are comparable to insertion loss values of previously reported conventional CPW lines. In addition, an InP RTD device was flip-chip bonded with a SiC substrate for the first time and its DC and RF performance was investigated through a comparison with the corresponding measures of a conventional InP RTD device.

2. Structure Design and Fabrication

2.1. Structure Design of a Flip-Chip μ-Bump Bonding Technology between InP and SiC Substrates

The designed structure of a flip-chip μ-bump bonding technology for a CPW interconnection between InP and SiC substrates is shown in Figure 1. To minimize the signal loss of the CPW line escaping from PAD metals (PAD_InP and PAD_SiC in Figure 1) to the substrates, semi-insulating InP and SiC substrates were chosen. An InP substrate, provided by JZ Nippon Mining & Metal Corporation, with a 3-inch diameter, a thickness of approximately 610 μm, and a resistivity of more than 1×10^7 Ω·cm was used. A 4H-SiC substrate with a 4-inch diameter, a thickness of approximately 510 μm, and a resistivity over 1×10^7 Ω·cm, provided by Synlight Crystal Co., Ltd. Hebei, China, was used. A dielectric layer of SiO_2 or BCB (Cyclotene 3022-46 resin) with a thickness of 2 μm was inserted between the PAD metals and the substrates for device passivation and planarization, considering the ultimate integration of transistors such as HEMTs and HBTs in the future, as well as for the minimization of the signal loss from the PAD metals to substrates, as shown in Figure 1b. A Ti/Au material was used as PAD metals with the thickness set to 0.8 μm, which corresponded to the maximum limit of the sputtering equipment used in this work. The μ-bump metal consisted of Cu/Ni/SnAg and its height was set to 20 μm or more to prevent any bonding failure caused by the fragile nature of the InP substrates. From an S-parameter simulation for the flip-chip-bonded CPW line between InP and SiC substrates using the advanced design system (ADS) momentum simulator, it was determined that the signal width (W) and the gap (G) of the CPW to have a characteristic impedance of 50 Ω in the mmW frequency range of more than 30 GHz were 60 and 40 μm, respectively. The length of the μ-bump pad ('a' in Figure 1) was set as 60 μm, the same as W. The size of the μ-bump ('b' in Figure 1) was set in the range of 25 to 40 μm by comprehensively considering several phenomena such as the alignment accuracy of the flip-chip bonder equipment, the SnAg overflow in the bonding process, and the increase in the insertion loss that occurs when the bump size is quite small compared to the bump-pad length.

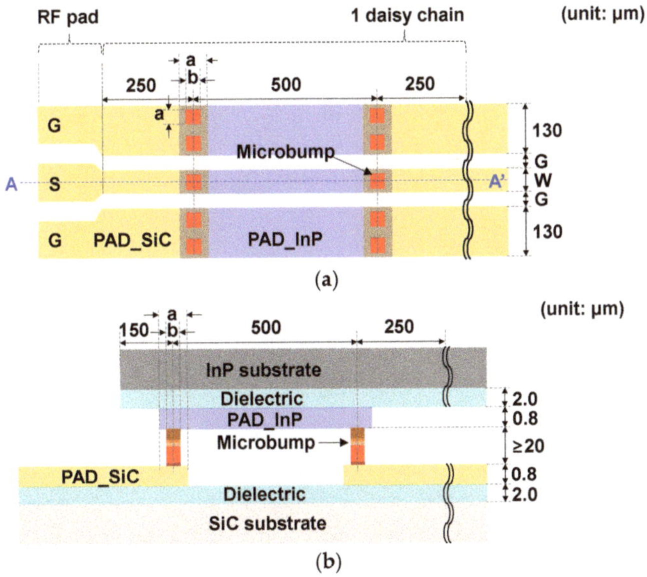

Figure 1. The structure of flip-chip microbump bonding technology for coplanar waveguide (CPW) interconnection between InP and SiC substrates: (**a**) a floor plan; (**b**) a cross-sectional view at point A–A' in Figure 1a.

2.2. Fabrication of a Flip-Chip μ-Bump Bonding Process between InP and SiC Substrates

Figure 2 shows a cross-sectional view of the process flow of the flip-chip μ-bump bonding for the CPW interconnection between InP and SiC substrates, which entails a dielectric layer deposition (Figure 2a), the formation of PAD metal (Figure 2b), formation of μ-bump (Figure 2c), dicing (Figure 2d), and flip-chip SnAg-to-Au solder bonding (Figure 2e).

The process for the deposition of the dielectric layer, seen in a conceptual diagram of Figure 2a, was conducted. To find the appropriate material for the dielectric layer, two kinds of materials with a low dielectric constant, SiO_2 and BCB, were tested. A 2.5 μm thick BCB was deposited on the InP substrate via spin-coating with spinner system equipment and cured at 210 °C in vacuum oven equipment with a N_2 atmosphere. Additionally, a 2 μm thick SiO_2 layer was deposited on the InP substrate using plasma-enhanced chemical vapor deposition (PECVD) at a temperature of 300 °C. After a 0.2 μm thick Ti/Au PAD metal for the CPW line formation was deposited on both the BCB-based and SiO_2-based InP substrates, as shown in the inset in Figure 3a, an S-parameter measurement was conducted at a frequency of 15 GHz. As a result, the insertion loss of the BCB-based and SiO_2-based CPW lines was 0.24 and 0.29 dB/mm at 15 GHz, respectively, as shown in Figure 3a. The lower insertion loss of the BCB-based CPW was attributed to the dielectric constant of the BCB (~2.5) being lower than that of the SiO_2 (~3.8). Even though the BCB layer was superior to the SiO_2 layer in terms of mmW performance, a problem was found, where the surface of the BCB layer was dirty enough to adversely affect the process yield, in contrast to the clean surface condition of the SiO_2 layer, as shown in Figure 3b. This yield issue of BCB was dependent on the size of the substrate, which occurred when the size of the substrate was increased to 3 inches or larger. It was determined that the cause of the issue was that the BCB solution sprayed through the dropper adhered to the wall of the spinner equipment during the spinning process, and then fell off and adhered to the wafer again. Consequently, to establish a high-yield process technology, the SiO_2 layer with a thickness of 2 μm was used as the dielectric layer. We noted that the yield issue of BCB may have been limited by our facility at this time, and could be sufficiently improved through the optimization of the spinner-based BCB process.

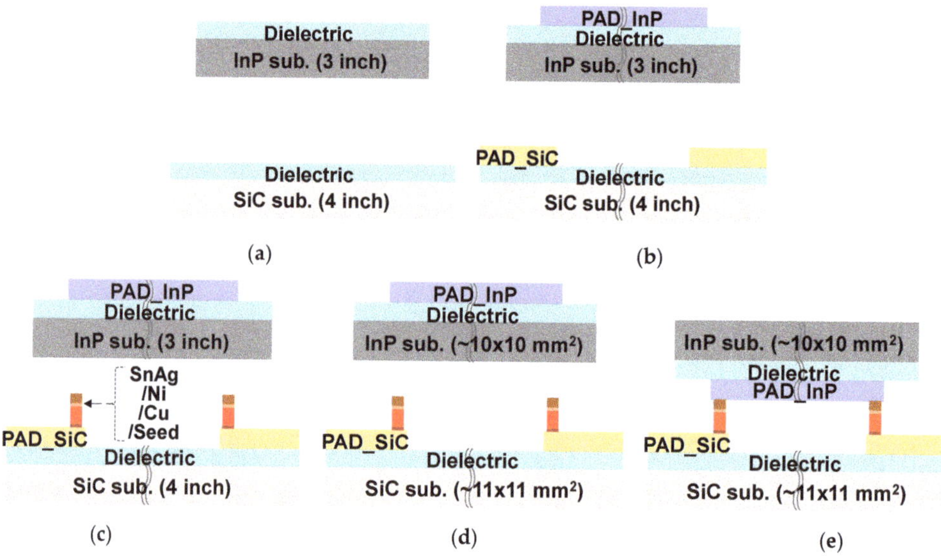

Figure 2. Cross-sectional view of the process flow for flip-chip microbump (μ-bump) bonding technology between InP and SiC substrates: (**a**) dielectric layer deposition; (**b**) formation of PAD metal; (**c**) formation of microbump metal; (**d**) dicing; (**e**) flip-chip SnAg-to-Au solder bonding.

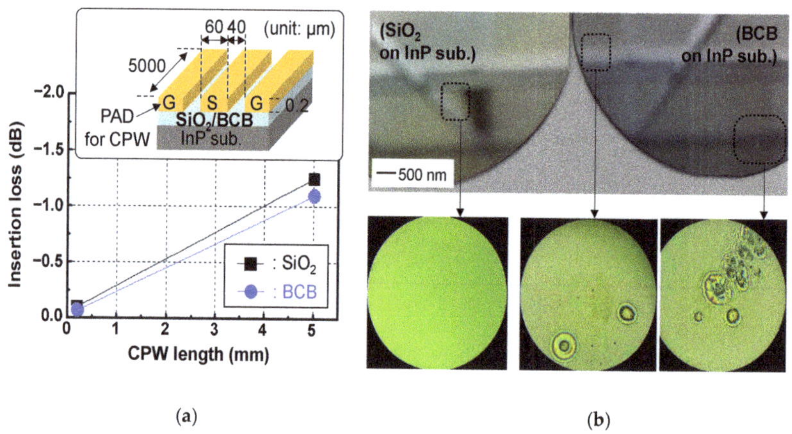

Figure 3. Implementation results of coplanar waveguides (CPWs) with different dielectric layers of SiO_2 and BCB: (**a**) measured insertion loss of fabricated SiO_2- and BCB-based coplanar waveguide (CPW) lines. The inset shows a structural diagram of the fabricated CPW lines; (**b**) microscope images immediately after deposition of SiO_2 and BCB layers on InP substrates.

The process for the formation of the PAD metal was conducted on both SiO_2-deposited InP and SiC substrates, as shown in Figure 2b. The process methodology of the evaporation and lift-off was first applied and the detailed process flow was as follows: A cleaning process was conducted using acetone/IPA/DI solutions. A photolithography process with a critical dimension (CD) of 60 μm and an undercut slope was performed using an NR93000PY negative photoresist (PR) and EVG640 contact aligner. The Ti/Au PAD metal with a thickness of 400/8000 Å was evaporated using the KVET-C500200 evaporator. A Ti/Au PAD metal pattern was formed by performing a lift-off process using an acetone solution. This evaporation and lift-off process methodology resulted in an adhesion problem between the PAD metals and the substrates, as shown in Figure 4a. To solve the problem, the process methodology of the sputtering and metal etching was finally applied, and the detailed process flow was as follows: The cleaning process was conducted as aforementioned. The Ti/Au PAD metal with a thickness of 400/8000 Å was then sputtered using sputter equipment with a predeposition of 10 s, bias power of 700 W, and operating pressure of 10^{-6} Torr. A photolithography process with the same CD of 60 μm was then performed using a positive AZ601 PR and EVG640 contact aligner equipment. The Ti/Au metal was then selectively wet-etched by immersing it in Ti and Au etchants for 50 and 30 s, respectively. The positive PR was removed with acetone-IPA-DI solutions. The PAD metal formed with the sputtering and metal etch process did not have any adhesion problems with the underlying substrate, as shown in Figure 4b.

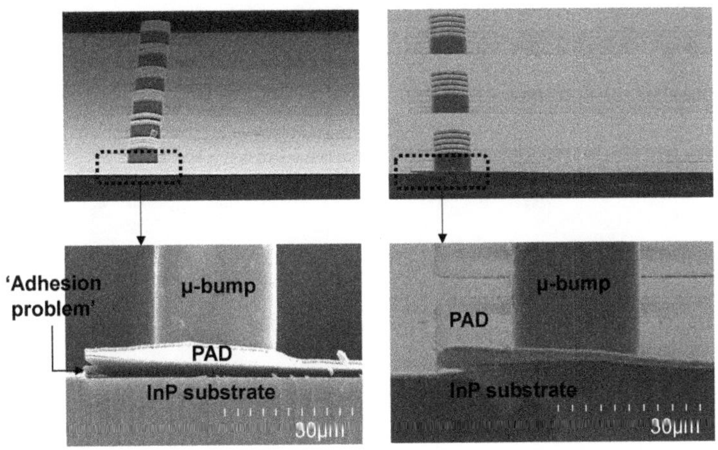

(a) (b)

Figure 4. SEM images of PAD metals: (**a**) PAD metal based on an evaporation and lift-off process; (**b**) PAD metal based on a sputtering and metal etching process.

The process for the formation of the μ-bump metal was performed on a SiC-substrate-based sample, as seen in the conceptual diagram in Figure 2c. First, a 300/2000 Å thick Ti/Au seed metal was deposited using the aforementioned sputtering equipment, and then a photolithography process for electroplating (EP) was performed using a negative JSR-126N PR, a contact aligner, and a hard-bake process (110 °C for 8 min), leading to a PR thickness of 30 μm or more. An EP process was then conducted, wherein Cu, Ni, and SnAg metals were sequentially deposited to form μ-bumps with a height of 20 μm or more. The thickness of Ni and SnAg metals was set to be more than 2 μm and 6 μm, respectively. The PR and seed metal were then removed by immersing in an STR2000 solution for 30 min at 40 °C and in Ti/Au etchants for a total of 80 s, respectively. It was paramount for the bump to have a flat-top shape to achieve a high-yield performance of the flip-chip bonding. In the EP process of the Cu metal, a copper sulfate solution and additives are generally used. According to our experimental results, when the additives were mixed with the copper sulfate solution, the μ-bump had a convex top shape with a height difference from the top center to the top edge of approximately 6 μm, as shown in Figure 5a, leading to bonding failure. Accordingly, the additives were not used in the Cu EP process to obtain a flat-top shape, as shown in Figure 5b. The copper sulfate solution was provided by ATOTECH.

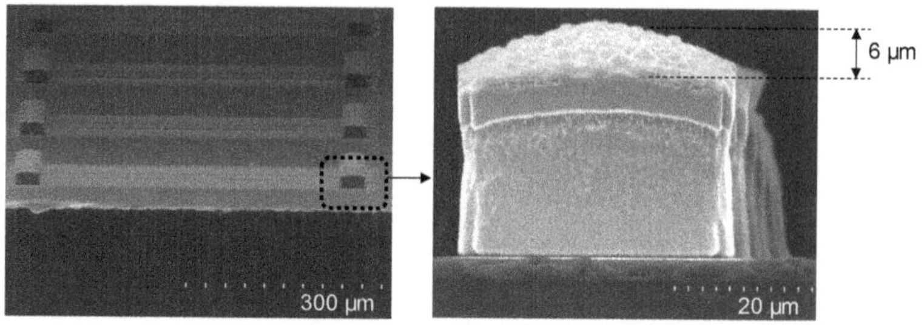

(a)

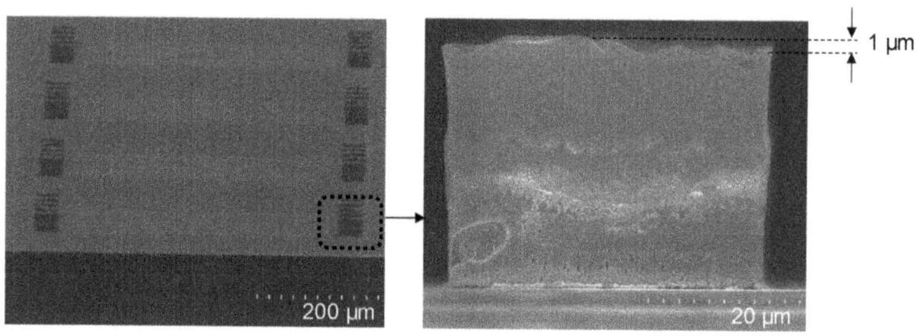

(b)

Figure 5. SEM images after formation of microbump (μ-bump) metal: (**a**) μ-bump metal that additives were used in in the Cu electroplating (EP) process; (**b**) μ-bump metal that additives were not used in in the Cu EP process.

A dicing process was conducted for the prefabricated InP-substrate-based and SiC-substrate-based samples, as seen in a conceptual diagram of Figure 2d. First, the two samples were protected with a AZ601 PR coating and soft-baking at 100° at 60 s to prevent wafer contamination by particles generated during the dicing process. Second, using the DISCO DFD640 dicing equipment and a KH5-1840 blade, an InP-substrate-based sample with a full size of 3 inches and a SiC-substrate-based sample with a full size of 4 inches were diced to a size of 1×1 cm^2 and 1.1×1.1 cm^2, respectively. Third, the PR was removed and the samples were cleaned using Acetone-IPA-DI solutions. As a result of the process, a peeling problem was found, where the SiO_2 dielectric layer was peeled off around the dicing lines on the InP substrate, as shown in Figure 6a. The peeling problem was solved by adding a dielectric removal process before the aforementioned dicing process, which selectively removed the dielectric layer around the dicing lines. The dielectric removal process was carried out in the order of a photolithography process using an S700 positive PR and an EVG contact aligner, a hard-baking process for 15 min at 150°, a wet-etching process for 90 min in BOE solution, and a PR removal process using Acetone-IPA-DI. Figure 6b shows an SEM image of the dicing process with the inclusion of the dielectric removal process.

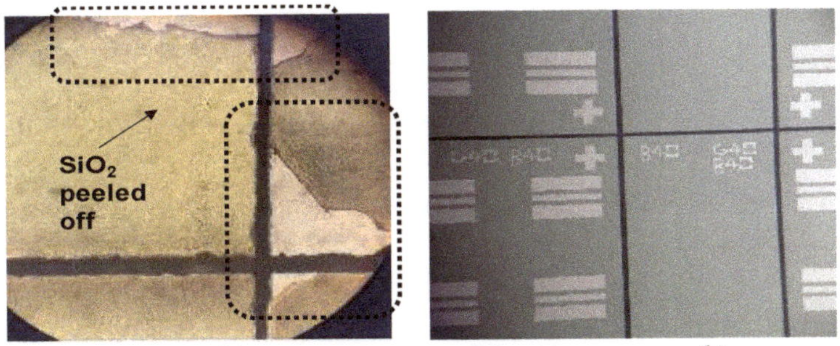

Figure 6. Microscope images after dicing process: (**a**) dicing process where a dielectric removal process was not included; (**b**) dicing process where a dielectric removal process was included.

A flip-chip SnAg-to-Au solder bonding process, as seen in the conceptual diagram of Figure 2e, was established by assembling the diced InP-substrate-based and SiC-substrate-based samples. The bonding process was conducted using the DFC-2000C flip-chip bonder equipment with conditions of a bonding pressure of 10 N and a total bonding time of 16.1 s. The bonding temperature was set to 300°, as the overflow phenomenon of SnAg material occurred above 350°, as shown in Figure 7. The alignment error of the flip-chip bonding was within 2 μm.

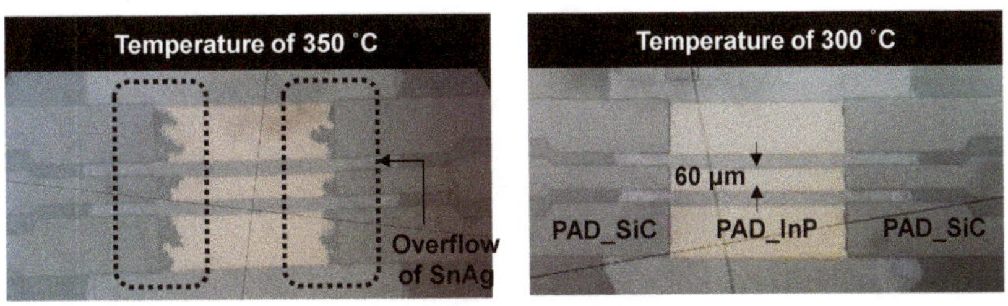

Figure 7. Microscope images of a flip-chip-bonded InP-to-SiC sample with different bonding temperatures of 350 and 300°.

3. Results and Discussion

3.1. Performance of Flip-Chip-Bonded InP-to-SiC CPW Lines Consisting of 10 Daisy Chains Interconnected by a Hundred μ-Bumps

By utilizing the flip-chip μ-bump bonding technology, InP-to-SiC CPW lines where both PAD metals on the InP and SiC substrates were interconnected through μ-bumps were implemented. To pursue the scale-up of the bonding technology for mmW application, an InP-to-SiC CPW line consisted of ten daisy chains interconnected by a hundred μ-bumps, and its length was as high as 10 mm, as shown in Figure 8a. Ten InP-to-SiC CPW lines were arranged in a flip-chip-bonded sample with an area of 11×10 mm^2, as shown in Figure 8b. Among ten CPW lines, the upper two lines and the lower two lines served as dummy patterns for achieving high-yield performance. Two identically designed flip-chip-bonded samples were fabricated. S-parameter data for all real CPW lines arranged in the two samples were measured using the N5225B PNA network analyzer (NA). Figure 9a,b show the measured results for the insertion and return losses of the CPW lines, respectively. The return loss was more than 15 dB over the frequency of 30 GHz from DC. The insertion loss was in the range of 2.24 to 2.71 dB at 30 GHz, and its average value was 0.25 dB/mm, which was comparable to the insertion loss values of previously reported conventional mmW CPW lines without any bonding technologies [19–21]. The deviation of the insertion loss for the twelve CPW lines was within ±10%, which verified that the flip-chip-bonded μ-bump process between the InP-to-SiC substrates was well established, exhibiting good uniformity.

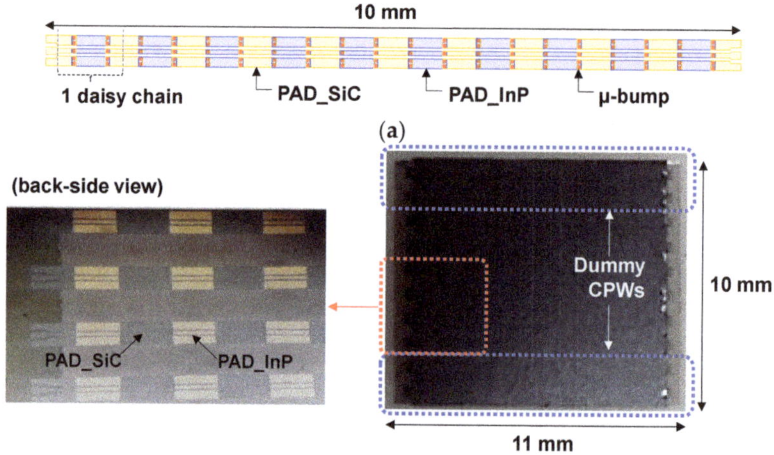

(b)

Figure 8. Implementation results of InP-to-SiC CPW lines with a hundred μ-bumps: (**a**) schematic diagram of a InP-to-SiC CPW line; (**b**) microscope images of a fabricated flip-chip-bonded sample with an area of 11 × 10 mm^2 arranged with ten CPW lines (six real and four dummy lines).

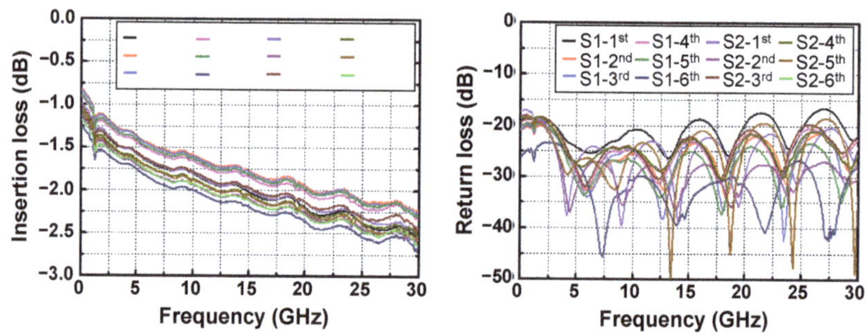

(a) (b)

Figure 9. Measured S-parameter results of twelve InP-to-SiC CPW lines with a hundred μ-bumps arranged in two flip-chip-bonded samples: (**a**) insertion loss; (**b**) return loss.

The RF modeling of fabricated μ-bumps was essential to utilize the flip-chip bonding technology for the mmW application. First, the RF pad region inserted for the RF measurement of CPW lines, as described in Figure 1a, was de-embedded using fabricated open and thru patterns. Figure 10a shows a modeled circuit diagram of a thru pattern with a length of 200 μm, which was drawn in the advanced design system (ADS) simulator. Shunt resistor (R_{PP}) and capacitor (C_{PP}) devices were used for the proper RF modeling of the pad region. Figure 10b shows measured and modeled results for the S-parameter of the thru pattern. When R_{PP} and C_{PP} were 40 Ω and 12 fF, respectively, the modeled results were in good agreement with the measured results. Next, the RF modeling of the fabricated μ-bump was carried out using measured S-parameter results of the flip-chip-bonded InP-to-SiC CPW line with 10 daisy chains. Figure 11a shows a circuit diagram used for the fabricated CPW lines. As an equivalent circuit model of the fabricated μ-bump, a pi model was used, which consisted of a series resistance (R_{BS}), a series inductance (L_{BS}), and two shunt capacitors (C_{BP}) [22]. Figure 11b shows measured and modeled results of the S-parameter of the fabricated CPW lines with 10 daisy chains. When R_{BS}, L_{BS}, and C_{BP} were 0.35 Ω, 50 pH, and 20 fF, respectively, the modeled results were well matched with the measured results.

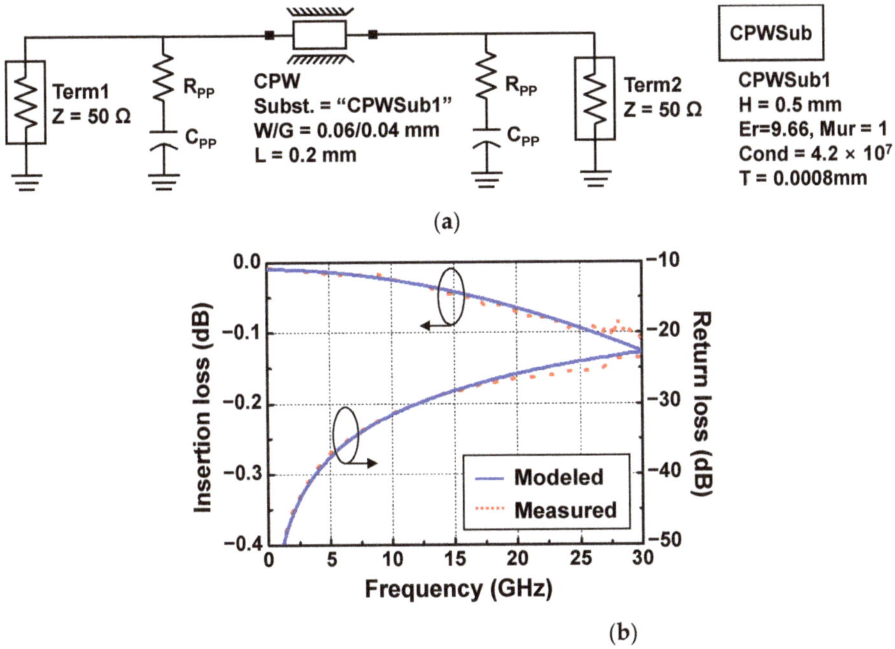

Figure 10. De-embedding for an RF pad region of CPW lines: (**a**) RF modelled circuit diagram of a thru pattern; (**b**) measured and modeled results for S-parameter of the thru pattern.

3.2. Application to mmW Device of the Flip-Chip μ-Bump Bonding Technology

To demonstrate the mmW application capability of the flip-chip bonding technology, an InP resonant tunneling diode (RTD), which is one of the semiconductor devices operating at mmW and terahertz (THz) frequencies [23,24], was flip-chip-bonded for the first time with the SiC substrate and its DC, and RF performance was investigated. Figure 12a shows a fabricated InP-substrate-based sample consisting of an RTD, CPW PAD metals for the flip-chip interconnection, and dummy PAD metals functioning as supporting pillars during the bonding process. The inset shows an SEM image before the device passivation process of the fabrication RTD. The epitaxial structure and process sequence of the RTD were described elsewhere [25]. Figure 12b shows a fabricated SiC-substrate-based sample consisting of CPW PAD and μ-bump metals for the flip-chip interconnection, and dummy PAD and μ-bump metals functioning as supporting pillars. Figure 12c shows a microscope image after the InP-substrate-based sample was flip-chip-bonded with the SiC-substrate-based sample.

A conventional InP RTD without a flip-chip bonding interconnection (C-RTD) and an InP-to-SiC RTD with a flip-chip bonding interconnection (F-RTD) were measured by being probed at measurement pads (seen in Figure 12a,b,) respectively, with respect to the DC and RF characteristics. Figure 13a shows a DC I-V curve of the two RTD devices, which was measured with the Keithley 4200-SCS/F semiconductor characterization system and Summit 11862B probe station. The two RTDs exhibited nearly the same peak and valley voltages of 0.3 and 0.75 V, respectively. The peak and valley currents of F-RTD were 3.21 and 0.24 mA, which were approximately 9 % higher than those (2.94 and 0.22 mA) of C-RTD. This current difference was attributed to the wet-chemical etching variation in the mesa process of the two RTDs [25]. Figure 13b shows measured S-parameter data of the two RTDs measured with the N5225B PNA network analyzer (NA) equipment. The RTDs were biased at 0.2 V. It was observed that the S_{11} value of F-RTD increased compared with that of C-RTD as the frequency increased. This S_{11} increase was mainly attributed to the high C_{BP} value of 20 fF. From the ADS simulation results, based on the aforementioned equivalent model of the bump, the S_{11} graph of F-RTD was the same as that of C-RTD when the C_{BP} value decreased to less than 10 fF from 20 fF, as shown in Figure 13b. C_{BP} was generated from the bump-pad region, corresponding to the region of bump-pad length of 'a' in Figure 1. In this work, the bump-pad length for F-RTD was as large as 80 μm, while the bump size for F-RTD ('b' in Figure 1) was 40 μm. Because the overflowed SnAg after the flip-chip bonding was present within 5 μm from the edge of the μ-bump metals and the alignment error of bonding equipment was within 2 μm, the bump-pad length for F-RTD could be reduced to less than 60 μm, corresponding to a C_{BP} of 10 fF.

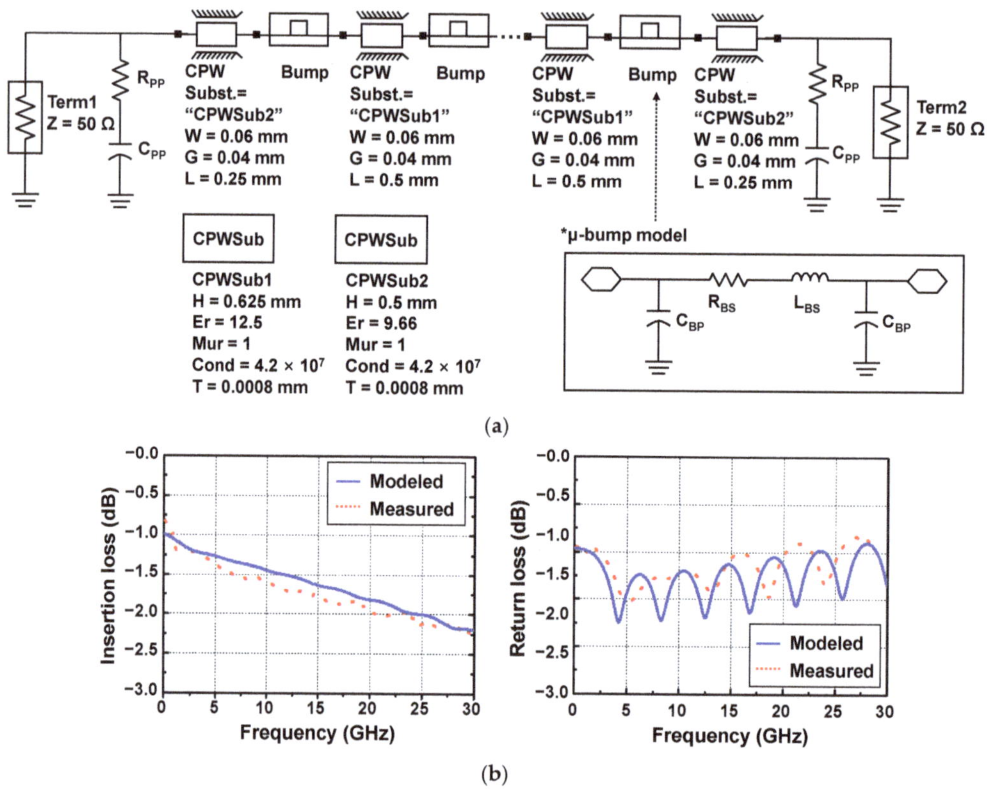

Figure 11. RF modeling of flip-chip-bonded InP-to-SiC CPW lines: (**a**) equivalent circuit diagram for RF modeling of CPW lines; (**b**) measured and modeled S-parameter results for a CPW line with 10 daisy chains.

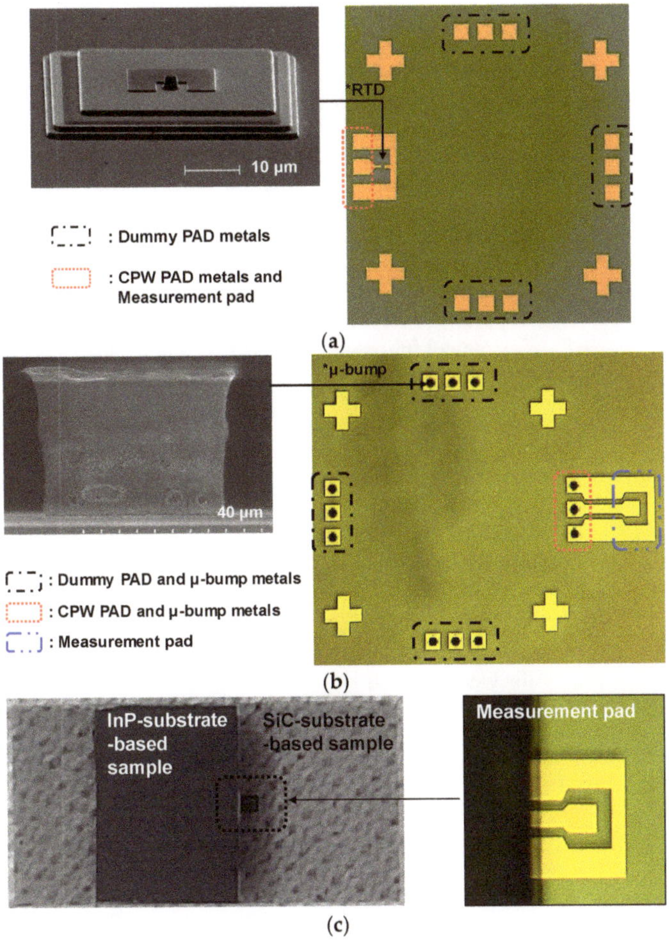

Figure 12. Implementation results of a flip-chip bonded InP-to-SiC resonant tunneling diode (RTD): (**a**) an InP-substrate-based sample, (**b**) a SiC-substrate-based sample, (**c**) a flip-chip-bonded sample with an InP-to-SiC RTD.

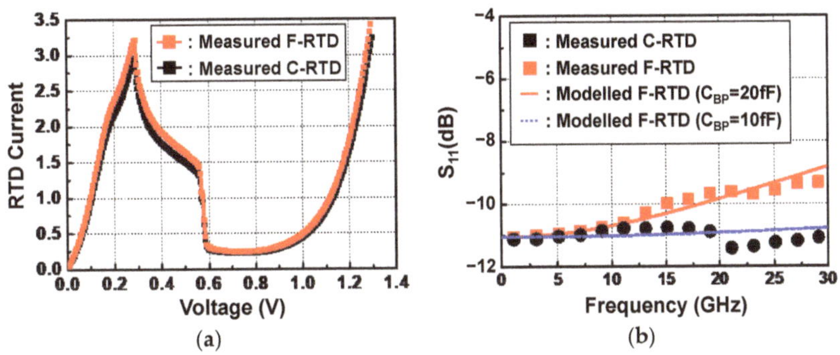

Figure 13. Measurement results of conventional InP resonant tunneling diode (C-RTD) and flip-chip-bonded InP-to-SiC resonant tunneling diodes (F-RTD): (**a**) DC I–V curve; (**b**) S-parameter data. C_{BP} denotes a parasitic shunt capacitance of μ-bump.

4. Conclusions

A process methodology for flip-chip μ-bump bonding between InP and SiC substrates for a mmW wireless communication application was proposed, consisting of a SiO_2-based dielectric passivation process, a sputtering-based pad metallization process, an EP bump process enabling a flat-top μ-bump shape, a dicing process without the peeling of the dielectric layer, and a SnAg-to-Au solder bonding process. By using the flip-chip bonding process, 10 mm long InP-to-SiC CPW lines with 10 daisy chains interconnected with a hundred μ-bumps were fabricated. All InP-to-SiC CPW lines placed on two samples, one of which had an area of approximately 11×10 mm^2, exhibited uniform performance with insertion loss deviation within ±10% along with an average insertion loss of 0.25 dB/mm, while achieving return losses of more than 15 dB at a frequency of 30 GHz, which were comparable to the insertion loss values of conventional CPW lines. In addition, an InP-to-SiC resonant tunneling diode device was fabricated for the first time and its DC and RF characteristics were investigated.

Author Contributions: Conceptualization, J.L.; methodology, J.L., J.Y.L., J.S. and G.S.; validation, J.L.; formal analysis, J.L.; investigation, J.L., J.Y.L., J.S. and G.S.; resources, J.S. and G.S.; data curation, J.L.; writing—original draft preparation, J.L.; writing—review and editing, J.L., H.K. and S.H.K.; visualization, J.L.; supervision, J.L., H.K. and S.H.K.; project administration, J.L.; funding acquisition, J.L. All authors have read and agreed to the published version of the manuscript.

Funding: This research was supported by the Civil-Military Technology Cooperation Program (no. 19-CM-BD-05).

Acknowledgments: The authors would like to express their gratitude for valuable help from S. Jo at SNI Co. Ltd. for the overall fabrication. The authors would also like to thank S. Kim at AJIN Electronics Co. Ltd. for conducting the flip-chip SnAg-to-Au solder bonding process and S. Kim at the Electronics and Telecommunications Research Institute for supporting the network analyzer equipment.

Conflicts of Interest: The authors declare no conflict of interest.

References

1. Mei, X.; Yoshida, W.; Lange, M.; Lee, J.; Zhou, J.; Liu, P.; Leong, K.; Zamora, A.; Padilla, J.; Sarkozy, S.; et al. First Demonstration of Amplification at 1 THz Using 25-nm InP High Electron Mobility Transistor Process. *IEEE Electron Device Lett.* **2015**, *36*, 327–329. [CrossRef]
2. Urteaga, M.; Griffith, J.; Young, J.; Pierson, R.; Rowell, P.; Seo, M.; Rodwell, M.J.W. A 130 nm InP HBT Integrated Circuit Technology for THz Electronics. In Proceedings of the 2016 IEEE International Electron Devices Meeting (IEDM), San Francisco, CA, USA, 3–7 December 2016; pp. 711–714. [CrossRef]
3. Romanczyk, B.; Wienecke, S.; Guidry, M.; Li, H.; Ahmadi, E.; Zheng, X.; Keller, S.; Mishra, U.K. Demonstration of Constant 8 W/mm Power Density at 10, 30, and 94 GHz in State-of-the-Art Millimeter-Wave N-Polar GaN MISHEMTs. *IEEE Trans. Electron Devices* **2018**, *65*, 45–50. [CrossRef]
4. Green, D.S.; Dohrman, C.L.; Demmin, J.; Zheng, Y.; Chang, T.-H. A Revolution on the Horizon from DARPA: Heterogeneous Integration for Revolutionary Microwave\/Millimeter-Wave Circuits at DARPA: Progress and Future Directions. *IEEE Microw. Mag.* **2017**, *18*, 44–59. [CrossRef]
5. Carter, A.; Urteaga, M. 3D integration unites InP, GaN and silicon CMOS. *Compd. Semicond.* **2018**, *24*, 50–53. Available online: https://compoundsemiconductor.net/article/104207/3D_Integration_Unites_InP_GaN_And_Silicon_CMOS (accessed on 15 June 2021).
6. Micromachines Lee, J.; Roh, K.; Lim, S.-K.; Kim, Y. Sidewall Slope Control of InP Via Holes for 3D Integration. *Micromachines* **2021**, *12*, 89. [CrossRef] [PubMed]
7. Scott, D.; Monier, C.; Wang, S.; Radisic, V.; Nguyen, P.; Cavus, A.; Deal, W.; Gutierrez-Aitken, A. InP HBT Transferred to Higher Thermal Conductivity Substrate. *IEEE Electron Device Lett.* **2012**, *33*, 507–509. [CrossRef]
8. Watanabe, M.; Yanagisawa, M.; Uesaka, K.; Ekawa, M.; Shoji, H. InP-DHBT Fabricated on High Heat Dissipation SiC Wafer Using Atomic Diffusion Bonding. *SEI Tech. Rev.* **2016**, *83*, 45–49.
9. Shiratori, Y.; Hoshi, T.; Ida, M.; Higurashi, E.; Matsuzaki, H. High-Speed InP/InGaAsSb DHBT on High-Thermal-Conductivity SiC Substrate. *IEEE Electron Device Lett.* **2018**, *39*, 807–810. [CrossRef]
10. Takenaka, M.; Takagi, S. InP-based photonic integrated circuit platform on SiC wafer. *Opt. Express* **2017**, *25*, 29993–30000. [CrossRef] [PubMed]

11. Wu, L.; JiaYun, D.; Cheng, W.; Kong, Y.; Chen, T.; Zhang, T. Heterogeneous Integration of InP DHBT and Si CMOS by 30μm Pitch Au-In Microbumps. In Proceedings of the 2021 Electron Devices Technology and Manufacturing Conference (EDTM), Chengdu, China, 8–11 April 2021; pp. 1–3. [CrossRef]
12. Geum, D.-M.; Kim, S.; Lee, S.; Lim, D.; Kim, H.-J.; Choi, C.; Kim, S.-H. Monolithic 3D Integration of InGaAs Photodetectors on Si MOSFETs Using Sequential Fabrication Process. *IEEE Electron Device Lett.* **2020**, *41*, 433–436. [CrossRef]
13. Li, J.; Zhang, Y.; Zhang, H.; Chen, Z.; Zhou, C.; Liu, X.; Zhu, W. The thermal cycling reliability of copper pillar solder bump in flip chip via thermal compression bonding. *Microelectron. Reliab.* **2020**, *104*, 113543. [CrossRef]
14. Koh, W.; Lin, B.; Tai, J. Copper Pillar Bump Technology Progress Overview. In Proceedings of the 2011 International Conference on Electronic Packaging Technology & High Density Packaging, Shanghai, China, 8–11 August 2011; pp. 1–5. [CrossRef]
15. Lee, J.; Lee, C.; Kim, C.; Kalchuri, S. Micro Bump System for 2nd Generation Silicon Interposer with GPU and High Bandwidth Memory (HBM) Concurrent Integration. In Proceedings of the 2018 IEEE 68th Electronic Components and Technology Conference, San Diego, CA, USA, 29 May 2018–1 June 2018; pp. 607–612. [CrossRef]
16. Mori, K.; Ono, Y.; Watanabe, S.; Ishikawa, T.; Sugiyama, M.; Imasu, S.; Ochiai, T.; Mori, R.; Kida, T.; Hashimoto, T.; et al. High Density and Reliable Packaging Technology with Non Conductive Film for 3D/TSV. In Proceedings of the 2013 IEEE International 3D Systems Integration Conference (3DIC), San Francisco, CA, USA, 2–4 October 2013; pp. 1–7. [CrossRef]
17. Urteaga, M.; Carter, A.; Griffith, Z.; Pierson, R.; Bergman, J.; Arias, A.; Rowell, P.; Hacker, J.; Brar, B.; Rodwell, M.J.W. THz Bandwidth InP HBT Technologies and Heterogeneous Integration with Si CMOS. In Proceedings of the 2016 IEEE Bipolar/BiCMOS Circuits and Technology Meeting (BCTM), New Brunswick, NJ, USA, 25–27 September 2016; pp. 35–41. [CrossRef]
18. Weimann, N.; Monayakul, S.; Sinha, S.; Schmückle, F.-J.; Hrobak, M.; Stoppel, D.; John, W.; Krüger, O.; Doerner, R.; Janke, B.; et al. Manufacturable Low-Cost Flip-Chip Mounting Technology for 300–500-GHz Assemblies. *IEEE Trans. Compon. Packag. Manuf. Technol.* **2017**, *7*, 494–501. [CrossRef]
19. Lahiji, R.R.; Sharifi, H.; Katehi, L.P.B.; Mohammadi, S. 3-D CMOS Circuits Based on Low-Loss Vertical Interconnects on Parylene-N. *IEEE Trans. Microw. Theory Tech.* **2010**, *58*, 48–56. [CrossRef]
20. Watanabe, A.O.; Ali, M.; Sayeed, S.Y.B.; Tummala, R.R.; Pulugurtha, M.R. A Review of 5G Front-End Systems Package Integration. *IEEE Trans. Compon. Package. Manuf. Technol.* **2021**, *11*, 118–133. [CrossRef]
21. Zhang, Q.X.; Yu, A.B.; Yang, R.; Li, H.Y.; Guo, L.H.; Liao, E.B.; Tang, M.; Lo, G.-Q.; Balasubramanian, N.; Kwong, D.-L. Integration of RF MEMS and CMOS IC on a Printed Circuit Board for a Compact RF System Application Based on Wafer Transfer. *IEEE Trans. Electron Devices* **2008**, *55*, 2484–2491. [CrossRef]
22. Staiculescu, D.; Sutono, A.; Laskar, J. Wideband Scalable Electrical Model for Microwave/Millimeter Wave Flip Chip Interconnects. *IEEE Trans. Adv. Packag.* **2001**, *24*, 255–259. [CrossRef]
23. Maekawa, T.; Kanaya, H.; Suzuki, S.; Asada, M. Oscillation up to 1.92 THz in resonant tunneling diode by reduced conduction loss. *Appl. Phys. Express* **2016**, *9*, 024101. [CrossRef]
24. Lee, J.; Kim, M.; Lee, J. 692 GHz High-Efficiency Compact-Size InP-Based Fundamental RTD Oscillator. *IEEE Trans. THz Sci. Technol.* **2021**, *11*, 716–719. [CrossRef]
25. Lee, J.; Kim, M.; Park, J.; Lee, J. 225 GHz triple-push RTD oscillator with 0.5 mW dc-power consumption. *IET Circuits Devices Syst.* **2020**, *14*, 209–215. [CrossRef]

micromachines

Article

Numerical Study on the Cavitation Characteristics of Micro Automotive Electronic Pumps under Thermodynamic Effect

Kaipeng Wu [1,†], Asad Ali [1,†], Changhong Feng [2,*], Qiaorui Si [1], Qian Chen [3] and Chunhao Shen [1]

1 Research Center of Fluid Machinery Engineering and Technology, Jiangsu University, Zhenjiang 212013, China; siqiaorui@ujs.edu.cn (Q.S.); 2222011062@stmail.ujs.edu.cn (K.W.); 5103190316@stmail.ujs.edu.cn (A.A.); shenchunhaoujs@163.com (C.S.)
2 Feilong Auto Components Co., Ltd., Nanyang 474500, China
3 School of Electrical and Information Engineering, Jiangsu University, Zhenjiang 212013, China; chenqian0501@ujs.edu.cn
* Correspondence: fengch@flacc.com; Tel.: +86-135-0390-8161
† These authors contributed equally to this work.

Abstract: In order to study the influence of thermodynamic effects on the cavitation performance of hydromechanics, the Singhal cavitation model was modified considering the influence of the thermo-dynamic effects, and the modified cavitation model was written into CFX using the CEL language. Numerical simulation of the cavitation full flow field at different temperatures (25 °C, 50 °C and 70 °C) was carried out with the automotive electronic water pump as the research object. The results show that the variation trend of the external characteristic simulation and experimental values is the same at all flow rates, and the calculation accuracy meets the subsequent cavitation demand. With the increase in temperature, the low-pressure area inside the automotive electronic pump's impeller decreases. NPSHr decreases and the cavitation resistance is enhanced. During the process of no cavitation to cavitation, the maximum pressure pulsation amplitude in the impeller channel gradually increases. The generation and collapse of cavitations cause the change of pressure pulsation in the internal flow field, causing pump vibration.

Keywords: micro automotive electronic pump; computational fluid dynamics; numerical simulation; experimental techniques; cavitation; thermodynamic effects

Citation: Wu, K.; Ali, A.; Feng, C.; Si, Q.; Chen, Q.; Shen, C. Numerical Study on the Cavitation Characteristics of Micro Automotive Electronic Pumps under Thermodynamic Effect. *Micromachines* **2022**, *13*, 1063. https://doi.org/10.3390/mi13071063

Academic Editor: Stelios K. Georgantzinos

Received: 3 June 2022
Accepted: 29 June 2022
Published: 1 July 2022

1. Introduction

Automotive electronic pumps are core components of internal combustion engine cooling systems for drainage and irrigation products, which need to be operated at a high temperature and high speed for a long time. Because the physical parameters of coolant change with temperature, it is easy to induce cavitation of the pump and then affect the stable operation of the system. Cavitation is a flow phenomenon involving a phase change process in which the thermodynamic effect exists during the occurrence, development, and collapse of the bubble [1–5]. As the vaporization medium needs to absorb heat from the surrounding liquid, the temperature will have a certain effect on the physical parameters of the medium, which in turn affects the process of heat exchange during the vaporization process. Therefore, the existing cavitation model should be appropriately modified to predict the cavitation phenomenon accurately, and the source term considering the thermodynamic effect should be added.

Many scholars have studied cavitation flow under thermodynamic effects. In terms of experiments, Fruman et al. [6] proposed a cavitation correction model considering cavity heat transfer by analyzing the heat transfer process in the cavitation process, using R114 as the test object, and found that the cavitation performance improved with the increase of temperature. Franc et al. [7] made a cavitation test of Freon R-114 flowing through the induced wheel at different temperatures and investigated the effect of temperature on the thermodynamic effect of significant fluid cavitation. By comparing the experimental phenomena of different temperature water around NACA0015 hydrofoil at the same cavitation number, Cervone [8] found that the difference in the water vacuole size was not significant at 293 K and 323 K; however, the vacuole length decreased significantly at 343 K and gave a reasonable explanation for this phenomenon. Yoshida et al. [9,10] systematically investigated the thermal effect on the cavitation performance and cavitation instability of the induced wheel by using liquid nitrogen as the working medium. Gustavsson et al. [11,12] carried out a cavitation flow test of water and fluorinated ketone around NACA0015 hydrofoil separately and found that the cavitation length of fluorinated ketone was shorter than that of water at the same cavitation number and the cavitation thermal inhibition effect was more obvious.

With the development of computer technology and computational fluid dynamics [13–15], numerical simulation has become an essential tool for studying cavitation problems. Deshpande et al. [16] built on the Navier-Stokes equation with compressibility and pseudo-time-step progression by coupling the energy equation, and developed a thermodynamic model applicable to the cavitation of low-temperature fluids. Franc et al. [17] performed an analysis founded on the Rayleigh-Plesset equation, modified the cavitation model with temperature as one of the influencing factors of cavitation, and validated it accordingly. Hosangadi et al. [18] described a compressible multiphase flow formulation to account for the energy balance and variable thermodynamic properties of the fluid, identified fundamental changes in the physical characteristics of the cavity when thermal effects are significant, and suggested that the heat transfer model assumed by Hord in the B-factor model variant is a poorer approximation. Watanabe et al. [19] simulated the heat transfer process between the bubble and the work mass using a one-dimensional unsteady-state thermal conductivity model. They analyzed the flow using the singularity method. Using an agent model, Goel et al. [20] synthesized the evaporation and condensation coefficients and the degree of influence of the thermophysical parameters on the objective function in the cavitation model. They gave optimal equation coefficients for the cavitation model. Tani et al. [21] applied the B-factor theory to correct the saturated vapor pressure variation brought about by temperature differences in the cavitation process and proposed a cavitation model considering thermodynamic effects. Tseng et al. [22] established an interface-based cavitation model, modified the model according to the low-temperature fluid medium, and investigated the thermodynamic effects by solving the energy equation in the full basin and combining the available physical parameters to verify the validity of the calculation method and the model. To study thermodynamic effects in different fluids at different temperatures, De et al. [23] proposed a stable semi-one-dimensional model based on the Rayleigh-Plesset equation to study the internal cavitation considering thermodynamic effects. Research on cavitation under thermodynamic effects has made some achievements, but it mainly focuses on inducer and airfoil flow cavitation. The research on the internal flow pattern and cavitation characteristics under a thermodynamic effect in small centrifugal pumps such as automobile electronic pumps remains to be further studied.

This study is focused on the cavitation flow field characteristics at different operating points of a high-speed automotive electronic water pump at a rated speed (5400 r/min), modifying the Singhal cavitation model considering the influence of thermodynamic effects, and carrying out numerical simulations of the cavitation full flow field at different temperatures (25 °C, 50 °C and 70 °C) based on the modified cavitation model to investigate the cavitation performance, bubble distribution law and pressure pulsation characteristics of an automotive electronic water pump at different temperatures, providing a theoretical basis for preventing and reducing the cavitation phenomenon of an automotive electronic pump.

2. Three-Dimensional Model and Grid Division

2.1. Model Pump Parameters

In this study, an automotive electronic water pump with a specific speed of 81 is used as the object. The working environment temperature of the pump is −40 °C to 80 °C, which is a micro high-speed pump driven by a DC brushless motor and controlled by an electronic unit. Table 1 shows the main parameters of the automotive electronic pump.

Table 1. Automotive electronic water pump parameters.

Parameters	Value
Rated flow rate, Q_d	1.25 m^3/h
Rated head, H_d	7 54 m
Rated rotating speed, n	5400 r/min
Specific speed, n_s	81
Blade number, Z	5
Inlet diameter of impeller, D_j	17 mm
Outlet diameter of impeller, D_2	47 mm

The automotive electronic water pump structure uses a closed centrifugal impeller and spiral pressurized volute as the main hydraulic components and an integral molding of the rotor and motor rotor. The whole pump body is divided into a pump barrel, impeller, and motor base, as shown in Figure 1.

(a) (b) (c)

Figure 1. Structure diagram of automotive electronic water pump: (**a**) pump barrel, (**b**) impeller, and (**c**) motor base.

2.2. Division of Computational Grid

Due to the complex structure, ICEM CFD software was used for unstructured meshing of the computational domain of the main overcurrent components of the automotive electronic water pump, as shown in Figure 2. To consider the calculation cycle and its reliability of numerical calculation, grid-independent analysis was carried out, as shown in Figure 3. Since the previous literature [24,25] showed that if the absolute error of the predicted value of the two heads before and after is within 2%, then the influence of the mesh elements can be ignored. From Figure 3, when the grid number is less than 1.72 million, the grid number has a large influence on the head. When the grid number is larger than 1.72 million, the numerical calculation of the head varies by 1% with the increase of the grid number, and the final determination of the numerical calculation grid number is around 1.72 million. Grid information for parts of automotive electronic pumps is shown in Table 2.

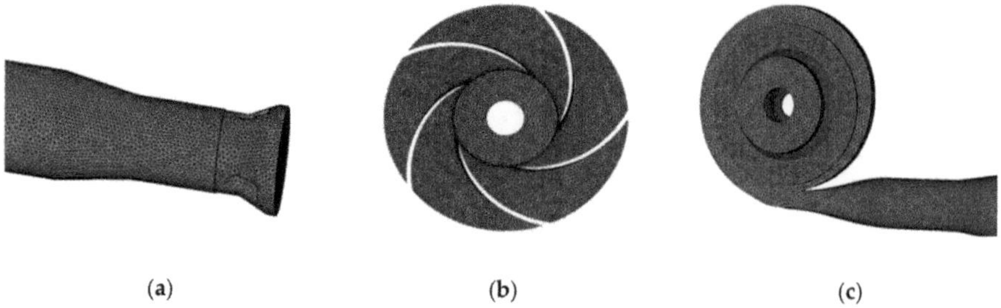

Figure 2. Grid of automotive electronic water pump: (**a**) inlet section, (**b**) impeller, and (**c**) volute and outlet.

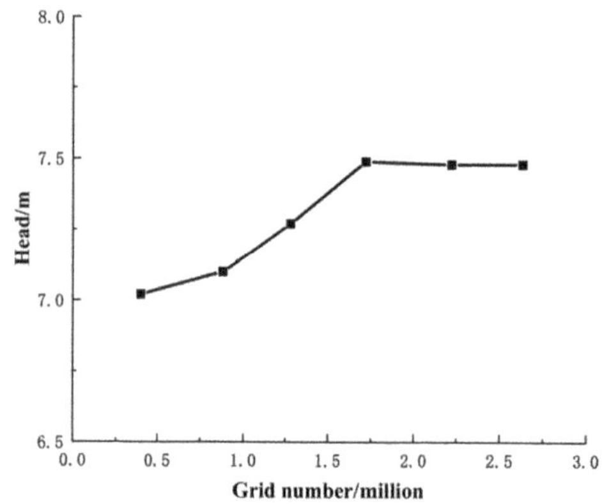

Figure 3. Grid independence analysis.

Table 2. Grid information of domains.

Domains	Inlet Section	Impeller	Volute and Outlet	Total
Grid number	328,403	562,914	833,778	1,725,095
Node number	57,584	96,006	143,978	297,568
Grid quality	0.34	0.32	0.26	/
y +	19.4145	38.9787	54.9447	/

3. Numerical Method

3.1. Turbulence Model

The corresponding Reynolds numbers at different temperatures in this study are 1.97×10^5 (25 °C), 1.95×10^5 (50 °C) and $1.93 \times 1\ 0^5$ (70 °C), which are high Reynolds number cases. Based on the research experience and the literature analysis [26], the RNG k-ε turbulence model is chosen to calculate the cavitation. Its control equation is:

$$\frac{\partial(\rho k)}{\partial t} + \nabla \cdot (\rho k u_i) = \nabla \cdot \left[\left(\mu + \frac{\mu_t}{\sigma_k}\right)\nabla k\right] + P_t - \rho\varepsilon \quad (1)$$

$$\frac{\partial(\rho\varepsilon)}{\partial t} + \nabla \cdot (\rho\varepsilon u_i) = C_{\varepsilon 1}\frac{\varepsilon}{k}P_t - C_{\varepsilon 2}\rho\frac{\varepsilon^2}{k} + \nabla \cdot [(\mu + \frac{\mu_t}{\sigma_\varepsilon})\nabla\varepsilon] \quad (2)$$

$$\mu_t = \frac{C_\mu \rho k^2}{\varepsilon} \quad (3)$$

where k is turbulent kinetic energy; ε is turbulent dissipation term; P_t is turbulent kinetic energy generation term; $C_{\varepsilon 1} = 1.42, C_{\varepsilon 2} = 1.68, C_\mu = 0.085, \sigma_k = 0.7179, \sigma_\varepsilon = 0.7179.$

3.2. Modification of Cavitation Model Considering Thermodynamic Effects

The commonly used cavitation simulation is based on the cavitation dynamics equation, namely the Rayleigh-Plesset equation [27]. The specific form is as follows:

$$\frac{p_B(t) - p_\infty(t)}{\rho_l} = R_B \frac{d^2 R_B}{dt^2} + \frac{3}{2}\left(\frac{dR_B}{dt}\right)^2 + \frac{4v_l}{R}\frac{dR_B}{dt} + \frac{2S}{\rho_l R} \tag{4}$$

The left side of the equation is the pressure-driven term, determined by the vacuum's boundary conditions. The right side of the equation shows the second-order motion term of the vacuole, the first-order motion term of the vacuole, the viscosity term, and the surface tension term, in that order, where R_B is the radius of the vacuole; $p_B(t)$ is the pressure inside the vacuole; $p_\infty(t)$ is the pressure at infinity; ρ_l is the density of the liquid; is the kinematic viscosity of the liquid phase; S is the surface tension coefficient.

Based on this equation, the Singhal cavitation model [28] is derived by neglecting the second-order term, the viscous term, and the effect of surface tension of the equation:

When $p < p_v$, liquid vaporizes into bubbles

$$m^+ = C_{vap} \frac{3\alpha_{nuc}(1 - \alpha_v)\rho_v}{R_B} \sqrt{\frac{2}{3}\frac{p_B(t) - p_\infty(t)}{\rho_l}} \tag{5}$$

When $p > p_v$, the cavity condenses into liquid

$$m^- = C_{con} \frac{3\alpha_v \rho_v}{R_B} \sqrt{\frac{2}{3}\frac{p_\infty(t) - p_B(t)}{\rho_l}} \tag{6}$$

where C_{vap} and C_{con} are the correction factors for the vaporization and condensation source terms, respectively, α_{nuc} is the volume fraction of the cavitation nucleus, and their values are $C_{vap} = 50$, $C_{con} = 0.01$, $\alpha_{nuc} = 5 \times 10^4$.

Singhal cavitation model is derived based on the isothermal assumption, ignoring the effect of thermodynamic effects in cavitation. When cavitation occurs, the liquid vaporization absorbs the latent heat of vaporization, resulting in a decrease in temperature near the vacuole, and a certain temperature difference is formed inside and outside the vacuole. The temperature difference affects the growth of the vacuole. Moreover, the Singhal cavitation model is modified in this paper to consider the thermodynamic effect of cavitation under different temperature conditions. The second-order term, viscous term, and surface tension of the Rayleigh-Plesset equation is neglected, and the Taylor series expansion of cavitation pressure is carried out to retain the first-order term. The second-order term and other terms after they are neglected [29], we can obtain the following:

$$\frac{dR_B}{dt} = \sqrt{\frac{2}{3}\left[\frac{p_B(t) - p_\infty(t)}{\rho_l} + \frac{dp_B}{dT}\frac{T_B - T_\infty}{\rho_l}\right]} \tag{7}$$

From the Clapeyron-Clausius equation, we can obtain:

$$\frac{dp_B}{dT} = \frac{\rho_l \rho_B}{T(\rho_l - \rho_B)} L \tag{8}$$

where L is the latent heat of vaporization.

Substituting Equation (8) into Equation (7) yields:

$$\frac{dR_B}{dt} = \sqrt{\frac{2}{3}\left[\frac{p_B(t) - p_\infty(t)}{\rho_l} + \left(\frac{\rho_l \rho_B}{T(\rho_l - \rho_B)} L\right)\frac{T_B - T_\infty}{\rho_l}\right]} \tag{9}$$

From Equations (5), (6) and (9), the condensation source term and the vaporization source term of the cavitation model considering thermodynamic effects are:

$$m^{+} = C_{vap}\frac{3\alpha_{nuc}(1 - \alpha_{v})\rho_{v}}{R_{B}}\sqrt{\frac{2}{3}\left[\frac{p_{B}(t) - p_{\infty}(t)}{\rho_{l}} + \left(\frac{\rho_{l}\rho_{B}}{T(\rho_{l} - \rho_{B})}L\right)\frac{T_{B} - T_{\infty}}{\rho_{l}}\right]} \quad (10)$$

$$m^{-} = C_{con}\frac{3\alpha_{v}\rho_{v}}{R_{B}}\sqrt{\frac{2}{3}\left[\frac{p_{\infty}(t) - p_{B}(t)}{\rho_{l}} + \left(\frac{\rho_{l}\rho_{B}}{T(\rho_{l} - \rho_{B})}L\right)\frac{T_{B} - T_{\infty}}{\rho_{l}}\right]} \quad (11)$$

3.3. Boundary Condition Setting

CFX software was used to simulate the flow field of the automotive electronic water pump at different temperatures with pure water as the medium for single-phase steady simulation and two-phase flow simulation with the addition of cavitation model. The RNG k-ε turbulence model was used for each phase of the simulation, the simulation speed was 5400 r/min, the inlet and outlet boundary conditions were set for the total pressure inlet and mass flow outlet, the cross-interface connection of GGI was used, and the convergence accuracy of all simulations was set to 1×10^{-4}.

For the single-phase stability simulation, the dynamic-static interface is set as the frozen rotor and the number of iterations is 1000. For steady cavitation simulation, the modified cavitation model is written into the CFX using the CEL language, pure water at the corresponding temperature is added as the working medium, and water vapor is added as the bubble generated during cavitation. The calculation sets the volume fraction of the liquid phase at the inlet boundary to 1 and the volume fraction of the vapor phase to 0. The inlet pressure is adjusted downwards by 0.2 atm for each simulation, and the change in the head is compared to determine the inlet pressure at which cavitation occurs. The physical parameters of pure water and water vapor at different temperatures are shown in Table 3. It can be seen that the temperature has a large effect on the physical parameters of water density, dynamic viscosity, and vaporization pressure, and the value of each physical parameter is smaller for water vapor compared with water.

Table 3. Physical parameters of water and water vapor at different temperatures.

Physical Parameters	25 °C		50 °C		70 °C	
	Water	Water Vapor	Water	Water Vapor	Water	Water Vapor
Density (kg/m^3)	997	0.02308	988.1	0.08302	977.8	0.1982
dynamic viscosity (10^5 Pa s)	8899	0.98626	54.94	1.002	40.16	1.0817
Constant pressure specific heat capacity (kJ/(kg K))	4.1817	1.9116	4.174	1.9343	4178	1962.7
Thermal conductivity(w/(m K))	0.6069	0.01854	0.6478	0.02182	0.6676	0.02346
Vaporization pressure (kPa)	3.1684	/	12.335	/	31.160	/

To obtain a stable and reliable solution, the unsteady simulation of the cavitation flow field takes its steady simulation results at the corresponding temperature as the initial conditions. The dynamic-static intersection is modified to a transient rotor stator in the calculation; the impeller rotation of 360° is taken as a calculation cycle, the time step is set to 6.17×10^{-5} s (calculated for every 2° rotation of the impeller), the number of inner loops at each time step is 20, and the total calculation time is set to 0.2 s (18 rotation periods). The Courant number is used as a criterion to determine whether the time step satisfies the periodic numerical simulation, and is defined as:

$$C_{0} = v\frac{\Delta t}{L} < 100 \quad (12)$$

where L is the smallest size of the grid cell, v is the biggest velocity of the main flow, and Δt is the time step. The maximum C_0 obtained by the calculation is 5.23, which satisfies the independence of the time step.

3.4. Monitoring Point Setting

To study the pressure variation inside the impeller of an automotive electronic water pump during cavitation under thermal effects, 12 monitoring points are set up at the impeller position of the pump, as shown in Figure 4.

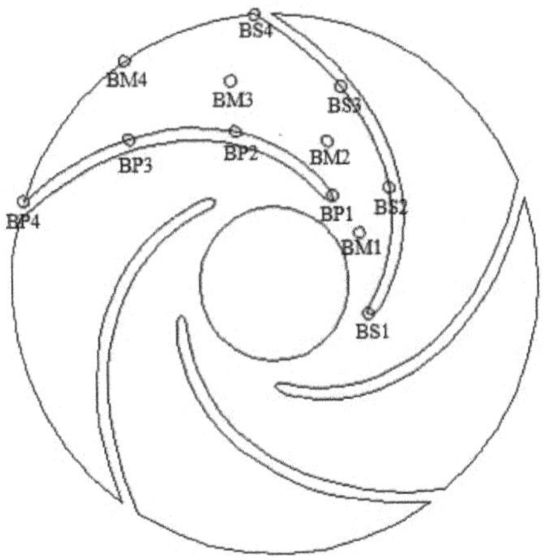

Figure 4. Monitoring points position in an automotive electronic water pump.

4. Analysis of Internal Flow of Automotive Electronic Water Pump under Non-Cavitation Condition

4.1. Verification of External Characteristics by Numerical Calculation

The external characteristic curve is the external expression of the internal flow characteristics of the automotive electronic water pump. Numerical calculations are performed for the automotive electronic water pump at different flow rates, and tests verify the reliability of the numerical calculation results. The performance test of an automotive electronic water pump is carried out concerning GB/T 3216–2016, and the test bench is shown in Figure 5.

Using the head coefficient ψ to calculate the external characteristics, the head coefficient can be calculated as the following equation:

$$\psi = \frac{gH}{u_2^2/2} \tag{13}$$

where u_2 is the circumferential exit speed of the impeller.

Figure 6 shows a comparison of the external characteristics results. As can be seen from the figure, the experimental head coefficient and efficiency are slightly lower than the numerical calculation due to the simplification of the model during the numerical calculation for the same flow conditions. At a small flow rate ($0.2Q_d$), the difference between the simulated and experimental values of the head coefficient is 0.06; at $0.8Q_d$~$1.6Q_d$, the two curves are in good agreement, and it can be considered that the numerical calculation method can accurately predict the external characteristics of the pump and ensure the accuracy of the subsequent analysis.

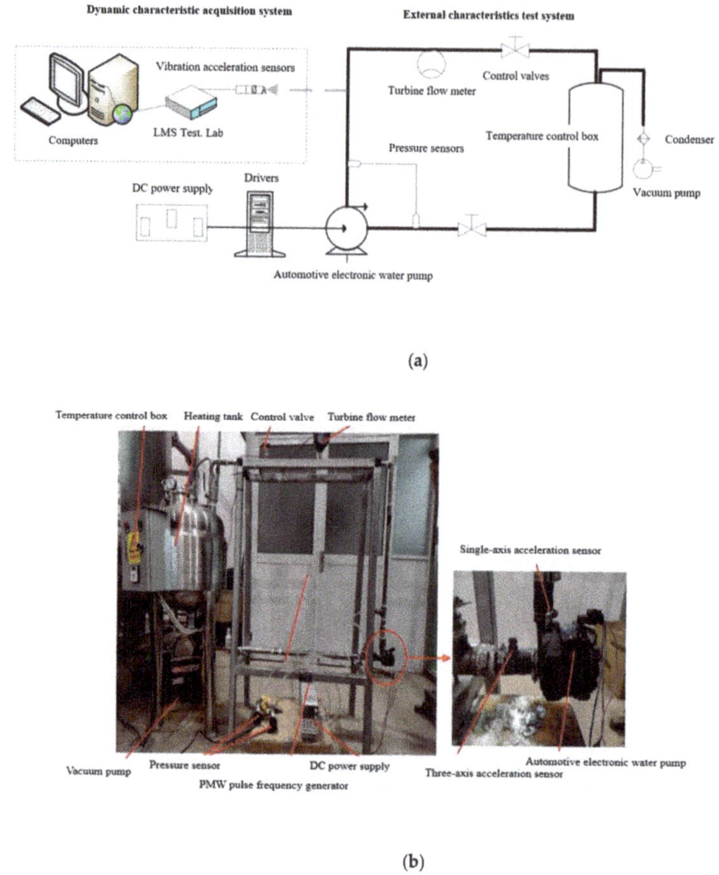

Figure 5. Test bench for an automotive electronic water pump: (**a**) experimental schematic, (**b**) experimental loop.

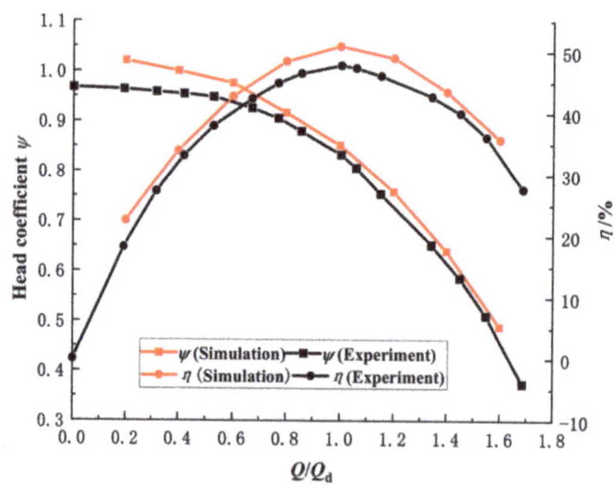

Figure 6. Comparison of pump performance between numerical and experimental results.

4.2. Pressure Field Distribution in the Pump at Different Temperatures

Figure 7 shows the static pressure distribution clouds of the impeller of the automotive electronic water pump under four different flow conditions at 25 °C, 50 °C, and 70 °C (speed of 5400 r/min). The static pressure distribution at different temperatures is similar; from the blade inlet to the blade outlet, the pressure in the impeller is of a gradient-type growth, and the pressure distribution is more uniform. As the flow rate increases, the blade suction surface pressure gradually decreases, and the low-pressure area gradually increases. Under the large flow condition ($1.2Q_d$), the area of the low-pressure area at the inlet gradually decreases as the temperature increases.

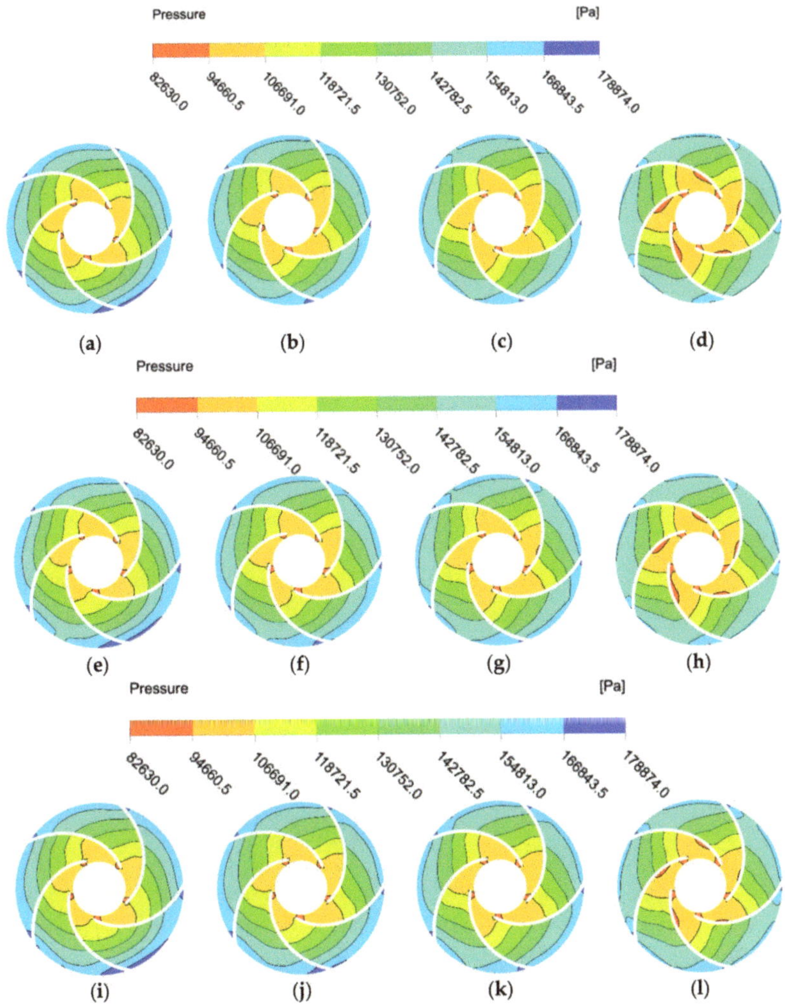

Figure 7. Pressure distribution of impeller under at different temperatures. (**a**) $0.6Q_d$ at 25 °C; (**b**) $0.8Q_d$ at 25 °C; (**c**) $1.0Q_d$ at 25 °C; (**d**) $1.2Q_d$ at 25 °C; (**e**) $0.6Q_d$ at 50 °C; (**f**) $0.8Q_d$ at 50 °C; (**g**) $1.0Q_d$ at 50 °C; (**h**) $1.2Q_d$ at 50 °C; (**i**) $0.6Q_d$ at 70 °C; (**j**) $0.8Q_d$ at 70 °C; (**k**) $1.0Q_d$ at 70 °C; (**l**) $1.2Q_d$ at 70 °C.

5. Automotive Electronic Water Pump Cavitation Performance Analysis

5.1. Computational Analysis of Cavitation Performance Considering Thermodynamic Effects

The net positive suction head is the surplus energy per unit weight of water at the pump suction inlet that exceeds the vaporization pressure and is expressed as follows:

$$\text{NPSHa} = \frac{p_1 - p_v}{\rho g} \tag{14}$$

where p_1 is the automotive electronic water pump inlet pressure; p_v is the saturation vapor pressure of water at that temperature; ρ is the fluid density.

Figure 8 shows the cavitation performance curves of the simulated and tested automotive electronic water pump at a temperature of 25 °C (flow rate of 1.25 m^3/h and speed of 5400 r/min). As can be seen from the figure, the head coefficient of the test value is smaller than the simulated value under the same inlet conditions, which are related to the existence of friction between the fluid and the pipe wall and energy loss between the motor and the impeller. Overall, the simulated and experimental cavitation performance curves have the same stagnation point position and trend. It can be considered that the numerical calculation method can accurately predict the cavitation performance of the pump and ensure the accuracy of subsequent cavitation calculations.

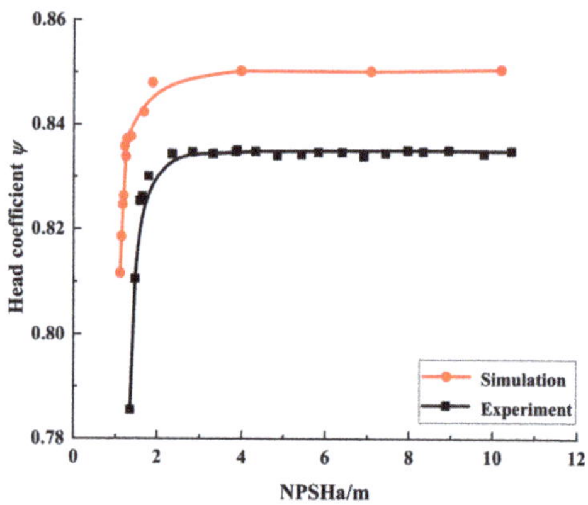

Figure 8. Comparison of cavitation performance curves between numerical and experimental results.

Figure 9 shows the cavitation performance curves of the simulated automotive electronic water pump at three different temperatures (flow rate of 1.25 m^3/h and speed of 5400 r/min). We define the automotive electronic water pump head drop of 1% as incipient cavitation, a head drop of 3% as cavitation, and the NPSH corresponding to the head drop of 3% as the required net positive suction head of the pump (NPSHr). The NPSHrs at three temperatures were 1.15 m at 25 °C, 1.01 m at 50 °C, and 0.91 m at 70 °C. As can be seen from the figure, with the decrease of NPSHa, the head coefficient maintains a level for a period of time and then rapidly decreases, which is because the flow pattern inside the automotive electronic water pump is turbulent and sensitive to cavitation, and the hydraulic performance deteriorates sharply once cavitation occurs; with the increase of temperature, NPSHr gradually decreases, and the anti-cavitation performance of the automotive electronic water pump gradually increases.

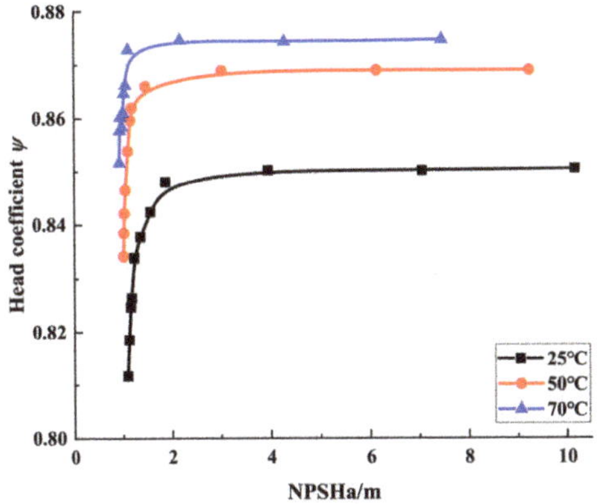

Figure 9. Cavitation performance curves with different temperatures.

5.2. Cavitation Steady Analysis Considering Thermodynamic Effects

5.2.1. Bubble Shape during Cavitation Formation

The bubble is the most visual product when cavitation occurs, and the volume fraction of the bubble can judge the degree of cavitation. Figure 10 shows the distribution of bubbles in the impeller at different NPSHa for a bubble volume concentration of 10% at a temperature of 25 °C (flow rate of 1.25 m^3/h and speed of 5400 r/min) for an automotive electronic water pump. It can be seen from the figure that when the NPSHa is larger, no bubble is generated in the impeller; when the NPSHa is reduced to 1.68 m, a small number of bubbles appears at the inlet side of the impeller near the suction surface of the blade, which is the cavitation inception stage. As the NPSHa gradually decreases, the distribution area of the bubble starts to spread from the suction surface to the pressure surface, and cavitation further develops; when the NPSHa decreases to NPSHr, the bubbles gradually increase and develop to the pressure surface and start to block the impeller flow channel, and the high-speed jet caused by the rupture of the bubbles will damage the impeller and produce cavitation, resulting in a serious drop in the head.

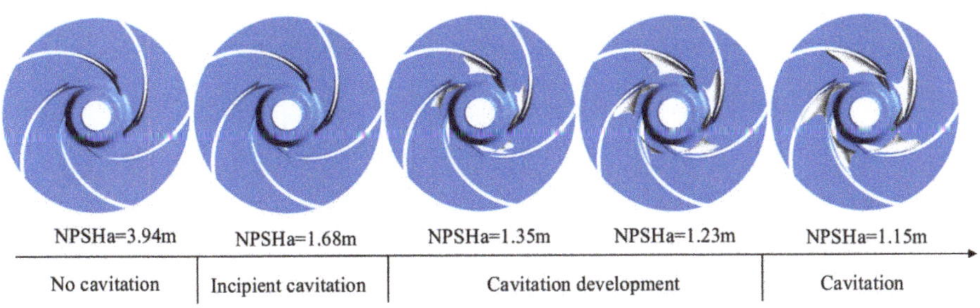

Figure 10. Bubble distribution during the cavitation formation process.

5.2.2. Analysis of the Pressure Field during Cavitation Formation

The occurrence of cavitation is pressure-dependent, and its distribution is influenced by the law of hydrostatic pressure distribution. Figure 11 shows cloud plots of static pressure distributions at different NPSHas of an automotive electronic water pump at a temperature of 25 °C (flow rate 1.25 m^3/h, speed 5400 r/min). As can be seen from the figure, from the blade inlet to the blade outlet, the pressure gradually increased; five blades on the static pressure distribution show a similar pattern in which the static pressure of the fluid along with the fluid flow direction increased. In the blade outlet pressure, it reached the maximum. The range of local low-pressure areas gradually expands with the decrease of NPSHa. In the no cavitation and incipient cavitation stage, the influence of NPSHa on the size of the low-pressure area range is small; from the cavitation development stage to the cavitation stage, the influence of NPSHa size on the size of the low-pressure area range at the impeller inlet of the automotive electronic water pump is large. Because of the low pressure at the impeller inlet, cavitation generally occurs first from the suction surface of the impeller inlet blade.

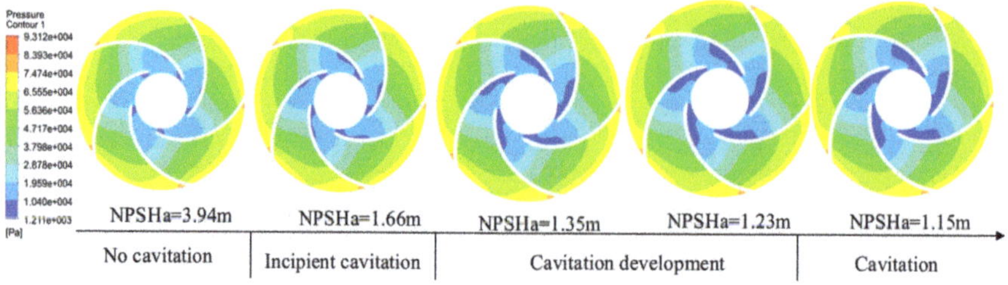

Figure 11. Pressure distribution of bubble formation process.

5.2.3. Cavitation Flow Field Analysis

Figure 12 shows the bubble distribution inside the impeller at three temperatures when the NPSHa is 1.15 m (flow rate 1.25 m^3/h, speed 5400 r/min). At 70 °C, when the inlet pressure is 0.407 atm, some bubbles are produced near the suction surface of some of the impeller's blades, and the pump head does not drop significantly at this time. At 50 °C, when the inlet pressure is 0.221 atm, the bubble area becomes larger, and all impeller blades produce bubbles and development to the middle of the radial direction of the impeller; at this time, the pump head drops by 1.5%. It can be seen that the bubbles generated at this time are enough to impact the performance of the pump. At 25 °C, when the inlet pressure is 0.131 atm, the bubbles cluster continues to increase and begins to block the flow channel, at which time the pump head decreases by 3%, and cavitation occurs. At 25 °C, the low-pressure area at the impeller inlet suction surface is larger, and as the temperature rises, the area of the low-pressure area becomes smaller and smaller, and the area of the corresponding bubbles also gradually decreases. In addition, both the low-pressure area and the bubble cluster are axially distributed.

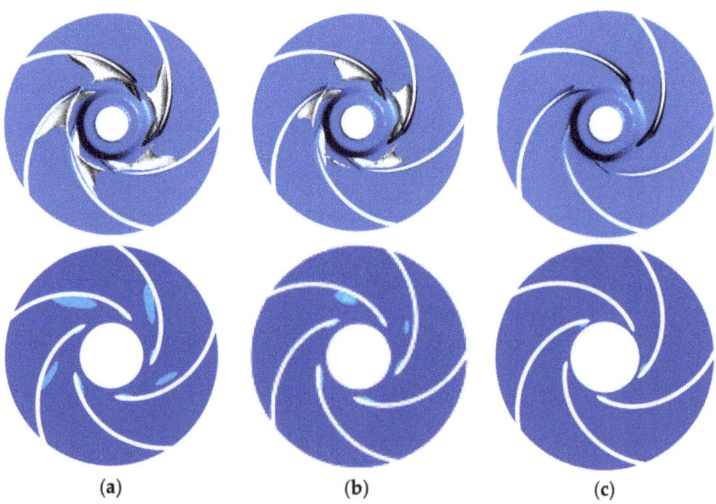

Figure 12. Bubble distribution with different temperatures based on the same. (**a**) T = 25 °C; (**b**) T = 50 °C; (**c**) T = 70 °C.

Combining Figures 8 and 9, it is found that the number of bubbles is higher during the stage of a sharp decline in the head, and the bubbles have a certain influence on the head; the bubbles are all generated from near the low-pressure area at the blade inlet and develop along with the impeller outlet, and the temperature has a certain influence on the distribution of bubble in the automotive electronic water pump. The bubble distribution inside the impeller is consistent with the cavitation performance curve: the more bubbles are generated, the more obvious the drop of its head.

5.3. Cavitation Transient Analysis Considering Thermodynamic Effects

5.3.1. Transient Bubble in the Impeller

Figure 13 shows the distribution of bubbles at a temperature of 25 °C and an inlet pressure of 0.131 atm for an impeller with a vapor volume concentration of 10% at six different moments in a cycle. It can be seen from the figure that at the moment t = 0, the suction surface in each flow channel adheres to a continuous bubble, and the shape of the bubble is of an irregular rugby ball with a wide middle and narrow front and back; as the blades are periodically arranged, with the impeller rotation, the bubble change in each flow channel has a similar pattern. Take runner No. 1 as an example: at $t = 2/6$ T, the bubble in the middle of the flow channel will be detached; at $t = 3/6$ T, the volume of the bubble in runner No. 1 is smaller than at $t = 2/6$ T. At this time, the bubble has detached and collapsed in the high-pressure area; with the development of time, at $t = 3/6$ T to $t = 5/6$ T, the bubble in runner No. 1 slowly develops toward the end and detaches. On the whole, the detachment and collapse of the bubble occurred in the middle and tail of the bubble cluster, where the middle bubble detached less and the tail detached more; the changes in the position and shape of the bubble were relatively small, indicating that the whole process of cavitation was relatively stable at this time.

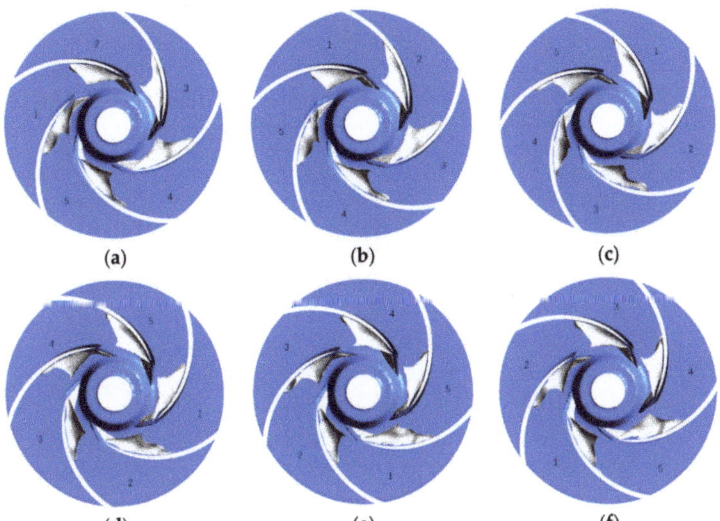

Figure 13. Bubble distribution in the impeller domain. (**a**) $t = 0$; (**b**) $t = 1/6$ T; (**c**) $t = 2/6$ T; (**d**) $t = 3/6$ T; (**e**) $t = 4/6$ T; (**f**) $t = 5/6$ T.

5.3.2. Comparative Analysis of Pressure Pulsation in the Frequency Domain

To analyze the relative pressure variation at each monitoring point within the impeller, we use a pressure coefficient expressed as:

$$C_p^* = \frac{p - \overline{p}}{0.5\rho u_2^2} \tag{15}$$

where p is the instantaneous pressure, $\overline{p}$ is the average pressure, ρ is the fluid density, and u_2 is the circumferential velocity at the impeller outlet. $\overline{p}$ is the average pressure of the last eight cycles simulated at the monitoring points in this study.

Figure 14 shows the frequency domain of the pressure pulsation at the monitoring point inside the impeller at 25 °C (flow rate 1.25 m^3/h, speed 5400 r/min) during a no cavitation operating condition of the automotive electronic water pump. As can be seen from the figure, along with the impeller runner from the blade inlet to the blade outlet, the pressure pulsation amplitude at each monitoring point gradually increases, the amplitude frequency is the axial frequency and its multiplier frequency, and the pressure pulsation amplitude reaches the maximum at the axial frequency. At the axial frequency, the pressure pulsation amplitude from the blade suction surface to the blade pressure surface pressure pulsation amplitude gradually increases; at the impeller runner inlet, the blade suction surface (BS1) and pressure surface (BP1) pressure pulsation amplitude is higher than the middle of the impeller runner (BM1), which is because at the impeller inlet, the three-jaw ring disturbs the incoming flow field and intensifies the pressure fluctuation at the impeller inlet.

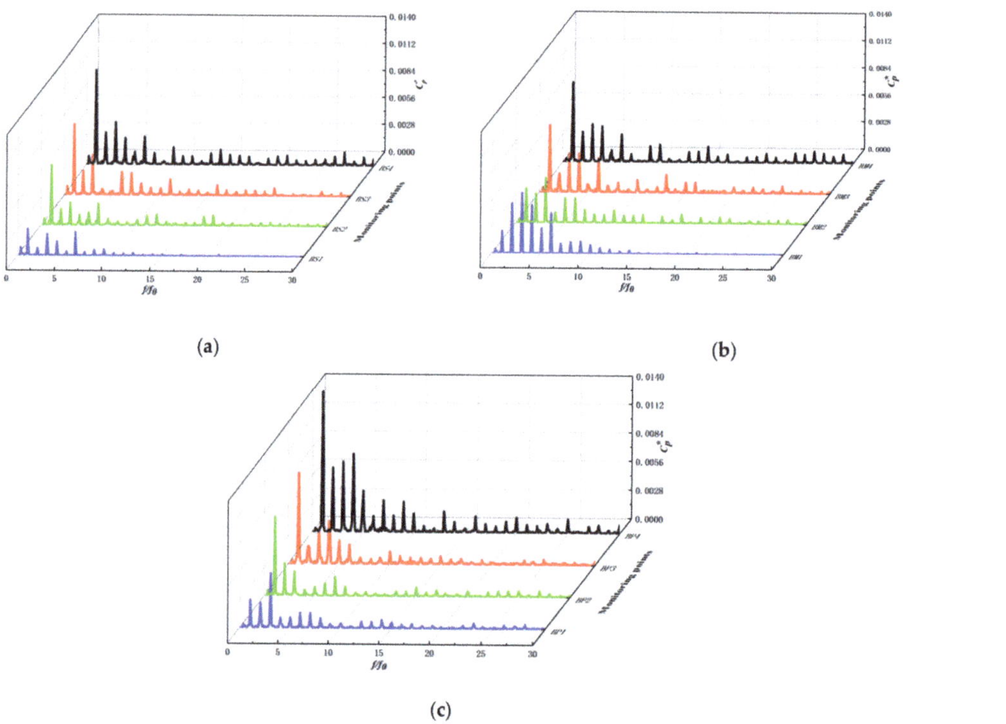

Figure 14. Frequency characteristic in the impeller. (**a**) Blade suction surface; (**b**) Middle of the flow channel; (**c**) Blade pressure surface.

Figure 15 shows the maximum values of pressure pulsations at the monitoring points inside the impeller during no cavitation, incipient cavitation, and cavitation of the automotive electronic water pump. The maximum amplitude value increases gradually along the fluid flow direction, except for individual monitoring points, and reaches the maximum value at the exit of each working surface. Overall, the maximum value of pressure pulsation during cavitation is higher on the blade working surface than no cavitation, which may be related to the generation and collapse of bubbles.

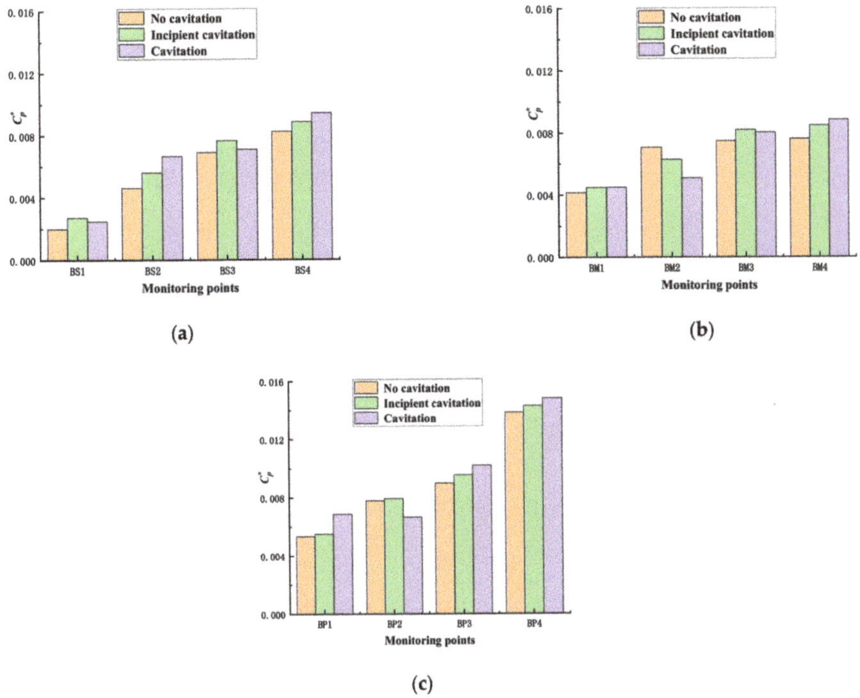

Figure 15. The maximum amplitude of pressure fluctuations in the impeller. (**a**) Blade suction surface; (**b**) Middle of the flow channel; (**c**) Blade pressure surface.

6. Conclusions

In this study, we have considered the influence of thermodynamic effect on the cavitation performance of hydromechanics, corrected the Singhal cavitation model considering a thermodynamic effect, written the corrected model into CFX software for numerical simulation of cavitation of automotive electronic water pump, explored the cavitation performance and the distribution pattern of cavitation bubble of automotive electronic water pump under different temperature (25 °C, 50 °C, 70 °C), and drawn the following conclusions:

1. The static pressure distribution of the automotive electronic water pump is similar at different temperatures. When operating under different flow conditions, there is a local low-pressure area on the suction surface of the blade that is more prone to cavitation, and the range of the low-pressure area gradually increases as the flow rate increases. Under the large flow condition ($1.2Q_d$), the area of the low-pressure area at the inlet gradually decreases as the temperature increases.
2. An automotive electronic water pump is sensitive to cavitation; once cavitation occurs will lead to a sharp deterioration in hydraulic performance. The required net positive suction head will decrease with the increase in temperature; the higher the temperature of the automotive electronic water pump's anti-cavitation performance, the less likely cavitation is.
3. When the inlet pressure of the automotive electronic water pump is large, no cavitation is generated inside the impeller. When the inlet pressure decreases to a certain value, the bubble starts to be generated on the suction surface of the impeller. It gradually increases with the decrease of the inlet pressure. The impeller's five blades in the low-pressure area and the bubble cluster show the same pattern, and the expansion of the bubble cluster and the expansion in the low-pressure area of the impeller show

the same trend. The pressure is closely related to the generation and development of cavitation.

4. Along with the impeller runner from the blade inlet to the blade outlet, each monitoring point's pressure pulsation amplitude gradually increased, and the amplitude frequency of the axial frequency and its multiplier frequency, as well as the pressure pulsation amplitude at the axial frequency, reached the maximum. In the high-frequency region, the pressure pulsation amplitude at the blade suction surface, the middle of the flow channel, and the blade pressure surface gradually increases along the inlet to the outlet. From no cavitation to cavitation, the maximum value of the pressure pulsation at the working surface gradually increases and reaches a maximum at the outlet.
5. This study is focused on the cavitation flow field characteristics at different operating points of a high-speed automotive electronic water pump at a rated speed (5400 r/min). In order to better prevent and reduce the cavitation phenomenon of the automotive electronic pump, our future research will focus on the cavitation monitoring technology, and we will establish a cavitation intelligent monitoring system to monitor the operating status of the automotive electronic water pump in real time.

Author Contributions: Data curation, K.W.; Validation, A.A. and C.S.; Formal analysis, Q.C.; Resources, Q.S.; Funding acquisition, Q.S.; Investigation, C.S and K.W.; Supervision, C.F.; Writing—original draft, K.W.; Writing—review and editing, A.A. All authors have read and agreed to the published version of the manuscript.

Funding: This research was funded by the National Natural Science Foundation of China (51976079, 12002136), the financial support of the National Key R&D Program of China (2020YFC1512403), and the Research Project of State Key Laboratory of Mechanical System and Vibration (MSV202201).

Institutional Review Board Statement: Not applicable.

Informed Consent Statement: Not applicable.

Data Availability Statement: The data supporting this study's findings are available within the article. The data pressented in this study are available on request from the corresponding author.

Acknowledgments: The authors gratefully acknowledge the National Natural Science Foundation of China (51976079, 12002136), the financial support of the National Key R&D Program of China (2020YFC1512403), and the Research Project of State Key Laboratory of Mechanical System and Vibration (MSV202201).

Conflicts of Interest: The authors declare no conflict of interest.

References

1. Utturkar, Y.; Wu, J.; Wang, G.; Shyy, W. Recent progress in modeling of cryogenic cavitation for liquid rocket propulsion. *Prog. Aerosp. Sci.* **2005**, *41*, 558–608. [CrossRef]
2. Li, Y.; Feng, G.; Li, X.; Si, Q.; Zhu, Z. An experimental study on the cavitation vibration characteristics of a centrifugal pump at normal flow rate. *J. Mech. Sci. Technol.* **2018**, *32*, 4711–4720. [CrossRef]
3. Tao, R.; Xiao, R.; Wang, F.; Liu, W. Cavitation behavior study in the pump mode of a reversible pump-turbine. *Renew. Energy* **2018**, *125*, 655–667. [CrossRef]
4. Fang, Y.; Zhang, J.; Xu, B.; Huang, C.S.; Guo, Z. A study on increasing the speed limit of axial piston pumps by optimizing the suction duct. *Chin. J. Mech. Eng.* **2021**, *34*, 1–10. [CrossRef]
5. Mao, Z.B.; Asai, Y.; Winranata, A.; Kong, D.Q.; Man, J. Eccentric actuator driven by stacked electrohydrodynamic pumps. *J. Zhejiang Univ. Sci. A* **2022**, *23*, 6. [CrossRef]
6. Fruman, D.H.; Reboud, J.-L.; Stutz, B. Estimation of thermal effects in cavitation of thermo sensible liquids. *Int. J. Heat Mass Transf.* **1999**, *42*, 3195–3204. [CrossRef]
7. Franc, J.-P.; Rebattet, C.; Coulon, A. An Experimental Investigation of Thermal Effects in a Cavitating Inducer. *J. Fluids Eng.* **2004**, *126*, 716. [CrossRef]
8. Cervone, A.; Testa, R.; Bramanti, C.; Rapposelli, E.; D'Agostino, L. Thermal Effects on Cavitation Instabilities in Helical Inducers. *J. Propuls. Power* **2005**, *21*, 893–899. [CrossRef]
9. Yoshida, Y.; Kikuta, K.; Hasegawa, S.; Shimagaki, M.; Tokumasu, T. Thermodynamic Effect on a Cavitating Inducer in Liquid Nitrogen. *J. Fluids Eng.* **2007**, *129*, 273. [CrossRef]

10. Kikuta, K.; Yoshida, Y.; Watanabe, M.; Hashimoto, T.; Nagaura, K.; Ohira, K. Thermodynamic Effect on Cavitation Performances and Cavitation Instabilities in an Inducer. *J. Fluids Eng.* **2008**, *130*, 111302. [CrossRef]
11. Gustavsson, J.P.R.; Denning, K.C.; Segal, C. Hydrofoil Cavitation Under Strong Thermodynamic Effect. *J. Fluids Eng.* **2008**, *130*, 091303. [CrossRef]
12. Gustavsson, J.P.R.; Denning, K.C.; Segal, C.; Dorney, D.J. Incipient Cavitation Studied Under Strong Thermodynamic Effect. *AIAA J.* **2009**, *47*, 710–716. [CrossRef]
13. Si, Q.; Asad, A.; Wang, B.; Peng, W.; Gerard, B.; Yuan, J.; Ameer, A. Numerical study on gas-liquid two phase flow characteristic of multistage electrical submersible pump by using a novel multiple-size group (MUSIG) model. *Phys. Fluids* **2022**, *34*, 063311. [CrossRef]
14. Ali, A.; Yuan, J.; Deng, F.; Wang, B.; Liu, L.; Si, Q.; Buttar, N.A. Research Progress and Prospects of Multi-Stage Centrifugal Pump Capability for Handling Gas-Liquid Multiphase Flow: Comparison and Empirical Model Validation. *Energies* **2021**, *14*, 896. [CrossRef]
15. Si, Q.; Ali, A.; Yuan, J.; Fall, I.; Muhammad Yasin, F. Flow-Induced Noises in a Centrifugal Pump: A Review. *Sci. Adv. Mater.* **2019**, *11*, 909. [CrossRef]
16. Deshpande, M.; Feng, J.; Merkle, C.L. Numerical Modeling of the Thermodynamic Effects of Cavitation. *J. Fluids Eng.* **1997**, *119*, 420. [CrossRef]
17. Franc, J.-P.; Pellone, C. Analysis of Thermal Effects in a Cavitating Inducer Using Rayleigh Equation. *J. Fluids Eng.* **2007**, *129*, 974. [CrossRef]
18. Hosangadi, A.; Ahuja, V. Numerical Study of Cavitation in Cryogenic Fluids. *J. Fluids Eng.* **2005**, *127*, 267. [CrossRef]
19. Watanabe, S.; Hidaka, T.; Horiguchi, H.; Furukawa, A.; Tsujimoto, Y. Steady Analysis of the Thermodynamic Effect of Partial Cavitation Using the Singularity Method. *J. Fluids Eng.* **2007**, *129*, 121. [CrossRef]
20. Goel, T.; Thakur, S.; Haftka, R.T.; Shyy, W.; Zhao, J. Surrogate model-based strategy for cryogenic cavitation model validation and sensitivity evaluation. *Int. J. Numer. Methods Fluids* **2008**, *58*, 969–1007. [CrossRef]
21. Tani, N.; Tsuda, S.; Yamanishi, N.; Yoshida, Y. Development and validation of new cryogenic cavitation model for rocket turbopump inducer. In Proceedings of the CAV2009, Ann Arbor, MI, USA, 17–22 August 2009; p. 63.
22. Tseng, C.-C.; Shyy, W. Modeling for isothermal and cryogenic cavitation. *Int. J. Heat Mass Transf.* **2010**, *53*, 513–525. [CrossRef]
23. De Giorgi, M.G.; Bello, D.; Ficarella, A. Analysis of Thermal Effects in a Cavitating Orifice Using Rayleigh Equation and Experiments. *J. Eng. Gas Turbines Power* **2010**, *132*, 092901. [CrossRef]
24. Ali, A.; Si, Q.; Wang, B.; Yuan, J.; Wang, P.; Rasool, G.; Shokrian, A.; Ali, A.; Zaman, M.A. Comparison of empirical models using experimental results of electrical submersible pump under two-phase flow: Numerical and empirical model validation. *Phys. Scr.* **2022**, *97*, 065209. [CrossRef]
25. Ali, A.; Si, Q.; Yuan, J.; Shen, C.; Cao, R.; Saad AlGarni, T.; Awais, M.; Aslam, B. Investigation of energy performance, internal flow and noise characteristics of miniature drainage pump under water-air multiphase flow: Design and part load conditions. *Int. J. Environ. Sci. Technol.* **2021**, 1–18. [CrossRef]
26. Deng, Y.; Feng, J.; Wan, F.; Shen, X.; Xu, B. Evaluation of the Turbulence Model Influence on the Numerical Simulation of Cavitating Flow with Emphasis on Temperature Effect. *Processes* **2020**, *8*, 997. [CrossRef]
27. Rayleigh, L. VIII. On the pressure developed in a liquid during the collapse of a spherical cavity. *Lond. Edinb. Dublin Philos. Mag. J. Sci.* **1917**, *34*, 94–98. [CrossRef]
28. Singhal, A.K.; Athavale, M.M.; Li, H.; Jiang, Y. Mathematical Basis and Validation of the Full Cavitation Model. *J. Fluids Eng.* **2002**, *124*, 617. [CrossRef]
29. Brennen, C.E. *Cavitation and Bubble Dynamics*; Oxford University Press: Oxford, UK, 1995.

micromachines

MDPI

Article

Electrochemical Milling of Deep-Narrow Grooves on GH4169 Alloy Using Tube Electrode with Wedged End Face

Zhisen Ye [1], Guilin Qiu [1] and Xiaolei Chen [1,2,*]

1 Guangzhou Key Laboratory of Nontraditional Machining and Equipment, Guangdong University of Technology, Guangzhou 510006, China; yezhisen_gdut@163.com (Z.Y.); ql2268554299@163.com (G.Q.)
2 State Key Laboratory of Precision Electronic Manufacturing Technology and Equipment, Guangdong University of Technology, Guangzhou 510006, China
* Correspondence: xlchen@gdut.edu.cn; Tel./Fax: +86-20-3932-2412

Abstract: Deep-narrow grooves (DNGs) of nickel-based alloy GH4169 are extensively used in aerospace industry. Electrochemical milling (EC-milling) can manufacture special structures including DNGs by controlling the moving path of simple tool, showing a flexible process with the advantages of high machining efficiency, regardless of material hardness, no residual stresses, burrs, and tool wear. However, due to the inefficient removal of electrolytic by-products in the inter-electrode gap (IEG), the machining accuracy and surface quality are always unsatisfactory. In this paper, a novel tube tool with wedged end face is designed to generate pulsating flow field in IEG, which can enhance the removal of electrolytic by-products as well as improve the machining quality of DNG. The flow field simulation results show that the electrolyte velocity in the IEG is changed periodically along with the rotation of the tube tool. The pulsating amplitude of electrolyte is changed by adjusting the wedged angle in the end face of the tube tool, which could affect the EC-milling process. Experimental results suggest that the machining quality of DNG, including the average width, taper of sidewall, and surface roughness, is significantly improved by using the tube tool with wedged end face. Compared with other wedged angles, the end face with the wedged angle of 40° is more suitable for the EC-milling process. DNG with the width of 1.49 mm ± 0.04 mm, taper of 1.53° ± 0.46°, and surface roughness (Ra) of 1.04 μm is well manufactured with the milling rate of 0.42 mm/min. Moreover, increasing the spindle speed and feed rate can further improve the machining quality of DNG. Finally, a complex DNG structure with the depth of 5 mm is well manufactured with the spindle speed of 4000 rpm and feed rate of 0.48 mm/min.

Keywords: deep-narrow groove; GH4169 alloy; electrochemical milling; wedged end face

Citation: Ye, Z.; Qiu, G.; Chen, X. Electrochemical Milling of Deep-Narrow Grooves on GH4169 Alloy Using Tube Electrode with Wedged End Face. *Micromachines* **2022**, *13*, 1051. https://doi.org/10.3390/mi13071051

Academic Editor: Stelios K. Georgantzinos

Received: 5 June 2022
Accepted: 29 June 2022
Published: 30 June 2022

1. Introduction

Due to their excellent fatigue resistance, corrosion resistance, radiation resistance, and high-temperature strength, Ni-based superalloys are used widely in die-casting molds, medical devices, and the aerospace sector as demand in the manufacturing field grows [1,2]. However, due to their material property of high-strength and poor heat conductivity, conventional machining often suffers from high tool wear, poor machining stability, and low process efficiency [3]. Nowadays, nontraditional processes with a low cost and high efficiency for manufacturing difficult-to-cut material are attracting more attention, including electrical discharge machining and laser machining. However, both cause heat affected layer and microcrack in machining surfaces, leading to lower machining quality [4,5].

Compared to other machining methods, electrochemical machining (ECM) has significant advantages for machining difficult-to-cut material, such as titanium alloys and Ni-based superalloys, due to the absence of tool wear, thermal and residual stresses, and cracks and burrs [6,7]. Electrochemical milling (EC-milling) combines the capabilities of ECM with the flexibility of numerical control (NC) technology, which can machine special structures, including deep-narrow grooves (DNGs), by controlling the moving path of the simple tool [8]. Natsu et al. fabricated three-dimensional complicated structures in SUS304 stainless using a tube electrode with an inner diameter of 0.26 mm by superimposing simple patterns [9]. Mitchell-Smith et al. produced different groove geometries by optimizing the bottom structure of a metallic nozzle to change current density distribution [10]. Liu et al. proposed a method of ultrasonic vibration-assisted electrochemical milling, which was helpful to improve the machined surface quality, and a three-dimensional structure with the total depth of 300 μm was machined [11]. Liu and Qu analyzed the anodic polarization curves of TB6 in $NaNO_3$ solution and the effect of current density on the morphology, and grooves and flat surfaces were machined successfully by electrochemical milling [12]. Zhu et al. established the electrochemical milling with nanosecond pulse model and machined 2D and 3D complex structures with good shape precision and surface quality [13]. Rathod et al. used sidewall insulation electrode to prevent the dissolution of the material along the sidewall of microgroove and reduced the taper angle from 58.39° to 25.20° for microgroove [14].

In the above studies, the electrochemical milling process is started from an initial gap between the bottom of the electrode and the surface of the workpiece, in which the electric field is mainly provided by the bottom of the electrode. With a motion along the feed direction, a very thin layer of material is removed from the surface of the workpiece. As a result, when a high-aspect-ratio structure is machined with this method, the tube tool is needed to multipass feed along the depth and length directions. Due to the stray corrosion, electric fields are repeatedly formed at the machined zone during the multipass feed, the machining accuracy and surface quality are always unsatisfactory. In an attempt to achieve better machining profile, Ghoshal and Bhattacharyya proved that the taper angle of grooves generated by sinking and milling method was far less than layer-by-layer method and then proposed a reversed taper tube tool for electrochemical milling to reduce the taper angle of groove [15]. This clearly showed that one-pass milling with the tube tool is more suitable for machining DNG structure.

However, the electric field for electrochemical anodic dissolution is provided by the sidewall of tool electrode in this method, in which large amounts of electrolytic by-products (sludge, gas bubbles, and heat) are generated in the inter-electrode gap (IEG). When the machining gap becomes small, it is difficult to remove them from the IEG. To maintain the stability of electrolyte flow during the whole machining process, a rapid removal of the electrolytic by-products is needed in the deep and narrow IEG. If the removal capacity is insufficient, the machining accuracy and surface quality will become poor. To obtain high machining accuracy and machining stability, Wang et al. suggested electrochemical machining with vibration superimposed to improve the removal of sludge and hydrogen bubbles [16]. Bilgi et al. proposed an ECM process with rotating electrode movement to enhance the uniformity of electrolyte flow and reduce or eliminate the flow field disrupting processes [17]. Moreover, Niu et al. proposed a flow channel structure with six slits in sidewall for improving the uniformity of flow field, and a thin-wall structure with the depth of 3 mm was machined [18]. However, the machining accuracy and surface quality in high-aspect ratio structures by EC-milling still need to be improved.

Pulsating flow is one of the unsteady flows that can change the characteristics of hydrodynamics and enhance mass transport [19,20]. This paper proposes a method for using a tube tool with wedged end face to generate pulsating flow field in the IEG to enhance the removal of electrolytic by-products, as well as improve the machining accuracy and surface quality in EC-milling. The parameters of pulsating flow field could be changed by adjusting the wedged angle in the end face of the tube tool, which would affect the EC-milling process. The flow field distributions of different wedged angles in the end face of the tube tool are simulated numerically. Experiments also are conducted to investigate the effects of different wedged angles in the end face of the tube tool on the machining quality of DNG.

2. Description of the Method and Numerical Simulation

2.1. Description of the Method

Figure 1 shows a schematic diagram of EC-milling of DNG with a wedged end face tube electrode. The milling procedure is divided into two processes. First, the tube electrode with wedged end face is fed along the *Z*-axis to the required machining depth through electrochemical drilling (Figure 1a). Second, the workpiece then is controlled to move along the specified rail in the *X*-*Y* plane (Figure 1b). The electrolyte flows from the bottom of the tube electrode into the IEG, thereby dissolving the workpiece material while simultaneously flushing out the electrolytic by-products.

As shown in Figure 1b, the front sidewall of the tube electrode provides an electric field for electrochemical anodic dissolution, and the amounts of electrolytic by-products are generated in the IEG. By using the wedged end face of the tube tool in the experiment, the quantity and flow rate of electrolyte flowing into IEG will be changed periodically with the rotation of the tube tool, and the pulsating flow field is generated, which could enhance the removal of electrolytic by-products. In addition, the pulsating parameters of electrolyte, including pulsating amplitude and pulsating frequency, can be changed by adjusting the wedged angle (α) and rotating speed of the tube tool, which will affect the EC-milling process.

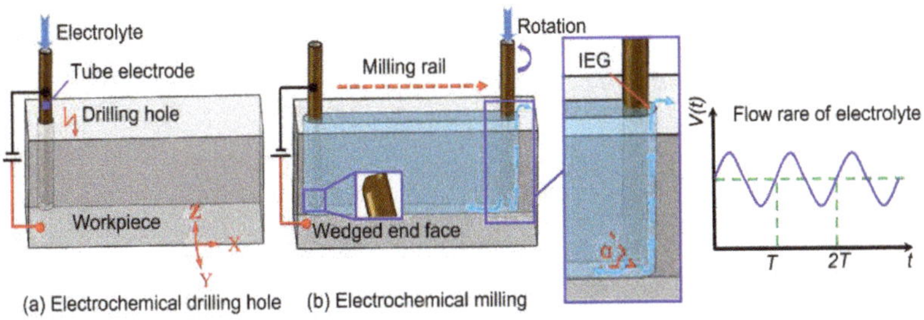

Figure 1. The schematic diagram of EC-milling of DNG with a wedged end face tube electrode. (**a**) Electrochemical drilling hole, (**b**) Electrochemical milling.

In the EC-milling process, the evolution process of DNG is illustrated in Figure 2. When the machining process is in a stable state, the front gap Δx in X direction can be described as [21]:

$$\Delta x = \frac{\eta \omega \kappa U_R}{v_x} \tag{1}$$

where η is the current efficiency, ω is the volumetric electrochemical equivalent of the metallic material, κ is the electric conductivity of electrolyte, U_R is the supply voltage, and v_x is the feed rate of the tube tool.

As shown in Figure 2a, when the tube tool is fed along the milling path in the plan *X*-*Y*, the relationship between the side gap and the *Y*-axis can be expressed as

$$\frac{dy}{dt} = \frac{\eta \omega \kappa U_R}{y} \tag{2}$$

At the initial time, $t = 0$ and $y = y_0$. Equation (2) can be integrated as follow:

$$\frac{y^2}{2} = \eta \omega \kappa U_R t + \frac{{y_0}^2}{2} \tag{3}$$

As the initial side gap, y_0 can be assumed to be the same with the front gap Δx, the machining side gap Δy in the Y direction can be expressed as:

$$\Delta y = \sqrt{\eta \omega \kappa U_R t + \Delta x^2} = \sqrt{\eta \omega \kappa U_R \frac{D}{v_x} + \Delta x^2} \tag{4}$$

where D is the outer diameter of the tube tool.

Thus, the groove width W can be described as:

$$W = D + 2\Delta y = D + 2\sqrt{\eta\omega\kappa U_R \frac{D}{v_x} + \Delta x^2} \quad (5)$$

As shown in Equation (5), it can be obtained that the width of DNG is affected by numerous parameters, including the outer diameter of the tube tool, feed rate, supply voltage, and electrolyte conductivity. During machining, the workpiece is dissolved. Meanwhile, hydrogen and oxygen are generated on the cathode and anode surfaces, respectively. The electrolyte in the IEG will be warmed by Joule heat. All these factors interactively influence the distribution of electrolyte conductivity along the flow path, resulting in a taper sidewall, as illustrated in Figure 2b. The relationship among the electrolyte conductivity κ, electrolyte temperature T, and gas void fraction β_{gas} can be described as follows:

$$\kappa = \kappa_0 (1 - \beta_{gas})^{bp} (1 + \alpha (T - T_0)) \quad (6)$$

where κ_0 is the initial electrolyte conductivity, T_0 is the initial electrolyte temperature, α is the degree of temperature dependence, and bp is Bruggeman's coefficient.

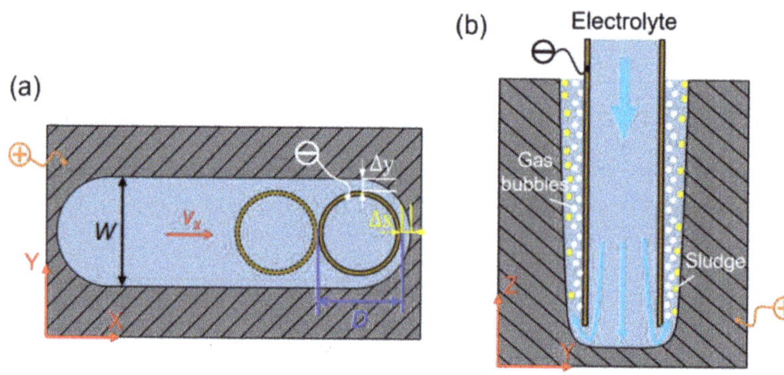

Figure 2. Diagram of the electrochemical milling process: (**a**) Sketch of the X-Y plane and (**b**) sketch of Y-Z plane.

Equation (6) shows that the efficient removal of electrolytic by-products (sludge, gas bubbles, and heat) is helpful to reduce the difference of electrolyte conductivity along the depth direction of DNG; thus, the taper of DNG's sidewall can be reduced.

2.2. Numerical Simulation

In this research, different wedged angles in the end face of the tube tool were used for generating different pulsating parameters for the electrolyte, and the flow field distribution in IEG was analyzed by computational fluid dynamics (CFD).

2.2.1. Model Building

Three-dimensional simulation model based on the flow field is established, as shown in Figure 3. The fluid domain includes the inner region of the tube electrode and DNG. Due to the rotation of the tube electrode, it needs to divide the fluid domain into a stationary zone and a rotation zone. The pink area represents the stationary zone, and the green area represents the rotation zone, which are transmitted data through the interface. In addition, the walls of the tube electrode are set as moving walls. Since the designed 3D model is relatively complex, tetrahedral meshes are used for the simulation model, and the mesh refinement is performed on the rotation zone and the around zone to ensure the accuracy of simulation results. In order to simplify the above model, this paper makes the following assumptions:

(1) Electrolyte is a continuous incompressible viscous fluid.
(2) The energy dissipation caused by the change of medium temperature and temperature difference is ignored in the machining process, and the flow is constrained by the conservation equation of mass and momentum.
(3) The flow field is not affected by bubbles or particles.

For incompressible viscous fluids, the fluid flow in the turbulent state is restricted by the Navier–Stokes equation:

$$\nabla \cdot \overline{v} = 0 \tag{7}$$

$$\frac{\partial \overline{v}}{\partial t} + (\overline{v} \cdot \nabla)\overline{v} = -\frac{1}{\rho}\nabla p + \mu \nabla^2 \overline{v} \tag{8}$$

where ρ is the electrolyte density, v is the flow velocity, p is the pressure, and μ is the dynamic viscosity of the electrolyte.

Based on the change of the flow field, the k-ε turbulence model in the standard equation is used to solve the turbulent energy k and the turbulent dissipation rate ε in the electrolyte flow process:

$$\frac{\partial(\rho k)}{\partial t} + \frac{\partial(\rho k u_i)}{\partial x_i} = \frac{\partial}{\partial x_j}\left[\left(\mu + \frac{\mu_t}{\sigma_k}\right)\frac{\partial k}{\partial x_j}\right] + G_k - \rho\varepsilon \tag{9}$$

$$\frac{\partial(\rho \varepsilon)}{\partial t} + \frac{\partial(\rho \varepsilon u_i)}{\partial x_i} = \frac{\partial}{\partial x_j}\left[\left(\mu + \frac{\mu_t}{\sigma_\varepsilon}\right)\frac{\partial k}{\partial x_j}\right] + \frac{C_{1\varepsilon}}{k}G_k - C_{2\varepsilon}\rho\frac{\varepsilon^2}{k} \tag{10}$$

where P_k is the generating term of turbulent energy, σ_k and σ_ε are Prandtl numbers corresponding to k and ε with values of 1.0 and 1.3, and $C_{1\varepsilon}$ and $C_{2\varepsilon}$ are model constants with values of 1.44 and 1.92.

All the models are solved by ANSYS FLUENT 19.2, and the simulation parameters are listed in Table 1.

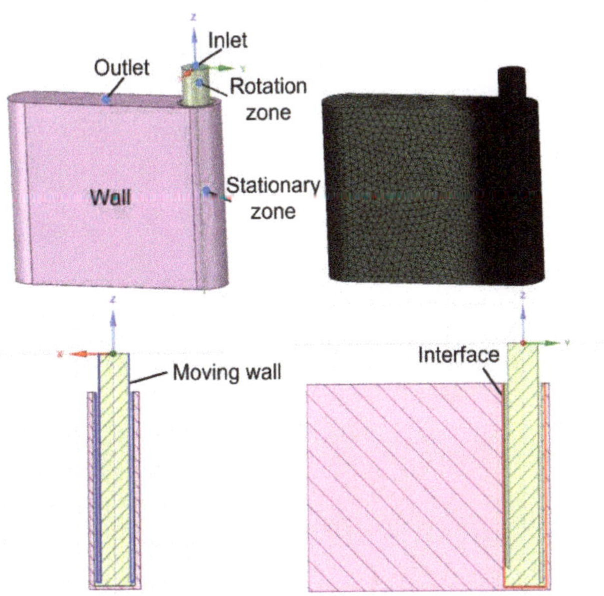

Figure 3. Building of three-dimensional model for the flow field simulation.

Table 1. The parameters for the simulation.

Parameter	Value
Inlet pressure, p_{in}	1.2 MPa
Outlet pressure, p_{out}	0 MPa
Rotational speed, ω	3000 rpm
Inter-electrode gap, Δ	0.25 mm
Dynamic viscosity of electrolyte, μ	1.003×10^{-3} Pa·s
Density of electrolyte, ρ	1100 kg/m^3
External diameter of tube electrode, D	1 mm
Internal diameter of tube electrode, d	0.8 mm
Wedged angles, α	0°, 40°, 50°, 60°
Groove length, L	5 mm
Groove depth, H	5 mm

2.2.2. Simulation Results

Figure 4 shows the contour of flow velocity on the cross section with different wedged angles under different rotational degrees. In the standard flat end face model (see Figure 4a), the velocity of electrolyte in the IEG is constant. In contrast, it can be found that the velocity of electrolyte in both front IEG and side IEG is changed in the wedged end face models under different rotation degrees (Figure 4b–d). The simulation results indicate that the pulsating flow field can be well generated with the wedged end face in tube electrode. Meanwhile, with the increase in the wedged angle, the velocity change of electrolyte in the IEG is more obvious. In the wedged angle of 40° model, the maximum and minimum velocities of electrolyte in the IEG are about 30 m/s and 8 m/s. In the wedged angle of 60° model, the maximum and minimum velocities of electrolyte in the IEG are about 32 m/s and 0 m/s.

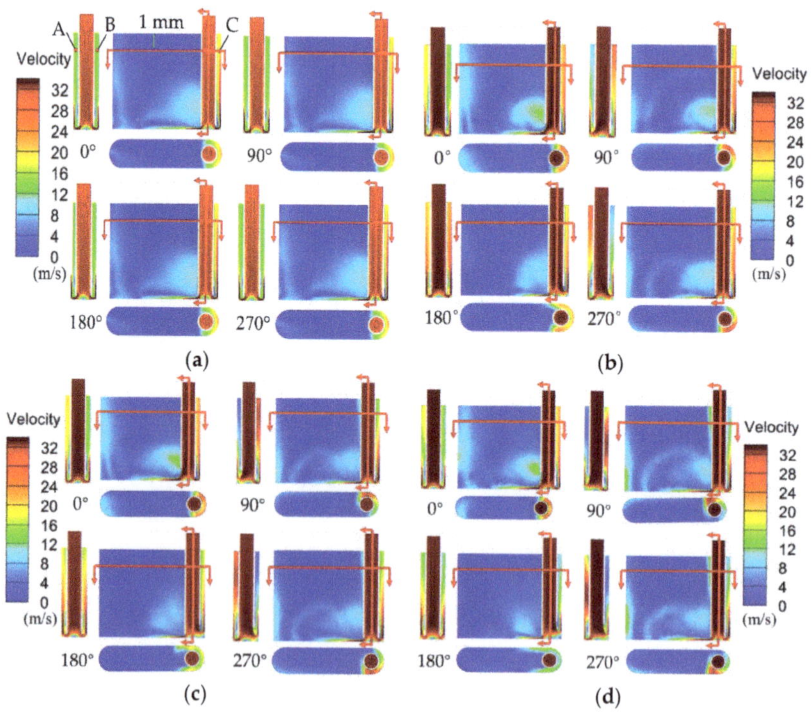

Figure 4. The contour of flow velocity on the cross section with different wedged angles in the end face of the tube tool. **(a)** 0°; **(b)** 40°; **(c)** 50°; **(d)** 60°.

In order to further analyze the change of electrolyte flow rate in the IEG, points A to C marked in Figure 4a are referenced to describe the specific electrolyte velocity in the IEG. Figure 5 shows the change of flow velocity at points A to C under different tube tools. In the standard flat end face model, the value of flow velocity at points A to C is constant. The velocity of electrolyte in the front gap and side gap is 23.8 m/s and 15.3 m/s, respectively.

The trends of the velocity of pulsating field with different wedged angles are the same. There are two peaks of electrolyte velocity in one cycle in the front machining gap, while the velocity of electrolyte changes like a sine wave in the side machining gap. With the increase in the wedged angle, the pulsating amplitude of electrolyte increases. When the end face with the wedged angle is 40°, the velocity of electrolyte at point A ranges from 6.7 to 30.6 m/s. When the wedged angle increases to 60°, the velocity of electrolyte ranges from 0.3 to 32.8 m/s.

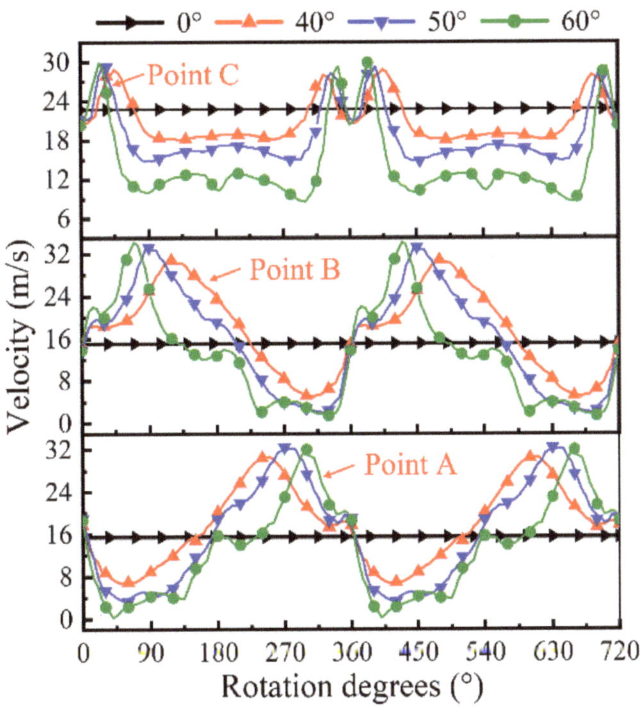

Figure 5. The variation of electrolyte velocity on points A to C in the IEG under kinds of tube tools.

3. Experimental Section

Figure 6 shows a schematic of EC-milling system with a tube electrode. The experimental setup included the pulse power supply, electrolyte supply unit, and electrolysis devices. The workpiece was installed on an *X-Y* stage. The electrode tool was carefully attached at the terminal of the spindle, and then the spindle was installed on a *Z*-axis and driven by an inverter. The speed of the spindle was adjusted by controlling the output frequency of the inverter. In addition, the function of internal flushing was achieved by using a rotating joint, which could transfer electrolyte from a pipeline into a rotating spindle. Figure 7 shows a photograph of the tube electrodes with different end faces, which were made from stainless steel 304, and the outer diameter and inner diameter were 1.0 mm and 0.8 mm, respectively.

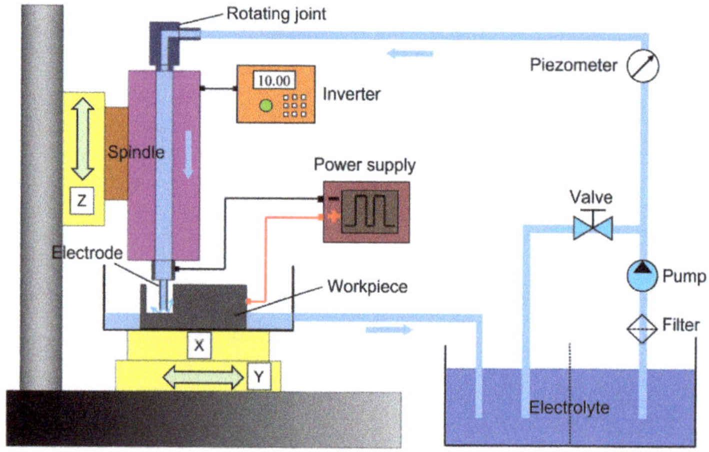

Figure 6. Schematic diagram of the experimental system.

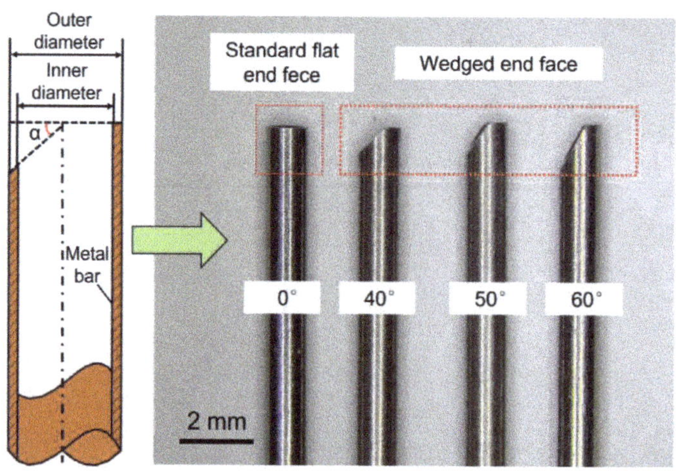

Figure 7. The photo of tube electrodes with different end faces.

In this work, the workpiece material was GH4169, and the machining depth and length of the electrode were 5 mm and 10 mm, respectively. The machining accuracy of DNG was investigated by detecting the width and taper. The size measurement of DNG is shown in Figure 8, where five points along the length of a DNG were measured, the width was the sectional width on the top, and the taper was the angle between the vertical line and the sectional sidewall. The average values of the width and taper were obtained, and the SD (standard deviation) was used to evaluate the dimensional uniformity. The morphologies of DNGs were examined using a scanning electron microscope. The profiles and surface roughness (Ra) of DNGs were measured using a confocal laser-scanning microscope (CLSM, Olympus LEXT OLS4100, Tokyo, Japan) and a Step profiler (Kosaka, ET-150, Tokyo, Japan).

Experiments were performed on each proposed tool to analyze its effect for machining accuracy and surface quality in the electrochemical milling process. Moreover, the influence of other parameters on machining quality was explored by single-factor experiment. The machining parameters are listed in Table 2.

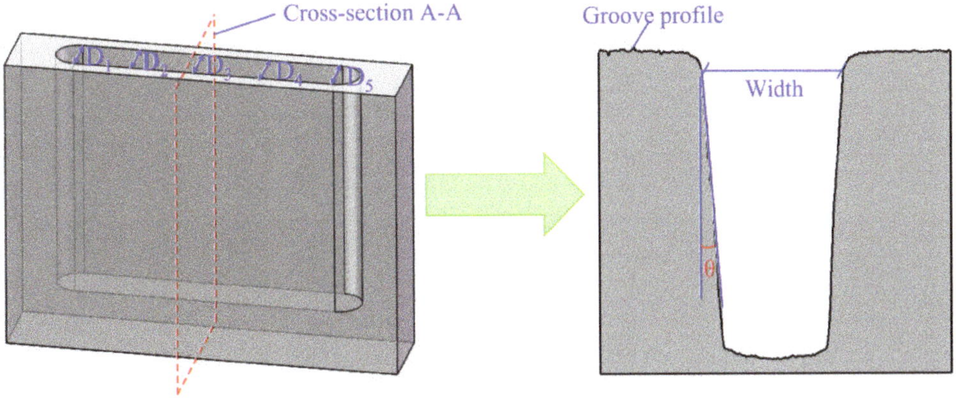

Figure 8. Schematic diagram of deep-narrow groove measurement.

Table 2. Machining parameters.

Parameters	Value
Electrolyte concentration	12% (wt.%), $NaNO_3$
Electrolyte temperature	25 °C
Electrolyte pressure	1.2 MPa
External diameter of tube electrode	1 mm
Internal diameter of tube electrode	0.8 mm
Wedge angle	0°, 40°, 50°, 60°
Spindle speed	2500, 3000, 3500, 4000 rpm
Feed rate	0.30, 0.36, 0.42, 0.48 mm/min
Applied voltage	25 V
Pulse frequency	1 kHz
Pulse duty cycle	50%
Machining depth	5 mm
Machining length	10 mm
Material of workpiece	GH4169

4. Results and Discussion

4.1. The Comparison of DNGs Generated with Different Wedged Angles

In order to compare the difference of DNGs generated with different wedged angles in the end face of the tube tool, comparative experiments were designed with the following machining parameters: the applied voltage of 25 V, pulse duty cycle of 50%, pulse frequency of 1 kHz, feed rate of 0.42 mm/min, and spindle speed of 3000 rpm.

Figures 9 and 10 show the SEM images, dimensions, and 3D profiles of DNGs generated with different wedged angles in the end face of the tube tool. Using standard flat end face with the wedged angle of 0°, the average width and taper of the grooves are 1.57 mm ± 0.06 mm (mean ± SD) and 2.2° ± 0.43°, respectively. When using the end face with wedged angle of 40°, the average width and taper of DNG decrease to 1.49 mm ± 0.04 mm and 1.53° ± 0.46°. However, with the increase in the wedged angle, the average width and taper of DNG gradually increase, reaching to 1.60 mm ± 0.04 mm and 2.3° ± 0.45°, respectively, when the wedged angle reaches to 60°.

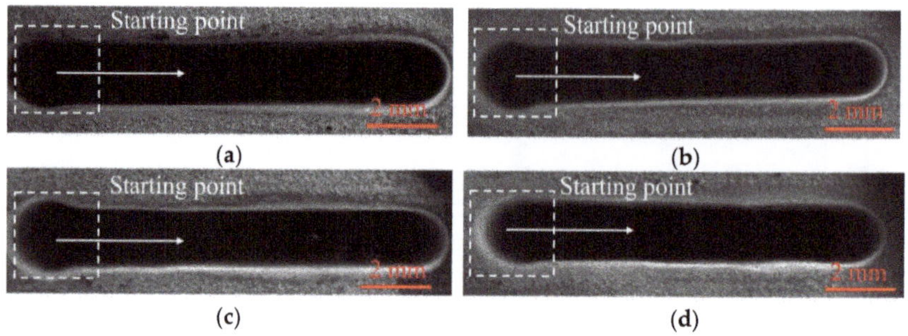

Figure 9. The SEM images of DNGs machined with different wedged angles in the end face of tube electrode. (**a**) Wedged angle = 0°; (**b**) wedged angle = 40°; (**c**) wedged angle = 50°; (**d**) wedged angle = 60°.

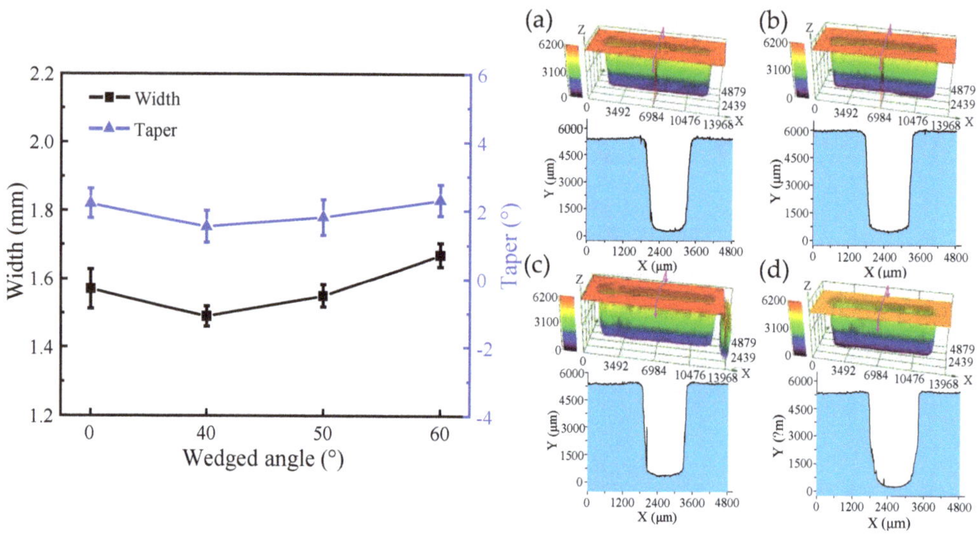

Figure 10. The dimension and 3D profile of DNGs milled with different wedged angles in the end face of the tube tool: (**a**) wedged angle = 0°; (**b**) wedged angle = 40°; (**c**) wedged angle = 50°; and (**d**) wedged angle = 60°.

Figure 11 shows the comparison of the SEM images and surface roughness (Ra) of DNGs milled with different wedged angles. The surface roughness (Ra) of sidewall is 1.387 μm under the standard flat end face. When the wedged angle is 40°, the surface roughness (Ra) suddenly drops to 1.09 μm. When the wedged angle varies from 50° to 60°, the surface roughness (Ra) increases to 1.65 μm. The reason is that a pulsating electrolyte is generated in the IEG by using the wedged end face of the tube tool, which occurs as flow separation resulting in a large number of vortices at the wall surface and increasing the turbulence and mixing of fluid. The boundaries of this flow are characterized by the creation and destruction of eddies of large turbulence energy and vortex shedding, which helps the removal of electrolytic by-products and reduces the difference of electrolyte conductivity in the IEG along the depth direction of DNG [19]. Hence, the material dissolution rate becomes uniform, reducing the taper of DNG and improving the surface quality. In addition, due to the reduction in the accumulation of electrolytic by-products in the IEG, the flow resistance decreases, avoiding the accumulation of electrolyte at the edge of DNG and reducing the stray corrosion at the upper edge of DNG. Thus, the upper width of DNG decreases. With the increase in the wedged angle, the machining quality becomes poor. As the wedged angle increases, the electric field distribution on the sidewall becomes nonuniform, which will lead to the nonuniform electrochemical dissolution. Thus, the taper of DNG increases. Meanwhile, combined with flow field simulation results, the pulsating amplitude of electrolyte increases but the minimum velocity of electrolyte decreases to about 0 m/s with the increase in the wedged angle. There is a low velocity zone of electrolyte in the IEG. Thus, anodic dissolution occurs in an instant static flow, the electrolytic by-products cannot be removed from IEG, which is not beneficial to the machining process. In fact, many studies have reported that a pulsating flow with proper pulsating parameters is helpful to the transfer process [22].

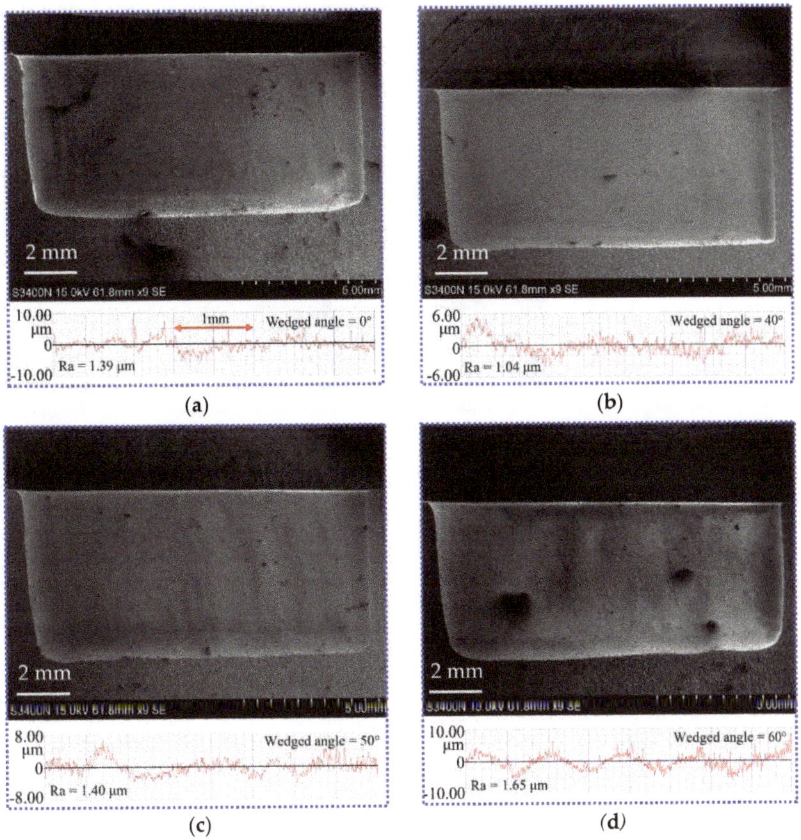

Figure 11. The SEM and surface roughness of sidewall of DNGs generated with different wedge angles in the end face of the tube tool: (**a**) wedged angle = 0°; (**b**) wedged angle = 40°; (**c**) wedged angle = 50°; and (**d**) wedged angle = 60°.

4.2. *The Effect of Spindle Speed on the Generation of DNGs*

In this subsection, single-factor experiments were performed to investigate the effect of spindle speed of 2500, 3000, 3500, and 4000 rpm on the dimension by using tube electrode with wedged angle of 40°. Additionally, other machining parameters were set as the applied voltage of 25 V, feed rate of 0.42 mm/min, pulse frequency of 1 kHz, and pulse duty cycle of 50%.

Figures 12 and 13 show the SEM images, dimensional change, and 3D profiles of DNGs generated with different spindle speeds. It can be observed that the average width of DNG decreases from 1.54 $\pm$ 0.04 mm to 1.49 $\pm$ 0.04 mm, and the average taper decreases from 1.62° ± 0.44° to 1.53° ± 0.46° when the spindle speed increases from 2500 rpm to 3000 rpm. As the spindle speed increases from 3000 to 4000 rpm, the average width maintains at about 1.5 mm, and the average taper gradually decreases from 1.53° ± 0.46° to 1.32° ± 0.32°. This result indicates that the spindle speed has no obvious influence on the width of DNG, while increasing spindle speed can improve the verticality of sidewall. High rotation is effective for the removal of electrolytic by-products from the IEG [23,24]. Meanwhile, the high pulsating frequency of electrolyte further improves the transfer of heat and electrolytic by-products, reducing the differences of electrolyte conductivity and material dissolution rate along the depth direction of DNG. Thus, the taper angle of DNG decreases. In addition, the spindle speed also affects the surface quality of DNGs. With the spindle speed increased, the pulsating frequency of electrolyte is increased, which could further enhance the removal of electrolytic by-products in IEG; thus, the electrochemical dissolution of material becomes more uniform, and the milled surface quality is improved. As shown in Figure 14, when the spindle speed increases from 2500 rpm to 4000 rpm, the surface roughness (Ra) of sidewall decreases from 1.14 μm to 0.92 μm.

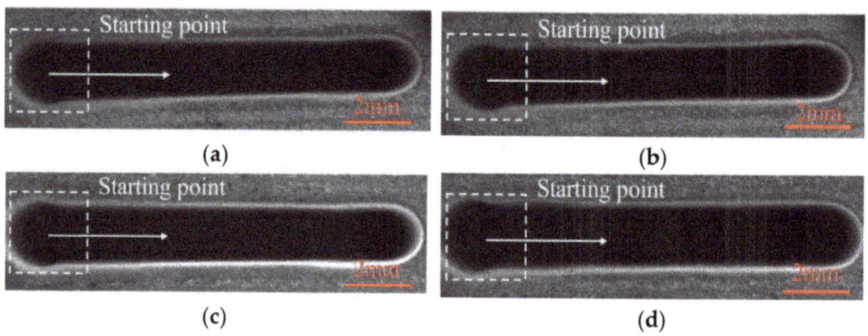

Figure 12. The SEM images of DNGs generated under different spindle speeds. (**a**) Spindle speed = 2500 rpm; (**b**) spindle speed =3000 rpm; (**c**) spindle speed = 3500 rpm; (**d**) spindle speed = 4000 rpm.

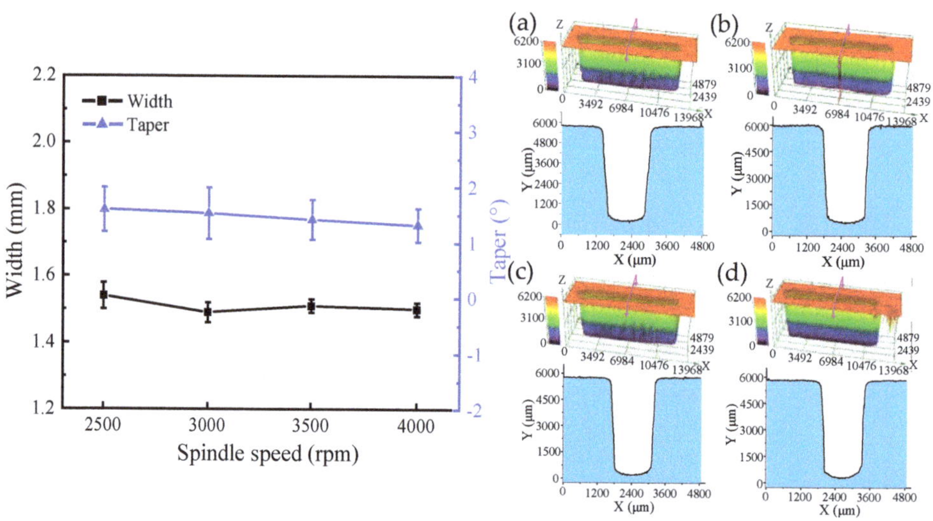

Figure 13. The dimension and 3D profile of DNGs milled with different spindle speeds: (**a**) spindle speed = 2500 rpm; (**b**) spindle speed = 3000 rpm; (**c**) spindle speed = 3500 rpm; and (**d**) spindle speed = 4000 rpm.

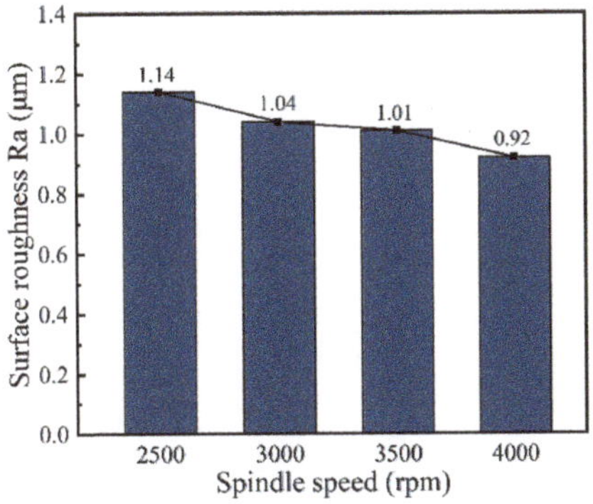

Figure 14. The surface roughness (Ra) of sidewall with different spindle speeds.

4.3. The Effect of Feed Rate on the Generation of DNGs

In order to investigate the effect of the tube tool feed rate on the machining quality, single-factor experiments with the feed rate at 0.30, 0.36, 0.42, and 0.48 mm/min and the spindle speed at 4000 rpm were conducted. The other machining parameters were set as the applied voltage of 25 V, wedged angle of 40°, rotational speed of spindle of 4000 rpm, pulse frequency of 1 kHz, and pulse duty cycle of 50%.

Figures 15 and 16 show the SEM images, dimensions, and 3D profiles of DNGs generated with different feed rates of tube electrode. As the feed rate increases from 0.30 to 0.48 mm/min, the average width of DNG decreases from 1.63 mm to 1.48 mm, and the average taper decreases from 2.66° to 1.32°. Moreover, the standard deviations of the width and taper of DNG decrease significantly with the increase in the feed rate. Thus, the consistency of the DNG enhances considerably.

According to Equation (1), with the increase in feed rate v_x of the tube tool, the front gap Δx decreases. As shown in Equations (4) and (5), with the decrease in the front gap Δx, the width of groove W decreases. In addition, when the workpiece is moving toward the electrode, the electric field is provided mostly by the front of the electrode. However, the milled surface of the groove is inevitably exposed to the back of the tube electrode, and a small amount of material is dissolved from the milled surface of the DNG due to corrosion by the stray current. Thus, with the increase in feed rate, the time of secondary stray dissolution to sidewall of DNG reduces, and the taper of sidewall and its standard deviation gradually decreases. At the same time, the reduction in secondary stray dissolution to sidewall improves the surface quality of sidewall. As shown in Figure 17, when the feed rate increases from 0.30 to 0.48 mm/min, the surface roughness (Ra) decreases from 1.41 μm to 0.85 μm.

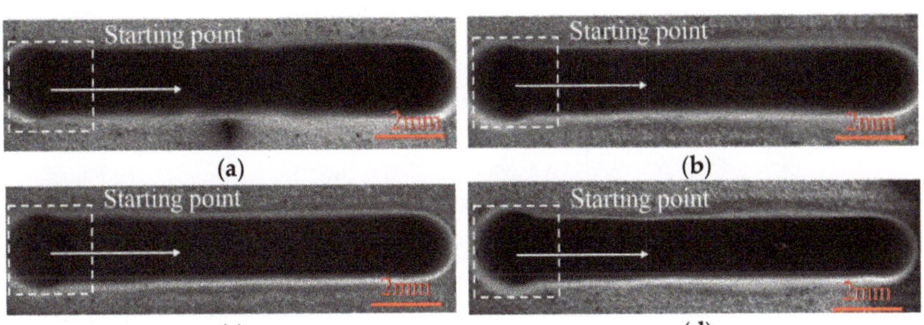

Figure 15. The SEM images of DNGs generated with varied feed rates. (**a**) Feed rate = 0.30 mm/min; (**b**) Feed rate = 0.36 mm/min; (**c**) Feed rate = 0.42 mm/min; (**d**) Feed rate = 0.48 mm/min.

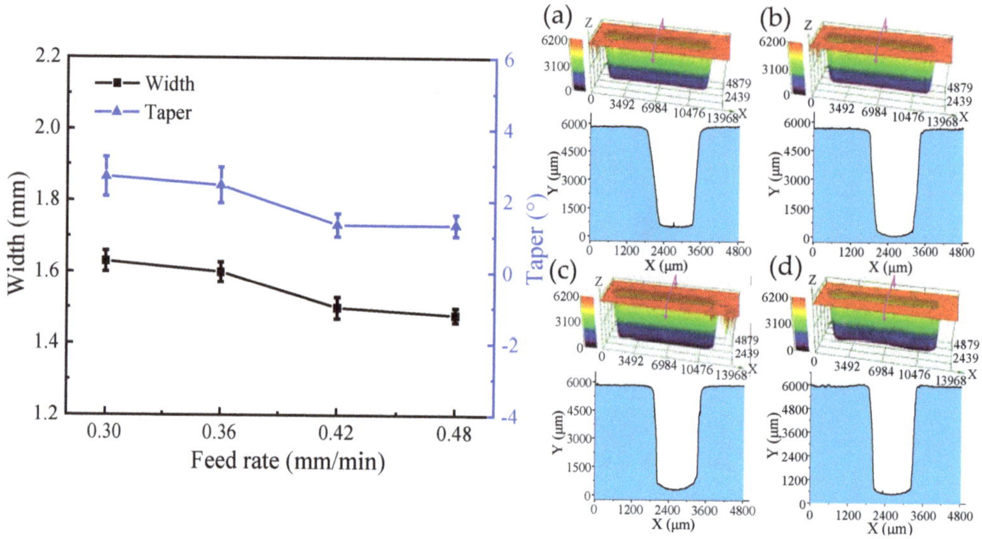

Figure 16. The dimension and 3D profile of DNGs milled with different feed rates: (**a**) feed rate = 0.30 mm/min; (**b**) feed rate = 0.36 mm/min; (**c**) feed rate = 0.42 mm/min; and (**d**) feed rate = 0.48 mm/min.

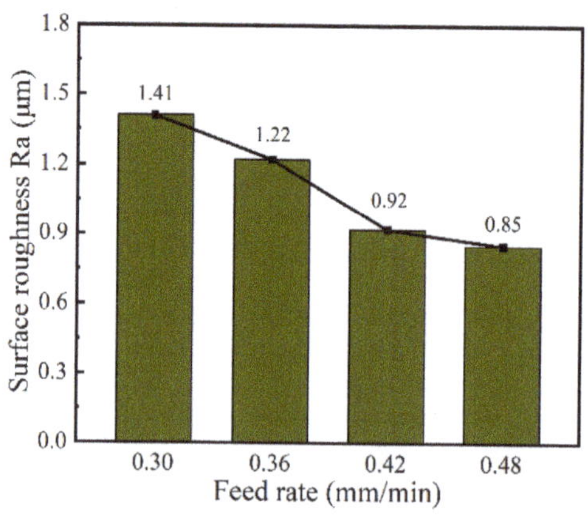

Figure 17. The surface roughness (Ra) of sidewall with varied feed rates.

4.4. EC-Milling of Complex Narrow Grooves by Using Wedged Tube Electrode

Based on the above research, the complex deep-narrow groove structures with machining depth of 5 mm are machined on GH4169 nickel-based alloy in one-pass feed by using a wedged end face tube electrode with wedged angle of 40°, as shown in Figure 18. The machining parameters are a feed rate of 0.48 mm/min, applied voltage of 25 V, spindle speed of 4000 rpm, pulse frequency of 1 kHz, and pulse duty of 50%.

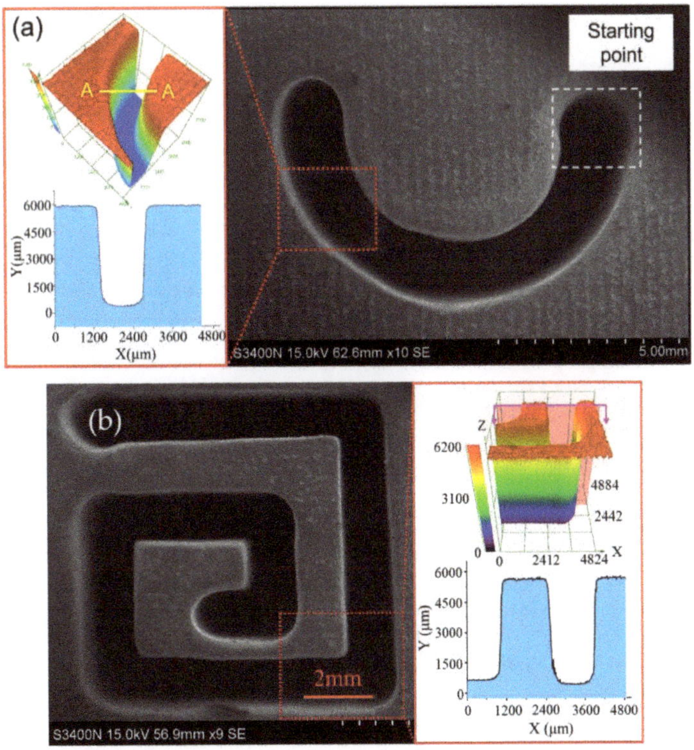

Figure 18. The SEM of complex DNG structures: (**a**) a semi-circular arc DNG structure and (**b**) a linear DNG structure.

5. Conclusions

In this study, a novel tube electrode with wedged end face has been proposed for electrochemical milling, deep-narrow groove on GH4160 alloy and to improve the machining accuracy and surface quality. Based on simulations and experiments, the main conclusions can be obtained as follows:

1. The simulation results indicated that the pulsating electrolyte could be generated in the inter-electrode gap by using a tube electrode with wedged end face. With the increase in wedged angle, the pulsating amplitude of electrolyte increased, but the minimum velocity electrolyte decreased.
2. Experiments verified that a pulsating electrolyte generated by the wedged angle of 40° was more suitable for the EC-milling process, and both the machining accuracy and surface quality were improved. Additionally, the average width and taper of DNG was 1.49 mm ± 0.04 mm and 1.53° ± 0.46°, respectively. The surface roughness (Ra) of the sidewall reduced to 1.04 μm at the same time.
3. The machining quality of DNG was improved by increasing the electrode rotational speed and feed rate. When the spindle speed was 4000 rpm and feed rate was 0.48 mm/min, the average width and taper of DNG was 1.48 mm and 1.32°, respectively. And the surface roughness (Ra) of sidewall was 0.85 μm.
4. A complex deep-narrow groove structure with the depth of 5 mm was fabricated stably on a GH4169 nickel-based alloy in one-pass feed by using a wedged end face tube electrode with wedged angle of 40° at a spindle speed of 4000 rpm and feed rate of 0.48 mm/min.

Author Contributions: Z.Y. and X.C. conceived and designed the experiments; Z.Y. and G.Q. performed the experiments; Z.Y. and X.C. analyzed the data; Z.Y. wrote the paper. All authors have read and agreed to the published version of the manuscript.

Funding: The work described in this study was supported by National Natural Science Foundation of China (Grant No. 52075105, 51705089).

Data Availability Statement: Not applicable.

Conflicts of Interest: The authors declare no conflict of interest.

References

1. Geng, P.H.; Qin, G.L.; Zhou, J. A computational modeling of fully friction contact-interaction in linear friction welding of Ni-based superalloys. *Mater. Des.* **2020**, *185*, 108244. [CrossRef]
2. Zhang, Z.W.; Fu, Q.; Wang, J.; Yang, R.; Xiao, P.; Ke, F.J.; Lu, C.S. Atomistic modeling for the extremely low and high temperature-dependent yield strength in a Ni-based single crystal superalloy. *Mater. Today Commun.* **2021**, *27*, 102451. [CrossRef]
3. Wang, Z.B.; Hou, G.S.; Sun, J.F.; Guo, J.Z.; Chen, W.Y. Characterization of residual stresses and grain structure in hot forging of GH4169. *Aerospace* **2022**, *9*, 92. [CrossRef]
4. Wang, M.L.; Qu, N.S. Improving performance of macro electrolyte jet machining of TC4 titanium alloy: Experimental and numerical studies. *Chin. J. Aeronant.* **2022**, *35*, 280–294. [CrossRef]
5. Xin, L.J.; Zhang, B.; Zhao, G.L.; Liu, H.L.; Yang, Y.F.; Li, L. Laser-induced oxidation-assisted micromilling of deep narrow microgroove on Inconel 718. *Int. J. Adv. Manuf. Technol.* **2021**, *114*, 173–184. [CrossRef]
6. Chen, X.L.; Zhu, J.J.; Xu, Z.Z.; Su, G.K. Modeling and experimental research on the evolution process of micro through-slit array generated with masked jet electrochemical machining. *J. Mater. Process. Technol.* **2021**, *298*, 117304. [CrossRef]
7. Li, H.S.; Fu, S.X.; Zhang, Q.L.; Niu, S.; Qu, N.S. Simulation and experimental investigation of inner-jet electrochemical grinding of GH4169 alloy. *Chin. J. Aeronant.* **2018**, *31*, 608–616. [CrossRef]
8. Zhang, C.Y.; Yao, J.B.; Zhang, C.Y.; Chen, X.L.; Liu, J.W.; Zhang, Y.J. Electrochemical milling of narrow grooves with high aspect ratio using a tube electrode. *J. Mater. Process. Technol.* **2020**, *282*, 116695. [CrossRef]
9. Natsu, W.; Ikeda, T.; Kunieda, M. Generating complicated surface with electrolyte jet machining. *Precis. Eng.* **2007**, *31*, 33–39. [CrossRef]
10. Mitchell-Smith, J.; Speridel, A.; Gaskell, J.; Clare, A.T. Energy distribution modulation by mechanical design for electrochemical jet processing techniques. *Int. J. Mach. Tools Manuf.* **2017**, *122*, 32–46. [CrossRef]
11. Liu, Y.; Chen, H.R.; Wnag, S.H.; Wang, K.; Li, M.H.; Peng, T.F. Micro electrochemical milling of micro metal parts with rotating ultrasonic electrode. *Sensors* **2020**, *20*, 6617. [CrossRef] [PubMed]
12. Liu, Y.; Qu, N.S. Electrochemical Milling of TB6 Titanium Alloy in $NaNO_3$ Solution. *J. Electrochem. Soc.* **2019**, *166*, 35–49. [CrossRef]
13. Liu, Y.; Jiang, Y.; Guo, C.S.; Deng, S.H.; Kong, H.H. Experimental Resarch on machining localization and surface quality in micro electrochemical milling of Nickel-Based Superalloy. *Micromachines* **2018**, *9*, 402. [CrossRef] [PubMed]
14. Rathod, V.; Doloi, B.; Bhattacharyya, B. Experimental investigations into machining accuracy and surface roughness of microgrooves fabricated by electrochemical micromachining. *Proc. Inst. Mech. Eng. B J. Eng. Manuf.* **2014**, *229*, 1781–1802.
15. Ghoshal, B.; Bhattacharyya, B. Investigation on profile of microchannel generated by electrochemical micromachining. *J. Mater. Process. Technol.* **2015**, *222*, 410–421. [CrossRef]
16. Wang, F.; Zhao, J.S.; Lv, Y.M.; Yang, Z.W.; Yao, J.; He, Y.F.; Tian, Z.J. Electrochemical machining of deep narrow slits on TB6 titanium alloys. *Int. J. Adv. Manuf. Technol.* **2017**, *92*, 3063–3071. [CrossRef]
17. Bilgi, D.S.; Jadhav, P.V. Enhancement of surface finish of pulse electrochemically machined (PECM) surface using rotationg electrode. *Int. J. Comput. Commun. Inform. Syst.* **2010**, *12*, 49–54.
18. Niu, S.; Qu, N.S.; Fu, S.X.; Fang, X.L.; Li, H.S. Investigation of inner-jet electrochemical milling of nickel-based alloy GH4169/Inconel 718. *Int. J. Adv. Manuf. Technol.* **2017**, *93*, 2123–2132. [CrossRef]
19. Qu, N.S.; Fang, X.L.; Zhang, X.D.; Zhu, D. Enhancement of surface roughness in electrochamical machining of Ti6Al4V by pulsating electrolyte. *Int. J. Adv. Manuf. Technol.* **2013**, *69*, 2703–2709. [CrossRef]
20. Yahya, A.; Hammouda, S.; Slimene, S.; Dhahri, H. Effect of cathode pulsating flow on mass transport and performance of solid oxide fuel cell. *Int. J. Therm. Sci.* **2022**, *174*, 107437. [CrossRef]
21. Zhu, D.; Wang, K.; Qu, N.S. Mircro wire electrochemical cutting by using in situ favricated wire electrode. *CIRP Ann.* **2007**, *56*, 241–244. [CrossRef]
22. Fang, X.L.; Qu, N.S.; Zhang, Y.D.; Xu, Z.Y.; Zhu, D. Effect of pulsating electrolyte flow in electrochemical machining. *J. Mater. Process. Technol.* **2014**, *214*, 36–43. [CrossRef]
23. Yahagi, Y.; Koyano, T.; Kunieda, M.; Yang, X.D. Micro drilling EDM with high rotation speed of tool electrode using the electrostatic induction feeding method. *Procedia Cirp* **2012**, *1*, 162–165. [CrossRef]
24. Feng, G.L.; Yang, X.D.; Chi, G.X. Experimental and simulation study on micro hole machining in EDM with high-speed tool electrode rotation. *Procedia Cirp* **2012**, *1*, 162–165. [CrossRef]

micromachines

MDPI

Article

Study on Solidification Process and Residual Stress of SiCp/Al Composites in EDM

Wenchao Zhang [1], Hao Chang [2] and Yu Liu [3,*]

[1] School of Mechanical Engineering and Automation, Dalian Polytechnic University, Dalian 116039, China; traum525@gmail.com
[2] Department of Mechanical Engineering, Zhengzhou Technical College, Zhengzhou 450121, China; changhaosc@163.com
[3] School of Mechanical Engineering, Dalian Jiaotong University, Dalian 116028, China
* Correspondence: liuyu_ly12@126.com

Abstract: To study the change of residual stress during heating and solidification of SiCp/Al composites, a one-way FSI (Fluid Structure Interaction) model for the solidification process of the molten material is presented. The model used process parameters to obtain the temperature distribution, liquid and solid-state material transformation, and residual stress. The crack initiated by the thermal stress in the recast layer was investigated, and a mathematical model of crack tip stress was proposed. The results showed a wide range of residual stresses from 44 MPa to 404 MPa. The model is validated using experimental data with three points on the surface layer.

Keywords: EDM; SiCp/Al; recast layer; residual stress; solidification

Citation: Zhang, W.; Chang, H.; Liu, Y. Study on Solidification Process and Residual Stress of SiCp/Al Composites in EDM. *Micromachines* **2022**, *13*, 972. https://doi.org/10.3390/mi13060972

Academic Editor: Stelios K. Georgantzinos

Received: 28 May 2022
Accepted: 16 June 2022
Published: 19 June 2022

Publisher's Note: MDPI stays neutral with regard to jurisdictional claims in published maps and institutional affiliations.

1. Introduction

SiCp/Al is a composite material made by adding silicon carbide particles with aluminum as a matrix and using specific technological means. This composite material has the characteristics of both aluminum and silicon carbide particles, such as electrical conductivity, low thermal expansion coefficient, low wear resistance, low corrosion resistance, etc., so it is widely used in precision fields such as aerospace, instrumentation, and optical components. However, due to the inhomogeneity of SiCp/Al composites, and containing a certain proportion of non-metallic particles with high strength, high hardness and high wear resistance, it is very difficult to process such materials by traditional machining methods. The main problems are as follows. The material processing is difficult and the processing quality is low. Especially during machining, the rake face is affected by the friction of the silicon carbide particles, resulting in severe tool wear, accumulation of debris, and low machining efficiency, making it challenging to meet machining requirements. Electrical discharge machining (EDM) relies on the spark discharge between two electrodes to remove the material. It is not limited by the strength and hardness of the material, and there is no physical contact between the tool and the workpiece during machining. Therefore, EDM is suitable for the machining of SiCp/Al composite materials.

Many scholars have conducted experiments on residual stress. Li et al. [1] experimented investigation on Cr12MoV Steel. The result showed that the depth of residual stress generated is much the same with different working liquids, but the working liquid with worse cooling capacity causes greater residual stress. Liu et al. [2] found the maximum value of average residual stress in the subsurface instead of the top surface owing to the high surface roughness. Pujari et al. [3] studied wire EDM parameters on residual stresses in the machining of aluminum alloy. The results obtained showed a wide range of residual stresses from 8.2 to 405.6 MPa. Mehmood et al. [4] found that the amount of residual stresses is proportional to the discharge current near the surface up to the depth of 75 μm. Ramulu et al. [5] studied the surface quality and subsequent performance of a 15% SiC particulate reinforced A356 Al. The results showed that the fatigue strength and residual stress of the material were significantly reduced using the tool with good conductivity. Batish et al. [6] analyzed the effect of parameters that induced residual stresses during electrical discharge machining of SiCp/Al. The pulse-off time was identified as the most significant factor in the formation of residual stresses. Additionally, better conductive electrode materials used during machining cause lower residual stress. Sidhu et al. [7] studied the influence of discharge parameters on the residual stress of the machined surface in EDM of SiCp/Al. The result showed that the workpiece, tool material properties, and pulse off time significantly contributed to the formation of residual stress. The concentration of reinforced particulates and matrix conductivity also play a vital role in the development of residual stress.

Many scholars have studied the changes of residual stress on the surface of SiCp/Al after EDM machining by experimental means. Still, few scholars have explored the change process of residual stress in the solidification process of molten material by both means of simulation and experiment. This paper linked the flow field with the structural field to simulate the change of residual stress during heating and solidification of SiCp/Al composites. The residual stress and the crack distribution were obtained. The discharge experiment was carried out on a self-built EDM machine to validate the simulation results, and the residual stress was measured by PROTO LXRD 3000.

2. Analysis of Residual Stress and Crack Formation

The solidification process of molten metal and molten particles mainly occurs in three phases. The first is that the molten material thrown into the working liquid solidifies when the working liquid cools and forms the debris. The second is that part of the molten material is thrown onto the corresponding electrode surface and cooled. Third, the residual materials that have not been thrown out of the pit cooled on the bottom of the pit. When new working liquid is introduced, a recast layer is formed on the surface, as shown in Figure 1.

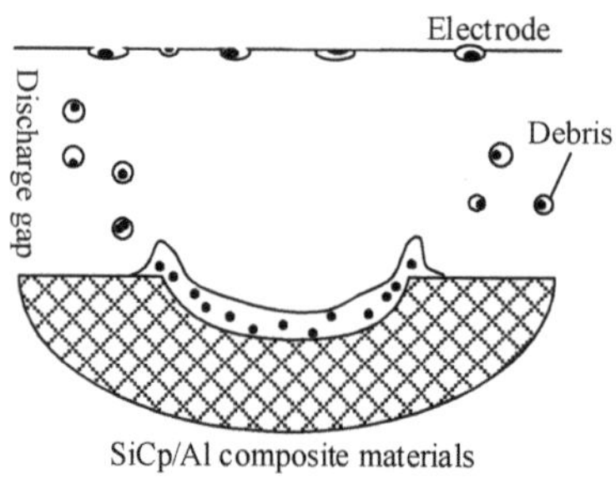

Figure 1. The process of solidification process for molten material.

The residual stresses are generated near the surface of the pit on both the tool and the workpiece, which mainly come from two aspects. Firstly, the heat source is the most critical factor affecting residual stress of discharge. The heat sources in EDM come from the discharge channels generated between the electrodes. When the heat source acts on the surface of the two electrodes, the heat on the surface of the tool and the workpiece promotes the melting and vaporization of the material by consuming its energy. In this process, a small amount of heat is still transferred to the tool and the workpiece. At this point, the physical properties of the material, such as the coefficient of expansion, will cause the material to expand and generate thermal stress [8]. For particle reinforced metal matrix composites, the coefficient of thermal expansion of matrix is different from that of particle. As the temperature rises, the internal stresses are released. Due to the difference in thermal expansion coefficient between the matrix material and the particles, the expansion of the matrix is much larger than that of the particles. In the expansion process, the matrix and the particles will inhibit each other and generate compressive stress, respectively. To relieve the stress, local microplastic deformation occurs in the material, and the stress below the yield strength is retained inside the workpiece as residual thermal stress [9,10].

The second is the phase transition stress caused by the material phase transition during rapid heating and cooling. As the temperature of the material decreases during cooling, the distance between the molecules will shorten, making the volume shrink. When the material is lower than its melting point, it will cause a phase transition from liquid to solid. The distance between molecules will be shortened again, generating phase transition stress [11].

Due to impurities, pores, and other defects as well as heating and cooling reasons, the workpiece will have an uneven distribution of stress during and after EDM machining. When the stress value exceeds the yield strength limit of the material, the microcrack forms. At this point, stress concentration occurs in the crack tip area. Although crack propagation can alleviate local stress, the stress concentration at the crack tip will lead to further crack propagation in both length and width, resulting in macroscopic crack formation.

3. Mathematical Model of Crack Tip Stress

When the residual stress is greater than the yield strength limit of the material, to alleviate the stress concentration, the local microplastic deformation of the material occurs. In this process, it is easy to promote the formation and expansion of the internal microcracks of the material. Generally, surface cracks are in opening mode after EDM machining [12]. Therefore, a crack model was established to analyze the residual stress at the crack tip, as shown in Figure 2.

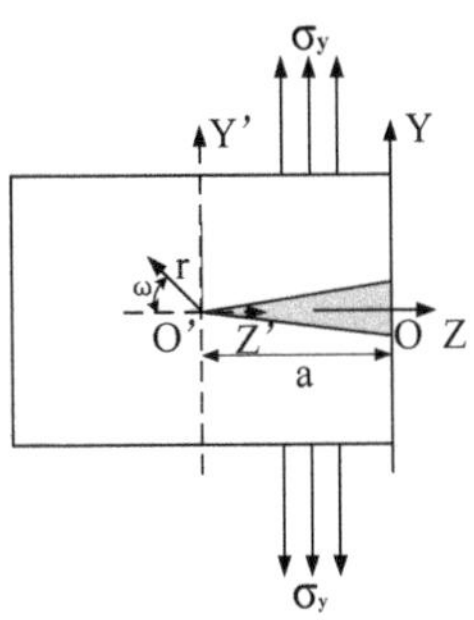

Figure 2. Schematic diagram of the crack model.

Select the minimum area of the pit surface. Firstly, a YOZ coordinate system was established in the plane to analyze the crack propagation, and the crack was placed in the negative direction of the Z-axis. Therefore, local boundary conditions on the pit surface were determined as follows:

$$\begin{array}{l} y = 0, -a < z < 0, \sigma_y = 0 \\ y = 0, z \to -\infty, \sigma_y \to \sigma \\ y = 0, z = -a, \sigma_y \to \infty \end{array} \tag{1}$$

where σ_y is the stress in the y-direction. Functions of a complex variable are usually used to solve the plane crack stress problem, and Westergaard is a common solution. The stress function is selected as:

$$\Psi = \mathrm{Re}\overline{\overline{\phi}}_I(\xi) + y\ \mathrm{Im}\ \overline{\phi}_I(\xi) \tag{2}$$

$$\sigma_x = \mathrm{Re}\phi_I(\xi) - y\ \mathrm{Im}\ \phi_I{}'(\xi) \tag{3}$$

$$\sigma_y = \mathrm{Re}\phi_I(\xi) + y\ \mathrm{Im}\ \phi_I{}'(\xi) \tag{4}$$

$$\phi_I(\xi) = \frac{\sigma}{\sqrt{1 + \frac{a}{\xi}}} \tag{5}$$

where Re represents the real part of functions of a complex; Im represents the imaginary part of functions of a complex; $\phi_I(\xi)$ is a widely used analytical function, $\overline{\phi}_I(\xi)$ and $\overline{\overline{\phi}}_I(\xi)$ are the primary and secondary integrals, respectively; $\xi = z + iy$.

The origin is moved to the crack tip, and the complex number is represented by trigonometric functions. Then, a point ξ (z, y) on the plane is transformed into η (r, ω), so $y = r \cdot \sin\omega$, $z + a = r \cdot \cos\omega$, then use the trigonometric function to express the complex number on the plane as follows:

$$\eta = \xi + a \tag{6}$$

The radius of stress field near the tip $r \to 0$, therefore $\eta = re^{i\omega} \to 0, \xi = \eta - a \to -a$. σ_y and σ_x are expressed as follows:

$$\sigma_x = \frac{\sigma\sqrt{a\pi}}{\sqrt{r\pi}} \cos\frac{\omega}{2}(1 - \sin\frac{\omega}{2}\sin\frac{3}{2}\omega) \tag{7}$$

$$\sigma_y = \frac{\sigma\sqrt{a\pi}}{\sqrt{\pi r}} \cos\frac{\omega}{2}(1 + \sin\frac{\omega}{2}\sin\frac{3}{2}\omega) \tag{8}$$

Let $K_I = \sigma\sqrt{\pi a}$, where K_I is called the fracture intensity factor, and the expression between stress and fracture intensity factor can be obtained as follows:

$$\sigma_y = \frac{K_I}{\sqrt{\pi r}} \cos\frac{\omega}{2}(1 + \sin\frac{\omega}{2}\sin\frac{3}{2}\omega) \tag{9}$$

According to the stress intensity criterion of fracture mechanics theory, when the stress intensity factor K_I reaches the material fracture toughness K_{IC}, the material fractures. Suppose the fracture intensity factor of Al alloy is K_{IC}, the critical stress σ_{ycr} at fracture can be obtained as follows:

$$\sigma_{ycr} = \frac{K_{IC}}{\sqrt{\pi r}} \cos\frac{\omega}{2}(1 + \sin\frac{\omega}{2}\sin\frac{3}{2}\omega) \tag{10}$$

4. Modeling of Melting, Throwing, and Solidification of Materials

4.1. Simulation Model of Melting and Throwing Processes

Figure 3 is the boundary conditions in the melting simulation model. The flow field geometric model was a rectangle with the length and height of 300 μm and 330 μm, respectively. The geometry models of tool, workpiece, and discharge channel were established in the flow field. The tool and workpiece were rectangular with a height of 100 μm, and the discharge channel height was 30 μm. To improve the calculation efficiency, a two-dimensional axisymmetric model was adopted in the simulation, and the left side of the flow field was set as the axisymmetric. Interface thermal resistance was selected for the contact surface between matrix and particles. The upper and lower sides and the right side of the flow field were set as walls. The materials and machining parameters are listed in Table 1. The tool was red copper, the workpiece was SiCp/Al, and the deionized water was used in the working liquid. DEFINE_PROFILE module was used to load the Gaussian heat flux function onto the tool and workpiece surfaces. The melting and solidification model was selected to simulate the melting and phase transition process of the material surface. The pulse-on was 20 μs in the simulation for single discharge in EDM. Because the simulation research focuses on the material surface temperature and the formation of the molten pool, the discharge area was finely meshed [13,14] in the mesh section, and the other area was coarsely meshed. The meshed model is shown in Figure 4. The generated mesh file was imported into Fluent software, boundary conditions and material properties were set, and after the simulation, the changes of the surface temperature and molten pool with time were observed.

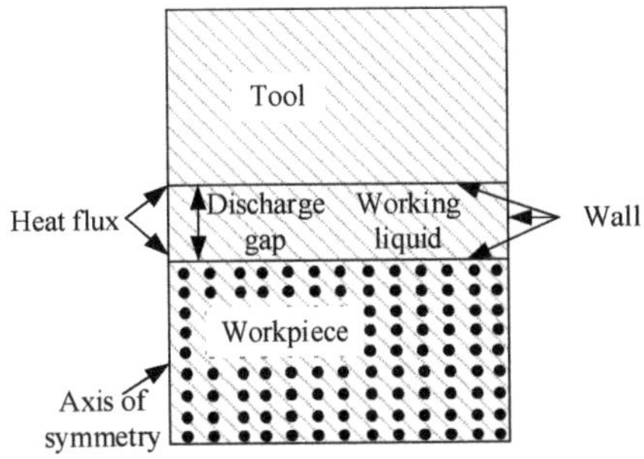

Figure 3. The boundary conditions in the melting simulation model.

Table 1. Simulation parameters in the melting simulation.

Parameter	Description
Tool material	Red copper
Workpiece material	SiCp/Al
Working liquid	Deionized water
Voltage U (V)	45
Current I (A)	20
Pulse-on T_{on} (μs)	20
Particle diameter	5 μm
Energy distribution (tool)	0.23
Energy distribution (workpiece)	0.30

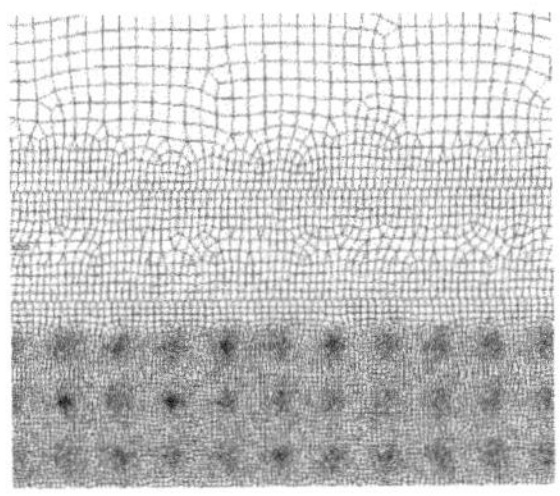

Figure 4. The meshed model.

The material throwing simulation was obtained by modifying boundary conditions based on melting results. The boundary between the lower surface of the tool and the upper surface of the workpiece was modified to interface, and the upper and lower surfaces of the discharge channel were set as velocity outlets. The boundary conditions in the throwing simulation model are shown in Figure 5.

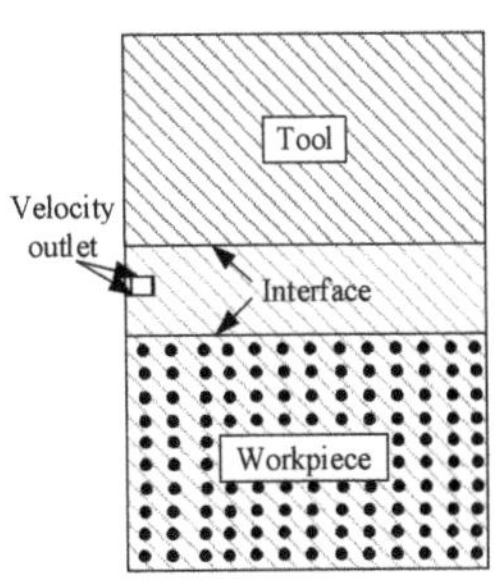

Figure 5. The boundary conditions in the throwing simulation model.

4.2. Modeling of Solidification Processes

The material solidification model was established by extracting the geometric model and modifying the boundary conditions based on the simulation results of the material throwing. The approach is as follows: Extract the pit surface contour curve from the throw simulation result file, output all points of the curve to the text file, and then import the file into ICEM to re-fit and draw a new two-dimensional pit model, as shown in Figure 6. The pit surface and other surfaces of the workpiece were set as the wall surface, the discharge channel was set as the velocity outlet, and the top of the flow field was set as the pressure outlet. Because the simulation time of the solidification process was long, a large time step was adopted in the simulation to shorten the calculation time.

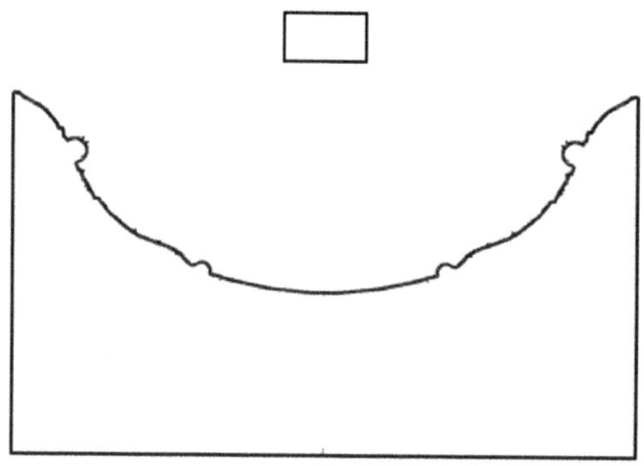

Figure 6. The 2D pit model after fitting.

4.3. Modeling of Residual Stress on Pit Surface

As a fluid simulation software, the latest version of Fluent is able to calculate the stress–strain field. Still, due to its weak analysis of structural mechanics, it can only calculate the linear stress–strain, which cannot meet the requirements of calculating the residual stress during the solidification process. In the simulation, the method of one-way fluid-solid coupling [15] was used to calculate the residual stress in the recast layer and the stress distribution at the crack tip after solidification. The flow field calculation results were taken as the initial conditions and introduced into the structural field for stress solution analysis. First, the flow field and structure field modules in Workbench were established. The flow field module calculated the melting, throwing, and solidification process of the material, and the structure field module calculated the residual stress and strain in the recast layer.

To study the influence of residual stress on crack formation, it is necessary to modify the workpiece geometry and add preset cracks on its surface, as shown in Figure 7a. The preset crack geometry model was re-meshed to study the stress at different crack tips. In the simulation, the main research object was the residual stress at the crack tip. To obtain more accurate calculation results and higher calculation efficiency, the mesh at the crack tip is locally fined in this paper. In other non-important calculation areas, the mesh was coarse, as shown in Figure 7b. After the calculation in the flow field was completed, the results were loaded into the structure field. The pressure field and temperature field calculation results were read, and then they were attached to the pit surface as the initial conditions for solving the residual stress. Transient analysis was adopted, and finally, the simulation calculation was carried out.

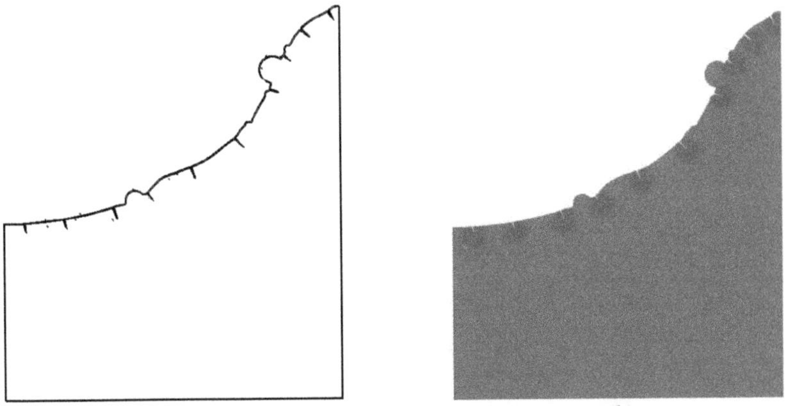

(a) (b)

Figure 7. Geometric and meshed model of preset crack. (**a**) A geometric model of a preset crack; (**b**) meshed model.

5. Simulation Results and Analysis

5.1. Analysis of Material Melting Process

Figure 8 shows the temperature field distribution on the tool and workpiece. The calculation error is 0.1%, so the values in the figures are all rounded. In Figure 8a, the maximum surface temperature of the workpiece at 5 μs is close to 14,100 K. This is mainly because the high-energy plasma in the discharge channel moves towards the electrodes at a very high speed. When it is bombarded on the surface of the electrodes, the kinetic energy of the plasma is converted into heat energy, which instantly forms a heat source with a small diameter and high temperature. As the heat source has a short action time, the heat does not transfer to other areas in time, so the heat mainly concentrates near the discharge point. However, at this time, the maximum temperature of the tool surface is about 8900 K, which is much lower than the surface temperature of the workpiece. This is because the presence of interface thermal resistance reduces the thermal conductivity of SiCp/Al composite material. The thermal conductivity of copper is much higher than that of copper, and the heat can be rapidly transferred to other locations, with less heat accumulation and less temperature rise. Figure 8b is the temperature field at 20 μs. As the heat transfers, the central temperature drops to 11,100 K, and the highest temperature on the tool surface drops to 7600 K. It can be seen from the figure that as the discharge time increases, the maximum temperature gradually decreases.

Figure 9 shows the curve of the maximum surface temperature of the tool and workpiece with time. In general, the maximum surface temperature of SiCp/Al is higher than that of copper electrode. From the curve, the slope of the curve becomes smaller, indicating that with the increase of discharge time, the decrease of the temperature reduction rate becomes slow. It is shown that the temperature changes violently at the beginning in EDM, and then the rate of temperature change decreases gradually.

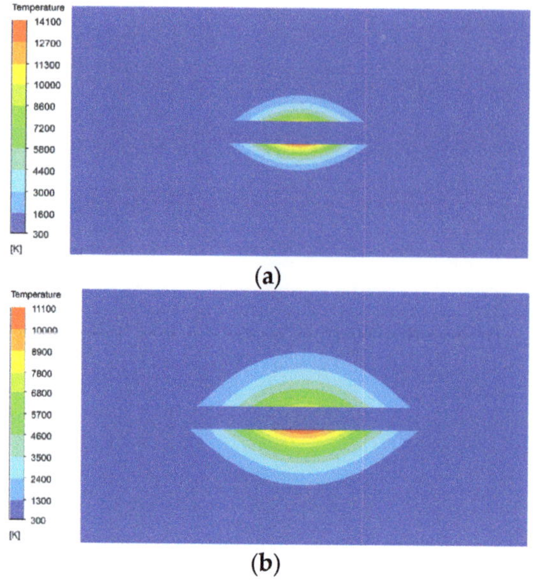

Figure 8. Temperature field distribution on the tool and workpiece at different discharge moments. (**a**) Discharge time at 5 μs; (**b**) Discharge time at 20 μs.

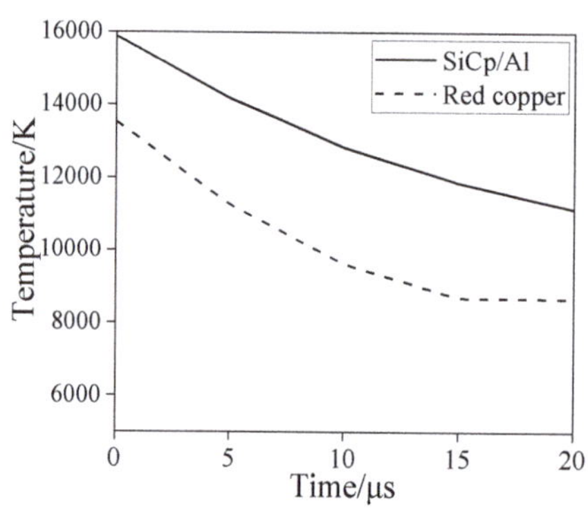

Figure 9. Curve of the maximum temperature of red copper and SiC_p/Al with time.

Figure 10a shows the molten pool at 5 μs. As can be seen from the figure, the molten pool is formed quickly at this time, and the material in the molten pool changes from the solid phase to liquid phase, and the radius is greater than the depth. This is because the heat generated by the discharge channel applied to the tool surface travels faster in the direction of a radius than in the direction of depth, so the radius of the temperature field is greater than the depth. When the material reaches the melting point, a phase transition occurs. The material changes from the solid phase to the liquid phase, and the metal matrix exists in the molten state locally, forming a molten pool. The melting point of SiC is 2300 K, which is much higher than that of Al, so when the Al matrix reaches the melting point, part of the SiC particles are still in the solid state. It can be seen from the figure that part of the SiC particles appears red, representing the liquid phase, while some SiC particles are blue, representing the solid phase. The SiC particles near the center of the workpiece surface have been completely melted, and those far away from the discharge center are still solid. In addition, there is also a melting pool on the surface of the tool. Compared with the melting pool of the workpiece, the volume of the melting pool on the tool is smaller than that of the workpiece, mainly because in the heating process, the thermal conductivity of the tool is greater than that of aluminum carbide, and the heat can be quickly transferred out. Hence, the local temperature is relatively low. In addition, the melting point of copper is 1350 K, which is much higher than that of Al, so the Al matrix melts first, and the molten pool volume is smaller than the workpiece. Figure 10b shows the molten pool at 20 µs. The pool increases significantly in depth and radius, mainly because the heat has transferred to the surrounding area, causing more material to melt.

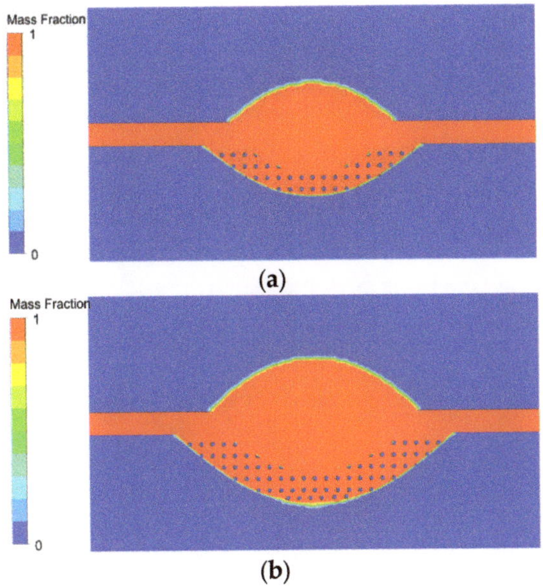

Figure 10. Molten pool on the tool and workpiece at different discharge moments. (**a**) Discharge time at 5 µs; (**b**) Discharge time at 20 µs.

5.2. Analysis of Material Solidification and Throwing Process

Figures 11 and 12 show the temperature field and the phase transition in the solidification process in the discharge area of the molten material from 500 µs to 700 µs. At 500 µs, as can be seen from Figure 11a, the surface temperature of the pit is higher than the melting point of the matrix, and the metal mainly exists on the surface in liquid form. Most of the molten metal with SiC particles was thrown out of the matrix, and only a small part of the metal materials and SiC particles remain in the discharge area. This part of the material is subjected to continuous shear stress and remains on the pit surface during the throwing process [16]. At 600 µs, the maximum temperature of molten Al decreases, and the solidified layer appears gradually during the phase transition from liquid to solid state. As can be seen from Figure 12b, the solidification rate at the bottom center of the pit was significantly faster than that of other positions. With the increase in time, the movement speed of the SiC particles in the molten Al slows down. At 700 µs, the residual molten Al has completely solidified, and the SiC particles are also solidified on the surface of the workpiece. At this time, the recast layer is formed, and the material enters the cooling stage after solidification.

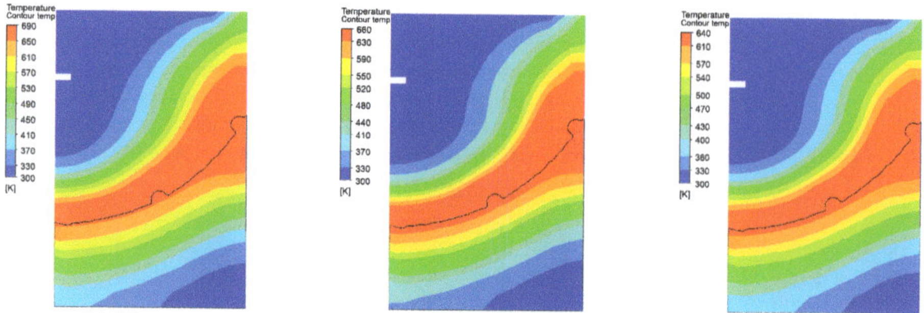

(**a**) 500 µs (**b**) 600 µs (**c**) 700 µs

Figure 11. The temperature field changes at different moments.

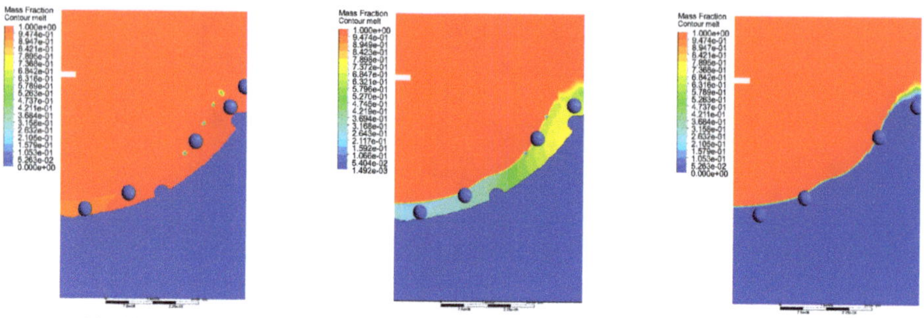

(**a**) 500 µs (**b**) 600 µs (**c**) 700 µs

Figure 12. Phase of solidification process of molten metal at different moments.

5.3. Analysis of the Residual Stress

The thermal expansion coefficient of particle and matrix is different. When the thermal stress exceeds the yield strength of the material, microplastic deformation occurs in local areas. Such microplastic deformation develops rapidly at the tip of the microcrack. The residual thermal stress below the yield strength of the matrix is preserved as the residual thermal stress. Figure 13 shows the equivalent stress field distribution after machining. It can be seen from the figure that the stress at the crack tip is relatively concentrated. As the temperature of the molten metal drops, the metal continues to solidify in the working fluid, and the state changes from liquid to solid. At this time, the distance between atoms is further shortened, and the volume of the metal is reduced. The material generates stress due to the shrinkage. The maximum stress in the figure is 404.49 MPa. The stress is mainly due to the phase transition stress during the transformation of the molten metal into solid metal. The result is close to the literature [17,18].

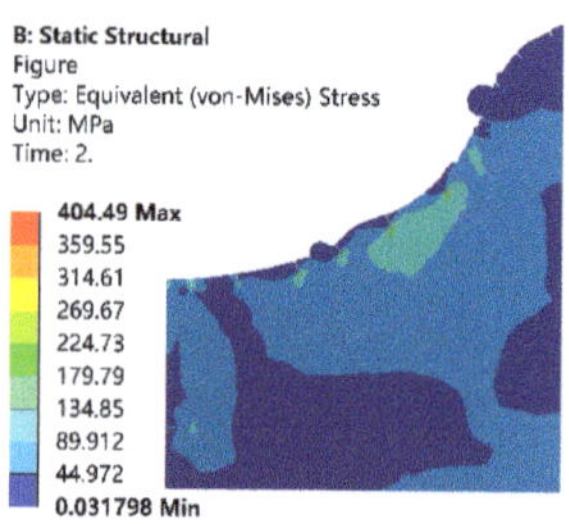

Figure 13. The residual stress after machining.

6. Surface Crack Observation Experiment

6.1. Experiment Setup

The experiment was carried out on a self-built EDM machine, as shown in Figure 14. 65 vol% SiCp/Al was selected as the workpiece, red copper with a diameter of 1 mm was selected as the tool, deionized water was used as the working liquid, RC power supply was used to provide discharge energy for EDM machining, and positive polarity machining was adopted. In the experiment, the discharge voltage of the power supply was 45 V, the peak current was 25 A, the discharge pulse width was 20 μs, and the discharge machining parameters were shown in Table 2. The surface of the tool and the workpiece were polished before machining to remove the oxide layer and other impurities. In the experiment, the single-pulse discharge was adopted. When a spark was generated between the tool and the workpiece, the control circuit immediately disconnected the power supply to realize a single effective discharge.

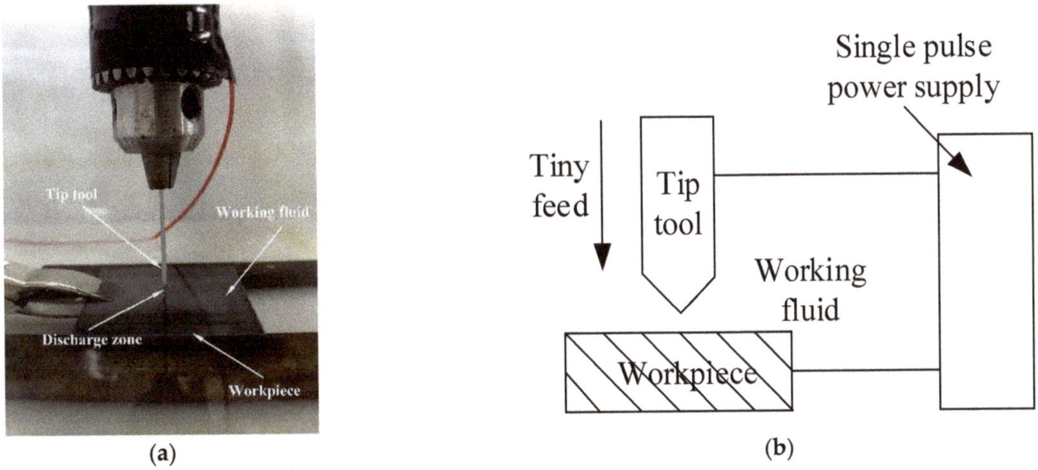

Figure 14. Experimental process of single pulse tip discharge and its schematic diagram. (**a**) Tip discharge with single pulse; (**b**) Tip discharge process.

Table 2. Discharge parameters in EDM machining.

Parameter	Description
Tool material	Red copper
Workpiece material	SiCp/Al
Working liquid	Deionized water
Tool diameter (mm)	1
Voltage U (V)	45
Current I (A)	25
Pulse-on T_{on} (μs)	20

JSM6360-LV high pressure vacuum scanning electron microscope (HVSEM) was used to observe the surface crack distribution of the pits after machining. The specific parameters of SEM are shown in Table 3. To study the influence of residual stress on surface cracks, three positions on the pit surface were selected for observation after the experiment. They are the center position of the pit, the middle position of the pit along the radius, and the position of the edge of the pit. The three positions A, B, C are shown in Figure 15. FV1000 electron microscope was used to take the overall photo of the pit, as shown in Figure 16.

Table 3. The parameters of SEM.

Parameter	Description
LV pressure P_L (Pa)	1~270
Accelerating voltage V_a (kV)	0.3~30
Specimen stage *s* (mm)	X: 125, Y: 100, Z: 5–80
Tilt R_l (°)	−10–+90
Rotation R_t (°)	360
Magnification *n*	×5–×100,000

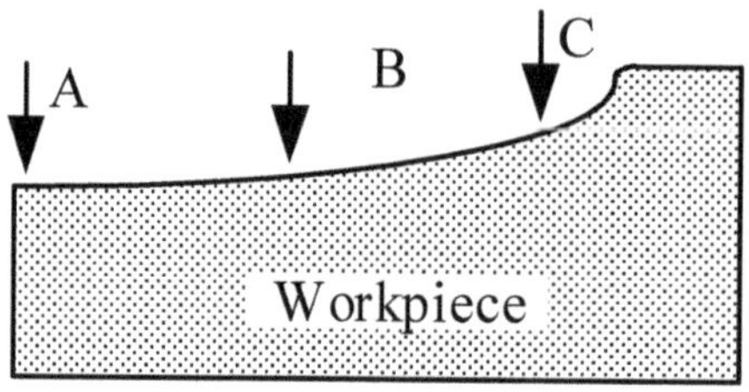

Figure 15. Positions on the surface of the pit.

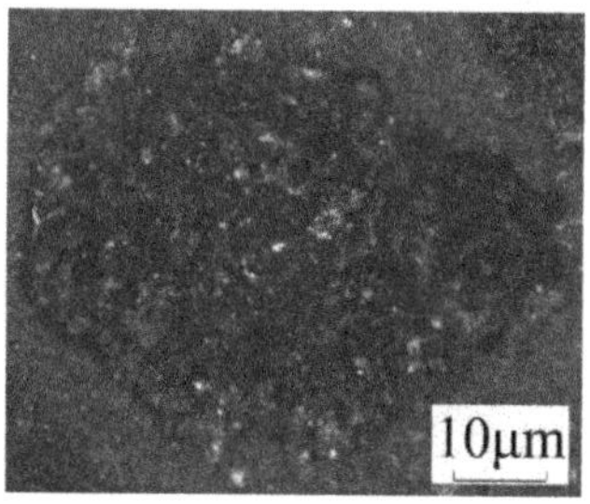

Figure 16. The overall photo of the pit.

6.2. Experimental Result and Analysis

Figure 17 shows the distribution of cracks on the surface of the pit at different positions observed by SEM. Figure 17a is the surface morphology at position A. As seen from the figure, there is no visible crack in this area. This is mainly because the stress in the central area of the pit is very small, the stress does not exceed the yield limit of the material, and no plastic deformation occurs in this area. It is difficult for the microcracks to form visible macroscopic cracks. Therefore, no cracks are observed here. Figure 17b is the surface morphology at position B. Cracks can be observed in this region, and the crack inside the square has a slender shape. This is mainly because during the heating process, the materials constrain each other due to thermal expansion, causing local stress between the two materials. In general, the thermal expansion coefficient of the metal matrix is large, but particles are much smaller. The difference of thermal expansion coefficient results in thermal mismatch stress, which leads to microplastic deformation of the material. This plastic deformation develops rapidly at the tip of the microcrack, and then promotes the initiation and propagation of the microcrack. Figure 17c is the surface morphology at position C. The crack in the block is obvious and has a slender shape. This is because the stress concentration here is larger than the yield strength limit of the material itself, and the plastic deformation is larger than that at position B, so the crack is further expanded in length and width [19].

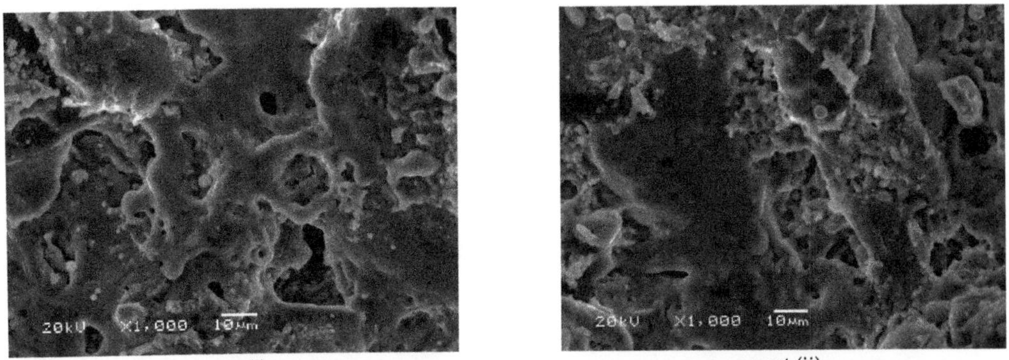

part (i) part (ii)

(**a**) Position A

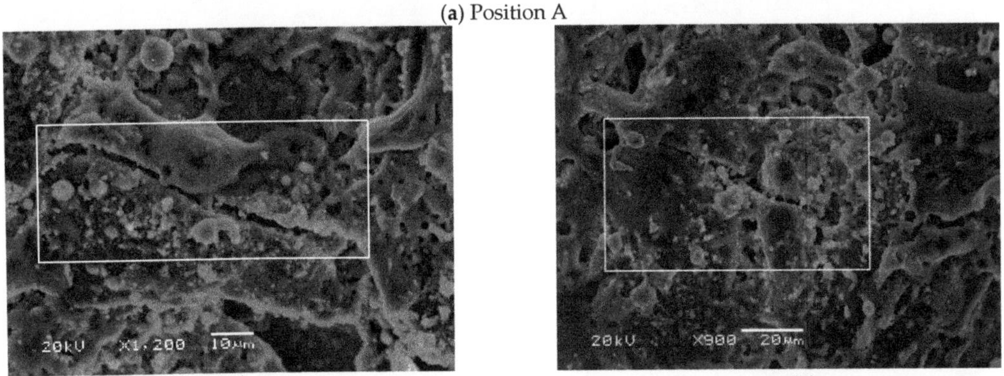

part (i) part (ii)

(**b**) Position B

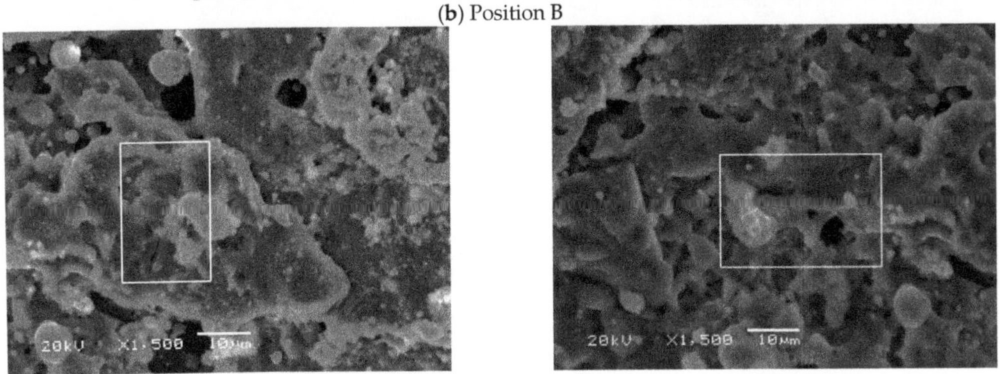

part (i) part (ii)

(**c**) Position C

Figure 17. The cracks on the pit surface.

The main reason for crack propagation in Figure 17b,c is that extremely high temperature is generated instantly on the tool and workpiece surface during EDM, and the temperature decreases rapidly after discharge, which causes the material to expand and contract in a short period and be subjected to the thermal shock of this cycle. Thermal stress and thermal fatigue occur inside the material. Due to defects such as pores, impurities and microcracks, the stress concentration at the microcrack tip easily results in the formation of macroscopic cracks when subjected to thermal impact. In the simulation, the tip stresses at position B and position C on the surface of the pit are larger. In the experiment, large cracks appear at position B and position C, indicating that a large residual stress is generated inside the pit and is greater than the yield strength of the material, resulting in microscopic plastic deformation in the material. The crack location and size in the experiment can indirectly verify the distribution of residual stress obtained in the simulation, which verifies the correctness of the simulation results.

In the experiment, the X-ray diffractometer can only measure the surface of the workpiece. Proto LXRD 3000 was used to measure the residual stress of the three points A, B, and C. The maximum stress exists on Position B, which is 164 MPa. The second largest stress is on position C, 131 MPa. The minimum stress is 86 MPa, which is on position A. The largest stress normally appears at the crack tip. In the simulation, a small area for each position is selected, and the average value is calculated. The simulation result at position A is 90.25 MPa, the result at position B is 168.06 MPa, and the result at position C is 110.63 MPa. The results are close to those results in literature [6,7]. Compared with the simulation results, the experimental results are similar to the simulation results and verify the distribution of the cracks on the surface of the pit. The simulation and experimental results are shown in Figure 18.

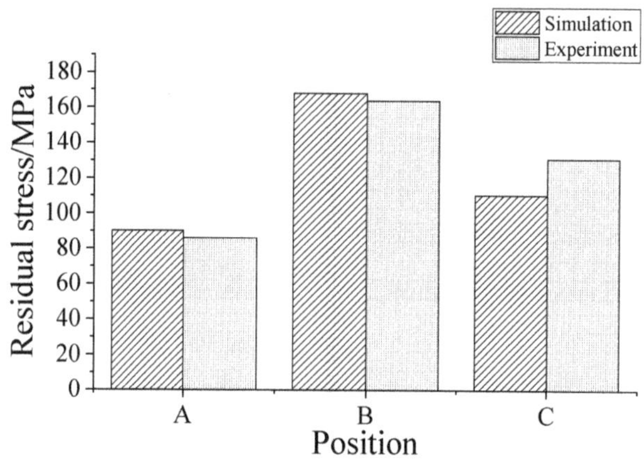

Figure 18. Residual stress at different positions.

After machining, to explore the elements on the surface, the spectrum analysis experiment was conducted. The impurities of the workpiece were removed by an ultrasonic cleaning machine. The types and contents of elements are shown in Table 4. The surface contains five elements, namely carbon, oxygen, aluminum, silicon, and copper. Oxygen comes from surface pollutants, while copper comes from the tool. When a discharge channel is established between the two electrodes, the electrode surface melts. Under the disturbance of working liquid, part of copper metal solidifies on the pit surface. There are complex chemical and physical changes in the discharge process, and new substances are produced. Some of these substances are free in the working fluid, and some are solidified on the surface of the pits to form a new recast layer.

Table 4. Elements and weights.

Elements	**C**	**O**	**Al**	**Si**	**Cu**
Contents (%)	6.25	18.08	14.13	16.58	44.96

7. Conclusions

In this paper, the one-way fluid-solid coupling method was adopted to establish the process of material melting, throwing, and solidification in the flow field. The residual material stress model in the recast layer was established in the structural field. The surface micromorphology and crack distribution state of the residual material after solidification were explored. The conclusions are as follows:

(1) At the initial moment of discharge, the largest temperature of the tool is 13,532 K and the largest temperature of the workpiece is 15,896 K. With the increase of discharge time, the maximum surface temperatures of the two electrodes present a downward trend, and the molten pool areas continue to expand. As the solidification process progresses, a new recast layer is formed on the surface of the pit.
(2) With the increase of time, the temperature of molten liquid on the surface of the pit decreases slowly. At 700 μs, the residual molten Al on the surface of the pit completed solidified. Meanwhile, the stress at the crack tip also decreases slowly. Micro-plastic deformation occurs in the material, and the residual stress is relaxed.
(3) In the surface crack observation experiment, the crack grows gradually from the center of the pit to the edge, indicating that the stress increases gradually along the radius direction, promoting the initiation and further expansion of microscopic cracks and the formation of macroscopic cracks.
(4) The smallest residual stress is 86 MPa and lies on the center of the pit. The highest stress is 164 MPa and lies in the middle of the bottom center to the edge of the pit, which is close to the simulation results.

In the modeling, the study conducted a one-way coupling fluid structure interaction method. In fact, during solidification, residual stresses are generated synchronously. In the future, the two-way coupling method will be considered in the simulation to improve the accuracy of the results.

Author Contributions: Conceptualization, W.Z. and Y.L.; methodology, W.Z.; software, W.Z. and H.C.; validation, Y.L.; formal analysis, W.Z. and Y.L.; investigation, W.Z.; resources, Y.L.; data curation, W.Z. and H.C.; writing—original draft preparation, W.Z.; writing—review and editing, Y.L. and H.C.; visualization, H.C.; supervision, Y.L.; project administration, Y.L.; funding acquisition, Y.L. and H.C. All authors have read and agreed to the published version of the manuscript.

Funding: This research was funded by National Natural Science Foundation of China (51875074), Natural Science Basic Research Project of Education Department of Liaoning Province (JDL2020009), and Science and Technology Research Project of Science and Technology Hall of Henan Province (212102210314).

Institutional Review Board Statement: Not applicable.

Informed Consent Statement: Not applicable.

Data Availability Statement: The data presented in this study are available on request from the corresponding author. The data are not publicly available due to further study.

Conflicts of Interest: The authors declare no conflict of interest.

References

1. Li, X.S.; Cai, A.H.; Zeng, J.J. Effect of EDM conditions on surface residual stress of Cr12MoV steel. *Mater. Sci. Forum* **2012**, *697*, 171–175. [CrossRef]
2. Liu, J.F.; Guo, Y.B. Residual stress modeling in electric discharge machining (EDM) by incorporating massive random discharges. *Procedia Cirp.* **2016**, *45*, 299–302. [CrossRef]
3. Pujari, S.R.; Koona, R.; Beela, S. Surface integrity of wire EDMed aluminum alloy: A comprehensive experimental investigation. *J. King Saud Univ.-Eng. Sci.* **2018**, *30*, 368–376. [CrossRef]
4. Butt, Z.; Mehmood, S.; Sultan, A.; Anjum, N.; Anwar, W. Determination of residual stress distribution in high strength aluminum alloy after EDM. *Adv. Sci. Technol. Res. J.* **2017**, *11*, 29–35. [CrossRef]
5. Ramulu, M.; Paul, G.; Patel, J. EDM surface effects on the fatigue strength of a 15 vol% SiCp/Al metal matrix composite material. *Compos. Struct.* **2001**, *1*, 79–86. [CrossRef]
6. Sidhu, S.S.; Batish, A.; Kumar, S. Neural network-based modeling to predict residual stresses during electric discharge machining of AlSiC metal matrix composites. *Proc. Inst. Mech. Eng. Part B J. Eng. Manuf.* **2013**, *11*, 1679–1692. [CrossRef]
7. Sidhu, S.S.; Batish, A.; Kumar, S. Analysis of residual stresses in particulate reinforced aluminium matrix composite after EDM. *Mater. Sci. Technol.* **2015**, *31*, 1850–1859. [CrossRef]

8. Rajendran, S.; Marimuthu, K.; Sakthivel, M. Study of crack formation and resolidified layer in EDM process on T90Mn2W50Cr45 tool steel. *Mater. Manuf. Process.* **2013**, *28*, 664–669.
9. Yilmaz, S.; Dunand, D.C. Finite-element analysis of thermal expansion and thermal mismatch stresses in a Cu–60vol% ZrW2O8 composite. *Compos. Sci. Technol.* **2004**, *64*, 1895–1898. [CrossRef]
10. Dutta, I. The nature and effect of thermal residual stresses in discontinuous fiber reinforced metal matrix composites. *Compos. Sci. Technol.* **1991**, *41*, 193–213. [CrossRef]
11. Guangwu, F.; Long, L.; Xiguang, G.; Yingdong, S. Finite Element Analysis of the Crack Deflection in Fiber Reinforced Ceramic Matrix Composites with Multilayer Interphase Using Virtual Crack Closure Technique. *Appl. Compos. Mater.* **2020**, *27*, 307–320. [CrossRef]
12. Rakotondrandisa, A.; Danaila, I.; Danaila, L. Numerical modelling of a melting-solidification cycle of a phase-change material with complete or partial melting. *Int. J. Heat Fluid Flow* **2019**, *76*, 57–71. [CrossRef]
13. Liu, Y.; Zhang, W.C.; Zhang, S.F.; Sha, Z.H. The Simulation Research of Tool Wear in Small Hole EDM Machining on Titanium Alloy. *Appl. Mech. Mater.* **2014**, *624*, 249–254. [CrossRef]
14. Ghalandari, M.; Koohshahi, E.M.; Mohamadian, F.; Shamshirband, S.; Chau, K.W. Numerical simulation of nanofluid flow inside a root canal. *Eng. Appl. Comput. Fluid Mech.* **2019**, *13*, 254–264. [CrossRef]
15. Salih, S.Q.; Aldlemy, M.S.; Rasani, M.R.; Ariffin, A.K.; Ya, T.M.Y.S.T.; Al-Ansari, N.; Yaseen, Z.M.; Chau, K.-W. Thin and sharp edges bodies-fluid interaction simulation using cut-cell immersed boundary method. *Eng. Appl. Comput. Fluid Mech.* **2019**, *13*, 860–877. [CrossRef]
16. Elger, D.F.; LeBret, B.A.; Crowe, C.T.; Roberson, J.A. *Engineering Fluid Mechanics*; John Wiley & Sons: Hoboken, NJ, USA, 2020.
17. Rajhi, W.; Alatawi, I.; Subhani, T.; Ayadi, B.; Al-Ghamdi, A.; Khaliq, A. A contribution to numerical prediction of surface damage and residual stresses on die-sinking EDM of Ti6Al4V. *J. Manuf. Process.* **2021**, *68*, 1458–1484. [CrossRef]
18. Bhattacharya, S.; Mishra, A.; Singh, T.; Bandyopadhyay, S.; Kumar, S.; Dey, G.; Kain, V. Simulation of residual stress using thermomechanical analysis on wire EDM cut surface. *Mater. Today Proc.* **2019**, *19*, 462–467. [CrossRef]
19. Li, Q.; Yang, X. Study on arc plasma movement and its effect on crater morphology during single-pulse discharge in EDM. *Int. J. Adv. Manuf. Technol.* **2020**, *106*, 5033–5047. [CrossRef]

micromachines

Review

Atomic Layer Assembly Based on Sacrificial Templates for 3D Nanofabrication

Guangzhou Geng [1], Zhongshan Zhang [1], Chensheng Li [1,2], Ruhao Pan [1], Yunlong Li [1], Haifang Yang [1,3,*] and Junjie Li [1,2,3,*]

1 Beijing National Laboratory for Condensed Matter Physics, Institute of Physics, Chinese Academy of Sciences, Beijing 100190, China; genggz@iphy.ac.cn (G.G.); zhangzs@iphy.ac.cn (Z.Z.); chenshengli@iphy.ac.cn (C.L.); panruhao@iphy.ac.cn (R.P.); liyl@cashq.ac.cn (Y.L.)
2 School of Physical Sciences, University of Chinese Academy of Sciences, Beijing 100049, China
3 Songshan Lake Materials Laboratory, Dongguan 523808, China
* Correspondence: hfyang@iphy.ac.cn (H.Y.); jjli@iphy.ac.cn (J.L.)

Abstract: Three-dimensional (3D) nanostructures have attracted widespread attention in physics, chemistry, engineering sciences, and biology devices due to excellent functionalities which planar nanostructures cannot achieve. However, the fabrication of 3D nanostructures is still challenging at present. Reliable fabrication, improved controllability, and multifunction integration are desired for further applications in commercial devices. In this review, a powerful fabrication method to realize 3D nanostructures is introduced and reviewed thoroughly, which is based on atomic layer deposition assisted 3D assembly through various sacrificial templates. The aim of this review is to provide a comprehensive overview of 3D nanofabrication based on atomic layer assembly (ALA) in multifarious sacrificial templates for 3D nanostructures and to present recent advancements, with the ultimate aim to further unlock more potential of this method for nanodevice applications.

Keywords: atomic layer deposition; sacrificial templates; 3D nanostructures

Citation: Geng, G.; Zhang, Z.; Li, C.; Pan, R.; Li, Y.; Yang, H.; Li, J. Atomic Layer Assembly Based on Sacrificial Templates for 3D Nanofabrication. *Micromachines* **2022**, *13*, 856. https://doi.org/10.3390/mi13060856

Academic Editor: Stelios K. Georgantzinos

Received: 11 May 2022
Accepted: 26 May 2022
Published: 30 May 2022

1. Introduction

3D nanostructures with highly-ordered, complex architectures have attracted significant interest in recent years; various devices based on 3D nanostructures have shown excellent functionalities that planar nanodevices cannot achieve [1,2], such as large specific surface area, more spatial dimensions, and multifunctional integration. Especially, these 3D nanostructures can provide additional accessibility and are open to multidimensional interactions, and can also be highly responsive to various external stimuli, which have the potential to develop newer and more innovative applications of all kinds. Due to the unique structural and functional advantages of 3D nanostructures, they are drawing more and more attention in optics [3], electronics [4], photocatalysis [5,6], piezoelectricity [7,8], energy storage devices [9,10], sensing detection [11], and other applications in nanoscience. To consummate and satisfy the kaleidoscopic needs for the fabrication and modulation of nanostructures, dedicated tools and special techniques are usually required, such as electron beam lithography (EBL), focused ion beam (FIB), induced coupled plasma-reactive ion etching (ICP-RIE), metal organic chemical vapor deposition (MOCVD), self-assembly, etc.

Among these conventional techniques, nanofabrication based on atomic layer deposition (ALD) has attracted more attention in recent years, due to its precise conformality and control over materials' thickness and composition. ALD is based on self-limiting reactions between two gaseous precursors, where the precursors are dosed successively over the growth surface, and achieves the deposition of thin films in a layer-by-layer fashion with atomic-level precision [12,13]. The cyclic and self-limiting nature of ALD enables extremely uniform and conformal films with excellent control over the thickness on virtually any complex substrates [14,15]. Uniformity of ALD film thickness extends not only over flat substrate surfaces but also into narrow holes and trenches. This ability of ALD to make conformal films is called "good step coverage" [16,17]. ALD has emerged as a powerful tool for nanostructure synthesis and functionalization offering unrivalled benefits, especially demonstrating its potential advantages over nanofabrication when combined with various templated structures. Furthermore, ALD enables deposition at low growth temperatures even down to room temperature, which makes ALD suitable for biological and polymer substrates. Up to now, hundreds of ALD processes have been developed to grow an enormous variety of materials, including oxides [18,19], nitrides [20,21], sulfides [22,23], and pure elements [24,25]. For example, representative ALD processes of titanium oxide (TiO_2), aluminum oxide (Al_2O_3), and zinc oxide (ZnO) could be conducted at typical temperature windows of 100–300 °C, and even down to room temperature and as high as 600 °C at certain conditions. Various homoleptic halides, alkoxides, and alkylamides have been widely employed as titanium, aluminum, and zinc precursors. Water is the most common oxygen source, while alternative oxygen sources, such as ozone or plasmas, are sometimes utilized to provide enhanced reactivities in the ALD process. As a result, suitable process parameters should be wisely chosen before utilizing the ALD process. All the advantages of ALD provide great opportunities that are often demanded for today's advanced devices with dimensions down-scaled to a nanometer level, such as magnetic recording, sensors, and drug delivery in medical treatment [26,27], and, of course, which also offer a new approach to nanofabrication.

The 3D nanostructures in this review infer to the intrinsic structures fabricated through sacrificial templates which are finally removed. Usually, it is quite complex and tough for conventional techniques to fabricate 3D nanostructures with giant controllability and extreme high aspect ratio. Due to the ability to coat or fill complex high-aspect-ratio structures with a wide range of materials and the excellent conformality, ALD has developed into an essential tool for the assembly of novel 3D nanostructures, resulting in an accurate replication of various rigid and soft sacrificial templates, which is collectively referred as the atomic layer assembly (ALA) fabrication method based on sacrificial templates. If 3D structures possess reasonable critical dimensions that are suitable for the ALD method after considering time consumption, and the composed material deposited by ALD is compatible with sacrificial templates, the ALA method is applicable to fabricate the 3D structures.

Various sacrificial templates have been investigated to fabricate 3D nanostructures through ALA method, and they could be generally divided into rigid and soft templates, as summarized in Figure 1. Here, rigid templates are hard structures, whose stable structures directly determine the size and morphology of 3D nanostructures. Rigid templates could be silicon (Si), anodic aluminum oxide (AAO), carbon, silica sphere, and even biological structures, etc. In the present context, the term "soft" is introduced mainly to differentiate the materials described here from "rigid". Soft templates could be electron or photoresist polymers, various assembled polymer fibers or membranes, polystyrene (PS) sphere, and so forth.

Generally, the ALA method based on sacrificial templates to obtain the 3D nanostructures includes three steps. Firstly, the sacrificial patterned templates can be fabricated using various advanced nano-techniques. The critical techniques for fabricating Si and AAO rigid templates commonly are lithography combined with etching (dry or wet) and anodization. For soft templates electron or photoresist polymers, electron beam lithography (EBL), direct laser writing (DLW), proximity field nanopatterning (PnP), and multibeam interference lithography (MBIL) are usually utilized. As for assembled polymer templates, block copolymer (BCP), electrospinning (ES), and self-assembly (SA) are regularly employed. Then the films with a certain thickness are deposited on sacrificial templates by ALD for 3D assembly to construct 3D nanostructures. Finally, the residual templates are selectively removed by etching or a heat treatment, and 3D nanostructures shaped by the sacrificial templates can be obtained. Through this ALA method based on sacrificial templates, many kinds of 3D nanostructures could be fabricated, like nanotubes, nanopillars, nanofibers, nanonetworks, inverse opals, and even replication of biological structures [28–31]. These 3D nanostructures possess full and precise control of size and shapes, considerable reproducibility, and flexible design, making them attractive for a broad range of practical applications.

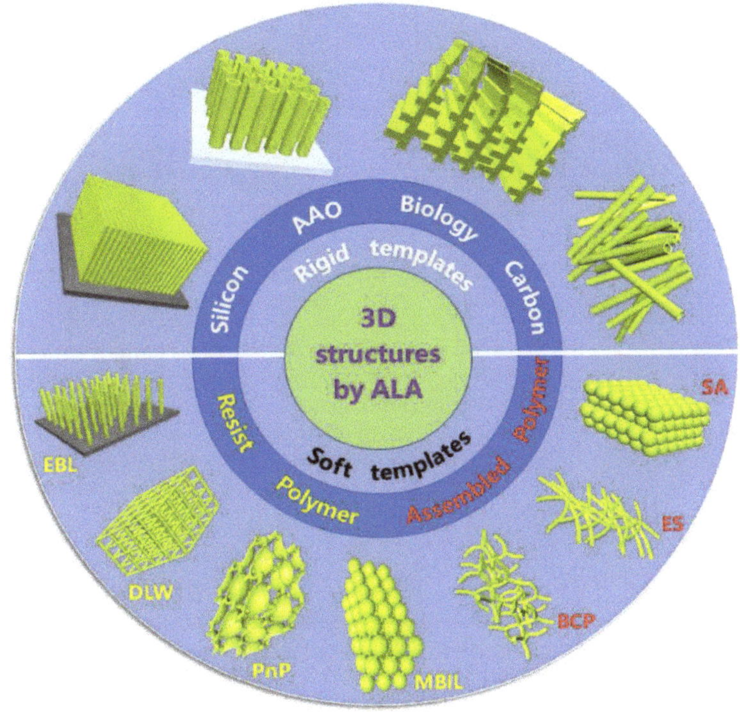

Figure 1. An illustrating summary of the sacrificial templates and obtained typical 3D structures fabricated by atomic layer assembly method. Generally, the sacrificial templates are divided into rigid and soft templates. Typically obtained 3D structures based on ALA method through different sacrificial templates are illustrated in the outer circle of the figure.

There are two main fabricating mechanisms for the 3D nanostructures based on sacrificial templates by ALA method. As shown in Figure 2a, the first case is that the films are directed coated onto the as-fabricated templates by ALD, fabricating the 3D nanostructures by full replication from the sacrificial templates after removing the templates. The obtained 3D nanostructures keep the same geometry with the outer contours of the templates, generally achieving hollow cross-linked nanostructures. The assembled polymer templates, carbon, biological templates, and resist polymer by PnP and MBIL are regularly geared to the first case, while another fabricating mechanism shows much more flexibilities in the morphology control and could achieve arbitrary custom patterns with separated pillars and tubes. 3D nanostructures fabricated from Si, AAO templates, and resist templates by EBL and DLW usually belong to this second fabricating mechanism. Different to the first mechanism, the caps of the films by ALD are generally eliminated followed by the removal of sacrificial templates, resulting in the separated pillars and tubes with user-defined patterns, as illustrated in Figure 2b. The advantages of the first mechanism are the mass productivity and easy operation, while the second mechanism shows superiority over the flexibility and controllability in custom patterns, showing plenty of applications in various active devices.

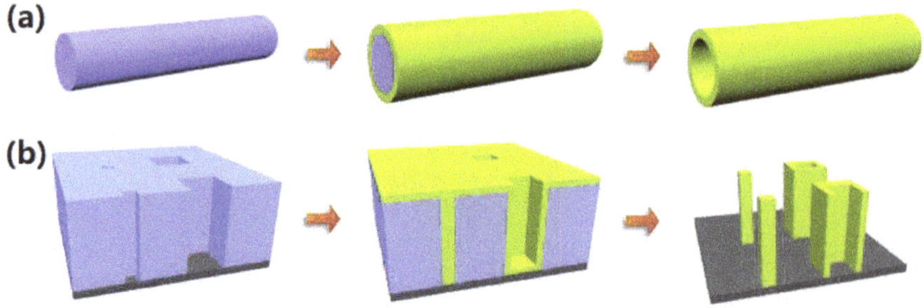

Figure 2. Schematic illustrations of two main fabricating mechanisms for 3D nanostructures based on sacrificial templates by ALA method. (**a**) Mechanism achieving hollow cross-linked 3D nanostructures. (**b**) mechanism achieving arbitrary custom patterns with separated pillars and tubes.

In this review, different types of sacrificial templates are highlighted to explore the fabrication of 3D nanostructures based on the ALA method, which consists of five sections. Following the introduction, 3D nanostructures based on rigid and soft templates are systematically categorized and introduced in Sections 2 and 3. The comparison and applications of 3D nanostructures using various templates are summarized in Section 4. Finally, the conclusion and outlook for the 3D nanostructures through this method are discussed in Section 5.

2. Rigid Templates for Fabricating 3D Nanostructures

Rigid templates are defined as opposed to soft templates, and are made of hard materials. Rigid templates have good chemical stability and mechanical rigidity, which are mostly used for the fabrication of nanostructure arrays. A wide choice of rigid templates is available, such as silicon, anodic aluminum oxide, carbon, silica spheres, biological structures, and so forth, which will be discussed later in detail. By an accurate replica of the rigid templates, the dimensions and specifications of 3D nanostructures can be modulated with higher resolution and high aspect ratio. Of course, both achieving and removing these rigid templates often requires a very complex process, and usually the removal process is environmentally unfriendly.

2.1. Si Templates

Perhaps the most well-known material to fabricate nanostructures, silicon may be the most widely investigated due to its mature fabrication process. Therefore, it is natural for researchers to fabricate 3D nanostructure replicas by utilizing silicon nanostructures as sacrificial templates. Figure 3a exhibits the fabrication routine of 3D nanostructures by ALA method through Si templates. Firstly, the Si templates are commonly prepared by lithography and etching process, resulting in templates with nanopillars or nanopores. Then the ALD process is applied sequentially and generally followed by a removal of the ALD cap on top of the structure. The last step is silicon template removal, which is done by dry or wet etching, leaving the high aspect ratio 3D nanostructures without distortion. In such a way, 3D nanostructures could be obtained, whose shape, size, and thickness can be modulated with higher resolution and high aspect ratio [32–34]. The success of this procedure is heavily dependent on the ability to etch silicon selectively and realize the perfect replication from photoresist patterns to Si templates, without destroying the ALD coatings, which also limits the diversity and resolution in nanostructure for the Si templates.

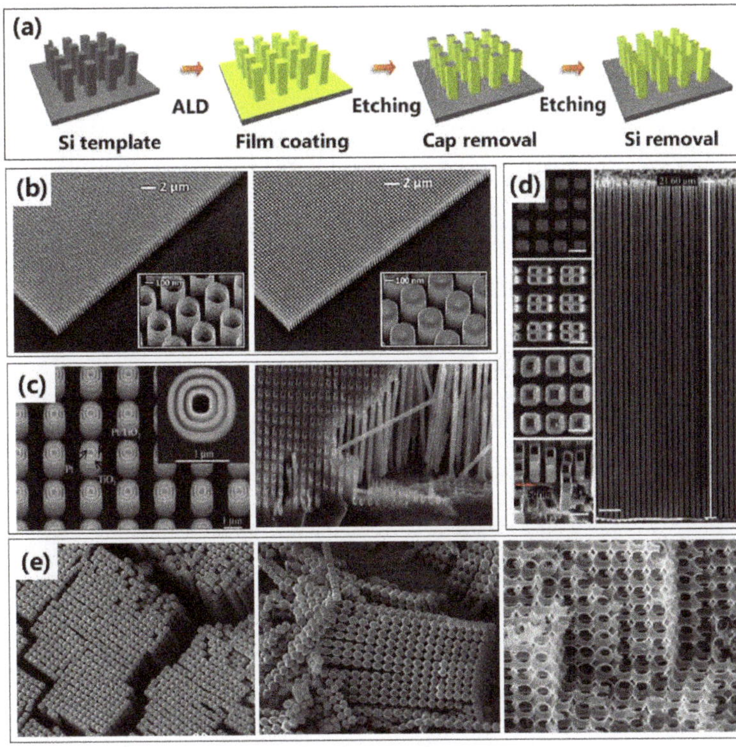

Figure 3. Representative 3D nanostructures fabricated by ALA method through Si templates. (**a**) Process routine of 3D nanostructures based on Si templates by ALA method. (**b**) Titled scanning electron microscope (SEM) images of the 3D metamaterial structures: Al-doped ZnO (AZO) nanotubes and nanopillars. The insets show an enlarged view of the metamaterials. Reprinted from [32], copyright 2017, with permission from Optical Society of America. (**c**) Multi-walled nested nanotube structures with alternating open annular layers after removal via chemical etching of two ALD Al_2O_3 sacrificial spacer layers, cleaved Si template highlighting that ALD Pt has infiltrated up to 90 µm deep into the bottom of the porous Si template. Reprinted with permission from Ref. [35]. Copyright 2011 Springer. (**d**) Top view SEM image of the Si nanostructures after Al_2O_3 film deposition by ALD, and SEM images of the top, cross-section, and tilted views of the generated Al_2O_3 nanotube arrays resulting from the sidewall transfer MacEtch process. Reprinted with permission from Ref. [36]. (**e**) An array of hollow TiO_2 micro-pearl chains viewed from the bottom and the side, and enlarged SEM images of TiO_2 porous network structures. Reprinted with permission from Ref. [37]. Copyright 2008 Springer.

As shown in Figure 3b, large-area 3D Al-doped ZnO (AZO) nanopillars and nanotubes using a combination of advanced reactive ion etching and ALD techniques were fabricated through sacrificial Si templates [32]. It was found that the last step of Si template removal process was heavily dependent on the selective chemistry of the SF_6 plasma. It was shown that silicon between AZO structures can be selectively removed with no observable influence on the ALD deposited coatings. Furthermore, complex multi-walled nested TiO_2–Pt nanotubes in series have been successfully fabricated using microporous Si templates, as shown in Figure 3c (left) [35]. Multilayered nested nanotubes separated by sacrificial spacer layers of TiO_2 (50 nm)/Al_2O_3 (120 nm)/Pt (25 nm)/TiO_2 (25 nm)/Al_2O_3 (120 nm)/Pt (50 nm) were firstly fabricated, followed by removal of the Al_2O_3 sacrificial spacer layers and the porous Si template by using NaOH solution, resulting in the novel multi-walled nested structures. These innovative nested nanostructures have the potential to produce novel multifunctional vertically-ordered 3D nanodevices in photovoltaic and sensing technologies. Especially, optimizing the ALD process by using an additional 10 s exposure time, as-formed Pt could infiltrate up to 90 μm deep into the bottom of the porous Si template, achieving extreme high aspect ratio Pt nanotubes in Figure 3c (right).

Furthermore, ultra-high aspect ratio Al_2O_3 nanotube arrays with line widths as small as 50 nm, heights of up to 21 μm, and an aspect ratio of up to 420:1 were successfully fabricated, utilizing reliable sidewall transfer low-temperature Au metal assisted chemical etching (MacEtch) process, as displayed in Figure 3d [36]. This technique combined the advantages of the high aspect ratio nanostructure capabilities of the MacEtch with the sidewall transfer process. Such a sidewall transfer MacEtch approach was compatible with well-established silicon planar processes, and has the benefits of possessing a fully controllable linewidth and height, high reproducibility, and flexible design, making it attractive for a broad range of practical applications. Similar high aspect ratio 3D nanostructures were investigated [34], while Bosch process was applied to fabricate the Si template with high aspect ratio nanoholes. As long as complex Si templates could be manufactured, more complex nanostructures could be realized by replication of the complex Si templates. As shown in Figure 3e, 3D macroporous silicon template was firstly fabricated by wet etching method [37], then hollow TiO_2 micropearl chains and networks were obtained by ALD after removal of Si template. For possible applications, the presented TiO_2 micropearls are well suited as carriers for drug delivery, whereas the network structures can work as highly porous photocatalytic material.

The advantages of Si templates are the mature fabrication process, custom flexible design, and easy implementation for array fabrication. While the removal of Si templates, in which wet etching is mostly utilized, is usually complicated and contaminative. Furthermore, it is a challenge to preserve exactly the morphology of the as-deposited nanostructures.

2.2. AAO Templates

Another widely investigated rigid template is the anodic aluminum oxide (AAO) template. AAO is an appealing template to fabricate highly ordered and regular nanostructures, showing advantages including low cost, good compatibility, easy scalability, and flexible controllability, and has been exploited for fabricating various nanostructure arrays [38,39]. As shown in Figure 4a, the fabrication process of 3D nanostructures based on AAO template is practically accordant with Si template, while the AAO templates are usually prepared by anodization method, resulting in templates with various nanopores. Once the AAO templates are prepared, the same following processes as ALD of certain materials and removal of cap and templates are conducted successively. An AAO template possesses significant advantages due to its uniform nanochannels size, tunable pore dimensions, good mechanical strength, and thermal stability. The morphological parameters of the 3D nanostructures, such as the packing density or the aspect ratio, are fully controlled by the geometrical parameters of the AAO template.

Diverse binary or multi nanostructure arrays with high degrees of controllability for each of the sub-components, including material, dimension and morphology, as shown in Figure 4b, were fabricated based on ALA method through an AAO template [40]. This binary nanostructure originated with a distinctive binary-pore AAO template that included two dissimilar sets of pores in one matrix, where the openings of the two sets of pores were towards opposite sides of the template. They also presented proof-of-principle photoelectrodes, transistors, and plasmonic devices made with binary nanostructure arrays using different combinations of materials and morphologies, and demonstrated superior performances compared to their single-component counterparts.

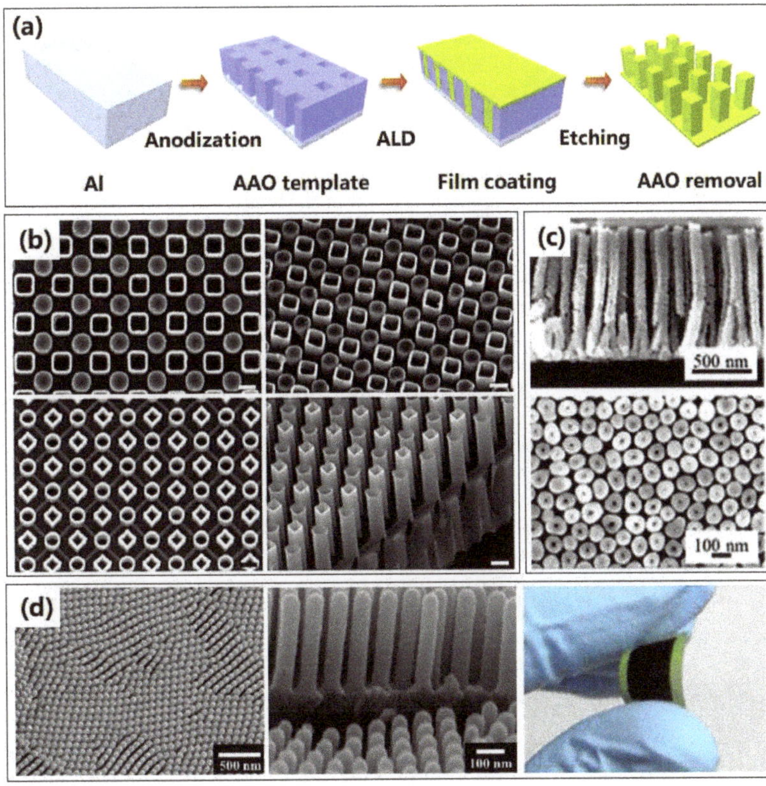

Figure 4. Typical 3D nanostructures fabricated by ALA method through AAO templates. (**a**) Process routine of 3D nanostructures based on AAO templates by ALA method. (**b**) SEM images of representative 3D binary nanostructure arrays. The upper two show nanostructures with nanopillars and nanotubes, and the lower two are nanostructures with all nanotubes. Reprinted with permission from Ref. [40], Copyright 2017 Macmillan Publishers Limited, part of Springer Nature. (**c**) SEM images of 3D ZnO nanowires for sensor applications. Reprinted with permission from Ref. [2], Copyright 2016 American Vacuum Society. (**d**) Overall and enlarged SEM images, and photograph of flexible 3D TiN nanotubes arrays. Reprinted with permission from Ref. [30].

Due to the high surface area, which will enlarge the contact area between structures and reactant, another important application of 3D nanostructures is sensors. A new low-temperature nanowire fabrication process using AAO sacrificial templates was presented [2], which allowed high aspect ratio nanowires to be readily integrated with microelectronic devices for sensor applications. As displayed in Figure 4c, the resulting ZnO nanowire array enhanced the sensor surface area by ~38 times over a conventional flat film, which was supposed to improve capacities of sensors, like sensitivity and response speed.

Flexible devices have achieved remarkable progress over the past years and have become increasingly important to many sectors. Even the AAO template is a rigid template, however, it can also be utilized to fabricate flexible devices. In 2020, the TiN nanotube arrays on flexible substrates were achieved by ALA method through AAO template [30], as shown in Figure 4d. The flexibility of the fabricated nanostructures was investigated using the cyclic bending and releasing method at high curvature. No noticeable cracks or delamination were detected upon operation after 1000 cycles. Meanwhile, delamination and faceted phenomena were not observed after 3000 cycles, implying that this method has a great potential for application in flexible systems as a 3D electrode.

For AAO templates, some problems of complicated post processes are inevitable, and the removal of the AAO template using chemical solvent is a big problem to be compatible with the fabrication of conventional Si-based integrated circuits. The compatibility with other common mature industrial processes needs to be developed further and the environmental concerns should be resolved.

2.3. Biological Templates

Apart from the artificial templates discussed above, the rigid templates even show enormous flexibilities in biological structure. Nature provides extensive micro-nanostructures, which could be utilized as templates to fabricate various 3D structures, such as photonic crystals, fiber networks, etc. Compared with other rigid templates, the bio-templates have the advantage of availability in a wide variety of sources, complex structure, nontoxicity, and easy removal, and they have great potential to be used as templates for the synthesis of other materials. A variety of biological templates have been utilized to fabricate 3D nanostructures, such as veritable biological templates of butterfly wings, fly eyes, cotton, and tubular trichome on legume, as shown in Figure 5. The first bio-inspired novel biostructures with the scale of a butterfly wing [31] were replicated through an Al_2O_3 coating by a low-temperature ALD process. An inverted 3D biostructure was achieved by removing the butterfly wing template at high temperature, forming a polycrystalline Al_2O_3 shell biostructure with precisely controlled thickness, as presented in Figure 5a. Other antireflection nanostructures were fabricated by replicating fly eyes using the same process, as shown in Figure 5b, indicating potential applications in optical coating, sensing, or lens arrays [41]. Furthermore, Tian et al. [42] deposited Al_2O_3 on the surface of low-cost cotton fibers, and hollow Al_2O_3 fibers were synthesized after the elimination of sacrificial cotton templates, and retained the micro- and macrostructures of the cotton. This low-weight Al_2O_3 fiber network with elastic and porous 3D structures overcomes the issues resulting from the 2D rigid Al_2O_3 layer and provides a low overpotential and dendrite-free growth of lithium metal, as shown in Figure 5c. Furthermore, a biomorphic mixed metal oxide framework through a legume bio-templated synthesis method by using a combination of ALD-in situ growth-calcination [43], showing potential applications in catalysis and adsorbents, was reported and investigated, as shown in Figure 5d.

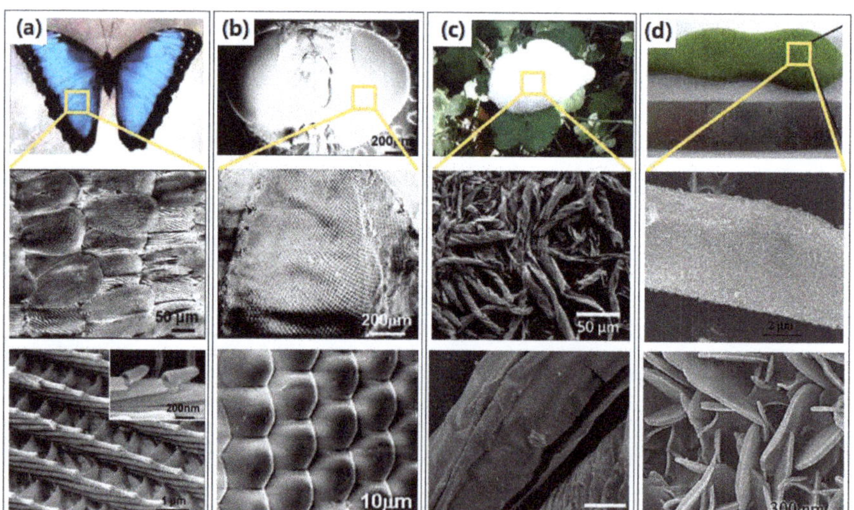

Figure 5. Attractive 3D nanostructures fabricated by ALA method through biological templates, with the photo image of veritable biological template, overall and enlarged SEM images in local area. (**a**) The

biological templates of butterfly wings, with overall and enlarged SEM images of the alumina replicas of the butterfly wings. Reprinted with permission from Ref. [31], Copyright 2006 American Chemical Society. (**b**) Veritable biological templates of fly eyes and SEM images of the alumina replica of a fly compound eye. Reprinted with permission from Ref. [41], Copyright 2008 IOP Publishing Ltd. (**c**) Veritable biological templates of cotton, overall, and enlarged SEM images of fibrous Al_2O_3 nanotube network. Reprinted with permission from Ref. [42], Copyright 2018 Wiley-VCH Verlag GmbH & Co. KGaA, Weinheim. (**d**) Veritable biological templates of tubular trichome on legume, with SEM images of mixed metal oxides framework obtained using a combination of ALD in situ growth calcination. Reprinted with permission from Ref. [43], Copyright 2009 American Chemical Society.

Other groups also investigated various nanostructures with remained biological templates, which enriched the feasibility and functionality of biological templates [44–46]. The advantages of biological templates are low cost, natural structural diversity, and highly accessible, which show enormous possibilities in commercial applications and deserve to be investigated further. However, the low chemical stability at high temperature, low tolerance for multistep fabrication processes, and weak reusability impede the development of the biological templates.

2.4. Other Rigid Templates

Except the templates discussed above, there are several special rigid templates that could be utilized to fabricate 3D nanostructures, such as cellulose, carbon coil, carbon spheres, and silica spheres, as schematically shown in Figure 6. Ortal et al. [47] deposited TiO_2 nanofilms by ALD onto cellulose microfibers, leading to the formation of chiral nanofilms with a spatial fibrous structure, which show potential applications in enantioselective areas. The obtained fibers on cellulose fibers and after cellulose extraction are displayed in Figure 6a. Natural cellulose was also investigated to prepare TiO_2, ZnO, and Al_2O_3 nanotube aerogels, resulting in an efficient humidity sensor with a fast response [48].

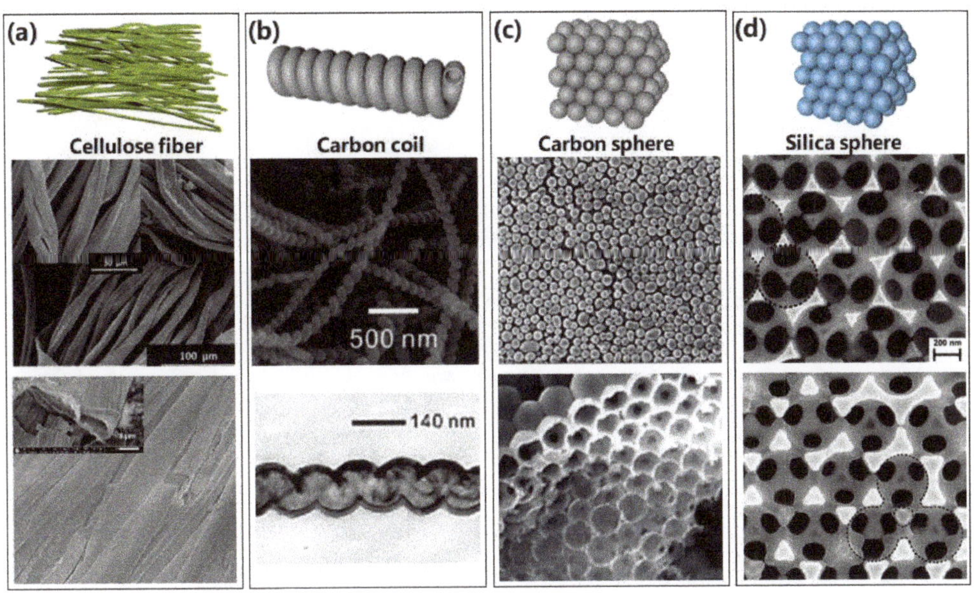

Figure 6. Several special 3D nanostructures fabricated by atomic layer assembly method through cellulose, carbon, and silica templates. (**a**) Schematic images of cellulose fiber templates, SEM images of TiO_2 fibers on cellulose fibers, and SEM images of Al_2O_3 nanofibers after cellulose extraction. Reprinted with permission from Ref. [47], Copyright 2017 American Chemical Society. (**b**) Schematic

images of carbon coil templates, SEM image of as-formed helical Al_2O_3 nanotube after removal of templates, and TEM image of TiO_2 helical nanotubes after removal of templates. Reprinted with permission from Ref. [49], Copyright 2010Wiley-VCH Verlag GmbH & Co. KGaA, Weinheim. (**c**) Schematic images of carbon sphere templates, and SEM images of veritable templates of carbon sphere opal and TiO_2 inverse opal. Reprinted with permission from Ref. [50], Copyright 2020 Elsevier. (**d**) Schematic images of silica spheres templates, and SEM images of TiO_2 close-packed inverse opals and non-close-packed inverse opals. Reprinted with permission from Ref. [51], Copyright 2006 WILEY-VCH Verlag GmbH & Co. KGaA, Weinheim.

A forest of 1D anatase TiO_2 nanoparticle chains was also synthesized based on ALA method using carbon nanotubes as sacrificial templates [52]. The resulting self-supported structure of fully crystallized nanoparticles offers a porous network with a large surface area. As shown in Figures 5c and 6b, carbon nanocoils [49] and nanospheres [50] were also investigated and used as sacrificial templates to prepare distinctive helical nanotubes and inverse opal photonic crystals, respectively, indicating potential applications in sensors, mechanical springs, and photocatalysis areas. Similarly, silica nanospheres [51] were utilized to fabricate inverse opal photonic crystals as shown in Figure 6d, which revealed a potential application in photocatalytic and solar cells.

However, the carbon and silica templates possess the disadvantages of inflexible structure and low variability, which restrict the realization of custom structures and functional expansion based on those templates. With the development of modern techniques, it is believed that more kinds of rigid templates will be fabricated and investigated to achieve 3D nanostructures based on ALA methods.

3. Soft Templates for Fabricating 3D Nanostructures

Just like rigid templates, there are numerous types of soft templates, including electron resist polymer, photoresist polymer, and various assembled polymers consisting of block polymer, fiber or membrane, polystyrene (PS) sphere, and so forth. These versatile soft templates can be used in the ALA method and have broad prospects for development in powerful fabrication of multiple nanostructures, which possess a lot of advantages, such as simple process, good flexibility, repeatable simplicity of the process, and environmentally friendly easy elimination of the templates, resulting in diversiform 3D nanostructures with numerous device applications.

3.1. Resist Polymer Templates

Due to the utilization of direct writing techniques, resist templates, including electron resists and photoresists, maybe the most flexible and controllable templates for 3D nanostructure fabrication based on ALA method. Nowadays, miniaturized conventional systems require the development of custom applications in imaging, displaying, and spectroscopy, especially for 3D nanosystems and 3D nanostructures. As a result, electron and photoresist templates, which could achieve 3D nanostructures by electron beam lithography (EBL) and direct laser writhing (DLW), are proposed and investigated in many fields, and their fabrication processes are illustrated in Figure 7a,d, respectively. Firstly, the electron or photoresists are patterned by EBL or DLW, and then ALD is applied to fill or coat the obtained templates to assemble 3D nanostructures which are confined by templates. Followed by the removal of the cap of the coated template and exposing residual resist polymer, the resist polymer is removed by O_2 plasma or remover.

In 2016, a novel process for fabricating dielectric metasurfaces through soft electron resist templates based on ALA method was first proposed, which could produce anisotropic, subwavelength-spaced 3D nanostructures with shape birefringence [53]. They patterned the electron resist using EBL firstly, resulting in templates with patterns in reverse of the final metasurface. Similar to the process to fabricate 3D nanostructures as discussed above, ALD-based TiO_2 films were coated on the template, followed by the removal of TiO_2 cap and residual resist, high performance 3D nanofin metasurface as a metalens, as shown in Figure 7b, was successfully fabricated and showed widespread applications in laser-based microscopy, imaging, and spectroscopy. This ALA process should be the simplest strategy to realize the complex 3D nanostructures of dielectric metamaterials up to now, which is also capable of achieving any high-efficiency metasurface optical element. Geng et al. developed the ALA method deeper to fabricate 3D nanostructures utilizing electron beam resist, resulting in large-scale arrays of multiple complex 3D nanostructures with high resolution down to nanometers and ultra-high aspect ratio of hundreds. Particularly, an extreme structural nanotube array with ultra-high aspect ratio of more than 80:1 (8 nm wall thickness and 650 nm height) is assembled successfully. The results are displayed in Figure 7c [54].

Creating lightweight, mechanically robust materials has long been an engineering pursuit. Many siliceous skeleton species, such as diatoms, sea sponges, and radiolarians, have remarkably high strengths compared with man-made materials of the same composition, yet are able to remain lightweight and porous [55,56]. As shown in Figure 7e [57], to investigate the lightweight nanostructures, creation of ceramic nanolattices begins with the design and writing of a negative photoresist template using two-photon lithography direct laser writing (DLW), achieving a polymer sacrificial template. Al_2O_3 film is then deposited onto the soft template by ALD, so that it coats the entire surface of this 3D skeleton. Then the internal polymer is etched away in O_2 plasma after the removal of the outermost sides of the coated structure by focused ion beam milling (FIB), resulting in 3D ceramic nanolattice consisting of a network of hollow tubes, which can recover their original shape after compressions in excess of 50% strain. This fabrication method enabled the creation of 3D structures with numerous geometries and custom designs, exhibiting a strong, ultralight, energy-absorbing, and recoverable metamaterial, called a mechanical metamaterial. Using a similar technique based on a photoresist template, gyroid photonic crystals were fabricated successfully, as shown in Figure 7f, which exhibited a complete bandgap in infrared spectroscopy measurements [58].

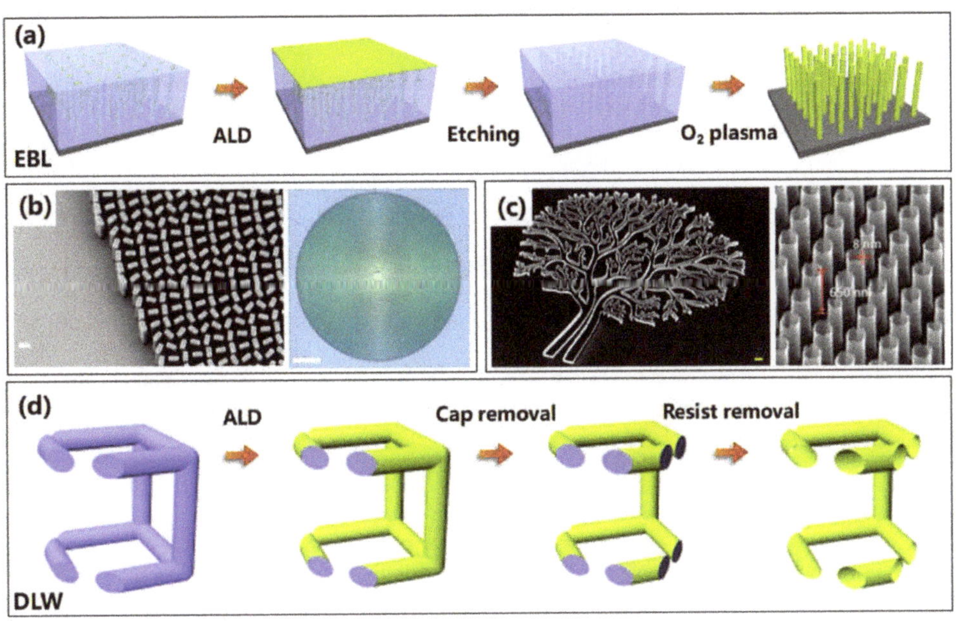

Figure 7. *Cont.*

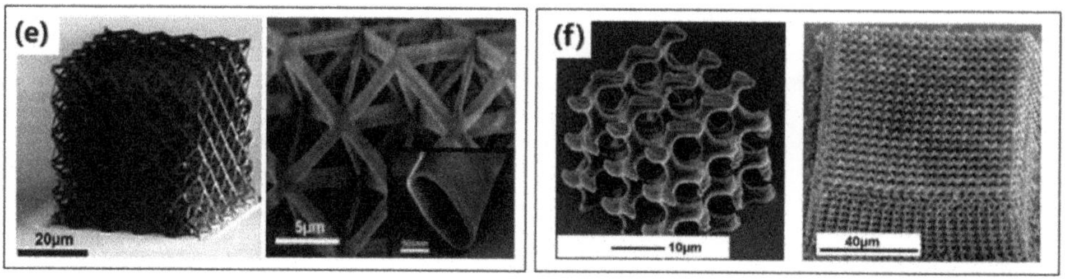

Figure 7. Resist polymer templates by EBL and DLW processes are used to fabricate various 3D nanostructures based on ALA method. (**a**) The illustrated process routine of 3D nanostructures based on electron resist templates through EBL technique. (**b**) SEM and optical images of the fabricated metalens. Reprinted with permission from Ref. [53], Copyright 2016 American Association for the Advancement of Science. (**c**) SEM images of the tree-shaped irregular nanostructures and nanotubes with ultra-high aspect ratio of more than 80:1, based on ALA method through electron resist templates by EBL technique. Reprinted with permission from Ref. [54], Copyright 2021 Elsevier. (**d**) The illustrated process routine of 3D nanostructures based on photoresist templates by DLW technique using ALA method. (**e**) Overall and enlarged SEM images of the overall alumina octet-truss nanolattice through photoresist by DLW technique based on ALA method. Reprinted with permission from Ref. [57], Copyright 2014 American Association for the Advancement of Science. (**f**) SEM images of two kinds gyroid photonic crystals through photoresist by DLW technique based on ALA method. Reprinted with permission from Ref. [58], Copyright 2016 American Chemical Society.

Due to the fact that it commonly relied on negative resists using two-photon lithography, the 3D nanostructure fabrications based on photoresist are limited [59,60], whose topography and size are restricted. Consequentially, electron resist shows more advantages than DLW photoresist in fabricating 3D nanostructures based on ALA method, which can maximize the virtues of following ALD process in flexible custom topography, precise size control, and more compatibility, further showing giant applications in various optical devices [61–68]. Despite the long-time consumption, the fabrication of 3D structures through resist polymers by ALA method shows incomparable flexibilities and great compatibility with conventional semiconductor processes, showing ability in achieving arbitrary custom pattern and complicated 3D structures with extreme geometries. New techniques combined with ALA method, like nanoprinting and multibeam exposing, should be developed to reduce the time consumption.

Apart from EBL and DLW, several other techniques could be utilized to pattern resist templates. As presented in Figure 8a, two feasible techniques to pattern photoresist templates are the proximity field nanopatterning (PnP) technique and multibeam interference lithography (MBIL).

PnP is a versatile 3D nanopatterning technique that creates highly ordered 3D nanostructures in photosensitive materials by capturing the 3D light distribution, generated by a phase mask with periodic relief structures. The 3D light intensity distribution, called the Talbot effect or self-imaging effect [69,70], has periodically repeated images generated by interferences of diffracted beams. This method can rapidly fabricate 3D nanostructures through only single exposure over a large area (>1 inch^2) [71]. As displayed in Figure 8b, hollow-tube-based 3D Al_2O_3 nanoarchitectures functionalized as lightweight materials were fabricated in large areas using PnP and ALA process [72]. The first two SEM images show nanoarchitectures taken during compression and after complete unloading, which revealed that the strengths of these nanoarchitecture materials were powerful. The zoomed-in image of the single unit cell is exhibited in the bottom of Figure 8b. Another group [73] reported a truss-like 3D hollow ZnO nanostructure using PnP method that exhibits a drastically improved elastic strain limit while maintaining a piezoelectric coefficient similar to that of single crystal ZnO, showing excellent potential application in enhanced haptic devices, flexible sensors, and energy harvesters. The SEM images of 3D ZnO hollow nanostructures after removal of the epoxy template are displayed in Figure 8c.

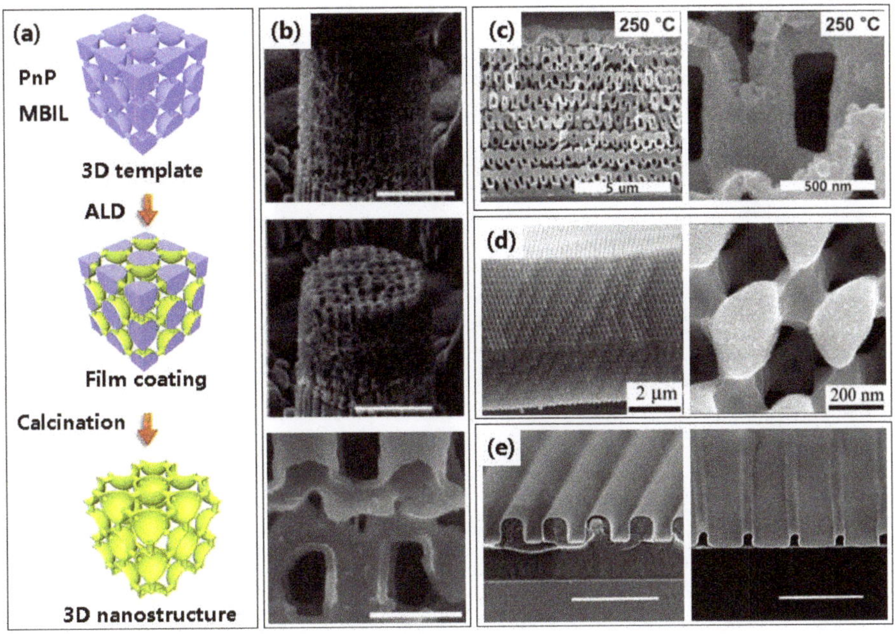

Figure 8. Other photoresist polymer templates by PnP and MBIL processes are used to fabricate 3D nanostructures by ALA methods. (**a**) Illustrated process routine of 3D nanostructures based on photoresist templates by PnP and MBIL techniques. The two techniques commonly share the same fabrication process, except the methods to fabricate resist templates. (**b**) SEM images of Al_2O_3 nano-architectures based on PnP technique taken before and after compression tests. And the enlarged SEM images of Al_2O_3 nano-architectures. Reprinted with permission from Ref. [72], Copyright 2018 WILEY-VCH Verlag GmbH & Co. KGaA, Weinheim. (**c**) SEM images of 3D ZnO hollow nanostructure deposited at 250 °C after removal of the epoxy template based on PnP method. Reprinted with permission from Ref. [73], Copyright 2020 Elsevier. (**d**) 3D TiO_2 photonic crystals after removal of the polymeric template by MBIL method. Reprinted with permission from Ref. [74], Copyright 2006 WILEY-VCH Verlag GmbH & Co. KGaA, Weinheim. (**e**) SEM images of polymer grating template by MBIL method with Platinum film by ALD; and free-standing Pt nano-accordions after removal of the template. Reprinted with permission from Ref. [75], Copyright 2016 The Royal Society of Chemistry.

Unlike PnP method, multibeam interference lithography (MBIL) is a maskless and practical fabrication technique of periodic microstructures over large areas [76,77]. This technique possesses advantages of low cost, no contamination, and fabricating pattern over a large area (up to mm in diameter), but the variety of the patterns is limited due to the exposure mechanism of MBIL, such as an anisotropic and aperiodic structure cannot be achieved by this method. By recording four-beam interference pattern into photoresist, a periodical pillar array is obtained as the sacrificial template [78], which can be used to form microlenses [79,80], optical biosensors [81], or microaxicons for beam generation [82]. The production and optical characterization of a high-quality, high-index 3D photonic crystal using MBIL and ALD was first demonstrated in 2006 [74]. A high-quality photonic crystal in amorphous TiO_2 was produced after removal of the polymer template, showing potential applications in all-optical circuits, microcavity-based light emitters, and other microstructured photonic devices, as shown in Figure 8d. Min et al. [75] also prepared complex Pt nano-accordion structures using a combination of MBIL and ALA; these metal nanostructures have good structural stability and electrical conductivity, as shown in Figure 8e, and their cross sectional profiles can be designed by specifying the template geometry, showing applications in stretchable electronics, photonics, and nanofluidics.

PnP and MBIL methods consume less time than EBL and DLW techniques, which reduces the fabrication cost greatly. Remarkably, there is also no anisotropic volume shrinkage occurring during the calcination to remove the template in the fabrication by PnP and MBIL, providing beneficial optical properties and mechanical stability, showing potential applications for highly efficient gas sensors by utilizing high porosity and large surface area. However, the simplex and inflexible morphologies by PnP and MBIL show less flexibility and controllability, which may limit the application area.

3.2. Assembled Polymer Templates

As well as resist polymers, alternative soft templates can be the variously assembled polymer templates, which generally include a wide variety of high polymers of electrospinning fibers (ESF), blocked polymer (BCP), and self-assembly polystyrene spheres (PSS), and so forth. Obviously, ESF were fabricated by electrospinning technique, and self-assembly technique is applied to obtain the BCP and PSS templates. The 3D nanostructure fabrication processes based on ALA method through ESF, BCP, and PSS templates are similar, which firstly are the preparation of templates by corresponding techniques, followed by film deposition by ALD, and removal of templates by calcination, as shown in Figure 9a.

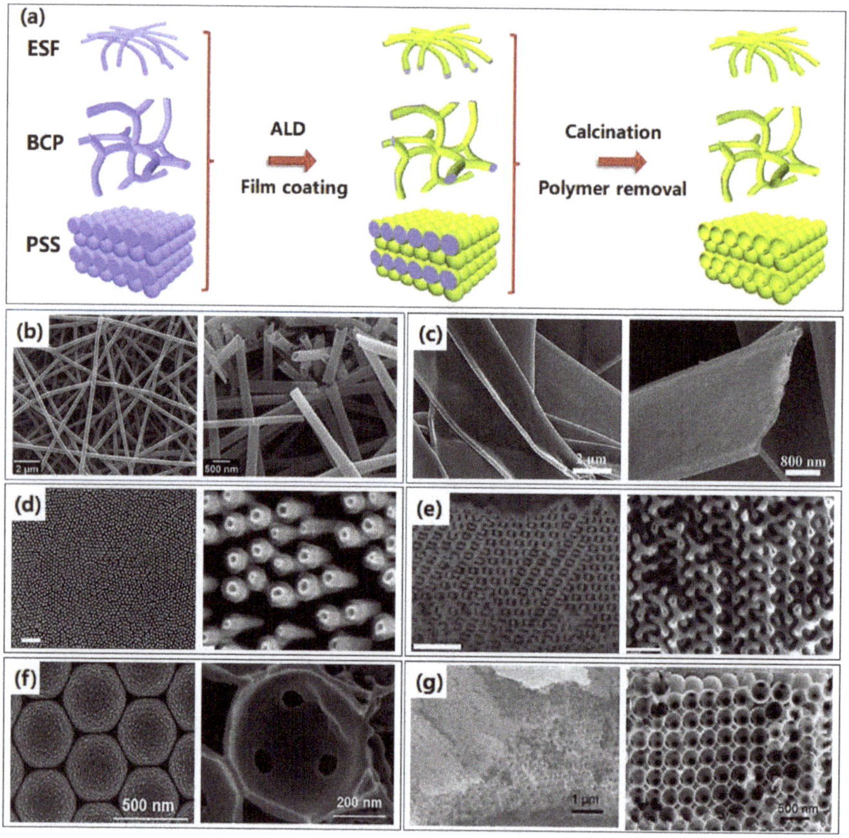

Figure 9. Typical assembled polymer templates are combined with ALA method to fabricate 3D nanostructures. (**a**) The illustrated process routines of 3D nanostructures based on assembled polymer

templates through electrospinning fibers (ESF), blocked polymer (BCP), and polystyrene spheres (PSS) templates. (**b**) Representative SEM images of AlN hollow nanofibers based on electrospinning after calcination. Reprinted with permission from Ref. [83], Copyright 2014, with permission from authors. (**c**) Overall and cross section SEM images of belt-like $BiVO_4$@ZnO heterojunction based on electrospinning after calcination. Reprinted with permission from Ref. [84], Copyright 2018 Elsevier. (**d**) Overall and enlarged SEM images of the nanorockets array based on BCP technique. Reprinted with permission from Ref. [6], Copyright 2017 WILEY-VCH Verlag GmbH & Co. KGaA, Weinheim. (**e**) SEM images of two kinds of mesoporous ZnO networks based on BCP technique. Reprinted with permission from Ref. [85], Copyright 2014 WILEY-VCH Verlag GmbH & Co. KGaA, Weinheim. (**f**) Overall and cross section SEM images of TiO_2 inverse opals based on self-assembled PS spheres. Reprinted with permission from Ref. [86], Copyright 2020 Elsevier. (**g**) SEM images of based TiO_2 inverse opals on self-assembled PS spheres. Reprinted with permission from Ref. [87], Copyright 2017 Elsevier.

The major advantages of the electrospinning process are easy operating and cost-effective to implement. The wide band gap semiconductors based on electrospinning through the ALA method, such as ZnO [88,89], SnO_2 [90,91], and TiO_2 [88,92], including CuO [93], depicted unique surface functionalities and higher sensitivities toward ethanol, O_2 [89,94], NO_2 [91,92], CO [92,94], NH_3 [88,90], H_2 [90,95], and so forth. As a result, the 3D nanostructures based on electrospinning toward sensors were widely investigated. Haider et al. [83] reported fabrication and characterization of aluminum nitride (AlN)/boron nitride (BN) bishell hollow nanofibers through successive ALD of AlN and sequential chemical vapor deposition (CVD) of BN on electrospinning polymeric nanofibrous sacrificial templates, which are exhibited in Figure 9b. This specific structure might find potential use in composite reinforcement, chemical sensing, and gas adsorption. In addition to sensors, nanofibers based on electrospinning and ALA also displayed enhanced photocatalytic activity in water splitting [96,97]. The fabrication of $BiVO_4$@ZnO heterojunction with a novel nanostructure for water splitting via electrospinning and ALA techniques were investigated [84], as shown in Figure 9c. The belt-like 3D structure exhibited enhanced photoelectrochemical performance and showed potential applications in solar cells. There are plenty of low-cost polymer materials to form polymer fiber templates using electrospinning; they could be polyacrylonitrile (PNA) [95,98], Nylon [99,100], polyvinyl acetate (PVA) [92,93], or Poly (vinylpyrrolidone) (PVP) [84,101], which show great opportunities in commercial applications.

Another important assembled polymer template for fabricating 3D nanostructures is the blocked polymer (BCP). The blocked copolymer template-assisted method, which is based on a self-assembly process, can generate well-ordered arrays with small feature sizes of 5–50 nm in dot, line, hole, or lamellar patterns, and has been extensively explored as a strategy to form periodic nanostructures [102,103]. BCP combined with ALA for wafer-scale to fabricate ultrasmall coaxial TiO_2/Pt nanotubes was reported, which is used as a catalytic rocket with a length below 150 nm and a tubular reactor size of only 20 nm, leading to the smallest man-made rocket engine reported to date, as shown in Figure 9d [6], demonstrating abilities to efficiently power the directional transport of significantly larger passive cargo. Two other different morphologies based on the self-assembly of BCP were synthesized to fabricate mesoporous ZnO networks, as shown in Figure 9e [85]. The manufacture of ZnO-based solar cells also showed the feasibility of the integration of the 3D mesostructured ZnO networks into photovoltaic devices.

3D nanostructure could also be formed by the inverse opal (IO) structures, which possess size adjustable pore structure, large surface area, and optical properties of photonic crystals. The inverse opal nanostructures are usually fabricated using a template of silica [50] or polymer micro-spheres by the self-assembly route. Polystyrene sphere (PSS) templates are most commonly used and easily removed by calcination. Inverse opal TiO_2 nanostructures [86] were synthesized by ALA using a template with ordered layers of PS spheres deposited on a Si substrate by spin coating, as illustrated in Figure 9f, where the IO nanostructures showed giant photocatalytic enhancement for the degradation of methylene blue than that of planar TiO_2 films. Similarly, a TiO_2/MoS_2 core/shell inverse opal structure was fabricated based on ALA on a self-assembled multilayer PSS template, as presented in Figure 9g. As a 3D photonic crystal, the TiO_2/MoS_2 inverse opal structure exhibited obvious stopband reflecting peaks, which can be adjusted through changing the opal diameters as well as the thickness of the MoS_2 layer [87]. The inverse opal structures could find plenty more applications in photocatalytic [104,105], solar cells [106,107], chemical sensors [108,109], surface enhanced Raman scattering (SERS) detection [110,111], etc.

The advantage of the ESF, BCP, and PSS template assisted methods should be that no complex and expensive equipment are applied to prepare the templates, as well as easy modification and feasibility of industrial-level high-volume mass production.

4. Comparison and Applications of 3D Nanostructures on Various Templates

The comparison of 3D nanostructures based on ALA through various rigid and soft templates are comprehensively summarized in Table 1. The templates' fabrication and elimination methods, obtained types of 3D structures, controllability, cost, productivity, flexibility, and applications are discussed for the different templates.

4.1. Comparison of 3D Nanostructures on Various Templates

Rigid and soft templates with certain topographies were firstly prepared through various fabrication techniques, which could be etching, anodization, lithography, self-assembly, and even native bio-structures, which result in nanowalls, nanopores, nanotubes, nanospheres, complex 3D nanostructures, and even some customized nanostructures. Generally, the fabrications of rigid templates are relatively complicated and need excessive fabrication processes. For example, wet or dry etching and anodization are utilized to fabricate the Si and AAO templates, respectively, and commonly only nanowalls and nanopores could be achieved in the rigid templates. However, the advantages of rigid templates are that extreme high aspect ratio (hundreds:one) of nanostructures could be achieved by Si or AAO templates, which is usually unreachable for soft templates. In contrast, direct writing lithography or self-assembly are regularly applied to achieve the soft templates with various nanostructures, which are easily conducted and show enormous controllability and flexibilities in size and morphology. However, direct writing lithography is relatively time consuming and expensive, and the self-assembly process lacks controllability and flexibility.

After the sacrificial templates are prepared, essential ALD process is applied to coat or fill the templates, followed by the removal of unwanted templates leaving 3D nanostructures. For the template elimination, dry or wet etching is applied for the Si and AAO templates, which is complex and environmentally unfriendly. Another convenient and low-cost method of calcination is also applied to remove the residual templates of nanostructures based on carbon, biological, and assembled polymer templates, showing potential applications in commercial devices.

Table 1. Comparison of nanostructures based on rigid and soft templates. [a] L = low, M = medium, H = high; [b] ● Optical devices, ★ Electronic devices, ■ Engineering devices, ◆ Sensing devices, ▲ Catalytic device.

Template		Template Fabrication Technique	Template Elimination Technique	Obtained 3D Nanostructures	Controllability [a]	Cost	Productivity	Flexibility	Application [b]
	Si	Etching (Dry, wet)	Etching (Dry, wet)	Nanotubes, Nanopillars	M	M	L	M	●★◆
	AAO	Anodization	Wet etching	Nanotubes, Nanopillars	M	M	M	M	●★◆
Rigid templates	Biological structure	Native	Calcination	Replica of biological structures	L	L	H	L	●◆▲
	Carbon	CVD	Calcination	Nanotubes Nanospheres	L	L	M	L	●▲
	Electron resist	EBL	O_2 plasma/Dissolution	Nanotubes, Nanopillars, Custom structures	H	H	L	H	●★■◆▲
	Photoresist	DLW/PnP/MBIL	O_2 plasma	Nanotubes, Nanopillars, Custom structures	H	M	M	H	●★■◆▲
	Fiber	ES	Calcination	Nanofibers	L	L	H	L	★◆▲
Soft templates	BCP	SA	Calcination	3D networks	M	L	H	M	●★■◆▲
	PS sphere	SA	Calcination	Inverse-opals	L	L	H	L	●◆▲

Various nanostructures could be achieved through the rigid and soft templates after the removal of the sacrificial templates. The electron and photoresist polymer templates exhibit the supreme controllability and flexibility over size and morphology, which could realize almost arbitrary customized nanostructures with high precision, including nanotubes, nanopillars, and custom structures. Especially, arbitrary custom 3D nanostructures with precise resolution and scale down to nanometers could be achieved based on ALA through electron resist, showing powerful applications in various metasurfaces. Therefore, the 3D nanostructures based on ALA method through resist polymer templates exhibit supreme controllability and flexibility. However, relatively low productivity increases the fabrication cost and inhibits the applications in commercial areas. The Si and AAO templates could also achieve various considerable nanostructures, with limited controllability and flexibility, which are generally limited to nanotubes and nanopillars. The integration with a modern conductor fabrication process demands to be developed further to fulfil modern industry need, as well as the exploitation of an environmentally friendly routine for removing the Si and AAO templates.

As to the nanostructures based on carbon and PS spheres, electrospinning fibers, BCP, and biological templates, which are usually achieved by repetition of the inflexible instinct structures, are relatively low cost and possess high productivity but lack high structure resolution, flexibility, and good controllability. As a result, it should be a comprehensive consideration to choose suitable templates and techniques for fabricating 3D nanostructures according to actual needs and conditions.

4.2. Applications of 3D Nanostructures on Various Templates

Along with innovating fabrication methods, expanding the application areas of 3D nanostructures is a growing demand of scientists. Due to the excellent controllability and flexibility of the ALA method based on electron resist templates, this approach provides a powerful platform to reconcile a new flexible 3D nanofabrication with versatile and multiple nanostructures towards broad potential applications in various custom applications in optical, electronic, and sensing fields, like nanophotonics, nanosensing, nanoelectronics, and nanobionics devices [40,54,112,113]. The 3D nanostructures based on AAO templates exhibit applications in plenty of different areas. For example, high aspect 3D nanostructures through AAO templates based on ALA of TiO_2 [114–117], ZnO [118–120], and HfO_2 [121,122] were achieved and further investigated, revealing potential applications in various advanced devices, including fluorescence, sensors, detectors, 3D batteries and supercapacitors, and so on.

Furthermore, the utilizations of 3D nanostructures fabricated by ALA method through other rigid and soft templates in catalytic, sensing, and engineering devices are attracting more attention, as illustrated in Table 1. For example, many 3D nanostructures based on ALA method through different sacrificial templates, with the materials of TiO_2, ZnO, SnO, et al., are photoactive, and thus their practical applications can be mostly found in the fields of photovoltaics [123,124], photoelectrolysis [115,125], photoelectrode [126], and photonic crystals [87,127]. Furthermore, recalling the well-known catalytic activity, these 3D nanostructures fabricated based on various templates, like biological, ES fibers, BCP, and PS sphere templates, provide us with an attractive means for fabricating efficient devices with high-surface-area surfaces for catalysis [128–130]. In addition, another field where high surface area is beneficial is sensor application. Various sensors with 3D nanostructures, like gas [131,132] and biological [133–135] sensors, have been investigated and exhibit excellent stability and response. Up to now, the applications of 3D nanostructures based on various rigid and soft templates have covered many other fields and attracted more and more researchers' attentions, like mechanical [57,136,137], electrical [40,138], optical [29,68], and flexible devices [139–141]. Certainly, all these applications are mass-produced only if reliable, controllable, and low-cost nanofabrication processes of 3D nanostructures are achieved and further improved, but they provide a promising routine for future commercial industry.

5. Conclusions and Outlook

Considering the specialty and importance of 3D nanostructures in excellent functionalities over planar nanostructures, 3D nanostructures have undergone rapid development in recent years, but no mature strategies for mass production exist thus far. ALD assisted 3D assembly through various sacrificial templates offers a promising approach to smoothly implementing the fabrication of 3D nanostructures. In this review, we have systemically addressed the fabrication strategies for 3D nanostructures based on the ALA method through various rigid and soft templates.

Many hard and soft templates with abundant geometries can be used by ALA method to fabricate most 3D nanostructures; key factors including reasonable critical dimension of ALD, the compatibility with the ALD process, and acceptable time consumption should be fully considered as golden rules to judge the feasibility of using the ALA method.

The goal for the development of 3D nanostructures by ALA method based on sacrificial templates is to fabricate commercial devices that will be available in the market, but this nanofabrication strategy still needs great improvement and advancement. 3D nanostructure-based devices still face many challenges, such as achieving a large scale, lost-cost, uniform, high-productivity and precise controllability. These challenges cannot be fully addressed with ALA methods based on various sacrificial templates thus far. Furthermore, integration with traditional semiconductor techniques is still the orientation of 3D nanostructure development. With the recent development of nanofabrication techniques, new fabrication strategies based on ALA method are expected to emerge; we are confident that such nanofabrication platforms with reduced cost, high productivity, precise controllability, and environmentally friendly for 3D nanostructure devices will eventually be commercially available, although great efforts are still needed.

Author Contributions: Original manuscript preparation, G.G., Z.Z. and C.L.; review and editing, R.P. and Y.L.; supervision, H.Y. and J.L. All authors have read and agreed to the published version of the manuscript.

Funding: The authors acknowledge the financial support received from the National Natural Science Foundation of China (Grant No. 12074420, U21A20140, and 61905274), the Beijing Municipal Science & Technology Commission, Administrative Commission of Zhongguancun Science Park (No. Z211100004821009), the Chinese Academy of Sciences through the Project for Young Scientists in Basic Research (YSBR-021), and the Key Deployment Project of Centre for Ocean Mega-Research of Science, Chinese Academy of Science (COMS2020J03). This work is also supported by the Synergic Extreme Condition User Facility.

Data Availability Statement: Not applicable.

Conflicts of Interest: The authors declare no conflict of interest.

References

1. Cui, A.; Li, W.; Shen, T.H.; Yao, Y.; Fenton, J.C.; Peng, Y.; Liu, Z.; Zhang, J.; Gu, C. Thermally Induced Shape Modification of Free-standing Nanostructures for Advanced Functionalities. *Sci. Rep.* **2013**, *3*, 2429. [CrossRef] [PubMed]
2. Clavijo, W.P.; Atkinson, G.M.; Castano, C.E.; Pestov, D. Novel low-temperature fabrication process for integrated high-aspect ratio zinc oxide nanowire sensors. *J. Vac. Sci. Technol. B Nanotechnol. Microelectron. Mater. Process. Meas. Phenom.* **2016**, *34*, 022203. [CrossRef]
3. Kelzenberg, M.D.; Boettcher, S.W.; Petykiewicz, J.A.; Turner-Evans, D.B.; Putnam, M.C.; Warren, E.L.; Spurgeon, J.M.; Briggs, R.M.; Lewis, N.S.; Atwater, H.A. Enhanced absorption and carrier collection in Si wire arrays for photovoltaic applications. *Nat. Mater.* **2010**, *9*, 239–244. [CrossRef] [PubMed]
4. Tomioka, K.; Yoshimura, M.; Fukui, T. A III–V nanowire channel on silicon for high-performance vertical transistors. *Nature* **2012**, *488*, 189–192. [CrossRef]
5. Kim, B.; Li, Y.; Jung, H.; Kim, J.Y.; Lee, D.-K.; Kim, B.; Son, H.J.; Kim, D.-H.; Ko, M.J. Enhanced interconnection of TiO_2 nanoparticles using atomic layer deposition for flexible dye-sensitized solar cells with plastic substrates. *Nano* **2014**, *9*, 1440011. [CrossRef]
6. Li, J.; Liu, W.; Wang, J.; Rozen, I.; He, S.; Chen, C.; Kim, H.G.; Lee, H.; Lee, H.; Kwon, S.; et al. Nanoconfined Atomic Layer Deposition of TiO_2/Pt Nanotubes: Toward Ultrasmall Highly Efficient Catalytic Nanorockets. *Adv. Funct. Mater.* **2017**, *27*, 1700598. [CrossRef]

7. Wu, W.; Wen, X.; Wang, Z.L. Taxel-Addressable Matrix of Vertical-Nanowire Piezotronic Transistors for Active and Adaptive Tactile Imaging. *Science* **2013**, *340*, 952–957. [CrossRef]
8. Pan, C.; Dong, L.; Zhu, G.; Niu, S.; Yu, R.; Yang, Q.; Liu, Y.; Wang, Z.L. High-resolution electroluminescent imaging of pressure distribution using a piezoelectric nanowire LED array. *Nat. Photon.* **2013**, *7*, 752–758. [CrossRef]
9. Banerjee, P.; Perez, I.; Henn-Lecordier, L.; Lee, S.B.; Rubloff, G. Nanotubular metal–insulator–metal capacitor arrays for energy storage. *Nat. Nanotechnol.* **2009**, *4*, 292–296. [CrossRef]
10. Liu, C.; Gillette, E.; Chen, X.; Pearse, A.J.; Kozen, A.C.; Schroeder, M.; Gregorczyk, K.E.; Lee, S.B.; Rubloff, G.W. An all-in-one nanopore battery array. *Nat. Nanotechnol.* **2014**, *9*, 1031–1039. [CrossRef]
11. Najiminaini, M.; Ertorer, E.; Kaminska, B.; Mittler, S.; Carson, J.J.L. Surface plasmon resonance sensing properties of a 3D nanostructure consisting of aligned nanohole and nanocone arrays. *Analyst* **2014**, *139*, 1876–1882. [CrossRef] [PubMed]
12. Puurunen, R.L. Surface chemistry of atomic layer deposition: A case study for the trimethylaluminum/water process. *J. Appl. Phys.* **2005**, *97*, 121301. [CrossRef]
13. Hämäläinen, J.; Ritala, M.; Leskelä, M. Atomic Layer Deposition of Noble Metals and Their Oxides. *Chem. Mater.* **2014**, *26*, 786–801. [CrossRef]
14. Tripathi, T.S.; Karppinen, M. Atomic Layer Deposition of p-Type Semiconducting Thin Films: A Review. *Adv. Mater. Interfaces* **2017**, *4*, 1700300. [CrossRef]
15. Miikkulainen, V.; Leskela, M.; Ritala, M.; Puurunen, R.L. Crystallinity of inorganic films grown by atomic layer deposition: Overview and general trends. *J. Appl. Phys.* **2013**, *113*, 021301. [CrossRef]
16. Park, T.; Kim, H.; Leem, M.; Ahn, W.; Choi, S.; Kim, J.; Uh, J.; Kwon, K.; Jeong, S.-J.; Park, S.; et al. Atomic layer deposition of Al_2O_3 on MoS_2, WS_2, WSe_2, and h-BN: Surface coverage and adsorption energy. *RSC Adv.* **2017**, *7*, 884–889. [CrossRef]
17. Yuan, G.; Wang, N.; Huang, S.; Liu, J. A brief overview of atomic layer deposition and etching in the semiconductor processing. In Proceedings of the 17th International Conference on Electronic Packaging Technology (ICEPT), Wuhan, China, 16–19 August 2016; pp. 1365–1368. [CrossRef]
18. Kim, D.H.; Lee, H.J.; Jeong, H.; Shong, B.; Kim, W.-H.; Park, T.J. Thermal Atomic Layer Deposition of Device-Quality SiO_2 Thin Films under 100 °C Using an Aminodisilane Precursor. *Chem. Mater.* **2019**, *31*, 5502–5508. [CrossRef]
19. Dominik, M.; Leśniewski, A.; Janczuk, M.; Niedziółka-Jönsson, J.; Hołdyński, M.; Wachnicki, Ł.; Godlewski, M.; Bock, W.J.; Śmietana, M. Titanium oxide thin films obtained with physical and chemical vapour deposition methods for optical biosensing purposes. *Biosens. Bioelectron.* **2017**, *93*, 102–109. [CrossRef]
20. Zhang, H.; Wang, B.; Brown, B. Atomic layer deposition of titanium oxide and nitride on vertically aligned carbon nanotubes for energy dense 3D microsupercapacitors. *Appl. Surf. Sci.* **2020**, *521*, 146349. [CrossRef]
21. Sowa, M.J.; Ju, L.; Kozen, A.C.; Strandwitz, N.C.; Zeng, G.; Babuska, T.F.; Hsain, Z.; Krick, B.A. Plasma-enhanced atomic layer deposition of titanium vanadium nitride. *J. Vac. Sci. Technol. A Vac. Surf. Film.* **2018**, *36*, 06A103. [CrossRef]
22. Meng, X.; Cao, Y.; Libera, J.A.; Elam, J.W. Atomic Layer Deposition of Aluminum Sulfide: Growth Mechanism and Electrochemical Evaluation in Lithium-Ion Batteries. *Chem. Mater.* **2017**, *29*, 9043–9052. [CrossRef]
23. Peters, A.W.; Li, Z.; Farha, O.K.; Hupp, J.T. Atomically Precise Growth of Catalytically Active Cobalt Sulfide on Flat Surfaces and within a Metal–Organic Framework via Atomic Layer Deposition. *ACS Nano* **2015**, *9*, 8484–8490. [CrossRef] [PubMed]
24. Kim, W.-H.; Shin, K.; Shong, B.; Godet, L.; Bent, S.F. Atomic Layer Deposition of Pt on the Surface Deactivated by Fluorocarbon Implantation: Investigation of the Growth Mechanism. *Chem. Mater.* **2020**, *32*, 9696–9703. [CrossRef]
25. Weber, M.J.; Mackus, A.J.M.; Verheijen, M.A.; Longo, V.; Bol, A.A.; Kessels, W.M.M. Atomic Layer Deposition of High-Purity Palladium Films from $Pd(hfac)_2$ and H_2 and O_2 Plasmas. *J. Phys. Chem. C* **2014**, *118*, 8702–8711. [CrossRef]
26. Pereira, A.; Palma, J.L.; DeNardin, J.C.; Escrig, J. Temperature-dependent magnetic properties of Ni nanotubes synthesized by atomic layer deposition. *Nanotechnology* **2016**, *27*, 345709. [CrossRef]
27. Escrig, J.; Daub, M.; Landeros, P.; Nielsch, K.; Altbir, D. Angular dependence of coercivity in magnetic nanotubes. *Nanotechnology* **2007**, *18*, 445706. [CrossRef]
28. Zhou, K.; Li, P.; Zhu, Y.; Ye, X.; Chen, H.; Yang, Y.; Dan, Y.; Yuan, Y.; Hou, H. Atomic Layer Deposition of ZnO on TiO_2 Nanofibers for Boosted Photocatalytic Hydrogen Production. *Catal. Lett.* **2020**, *151*, 78–85. [CrossRef]
29. Wei, Z.; Ruisheng, Y.; Guangzhou, G.; Yuancheng, F.; Xuyue, G.; Peng, L.; Quanhong, F.; Fuli, Z.; Changzhi, G.; Junjie, L. Titanium dioxide metasurface manipulating high-efficiency and broadband photonic spin Hall effect in visible regime. *Nanophotonics* **2020**, *9*, 4327–4335.
30. Yun, S.; Kim, S.-J.; Youn, J.; Kim, H.; Ryu, J.; Bae, C.; No, K.; Hong, S. Flexible 3D Electrodes of Free-Standing TiN Nanotube Arrays Grown by Atomic Layer Deposition with a Ti Interlayer as an Adhesion Promoter. *Nanomaterials* **2020**, *10*, 409. [CrossRef]
31. Huang, J.; Wang, X.; Wang, Z.L. Controlled Replication of Butterfly Wings for Achieving Tunable Photonic Properties. *Nano Lett.* **2006**, *6*, 2325–2331. [CrossRef]
32. Shkondin, E.; Takayama, O.; Panah, M.E.A.; Liu, P.; Larsen, P.V.; Mar, M.D.; Jensen, F.; Lavrinenko, A.V. Large-scale high aspect ratio Al-doped ZnO nanopillars arrays as anisotropic metamaterials. *Opt. Mater. Express* **2017**, *7*, 1606–1627. [CrossRef]
33. Tian, J.-L.; Wang, G.-G.; Zhang, H.-Y.; Han, J.-C. Nanopore diameter-dependent properties of thin three-dimensional ZnO layers deposited onto nanoporous silicon substrates. *Ceram. Int.* **2017**, *43*, 5173–5181. [CrossRef]
34. Shkondin, E.; Takayama, O.; Lindhard, J.M.; Larsen, P.V.; Mar, M.D.; Jensen, F.; Lavrinenko, A.V. Fabrication of high aspect ratio TiO_2 and Al_2O_3 nanogratings by atomic layer deposition. *J. Vac. Sci. Technol. A Vac. Surf. Film.* **2016**, *34*, 031605. [CrossRef]

35. Gu, D.; Baumgart, H.; Tapily, K.; Shrestha, P.; Namkoong, G.; Ao, X.; Müller, F. Precise control of highly ordered arrays of nested semiconductor/metal nanotubes. *Nano Res.* **2010**, *4*, 164–170. [CrossRef]
36. Li, H.; Xie, C. Fabrication of Ultra-High Aspect Ratio (>420:1) Al_2O_3 Nanotube Arraysby Sidewall TransferMetal Assistant Chemical Etching. *Micromachines* **2020**, *11*, 378. [CrossRef] [PubMed]
37. Langner, A.; Knez, M.; Müller, F.; Gösele, U. TiO_2 microstructures by inversion of macroporous silicon using atomic layer deposition. *Appl. Phys. A* **2008**, *93*, 399–403. [CrossRef]
38. Wen, L.; Wang, Z.; Mi, Y.; Xu, R.; Yu, S.-H.; Lei, Y. Designing Heterogeneous 1D Nanostructure Arrays Based on AAO Templates for Energy Applications. *Small* **2015**, *11*, 3408–3428. [CrossRef]
39. Zhou, Z.; Nonnenmann, S.S. Progress in Nanoporous Templates: Beyond Anodic Aluminum Oxide and Towards Functional Complex Materials. *Materials* **2019**, *12*, 2535. [CrossRef]
40. Wen, L.; Xu, R.; Mi, Y.; Lei, Y. Multiple nanostructures based on anodized aluminium oxide templates. *Nat. Nanotechnol.* **2017**, *12*, 244–250. [CrossRef]
41. Huang, J.; Wang, X.; Wang, Z.L. Bio-inspired fabrication of antireflection nanostructures by replicating fly eyes. *Nanotechnology* **2007**, *19*, 025602. [CrossRef]
42. Tian, R.; Feng, X.; Duan, H.; Zhang, P.; Li, H.; Liu, H.; Gao, L. Low-Weight 3D Al_2 O_3 Network as an Artificial Layer to Stabilize Lithium Deposition. *ChemSusChem* **2018**, *11*, 3243–3252. [CrossRef] [PubMed]
43. Zhao, Y.; Wei, M.; Lu, J.; Wang, Z.L.; Duan, X. Biotemplated Hierarchical Nanostructure of Layered Double Hydroxides with Improved Photocatalysis Performance. *ACS Nano* **2009**, *3*, 4009–4016. [CrossRef] [PubMed]
44. Ding, Y.; Xu, S.; Zhang, Y.; Wang, A.C.; Wang, M.H.; Xiu, Y.; Wong, C.P.; Wang, Z.L. Modifying the anti-wetting property of butterfly wings and water strider legs by atomic layer deposition coating: Surface materials versus geometry. *Nanotechnology* **2008**, *19*, 355708. [CrossRef] [PubMed]
45. Rodríguez, R.E.; Agarwal, S.P.; An, S.; Kazyak, E.; Das, D.; Shang, W.; Skye, R.; Deng, T.; Dasgupta, N.P. Biotemplated Morpho Butterfly Wings for Tunable Structurally Colored Photocatalysts. *ACS Appl. Mater. Interfaces* **2018**, *10*, 4614–4621. [CrossRef]
46. Liu, F.; Liu, Y.; Huang, L.; Hu, X.; Dong, B.; Shi, W.; Xie, Y.; Ye, X. Replication of homologous optical and hydrophobic features by templating wings of butterflies Morpho menelaus. *Opt. Commun.* **2011**, *284*, 2376–2381. [CrossRef]
47. Lidor-Shalev, O.; Pliatsikas, N.; Carmiel, Y.; Patsalas, P.; Mastai, Y. Chiral Metal-Oxide Nanofilms by Cellulose Template Using Atomic Layer Deposition Process. *ACS Nano* **2017**, *11*, 4753–4759. [CrossRef]
48. Korhonen, J.T.; Hiekkataipale, P.; Malm, J.; Karppinen, M.; Ikkala, O.; Ras, R.H.A. Inorganic Hollow Nanotube Aerogels by Atomic Layer Deposition onto Native Nanocellulose Templates. *ACS Nano* **2011**, *5*, 1967–1974. [CrossRef]
49. Qin, Y.; Kim, Y.; Zhang, L.; Lee, S.-M.; Bin Yang, R.; Pan, A.; Mathwig, K.; Alexe, M.; Gösele, U.; Knez, M. Preparation and Elastic Properties of Helical Nanotubes Obtained by Atomic Layer Deposition with Carbon Nanocoils as Templates. *Small* **2010**, *6*, 910–914. [CrossRef]
50. Bakos, L.P.; Karajz, D.; Katona, A.; Hernadi, K.; Parditka, B.; Erdélyi, Z.; Lukács, I.; Hórvölgyi, Z.; Szitási, G.; Szilágyi, I.M. Carbon nanosphere templates for the preparation of inverse opal titania photonic crystals by atomic layer deposition. *Appl. Surf. Sci.* **2020**, *504*, 144443. [CrossRef]
51. Graugnard, E.; King, J.S.; Gaillot, D.P.; Summers, C.J. Sacrificial-Layer Atomic Layer Deposition for Fabrication of Non-Close-Packed Inverse-Opal Photonic Crystals. *Adv. Funct. Mater.* **2006**, *16*, 1187–1196. [CrossRef]
52. Deng, S.; Verbruggen, S.W.; He, Z.; Cott, D.J.; Vereecken, P.M.; Martens, J.A.; Bals, S.; Lenaerts, S.; Detavernier, C. Atomic layer deposition-based synthesis of photoactive TiO_2 nanoparticle chains by using carbon nanotubes as sacrificial templates. *RSC Adv.* **2014**, *4*, 11648–11653. [CrossRef]
53. Khorasaninejad, M.; Chen, W.T.; Devlin, R.C.; Oh, J.; Zhu, A.Y.; Capasso, F. Metalenses at visible wavelengths: Diffraction-limited focusing and subwavelength resolution imaging. *Science* **2016**, *352*, 1190–1194. [CrossRef] [PubMed]
54. Geng, G.; Zhu, W.; Pan, R.; Zhang, Z.; Gu, C.; Li, J. Precise tailoring of multiple nanostructures based on atomic layer assembly via versatile soft-templates. *Nano Today* **2021**, *38*, 101145. [CrossRef]
55. Kröger, N. Prescribing diatom morphology: Toward genetic engineering of biological nanomaterials. *Curr. Opin. Chem. Biol.* **2007**, *11*, 662–669. [CrossRef]
56. Yang, W.; Chen, I.H.; Gludovatz, B.; Zimmermann, E.A.; Ritchie, R.O.; Meyers, M.A. Natural Flexible Dermal Armor. *Adv. Mater.* **2013**, *25*, 31–48. [CrossRef]
57. Meza, L.R.; Das, S.; Greer, J.R. Strong, lightweight, and recoverable three-dimensional ceramic nanolattices. *Science* **2014**, *345*, 1322–1326. [CrossRef]
58. Peng, S.; Zhang, R.; Chen, V.H.; Khabiboulline, E.T.; Braun, P.; Atwater, H.A. Three-Dimensional Single Gyroid Photonic Crystals with a Mid-Infrared Bandgap. *ACS Photon.* **2016**, *3*, 1131–1137. [CrossRef]
59. Frölich, A.; Fischer, J.; Zebrowski, T.; Busch, K.; Wegener, M. Titania Woodpiles with Complete Three-Dimensional Photonic Bandgaps in the Visible. *Adv. Mater.* **2013**, *25*, 3588–3592. [CrossRef]
60. Kawata, S.; Sun, H.; Tanaka, T.; Takada, K. Finer features for functional microdevices. *Nature* **2001**, *412*, 697–698. [CrossRef]
61. Shi, Z.; Khorasaninejad, M.; Huang, Y.-W.; Roques-Carmes, C.; Zhu, A.Y.; Chen, W.T.; Sanjeev, V.; Ding, Z.-W.; Tamagnone, M.; Chaudhary, K.; et al. Single-Layer Metasurface with Controllable Multiwavelength Functions. *Nano Lett.* **2018**, *18*, 2420–2427. [CrossRef]

62. Devlin, R.C.; Ambrosio, A.; Wintz, D.; Oscurato, S.L.; Zhu, A.Y.; Khorasaninejad, M.; Oh, J.; Maddalena, P.; Capasso, F. Spin-to-orbital angular momentum conversion in dielectric metasurfaces. *Opt. Express* **2017**, *25*, 377–393. [CrossRef] [PubMed]
63. Khorasaninejad, M.; Chen, W.T.; Zhu, A.Y.; Oh, J.; Devlin, R.C.; Rousso, D.; Capasso, F. Multispectral Chiral Imaging with a Metalens. *Nano Lett.* **2016**, *16*, 4595–4600. [CrossRef] [PubMed]
64. Liang, H.; Lin, Q.; Xie, X.; Sun, Q.; Wang, Y.; Zhou, L.; Liu, L.; Yu, X.; Zhou, J.; Krauss, T.F.; et al. Ultrahigh Numerical Aperture Metalens at Visible Wavelengths. *Nano Lett.* **2018**, *18*, 4460–4466. [CrossRef] [PubMed]
65. Devlin, R.C.; Khorasaninejad, M.; Chen, W.T.; Oh, J.; Capasso, F. Broadband high-efficiency dielectric metasurfaces for the visible spectrum. *Proc. Natl. Acad. Sci. USA* **2016**, *113*, 10473–10478. [CrossRef]
66. Mueller, J.P.B.; Rubin, N.A.; Devlin, R.C.; Groever, B.; Capasso, F. Metasurface Polarization Optics: Independent Phase Control of Arbitrary Orthogonal States of Polarization. *Phys. Rev. Lett.* **2017**, *118*, 113901. [CrossRef]
67. Wen, D.; Cadusch, J.J.; Meng, J.; Crozier, K.B. Multifunctional Dielectric Metasurfaces Consisting of Color Holograms Encoded into Color Printed Images. *Adv. Funct. Mater.* **2020**, *30*, 1906415. [CrossRef]
68. Zhang, C.; Divitt, S.; Fan, Q.; Zhu, W.; Agrawal, A.; Lu, Y.; Xu, T.; Lezec, H.J. Low-loss metasurface optics down to the deep ultraviolet region. *Light. Sci. Appl.* **2020**, *9*, 1–10. [CrossRef]
69. Ahn, C.; Park, J.; Kim, D.; Jeon, S. Monolithic 3D titania with ultrathin nanoshell structures for enhanced photocatalytic activity and recyclability. *Nanoscale* **2013**, *5*, 10384–10389. [CrossRef]
70. Latimer, P.; Crouse, R.F. Talbot effect reinterpreted. *Appl. Opt.* **1992**, *31*, 80–89. [CrossRef]
71. Park, J.; Wang, S.; Li, M.; Ahn, C.; Hyun, J.K.; Kim, D.S.; Kim, D.K.; Rogers, J.A.; Huang, Y.; Jeon, S. Three-dimensional nanonetworks for giant stretchability in dielectrics and conductors. *Nat. Commun.* **2012**, *3*, 916. [CrossRef]
72. Na, Y.-E.; Shin, D.; Kim, K.; Ahn, C.; Jeon, S.; Jang, D. Emergence of New Density-Strength Scaling Law in 3D Hollow Ceramic Nanoarchitectures. *Small* **2018**, *14*, 1802239. [CrossRef] [PubMed]
73. Kim, H.; Yun, S.; Kim, K.; Kim, W.; Ryu, J.; Nam, H.G.; Han, S.M.; Jeon, S.; Hong, S. Breaking the elastic limit of piezoelectric ce-ramics using nanostructures: A case study using ZnO. *Nano Energy* **2020**, *78*, 105259. [CrossRef]
74. King, J.S.; Graugnard, E.; Roche, O.M.; Sharp, D.N.; Scrimgeour, J.; Denning, R.G.; Turberfield, A.J.; Summers, C.J. Infiltration and Inversion of Holographically Defined Polymer Photonic Crystal Templates by Atomic Layer Deposition. *Adv. Mater.* **2006**, *18*, 1561–1565. [CrossRef]
75. Min, J.-H.; Bagal, A.; Mundy, J.Z.; Oldham, C.J.; Wu, B.-I.; Parsons, G.N.; Chang, C.-H. Fabrication and design of metal nano-accordion structures using atomic layer deposition and interference lithography. *Nanoscale* **2016**, *8*, 4984–4990. [CrossRef]
76. Stankevičius, E.; Daugnoraitė, E.; Račiukaitis, G. Mechanism of pillars formation using four-beam interference lithography. *Opt. Lasers Eng.* **2019**, *116*, 41–46. [CrossRef]
77. Gedvilas, M.; Indrišiūnas, S.; Voisiat, B.; Stankevičius, E.; Selskis, A.; Račiukaitis, G. Nanoscale thermal diffusion during the laser interference ablation using femto-, pico-, and nanosecond pulses in silicon. *Phys. Chem. Chem. Phys.* **2018**, *20*, 12166–12174. [CrossRef]
78. Stankevicius, E.; Balciunas, E.; Malinauskas, M.; Raciukaitis, G.; Baltriukiene, D.; Bukelskiene, V. Holographic lithography for biomedical applications. *Proc. SPIE* **2012**, *8433*, 843312. [CrossRef]
79. Wang, Z.; Zhao, G.; Zhang, W.; Feng, Z.; Lin, L.; Zheng, Z. Low-cost micro-lens arrays fabricated by photosensitive sol–gel and multi-beam laser interference. *Photon. Nanostructures Fundam. Appl.* **2012**, *10*, 667–673. [CrossRef]
80. Stankevicius, E.; Gedvilas, M.; Raciukaitis, G. Investigation of laser-induced polymerization using a smoothly varying intensity distribution. *Appl. Phys. B* **2015**, *119*, 525–532. [CrossRef]
81. Yuan, L.; Herman, P.R. Laser Scanning Holographic Lithography for Flexible 3D Fabrication of Multi-Scale Integrated Nano-structures and Optical Biosensors. *Sci. Rep.* **2016**, *6*, 22294. [CrossRef]
82. Stankevičius, E.; Garliauskas, M.; Raciukaitis, G. Bessel-like Beam Array Generation Using Round-tip Micro-structures and Their Use in the Material Treatment. *J. Laser Micro Nanoeng.* **2016**, *11*, 352–356.
83. Haider, A.; Ozgit-Akgun, C.; Kayaci, F.; Okyay, A.K.; Uyar, T.; Biyikli, N. Fabrication of AlN/BN bishell hollow nanofibers by electrospinning and atomic layer deposition. *APL Mater.* **2014**, *2*, 096109. [CrossRef]
84. Hou, H.; Liu, H.; Gao, F.; Shang, M.; Wang, L.; Xu, L.; Wong, W.-Y.; Yang, W. Packaging $BiVO_4$ nanoparticles in ZnO microbelts for efficient photoelectrochemical hydrogen production. *Electrochim. Acta* **2018**, *283*, 497–508. [CrossRef]
85. Kim, E.; Vaynzof, Y.; Sepe, A.; Guldin, S.; Scherer, M.R.J.; Cunha, P.; Roth, S.V.; Steiner, U. Gyroid-Structured 3D ZnO Networks Made by Atomic Layer Deposition. *Adv. Funct. Mater.* **2013**, *24*, 863–872. [CrossRef]
86. Birnal, P.; Lucas, M.M.D.; Pochard, I.; Domenichini, B.; Imhoff, L. Photocatalytic properties of atomic layer deposited TiO_2 inverse opals and planar films for the degradation of dyes. *Appl. Surf. Sci.* **2020**, *512*, 145693. [CrossRef]
87. Li, X.; Puttaswamy, M.; Wang, Z.; Tan, C.K.; Grimsdale, A.C.; Kherani, N.P.; Tok, A.I.Y. A pressure tuned stop-flow atomic layer deposition process for MoS_2 on high porous nanostructure and fabrication of TiO_2/MoS_2 core/shell inverse opal structure. *Appl. Surf. Sci.* **2017**, *422*, 536–543. [CrossRef]
88. Boyadjiev, S.I.; Kéri, O.; Bárdos, P.; Firkala, T.; Gáber, F.; Nagy, Z.K.; Baji, Z.; Takács, M.; Szilágyi, I.M. TiO_2/ZnO and ZnO/TiO_2 core/shell nanofibers prepared by electrospinning and atomic layer deposition for photocatalysis and gas sensing. *Appl. Surf. Sci.* **2017**, *424*, 190–197. [CrossRef]
89. Park, J.Y.; Choi, S.-W.; Lee, J.-W.; Lee, C.; Kim, S.S. Synthesis and Gas Sensing Properties of TiO_2-ZnO Core-Shell Nanofibers. *J. Am. Ceram. Soc.* **2009**, *92*, 2551–2554. [CrossRef]

90. Kim, W.-S.; Lee, B.-S.; Kim, D.-H.; Kim, H.-C.; Yu, W.-R.; Hong, S.-H. SnO_2 nanotubes fabricated using electrospinning and atomic layer deposition and their gas sensing performance. *Nanotechnology* **2010**, *21*, 245605. [CrossRef]
91. Choi, S.-W.; Park, J.Y.; Kim, S.S. Synthesis of SnO_2–ZnO core-shell nanofibers via a novel two-step process and their gas sensing properties. *Nanotechnology* **2009**, *20*, 465603. [CrossRef]
92. Katoch, A.; Kim, J.-H.; Kim, S.S. TiO_2/ZnO Inner/Outer Double-Layer Hollow Fibers for Improved Detection of Reducing Gases. *ACS Appl. Mater. Interfaces* **2014**, *6*, 21494–21499. [CrossRef] [PubMed]
93. Katoch, A.; Choi, S.-W.; Sun, G.-J.; Kim, H.W.; Kim, S.S. Mechanism and prominent enhancement of sensing ability to reducing gases in p/n core-shell nanofiber. *Nanotechnology* **2014**, *25*, 175501. [CrossRef] [PubMed]
94. Katoch, A.; Choi, S.-W.; Kim, S.S. Effect of the wall thickness on the gas-sensing properties of ZnO hollow fibers. *Nanotechnology* **2014**, *25*, 455504. [CrossRef]
95. Lee, B.-S.; Kim, W.-S.; Kim, D.-H.; Kim, H.-C.; Hong, S.-H.; Yu, W.-R. Fabrication of SnO_2 nanotube microyarn and its gas sensing behavior. *Smart Mater. Struct.* **2011**, *20*, 105019. [CrossRef]
96. Khalily, M.A.; Patil, B.; Yilmaz, E.; Uyar, T. Atomic layer deposition of Co_3O_4 nanocrystals on N-doped electrospun carbon nanofibers for oxygen reduction and oxygen evolution reactions. *Nanoscale Adv.* **2019**, *1*, 1224–1231. [CrossRef]
97. Du, Q.; Wu, J.; Yang, H. Pt@Nb-TiO_2 Catalyst Membranes Fabricated by Electrospinning and Atomic Layer Deposition. *ACS Catal.* **2014**, *4*, 144–151. [CrossRef]
98. Kayaci, F.; Vempati, S.; Ozgit-Akgun, C.; Biyikli, N.; Uyar, T. Enhanced photocatalytic activity of homoassembled ZnO nanostructures on electrospun polymeric nanofibers: A combination of atomic layer deposition and hydrothermal growth. *Appl. Catal. B Environ.* **2014**, *156*, 173–183. [CrossRef]
99. Donmez, I.; Kayaci, F.; Ozgit-Akgun, C.; Uyar, T.; Biyikli, N. Fabrication of hafnia hollow nanofibers by atomic layer deposition using electrospun nanofiber templates. *J. Alloy. Compd.* **2013**, *559*, 146–151. [CrossRef]
100. Ozgit-Akgun, C.; Kayaci, F.; Vempati, S.; Haider, A.; Celebioglu, A.; Goldenberg, E.; Kizir, S.; Uyar, T.; Biyikli, N. Fabrication of flexible polymer–GaN core–shell nanofibers by the combination of electrospinning and hollow cathode plasma-assisted atomic layer deposition. *J. Mater. Chem. C* **2015**, *3*, 5199–5206. [CrossRef]
101. Hu, M.; Teng, F.; Chen, H.; Jiang, M.; Gu, Y.; Lu, H.; Hu, L.; Fang, X. Novel Ω-Shaped Core-Shell Photodetector with High Ultraviolet Selectivity and Enhanced Responsivity. *Adv. Funct. Mater.* **2017**, *27*, 1704477. [CrossRef]
102. Ku, S.J.; Jo, G.C.; Bak, C.H.; Kim, S.M.; Shin, Y.R.; Kim, K.H.; Kwon, S.H.; Kim, J.-B. Highly ordered freestanding titanium oxide nanotube arrays using Si-containing block copolymer lithography and atomic layer deposition. *Nanotechnology* **2013**, *24*, 85301. [CrossRef] [PubMed]
103. Jung, Y.S.; Chang, J.B.; Verploegen, E.; Berggren, K.K.; Ross, C.A. A Path to Ultranarrow Patterns Using Self-Assembled Lithography. *Nano Lett.* **2010**, *10*, 1000–1005. [CrossRef] [PubMed]
104. Jiao, J.; Wei, Y.; Chi, K.; Zhao, Z.; Duan, A.; Liu, J.; Jiang, G.; Wang, Y.; Wang, X.; Han, C.; et al. Platinum Nanoparticles Supported on TiO_2 Photonic Crystals as Highly Active Photocatalyst for the Reduction of CO_2 in the Presence of Water. *Energy Technol.* **2017**, *5*, 877–883. [CrossRef]
105. Qi, D.; Lu, L.; Xi, Z.; Wang, L.; Zhang, J. Enhanced photocatalytic performance of TiO_2 based on synergistic effect of Ti_3+ self-doping and slow light effect. *Appl. Catal. B Environ.* **2014**, *160*, 621–628. [CrossRef]
106. Lee, J.W.; Lee, J.; Kim, C.; Cho, C.-Y.; Moon, J.H. Facile fabrication of sub-100 nm mesoscale inverse opal films and their appli-cation in dye-sensitized solar cell electrodes. *Sci. Rep.* **2014**, *4*, 6804. [CrossRef]
107. Kim, H.-N.; Moon, J.H. Enhanced Photovoltaic Properties of Nb_2O_5-Coated TiO_2 3D Ordered Porous Electrodes in Dye-Sensitized Solar Cells. *ACS Appl. Mater. Interfaces* **2012**, *4*, 5821–5825. [CrossRef]
108. Li, J.; Zheng, T. A comparison of chemical sensors based on the different ordered inverse opal films. *Sens. Actuators B Chem.* **2008**, *131*, 190–195. [CrossRef]
109. Chiang, C.-C.; Tuyen, L.D.; Ren, C.-R.; Chau, L.-K.; Wu, C.Y.; Huang, P.-J.; Hsu, C.C. Fabrication of titania inverse opals by multi-cycle dip-infiltration for optical sensing. *Photonics Nanostruct.* **2016**, *19*, 48–54. [CrossRef]
110. Li, X.; Wu, Y.; Shen, Y.; Sun, Y.; Yang, Y.; Xie, A. A novel bifunctional Ni-doped TiO_2 inverse opal with enhanced SERS perfor-mance and excellent photocatalytic activity. *Appl. Surf. Sci.* **2018**, *427*, 739–744. [CrossRef]
111. Qi, D.; Yan, X.; Wang, L.; Zhang, J. Plasmon-free SERS self-monitoring of catalysis reaction on Au nanoclusters/TiO_2 photonic microarray. *Chem. Commun.* **2015**, *51*, 8813–8816. [CrossRef]
112. Zhao, R.; Geng, G.; Wei, Q.; Liu, Y.; Zhou, H.; Zhang, X.; He, C.; Li, X.; Li, X.; Wang, Y.; et al. Controllable Polarization and Diffraction Modulated Multi-Functionality Based on Metasurface. *Adv. Opt. Mater.* **2022**, *10*, 2102596. [CrossRef]
113. Wang, F.; Geng, G.; Wang, X.; Li, J.; Bai, Y.; Li, J.; Wen, Y.; Li, B.; Sun, J.; Zhou, J. Visible Achromatic Metalens Design Based on Artificial Neural Network. *Adv. Opt. Mater.* **2022**, *10*, 2101842. [CrossRef]
114. Yao, Z.; Wang, C.; Li, Y.; Kim, N.-Y. AAO-assisted synthesis of highly ordered, large-scale TiO_2 nanowire arrays via sputtering and atomic layer deposition. *Nanoscale Res. Lett.* **2015**, *10*, 166. [CrossRef] [PubMed]
115. Liang, Y.-C.; Wang, C.-C.; Kei, C.-C.; Hsueh, Y.-C.; Cho, W.-H.; Perng, T.-P. Photocatalysis of Ag-Loaded TiO_2 Nanotube Arrays Formed by Atomic Layer Deposition. *J. Phys. Chem. C* **2011**, *115*, 9498–9502. [CrossRef]
116. Chang, W.-T.; Hsueh, Y.-C.; Huang, S.-H.; Liu, K.-I.; Kei, C.-C.; Perng, T.-P. Fabrication of Ag-loaded multi-walled TiO_2 nanotube arrays and their photocatalytic activity. *J. Mater. Chem. A* **2013**, *1*, 1987–1991. [CrossRef]

117. Al-Haddad, A.; Wang, Z.; Zhou, M.; Tarish, S.; Vellacheri, R.; Lei, Y. Constructing Well-Ordered CdTe/TiO_2 Core/Shell Nanowire Arrays for Solar Energy Conversion. *Small* **2016**, *12*, 5538–5542. [CrossRef]
118. Zhang, Y.; Liu, M.; Ren, W.; Ye, Z.-G. Well-ordered ZnO nanotube arrays and networks grown by atomic layer deposition. *Appl. Surf. Sci.* **2015**, *340*, 120–125. [CrossRef]
119. Norek, M.; Zaleszczyk, W.; Łuka, G.; Budner, B.; Zasada, D. Tailoring UV emission from a regular array of ZnO nanotubes by the geometrical parameters of the array and Al_2O_3 coating. *Ceram. Int.* **2017**, *43*, 5693–5701. [CrossRef]
120. Abdel-Fattah, T.M.; Gu, D.; Baumgart, H.; Namkoong, G. Optimizing the Release of ALD Grown ZnO Nanotubes from Anodic Aluminum Oxide Templates. *ECS Trans.* **2009**, *25*, 315–320. [CrossRef]
121. Gu, D.; Baumgart, H.; Abdel-Fattah, T.M.; Namkoong, G. Synthesis of Nested Coaxial Multiple-Walled Nanotubes by Atomic Layer Deposition. *ACS Nano* **2010**, *4*, 753–758. [CrossRef]
122. Gu, D.; Baumgart, H.; Namkoong, G.; Abdel-Fattah, T. Atomic Layer Deposition of ZrO_2 and HfO_2 Nanotubes by Template Replication. *Electrochem. Solid State Lett.* **2009**, *12*, K25. [CrossRef]
123. Tian, L.; Soum-Glaude, A.; Volpi, F.; Salvo, L.; Berthomé, G.; Coindeau, S.; Mantoux, A.; Boichot, R.; Lay, S.; Brizé, V.; et al. Undoped TiO_2 and nitrogen-doped TiO_2 thin films deposited by atomic layer deposition on planar and architectured surfaces for photovoltaic applications. *J. Vac. Sci. Technol. A Vac. Surf. Film.* **2015**, *33*, 01A141. [CrossRef]
124. Liu, L.; Karuturi, S.K.; Su, L.T.; Tok, A.I.Y. TiO_2 inverse-opal electrode fabricated by atomic layer deposition for dye-sensitized solar cell applications. *Energy Environ. Sci.* **2011**, *4*, 209–215. [CrossRef]
125. Singh, R.; Bapat, R.; Qin, L.; Feng, H.; Polshettiwar, V. Atomic Layer Deposited (ALD) TiO_2 on Fibrous Nano-Silica (KCC-1) for Photocatalysis: Nanoparticle Formation and Size Quantization Effect. *ACS Catal.* **2016**, *6*, 2770–2784. [CrossRef]
126. You, J.; Dou, L.; Yoshimura, K.; Kato, T.; Ohya, K.; Moriarty, T.; Emery, K.; Chen, C.-C.; Gao, J.; Li, G.; et al. A polymer tandem solar cell with 10.6% power conversion efficiency. *Nat. Commun.* **2013**, *4*, 1446. [CrossRef] [PubMed]
127. Yip, C.T.; Huang, H.; Zhou, L.; Xie, K.; Wang, Y.; Feng, T.; Li, J.; Tam, W.Y. Direct and Seamless Coupling of TiO_2 Nanotube Photonic Crystal to Dye-Sensitized Solar Cell: A Single-Step Approach. *Adv. Mater.* **2011**, *23*, 5624–5628. [CrossRef]
128. Kim, H.J.; Jackson, D.H.K.; Lee, J.; Guan, Y.; Kuech, T.F.; Huber, G.W. Enhanced Activity and Stability of TiO_2-Coated Cobalt/Carbon Catalysts for Electrochemical Water Oxidation. *ACS Catal.* **2015**, *5*, 3463–3469. [CrossRef]
129. Sridharan, K.; Jang, E.; Park, Y.M.; Park, T.J. Superior Photostability and Photocatalytic Activity of ZnO Nanoparticles Coated with Ultrathin TiO_2 Layers through Atomic-Layer Deposition. *Chem. A Eur. J.* **2015**, *21*, 19136–19141. [CrossRef]
130. Kurttepeli, M.; Deng, S.; Verbruggen, S.W.; Guzzinati, G.; Cott, D.J.; Lenaerts, S.; Verbeeck, J.; Tendeloo, G.V.; Detavernier, C.; Bals, S. Synthesis and Characterization of Photoreactive TiO_2–Carbon Nanosheet Composites. *J. Phys. Chem. C* **2014**, *118*, 21031–21037. [CrossRef]
131. Park, S.-G.; Jeon, T.Y.; Jeon, H.C.; Yang, S.-M.; Kwon, J.-D.; Mun, C.-W.; Cho, B.; Kim, C.S.; Kim, D.-H. Fabrication of 3D ZnO hollow shell structures by prism holographic lithography and atomic layer deposition. *J. Mater. Chem. C* **2014**, *2*, 1957–1961. [CrossRef]
132. Marichy, C.; Donato, N.; Latino, M.; Willinger, M.G.; Tessonnier, J.-P.; Neri, G.; Pinna, N. Gas sensing properties and p-type response of ALD TiO_2 coated carbon nanotubes. *Nanotechnology* **2014**, *26*, 24004. [CrossRef] [PubMed]
133. Baranowska-Korczyc, A.; Sobczak, K.; Dłużewski, P.; Reszka, A.; Kowalski, B.J.; Kłopotowski, Ł.; Elbaum, D.; Fronc, K. Facile synthesis of core/shell ZnO/ZnS nanofibers by electrospinning and gas-phase sulfidation for biosensor applications. *Phys. Chem. Chem. Phys.* **2015**, *17*, 24029–24037. [CrossRef] [PubMed]
134. Purniawan, A.; Pandraud, G.; Moh, T.S.Y.; Marthen, A.; Vakalopoulos, K.A.; French, P.J.; Sarro, P.M. Fabrication and optical measurements of a TiO_2-ALD evanescent waveguide sensor. *Sens. Actuator A Phys.* **2012**, *188*, 127–132. [CrossRef]
135. Shkondin, O.T.E.; Panah, M.E.A.; Liu, P.; Larsen, P.V.; Mar, M.D.; Jensen, F.; Lavrinenko, A.V. High aspect ratio titanium nitride trench structures as plasmonic biosensor. *Opt. Mater. Express* **2017**, *7*, 4171–4182. [CrossRef]
136. Jang, D.; Meza, L.; Greer, F.; Greer, J.R. Fabrication and deformation of three-dimensional hollow ceramic nanostructures. *Nat. Mater.* **2013**, *12*, 893–898. [CrossRef]
137. Meza, L.R.; Greer, J.R. Mechanical characterization of hollow ceramic nanolattices. *J. Mater. Sci.* **2014**, *49*, 2496–2508. [CrossRef]
138. Yoon, J.; Kim, S.; Kim, D.; Kim, I.-D.; Hong, S.; No, K. Fabrication of Highly Ordered and Well-Aligned $PbTiO_3$/TiN Core-Shell Nanotube Arrays. *Small* **2015**, *11*, 3750–3754. [CrossRef]
139. Park, J.; Kim, K.-I.; Kim, K.; Kim, D.-C.; Cho, D.; Lee, J.H.; Jeon, S. Rapid, High-Resolution 3D Interference Printing of Multilevel Ultralong Nanochannel Arrays for High-Throughput Nanofluidic Transport. *Adv. Mater.* **2015**, *27*, 8000–8006. [CrossRef]
140. Yang, J.; Mi, H.; Luo, S.; Li, Y.; Zhang, P.; Deng, L.; Sun, L.; Ren, X. Atomic layer deposition of TiO_2 on nitrogen-doped carbon nanofibers supported Ru nanoparticles for flexible Li-O_2 battery: A combined DFT and experimental study. *J. Power Sources* **2017**, *368*, 88–96. [CrossRef]
141. Vempati, S.; Ranjith, K.S.; Topuz, F.; Biyikli, N.; Uyar, T. Electrospinning Combined with Atomic Layer Deposition to Generate Applied Nanomaterials: A Review. *ACS Appl. Nano Mater.* **2020**, *3*, 6186–6209. [CrossRef]

micromachines

MDPI

Review

Design Aspects of Additive Manufacturing at Microscale: A Review

Nikolaos Rogkas [1], Christos Vakouftsis [1], Vasilios Spitas [1], Nikos D. Lagaros [2] and Stelios K. Georgantzinos [3,*]

[1] Laboratory of Machine Design, National Technical University of Athens, 9 Iroon Polytechniou, 15780 Zografou, Greece; nrogkas@mail.ntua.gr (N.R.); cvakouftsis@mail.ntua.gr (C.V.); vspitas@central.ntua.gr (V.S.)
[2] Institute of Structural Analysis and Antiseismic Research, School of Civil Engineering, National Technical University of Athens, 9 Iroon Polytechniou, 15780 Zographou, Greece; nlagaros@central.ntua.gr
[3] Laboratory for Advanced Materials, Structures and Digitalization, Department of Aerospace Science and Technology, National and Kapodistrian University of Athens, Evripus Campus, 34400 Psachna, Greece
* Correspondence: sgeor@uoa.gr

Abstract: Additive manufacturing (AM) technology has been researched and developed for almost three decades. Microscale AM is one of the fastest-growing fields of research within the AM area. Considerable progress has been made in the development and commercialization of new and innovative microscale AM processes, as well as several practical applications in a variety of fields. However, there are still significant challenges that exist in terms of design, available materials, processes, and the ability to fabricate true three-dimensional structures and systems at a microscale. For instance, microscale AM fabrication technologies are associated with certain limitations and constraints due to the scale aspect, which may require the establishment and use of specialized design methodologies in order to overcome them. The aim of this paper is to review the main processes, materials, and applications of the current microscale AM technology, to present future research needs for this technology, and to discuss the need for the introduction of a design methodology. Thus, one of the primary concerns of the current paper is to present the design aspects describing the comparative advantages and AM limitations at the microscale, as well as the selection of processes and materials.

Keywords: additive manufacturing; microscale; design; materials; processes; applications

Citation: Rogkas, N.; Vakouftsis, C.; Spitas, V.; Lagaros, N.D.; Georgantzinos, S.K. Design Aspects of Additive Manufacturing at Microscale: A Review. *Micromachines* **2022**, *13*, 775. https://doi.org/10.3390/mi13050775

Academic Editors: David J. Collins and Nam-Trung Nguyen

Received: 25 March 2022
Accepted: 12 May 2022
Published: 15 May 2022

1. Introduction

Current advancements in design processes and procedures for additive manufacturing technology require the sufficient capturing of the scale aspect from a perspective that incorporates the evaluation of the overall dimensions (macro-aspect) and the local features (micro-aspect). This includes the different functionalities and/or physical laws that govern them, as well as their synergy, including individual features and macroscopic tolerances and sensitivity analysis. In order to take into account the scale aspect, modeling, analysis, and simulation tools, as well as manufacturing and experimental methods, have been employed that can treat micro- and nanoscale problems efficiently, treating them either as individual entities or combined [1–3] (multi-scale approach). The efficient capturing of the underlying physics (i.e., that governing physical laws), the level of material modeling, the manufacturing methods employed, and the metrological considerations for the evaluation of the product performance are of major importance during both the design and the manufacturing phase, which, at the microscale, suffers from a major dilemma. The above concerns have been heavily discussed and investigated over the past few years at the meso/macroscale [4,5] and, during the last two decades, at the nanoscale [6,7]. However, although the distinction between macro- and nanoscale tools is well established, the tools and methodologies of these scales can be applied and frequently find a use in research into microscale applications, by underutilizing or fitting these methodologies to various microscale aspects. Sometimes, this may lead to the treatment of microscale as a fuzzy boundary between meso and nanoscale, rendering both approaches valid and invalid at the same time without providing any clear distinction between them. For instance, the case of the contact between asperities in tribological applications can be considered as macro-, micro- or nanoscale, depending on the number of asperities that are in contact [8]. Nanoscale corresponds to single-asperity contact, while macroscale corresponds to contact between millions of asperities. However, microscale is represented as contact between a few asperities; thus, it is not accurately or quantitatively defined.

Design at the nanoscale follows the physics prevailing at the nanoscale level (i.e., hydrogen bonds, van der Waals forces, London forces, Pauli repulsions, etc.) along with the manufacturing considerations for scale-specific methodologies (i.e., laser ablation [9,10], electrochemical machining [11–13], photo-etching [14], chemical vapor decomposition [15] or physical vapor decomposition [16]) either for material removal (machining) or for material addition (additive manufacturing—AM). Macroscale design follows the laws of physics prevailing at the macroscale (i.e., classical continuum mechanics, flow dynamics, the isotropy of material properties, etc.) and conventional manufacturing methods, which are validated for the forming of small-scale structures (micromachining [17]) or the control of the texture (surface roughness) using superfinishing, lapping, etc.

Therefore, it is considered that the clear and concrete definition of microscale from a perspective that includes the analysis and simulation tools and the manufacturing methods, is a necessary step before proceeding to the design problem. The same applies to Additive Manufacturing (AM) which was originally used for the manufacturing of components in applications on a macro scale but, in recent years, this expanded to micro- and even nanoscale, either by creating new AM manufacturing processes or by adjusting the existing processes to fit smaller scales. Therefore, the tools and methodologies originally developed for macro-scale manufacturing were usually implemented, with adjuncts, to study and evaluate smaller scales. Moreover, AM, as a relatively new fabrication method that is completely different from the traditional subtractive manufacturing methods, poses a challenge in creating a design methodology fitted to its unique capabilities and strict limitations, to fully exploit its potential and revolutionize the manufacturing industry.

Today, the benefits of AM are well-established: design freedom, partial consolidation, no tooling being necessary, just-in-time inventory, faster production, easy accessibility, cost-effectiveness, tangible and creative designs, unlimited shapes and geometry, a variety of raw materials, less waste production, risk reduction, and others [18]. AM or three-dimensional printing (3DP) was developed in the 1980s as a rapid prototyping method and is now considered to be a manufacturing process in the same genre as conventional manufacturing processes. AM has generated an impact in all industry areas, including aerospace [19,20], automobile manufacture [21], construction [22], and medical and military applications [23]. It offers flexibility in the product design process and a considerable reduction in material consumption, as well as making product personalization affordable [24]. ASTM International categorizes AM processes into 7 groups [25]: i.e., material extrusion (e.g., fused deposition modeling—FDM), powder bed fusion, vat photopolymerization (stereolithography—SLA), material jetting (e.g., PolyJet), sheet lamination, directed energy deposition, and binder jetting, each typified by the principle according to which the resulting matter is formed.

Although AM techniques have progressed greatly at the macro-scale, many challenges remain to be addressed, mainly at micro- and lower scales. AM fabrication at these scales is associated with certain constraints, such as feature size limitations, the expansion of the range of materials used in order to include alternative ones (i.e., non-metallics, ceramics, composites, etc.), the improvement of surface quality and the minimization of porosity, as well as other geometrical defects. The rapid evolution of the current AM fabrication technologies, in addition to those newly introduced, does not give enough time for the introduction of a design methodology that addresses all the aforementioned limitations.

Therefore, AM design methodologies need to involve and continuously adjust, based on the rapid evolution of AM technologies and processes.

This paper reviews the existing AM processes, their underlying techniques, commercial systems, and the materials used in AM fabrication at the microscale, together with applications in the production of microscale actuators, soft robotics, as well as biomedical and microfluidic devices. The primary concern is the presentation of design aspects describing the comparative advantages and AM limitations at the microscale, as well as the selection of processes and materials. The goal of this paper is to introduce the main design aspects of microscale AM, as described in the literature over the last five to ten years, and adumbrate a design methodology that is better fitted for microscale AM.

2. Technologies and Materials for AM at the Microscale

Additive manufacturing is a relatively new manufacturing method with increased popularity, aiming to expand the manufacturing capabilities of functional components. Among other applications, AM is used for the fabrication of products at micro- and nanoscale, with various degrees of geometrical and functional complexity and ever-increasing market infiltration into these areas. ISO/ASTM 52910-17 [25] is a standard that tries to set some guidelines regarding common industry practices in the field of AM at macro/meso scale; its extension to micro- or nanoscale is both an opportunity and a challenge. Although micro- and nanoscale AM is a relatively new field of application, macro-scale technologies are still of use to fabricate complex parts at these smaller scales. Many researchers, such as Vaezi et al., Paul et al., Behera et al., and Chizari et al. [26–30] proposed different classification categories for the application of these technologies at such scales, based on the production equipment used, the materials, the dimensions and the required tolerances of critical features, as well as other product attributes (i.e., intended use, texture, color, and strength). The most popular classification takes into consideration well-established macro-AM technologies, including 2D ink printing and other technologies fitted to the micro- and nanoscales. These can be divided into three main groups: macro or scalable additive manufacturing, 2D ink writing, or 3D direct writing processes, as well as hybrid ones.

2.1. Macro-AM Processes Fitted to Microscale Fabrication

The first group consists of traditional macro-AM processes fitted to micro- and nanoscale fabrication. Among them are the technologies widely used for macro-AM, such as Stereolithography, Selective Laser Sintering (SLS), Fused Deposition Modeling (FDM), Laminated Object Manufacturing (LOM), and inkjet printing processes. These scalable processes are adapted to microscale fabrication (micro-stereolithography or MSL, micro-laser sintering or MLS) but still face many difficulties and limitations. Modern micro-stereolithography (μ-SLA) pushes the resolution limits down to the sub-100 μm range with precision optics. The most common materials used are SL resins, hydrogels, biocompatible materials, and bioactive agents for a variety of bio-functional, implantable tissue-engineering applications, including nerve regeneration and guided angiogenesis [31–36]. Two-photon photopolymerization/lithography (TPL) is a photopolymerization-based technique that is primarily applied for the printing of polymer materials. Other materials are biocompatible and organic, mixing additives into the resist blend to generate composite structures, such as electrically conductive polymer microstructures loaded with carbon nanotubes [37–44]. Micro-SLS uses a laser to sinter small particles, consolidating powders in a layer-by-layer manner. The commercially available materials used in SLS come in powder form and include, but are not limited to, polymers such as polyamides (PA), polystyrenes (PS), thermoplastic elastomers (TPE), and polyaryletherketones (PAEK) [45–47]. Laminated object manufacturing is older and is slightly different from the known technologies and processes. LOM technology is based on the layer-by-layer fabrication of parts using sheets of various materials. Each sheet is cut into the desired geometry and used as a layer on top of the previous materials. The use of a binding substance ensures adhesion and creates the final functional component. Ceramics and metals, such as 316 L stainless steel, are among the most popular materials for LOM. Fused deposition modeling (FDM) is the most widely known and used AM process in macroscale fabrication. Nevertheless, it is fitted into producing components at the microscale as well. New developments help adapt to microscale needs and overcome limitations in terms of shape, resolution, and material usage. Polymers, metals, composites, and even biomaterials for tissue engineering are some of the most popular FDM materials.

2.2. *Two-Dimensional Ink Writing Technologies*

The second group of AM processes is based on 2D Ink writing technologies. These processes, traditionally used in the past for ink writing, are fitted for the fabrication of 3D microstructures. One of the most popular 2D-based methods is chemical vapor deposition (CVD). CVD creates a coating induced by a chemical reaction at the surface of a heated material. Laser chemical vapor deposition (LCVD) is a modified CVD process for the deposition of thin films. The two main categories of LCVD are photolytic LCVD and pyrolytic LCVD. In the first subcategory, the energy of the focused laser beam is absorbed by reagent gases, leading to the decomposition of gas molecules and the formation of a thin solid film on the substrate. In the second subcategory, the laser beam is focused on the locations to be deposited. This way, the temperature locally increases on the substrate until it reaches the threshold required. This leads to the deposition of a thin solid film on the substrate. Any material electroplated with nickel/chrome or stainless steel best supports PVD coatings. The most common chrome-plated materials are brass, zinc, steel, aluminum, and ABS plastic. Focused ion-beam direct writing (FIBDW) is another multi-material AM method that can use metallic, ceramic, and polymer inks for the fabrication of microscale structures. All inks must have specific rheological properties in order to be able to flow continuously through the printing nozzle and form a continuous filament, similar to that in FDM printing. This flow must be continuous, consistent, and able to create a discrete shape and form of layer [27,48–50]. Laser-induced forward transfer (LIFT) has been used to deposit a variety of materials, such as metals (Cu, Ag, Au, Pt, Cr, Al), semiconducting materials such as Ge and Se, oxide layers, nanocomposites, conductive polymers (PEDOT-PSS), biomaterials, and superconductors, among others. Donor materials with viscosities ranging from 10 to 100,000 cP have been printed with LIFT [51–61]. Another popular process, electrohydrodynamic (EHD) printing is a spray-based printing process that can pattern functional materials. EHD printing has been used to deposit metallic, carbon-based, ceramic, and polymer-based conductive materials, semiconducting nanoparticles (quantum dots), biomaterials, and molten metals on a wide range of substrates [62–71].

2.3. *Hybrid Processes*

The last category consists of methods combining additive and subtractive processes for micro-3D fabrication. Some typical methods are shape deposition modeling (SDM) and electrochemical fabrication (EFAB). SDM processes utilize additive and subtractive processes sequentially to produce 3D structures, but their use in micro-AM is limited. The EFAB process is based on the multilayer electrodeposition and planarization of at least two metals: one structural material and one sacrificial material. This process is capable of manufacturing microdevices with features as small as 20 μm and tolerances of ±2 μm. It is a popular method of manufacturing complex mechanisms without the need for assembly, which is favorable for medical devices. Common materials are Val-loy-120 (Ni–Co alloy), Edura-180 (electroplated Rh), and palladium [72].

From the above categorization, which is explored in more detail by the authors of [26], it can be deduced that the field of micro- and nano-AM both borrows existing practices from the mesoscale, expanding them into smaller sizes, and is based on either existing or new dedicated techniques at the micro- and nanoscales. The ability to control the geometry and tolerances at such a small scale is clearly one of the major issues, which, however, is at odds with the productivity of the more accurate yet slower dedicated techniques. Furthermore, the suitability of the proposed techniques for the desired material, size, accuracy, and productivity must be thoroughly assessed before the selection of the appropriate method.

3. AM Applications at the Microscale

Microscale AM processes have been employed in many technological fields for the fabrication of miniature devices. This paper addresses mainly recently published articles in the field of the design of microscale actuators and biomedical and microfluidic devices.

3.1. Actuator Applications

In the field of micro and miniature actuators, microscale AM is indicated as a promising fabrication solution, and it promotes the production of micromachines with complex geometry using monolithic approaches, which would otherwise require a combination of advanced micro-subtractive manufacturing methods and, usually, assemblies with a large number of components. Recent review papers present and discuss the use of AM processes in the design and fabrication of microelectromechanical systems (MEMS) actuators, biohybrid actuators, and piezoelectric systems as relevant applications. In [73,74], the authors investigated the recent developments and achievements regarding the most widely used 3D printing technologies for MEMS fabrication and discussed their challenges and potential. Several papers [75,76] presented recent advancements in the field of small-scale soft robotics and actuators using AM, while other researchers [77,78] examined the application of 3D printing for the fabrication of piezoelectric actuators; finally, the authors of [79,80] discussed an approach regarding the fabrication of biohybrid actuators using AM.

The fabrication of micro-grippers is an indicative example revealing the importance of micro–AM. Accurate tip displacements, which are as small as 20 µm, are necessary for handling and pick-and-place, and the sterile handling of sensitive parts, which is a common procedure in the biology and health sector environments. The actuation principle may be piezoelectric, magnetic, or electrothermal. Shao et al. fabricated magnetically active 3D microstructures using a high-resolution micro-continuous liquid interface production process (µCLIP), combining 3D-printed centimeter-sized samples with sub-75 µm fine features [81]. The magnetic photopolymerizable resin that was used maintains high solid loading (30 wt % Fe_3O_4 nanoparticles), improves the surface properties by reducing the stair-like surface roughness, and accelerates the fabrication process. In another study [82], the authors used the same method (µCLIP) to fabricate a 3D printed magnetically driven triple-finger micro-gripper (Figure 1C), and tested its efficiency using a 300 µm diameter microsphere, both in air and in deionized water. The printing process involved the soaking of the part in acetone to remove the residual liquid resin (2 min), then its transfer into ethanol and ultrasonic cleaner (5 min), and finally, after drying (30 min), the specimens were post-cured in UV light of a 405 nm wavelength (10 min). Daniel et al. [83] fabricated a chevron-type electrothermal actuator, using the material extrusion-based manufacturing of a shape memory polymer composite. Using a resistivity of 1.8 Ωcm and an operational voltage as low as 3 V, they accomplished 100 µm tip displacement, which was computationally and experimentally investigated. Their main computational finding was that the grippers can be actuated quickly (3–5 s) with voltages as low as 5 V, but they recover slowly (60–100 s). Experimentally, higher voltages were required for actuation; a tip displacement of up to 77–117 µm was achieved in 5 s with an operational voltage of 17.5–19.5 V. In [84], Tyagi et al. used a custom-built syringe-based extrusion 3D printer to fabricate bilayer micro-actuators, driven by hydrogels, down to a size of 300 × (1000 ÷ 5000) µm^2, with a minimum thickness of 30 µm. The printing resolution was 25 µm in the *x*-*y* plane; the rate of the lateral motion of the stage was ~2.5 mm/s and the air-dispensing pressure was 50–65 psi. The printing ink consisted of dissolved Hydromend D4 (hydrogel) in ethanol at a concentration of 20%. Lantada et al. [85] presented the development process of geometrically complex micro-vascular shape-memory polymer actuators by laser SLA, using a shape-memory epoxy that could change its shape as an effect of temperature increase. They presented two proof-of-concept applications: an active micro-claw with inner vasculatures of different cross-sections and an active spring with inner vasculatures of different cross-sections. In order to assess the effect of temperature on the closing of the gripper and the compression of the spring, they heated the prototypes with water flow (80 °C) running through the micro-vasculatures. In [86], Kozaki et al. presented the design of a microgripper for handling spheroid microstructures, mounted on a glass capillary (Figure 1A). They used a top-down micro-stereolithography setup, based on a 405-nm blue laser developed in their previous study [87]. The photo-curable polymer used is a mixture of acrylate resin and a photopolymerization initiator, polymerization inhibitor, and blue light absorber (wavelength 405 nm). The mixture was mixed, degassed (2000 rpm and 5 min for each mode), and stirred for 24 h at 60 rpm in a ball mill. The nominal diameter of the micro-gripper tip was 300 μm, while the effective force could reach values of between 0.01 and 0.04 N and the tip displacement varied between 20 and 80 μm, respectively. Alblalaihid et al. [88] demonstrated the application of a sputter-coating process for the deposition of metallic layers on polymer components and validated their approach for the fabrication of a micro-gripper device (Figure 1C). They used a 3D projection micro-stereolithography (PMSL) system. The gripper was thermally actuated and the tip displacement, in this case, was in the range of 10–180 μm, depending on the applied potential.

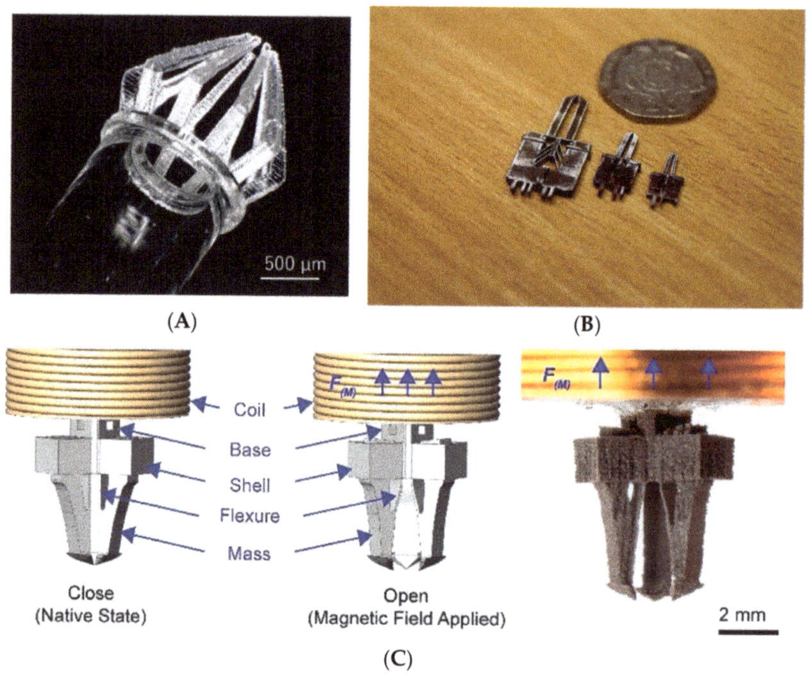

Figure 1. Examples of fabricated micro-grippers: (**A**) Micro-gripper for handling microstructure spheroids, mounted on a glass capillary. Reprinted with permission from Ref. [86]. Copyright 2020 MDPI. (**B**) Fabrication of a thermally actuated micro-gripper using a sputter-coating process for the deposition of metallic layers on polymer parts. Reprinted with permission from Ref. [88]. Copyright 2017 IOP Publishing. (**C**) Magnetically actuating micro-gripper for operation in air and water. Reprinted with permission from Ref. [82]. Copyright 2021 Elsevier.

3.2. Soft Robotics Applications

Besides micro-grippers, which, in most cases, maintain a rigid-type behavior during their operation, flexible actuators or soft robotics yield another application of microscale AM. Almeida et al. [89] designed an actuation mechanism for robotic micro-tweezers, based on a 3D-printed nylon flexure and a piezo-bimorph actuator, targeting the desired manipulation range from 100 μm to 1 mm. Bas et al. [90] designed miniature inflatable bending actuators, consisting of ultra-fine fibers (diameter of between 1 and 50 μm) and a soft elastomer matrix able to exhibit diverse movements. They used melt electro-writing (MEW) technology to create the prototypes (Figure 2B). Their actuators, with a length of 10–15 mm and an inner diameter of 1 mm, can reach their full range of motion within ~20 ms without exploiting snapping instabilities or material non-linearities. Joyee and Pan [91] fabricated a fully 3D-printed multi-material, multi-modal functional soft monolithic robot, composed of polymer and magnetic particle-polymer composites. The fabrication process was magnetic field-assisted projection stereolithography (M-PSL), capable of fabricating smart particle-polymer composites layer by layer. A photocurable flexible resin was used as the base material for 3D printing, while the magnetic nanoparticles (10 nm in nominal diameter) contained 60–80 wt % iron oxide. The maximum bending deformation was 5.2 mm on the z-axis and the maximum deflection in the xy plane was 146°. Schaffner et al. [92] reported a 3D-printing platform for the seamless digital fabrication of pneumatic silicone actuators, exhibiting programmable bioinspired architectures and motions with spatial resolutions in the range of 300 μm. They used viscoelastic silicone inks, resulting in elastomers with variable stiffness after polymerization. Sinatra et al. [93] introduced a novel fabrication strategy for nanofiber-reinforced soft micro-actuators with 30 μm feature sizes. The design and manufacturing of composite polydimethylsiloxane (PDMS)/nanofiber actuators using soft lithography and rotary jet spinning are described. Among the examined parameters were the lamina design and fiber orientation on the actuator curvature, mechanical properties, and pressurization range. Composite actuators displayed a 25.8% higher maximum pressure than pure PDMS devices. Furthermore, the best nanofiber-reinforced laminates tested were 2.3 times tougher than the control PDMS material, while maintaining comparable elongation. Xavier et al. [94] presented the design and direct 3D printing of novel omnidirectional soft pneumatic actuators using SLA (Figure 2A). They used an elastic resin and FDM with a soft thermoplastic polyurethane (TPU), achieving multimodal actuation including bending, extension, and contraction motions under positive, negative, or differential pressures. The printing time for a single actuator using the SLA method was 6 h and 40 min while the printing time using the FDM method was approximately 29 h and 20 min. In [95], Zhang et al. presented a generic process flow for the systematic and efficient tailoring of the material formulation and key processing parameters for the digital light processing-based 3D printing of miniature pneumatic actuators for soft robots. They printed various miniature pneumatic robots with an overall size of 2–15 mm and a feature size of 150–350 μm. They used a commercially available UV-curable elastomer, to which was added 30 wt % epoxy aliphatic acrylate (EAA), leading to a reduction in Young's modulus and an increase in failure strain. All the specimens were post-cured for 10 min. Ge et al. [96] presented the design of a bottom-up digital light processing (DLP) 3D printer system (385 nm UV light source, 50 μm normal resolution) and the fabrication of multiple-size soft pneumatic actuators integrally, with fast speed and high precision. Their experiments demonstrated that the printer could print objects with features as small as 87.5 μm. They also presented the design and fabrication of a soft pneumatic gripper containing three micro pneumatic actuators with 0.4-mm-wide square air channels, as well as 0.2-mm-thick chamber walls.

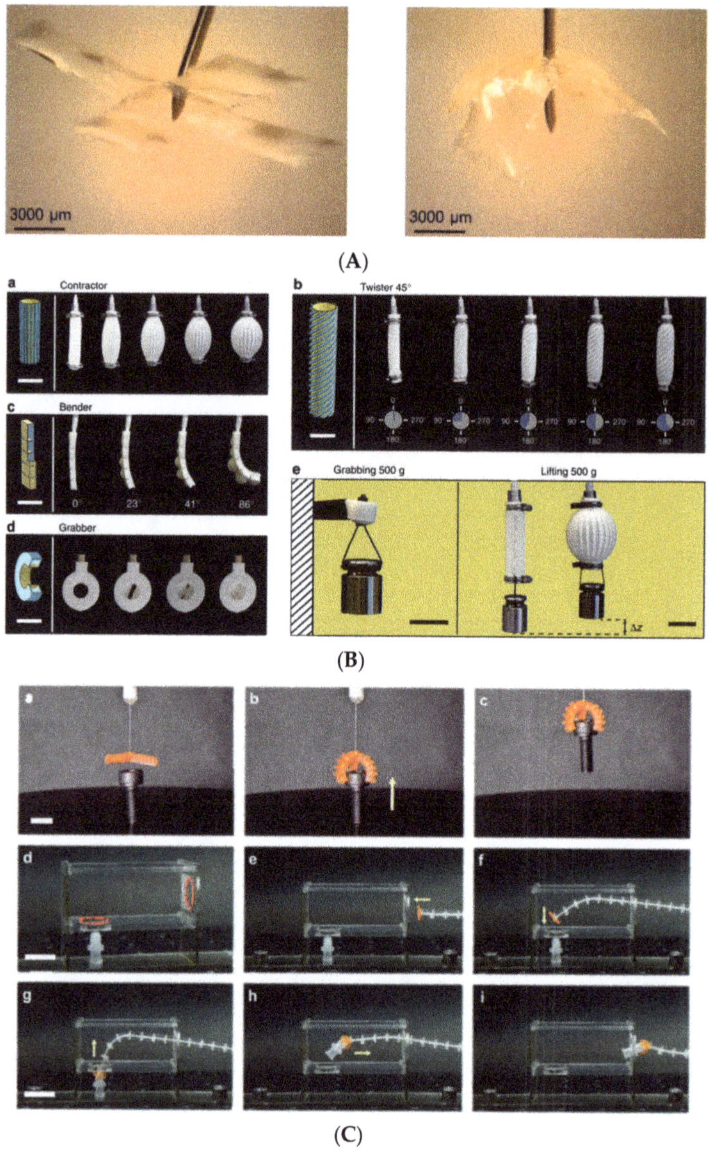

Figure 2. Examples of fabricated soft robotics: (**A**) Omnidirectional soft pneumatic actuator. Reprinted with permission from Ref. [84]. Copyright 2020 IOP Publishing. (**B**) Examples of soft actuators with programmed motion modes, fabricated from soft silicones. Reprinted with permission from Ref. [92]. Copyright 2018 Nature Communications, (**C**) Operation of the fabricated soft pneumatic gripper: the debris removal process within a confined space. Reprinted with permission from Ref. [95]. Copyright 2019 Wiley Online Library.

3.3. Biomedical and Microfluidics Applications

Microscale AM has been efficiently used in biomedical engineering, including many microfluidic applications, which can also be treated as a separate category. The methods, potential, challenges, and limitations of microscale AM in biomedical engineering have been reported in recent review studies. The applications of 3D printing in the health and pharmaceutical sectors have been thoroughly investigated over the last few years and can be tracked in the following review papers [97–99]. According to the authors of [99], the applications can be divided into the following categories: 3D disease modeling, pharmaceutical products, organ printing, and patient-specific in situ implants. Other possible applications include drug-delivery devices [100–102], the fabrication of microneedles [103,104], microfluidic devices and biomedical micro-devices [105,106], and the fabrication of tissues [107,108].

Drug delivery applications incorporate design solutions characterized by microscale features where AM has been successfully incorporated. In [109], Joyee and Pan proposed the design of a 3D-printed soft robot capable of multimodal locomotion. Utilizing computer aided design and computer aided engineering (CAD-CAE) tools for the design, they printed the robot via a novel magnetic field-assisted projection stereolithography (M-PSL) technique (Figure 3A). This soft robot is capable of bi-directional bending in the *xy* plane and *z*-direction and consists of anterior and posterior legs that contain a drug. The maximum dimensions of the robot in width and height is 5 mm × 5.5 mm, while the drug is released from a 200 μm hole.

In the field of microfluidics, Coltelli et al. [110] combined microfluidics, AM, and electrostatic actuation to design artificial muscles capable of generating up to 33 Mpa stress and 10–20% strain. Their design consists of arrays of rectangular cavities arranged accordingly, filled with conducting material inside a bulk dielectric volume. They suggest that the microfluidic devices are AM-fabricated in such a way that the channels would form wiring when filled with conducting fluid, while the bulk core would serve as the dielectric and as the force-transfer medium. The non-flexed lateral size of the electrode plates was kept at 400 μm × 400 μm. The non-flexed plate thickness was kept at 100 μm for each plate and the non-flexed separation between paired plates within the same micro-capacitor was kept at 100 μm. The accuracy of the SLS printing was kept at 100 μm. In [111], the authors illustrated the direct fabrication of a 3D complex microchannel design using AM, for the continuous mixing of micro/nano-particles with biomolecules. The fabrication process was conducted using the DLP method. After the 3D printing stage, the part was removed and washed with IPA (70% ethanol and water), blow-dried with pressurized air, and, finally, cured under UV light for 120 s. The cross-section of the trapezoidal channel had a width of 600 μm and heights of 80 and 130 μm. Another example of a design of microfluidic MEMS was presented in [112], where the authors proposed a micro-extrusion 3D printing system that contained integrated pick-and-place functionality. The case study was the fabrication of microfluidic-based 3D MEMS (three-dimensional microelectromechanical systems) that contain orthogonal out-of-plane piezoelectric sensors and actuators, using additive manufacturing.

Miniature pumps are very critical components in the health sector. In [113], Thomas et al. fabricated a 3D-printed electromagnetically actuated microfluidic pump, capable of generating a 2.2 μL/min flow rate of biofluid (Figure 3B). An FDM process with 100 μm-layer resolution was used to deposit polylactic acid on a plastic filament. Taylor et al. [114] fabricated a multi-material miniature diaphragm pump for the creation and maintenance of a low vacuum from atmospheric conditions, using PolyJet printing. The output surface was assessed in terms of roughness, giving values of R_a in the order of some microns (~2–3 μm), while the R_z values were close to layer thickness (~16–18 μm), which was considered acceptable. The stroke of the pump was 2.5 μm. In [115], a low-cost (~$120), open-source peristaltic pump was constructed with a combination of 3D-printed parts and common hardware. The pump was capable of producing flow rates of up to 1.6 mL min^{-1}.

In the field of microneedles (MNs), Economidou et al. [116] fabricated a hollow MN MEMS system for controlled transdermal drug delivery. They fabricated hollow cone-shaped MNs with a base diameter of 1000 μm, a tip diameter of 100 μm, and a height of 1000 μm using SLA and, afterward, integrated the MNs onto the MEMS. The hollow cones featured a wall thickness of 100 μm and the internal bores had a diameter of 800 μm at the cone base. The MNs were fabricated using an SLA 3D printer, followed by curing for 60 min under 40 °C UV radiation. The authors observed smooth surfaces on the MNs (no "stair-stepping" effect), as an outcome of the printing method they selected. In [117], the authors provided the capabilities of FDM low-budget printers (using PLA printing material) to print the non-transparent and closed internal microfeatures of in-plane linear, curved, and spiral microchannels with a diameter of less than 0.5 mm (i.e., linear, curved, and spiral channel profiles) and varying cross-sections. The surface roughness of each microchannel configuration was measured and was found to be in the order of some microns (~0.5–3 µm). In addition, each configuration was tested in terms of leakage flow. Caudill et al. [118] designed and printed microneedle arrays utilizing a three-dimensional (3D)-printing technique called continuous liquid interface production (CLIP). Besides pyramidal MNs, the design involved faceted MNs with horizontal grooves, leading to an increase in surface area and, thus, better vaccination properties. The MNs were 700 µm in height and 500 µm in width and were printed in a 10 × 10 array on a 10 mm × 10 mm patch for vaccine delivery. Chen et al. [119] proposed a novel 3D AM method, known as magnetorheological drawing lithography (MRDL), to efficiently fabricate bio-inspired MNs imitating the honeybee's stinger. With the assistance of an external magnetic field, a parent MN (20 µm tip width) was directly drawn on the pillar tip, and tilted micro barbs (5 µm tip width) were subsequently formed on the four sides of the parent MN. The fabrication process of the parent MN was conducted by means of insertion and, afterward, the removal of a copper pillar inside a pool filled with curable magneto-rheological fluid (CMRF) under an external magnetic field. Micro barbs were formed later, on the curved surface of the parent MN. Compared with a barbless microneedle, the micro-structured barbs enabled the bio-inspired microneedle to be easily inserted into the skin, with difficult removal. In [120], the authors used a commercially available stereolithographic 3D printing, which was assessed regarding its microscale fabrication properties, in order to fabricate sharp MNs (12 × 12 array, in total, 144; 30 min per patch) with a tip radius of approximately 15 µm. In another study [121], a microneedle mold fabrication technique using a low-cost desktop SLA 3D printer was presented, and the fabrication of needles with high-aspect ratios and tip radii of 20–40 µm took place.

Examples of different biomedical devices produced via AM are presented in Figure 3.

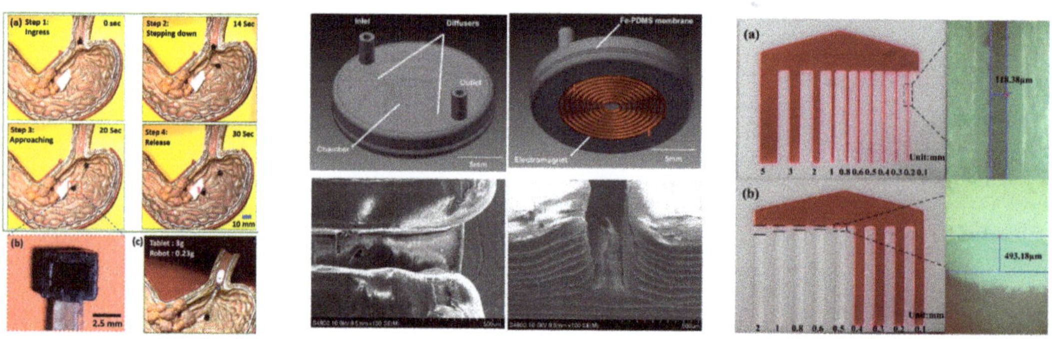

(A) (B) (C)

Figure 3. *Cont.*

(D)

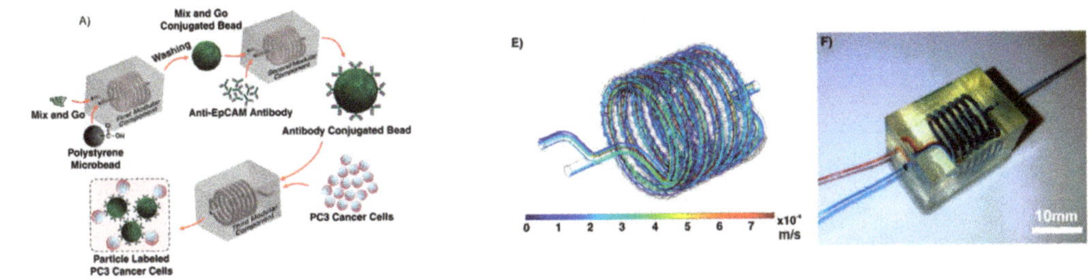

(E)

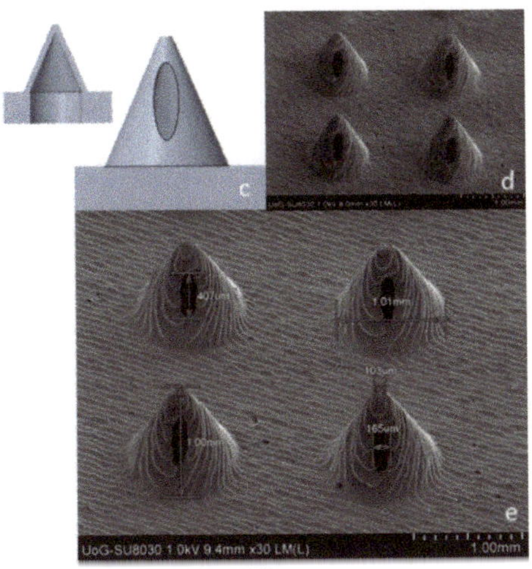

(F)

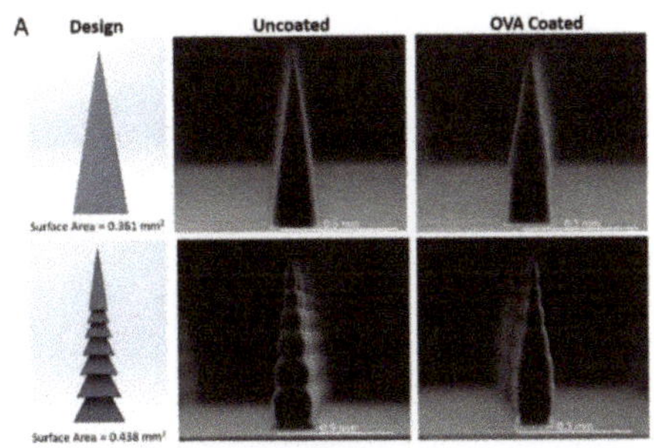

(G)

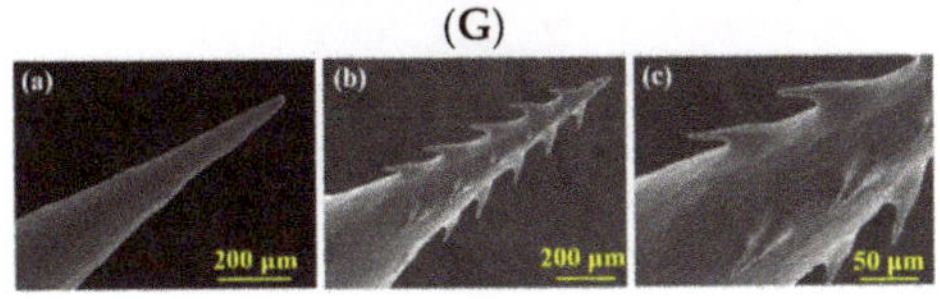

(H)

Figure 3. Examples of biomedical devices: drug delivery applications, microchannels, and microneedles: (**A**) Presentation of the operation of the multi-material soft robot. Movement of the actuator inside an anatomical stomach model with cancer tissue (target) and release of the drug at the targeted tumor location. Reprinted with permission from Ref. [109]. Copyright 2020 Elsevier. (**B**) Electromagnetically actuated microfluidic pump. Reprinted with permission from Ref. [113]. Copyright 2016 Elsevier. (**C**) Fabricated hydrophilic channels and hydrophobic chambers. Reprinted with permission from Ref. [122]. Copyright 2016 MDPI. (**D**) 3D printer head for additive manufacturing of sugar glass for tissue engineering applications. Reprinted with permission from Ref. [123]. Copyright 2017 Elsevier. (**E**) The design of microchannels for mixing: concentration distribution, streamlined simulations, and fabricated micromixer. Reprinted with permission from Ref. [111]. Copyright 2020 Elsevier. (**F**) CAD images of the microneedle designs and respective cross-sections, along with SEM images. Reprinted with permission from Ref. [116]. Copyright 2021 Elsevier. (**G**) Design and environmental scanning electron microscope (ESEM) images of printed microneedles, pyramidal, and faceted designs. Reprinted with permission from Ref. [118]. Copyright 2021 National Academy of Science (**H**) Micro barb features of printed microneedles with SEM images. Reprinted with permission from Ref. [119]. Copyright 2018 American Chemical Society.

4. Design Considerations

Additive manufacturing allows the production of parts with almost no geometry restrictions, even at a microscale. This manufacturing method can produce freeform, topology-optimized geometries (structures, arrays, patterns, small assemblies, micromachines, etc.), controlling the micro-structure of the component's material, and implementing the use of lattice structures, trusses, and multi-material fabrication. This way the engineer is able to "design" the structure of the material in macro- and even microscale, controlling the material properties based on the functionality of the component.

One may consider that the design for AM at the microscale deviates from the design protocols and approaches at the macroscale. From a general perspective, design at the macroscale is based on the principle of integrating parts into assemblies, while the interconnection or mating is achieved via the use of fundamental machine elements and machine-design methods, such as screws, tight fits, weldments, etc. Therefore, the functionality and the operational precision of a complex assembly are limited by the number and the attributes of the elements comprising it, since the rule of thumb is that as the number of the parts increases, precision drops. AM can be seen as an approach for creating more complex forms of parts-like assemblies, which is critical for applications at the microscale since many of the available micromachining techniques are mere extensions of their macroscale equivalents. This might attract questions and ambiguity regarding the appropriateness of the physics and the operational principle of the method due to the scale aspect; nevertheless, the alternative approach of fabricating micro machines with increased functionality and fewer but more complex parts is intriguing. For instance, multiple degrees of freedom in robotic arms can be achieved via the use of joints that are translated into several components (motors, bearings, screws, etc.) but in small-scale soft robotics, a single monolithic part made of soft material is able to derive controllable motions. Thus, AM is an alternative approach to designing at microscale, with an entirely different basis.

In Figure 4, joints in the macro- and microscales are presented. Specifically, Figure 4a shows the testing of a wire-driven continuum robot arm [124], while Figure 4b experiments with bending deformation due to the applied pressure of an omnidirectional soft pneumatic actuator [94].

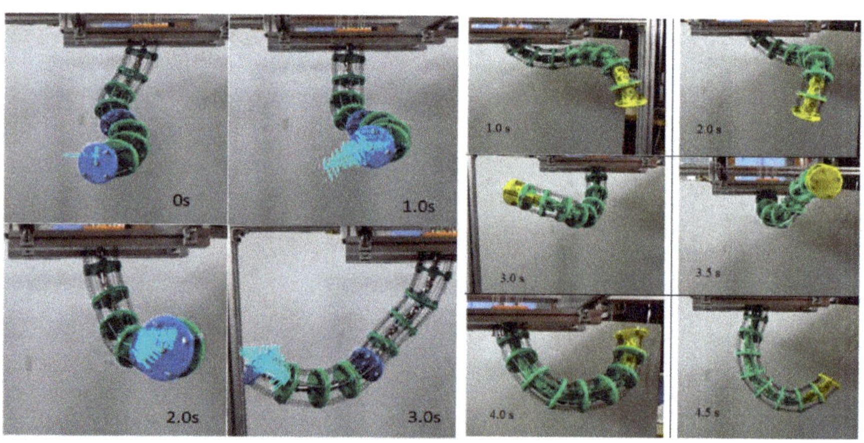

(a)

Figure 4. *Cont.*

(b)

Figure 4. Joints in macro and microscale: (**a**) Testing of the wire-driven robot arm. Reprinted with permission from Ref. [124]. Copyright 2019 MDPI. (**b**) Experiments on bending deformation due to applied pressure from an omnidirectional soft pneumatic actuator. Reprinted with permission from Ref [94]. Copyrights 2021 Elsevier.

In order to fully utilize the capabilities of AM and reduce design and fabrication defects, it is critical to quantify each process's parameters, limitations, and repeatability. For this purpose, numerous analytical methods were developed, along with many computational tools. Despite these methods and tools, experimental evaluation remains crucial for the optimization of AM processes.

4.1. Comparative Advantages and AM Limitations

AM enables the fabrication of complex, freeform, and smart structures [125]. Among their other unique capabilities, the use of lattice structures for the topology optimization of structures, and the design of lightweight components is possible. Nature-inspired design for the mimetics of complex nature layouts is also feasible via AM. The aim of these approaches is usually to yield controllable mechanical properties that can be tuned according to the requirements of the application. Lightweight structures, energy absorption, the fabrication of nature-inspired micro-patterns, and modeling and simulation techniques are important state-of-the-art aspects in the field of cellular micro-lattice architectures. Moreover, AM enables the control of the microstructure of the material, allowing the design of desired component properties [126]. In this way, AM could be used for the on-demand production of metamaterials. Metamaterials are ordered composites that have material properties not usually found in nature. The use of auxetic and custom infill patterns that are directly optimized to transfer energy absorption properties or dumping capabilities to AM components leads to unique smart materials and highly efficient components. Multi-material fabrication poses new challenges in the design and fabrication of smart components [26]. Recent advancements in the field of design, modeling and simulation, fabrication, and testing of lattice structures can be found in the following review papers [127–136]. Deriving the effective properties of additively manufactured micro-lattice structures is an important tool in the hands of designers for performing fast simulations at a low computational cost [126,137–139]. Souza et al. [139] derived a closed-form analytical solution of lattice structures fabricated by selective laser melting, using beam models. Athanasiadis et al. [140], in work based on fracture mechanics theory, investigated the potential of lattice structures to replace adhesives in sandwich-type structures, using both analytical and FEA calculations. Kenel et al. [141], using 3D ink extrusion, fabricated CoCrFeNi micro-lattices with strut diameters as narrow as 100 µm, and tested their compression and tension properties at ambient and cryogenic temperatures. Boulvert et al. [142] tested the acoustic behavior of 3D-printed micro-lattices in order to extract conclusions about the defects of FDM. The size of their samples was in the order of 200 µm, and defects included the presence of micro-grooves on the lattices' surfaces in the order of 10 µm. Studies regarding the defects of lattices were also conducted in [143–146]. In [147], McGregor et al. conducted a statistical study in order to assess the geometric quality of 2D and 3D micro-lattice structures. At the same time, as lattice structures outstripped bulk cores in many technological applications, the introduction of artificial, additively manufactured, textures come forward as an efficient tool for the fabrication of surfaces. Additive texturing is a state-of-the-art approach for fabricating surfaces with superior tribological, wetting, and wear characteristics [148]. In [149], Wang et al. used the selective laser melting of ink-printed copper nanoparticles (SLM-IP Cu NPs) in order to fabricate a friction-reduced surface for operating in mixed-lubrication conditions (Figure 5A). The patterns investigated were concave and convex, squared and fully sintered Cu film; the height of the features was 20 μm. Mekhie et al. [150] printed metallic hierarchical micro-features (pillars, channels, etc.) using selective laser melting (Figure 5B) for the wetting control, achieving hydrophobic surfaces with a contact angle greater than 140°.

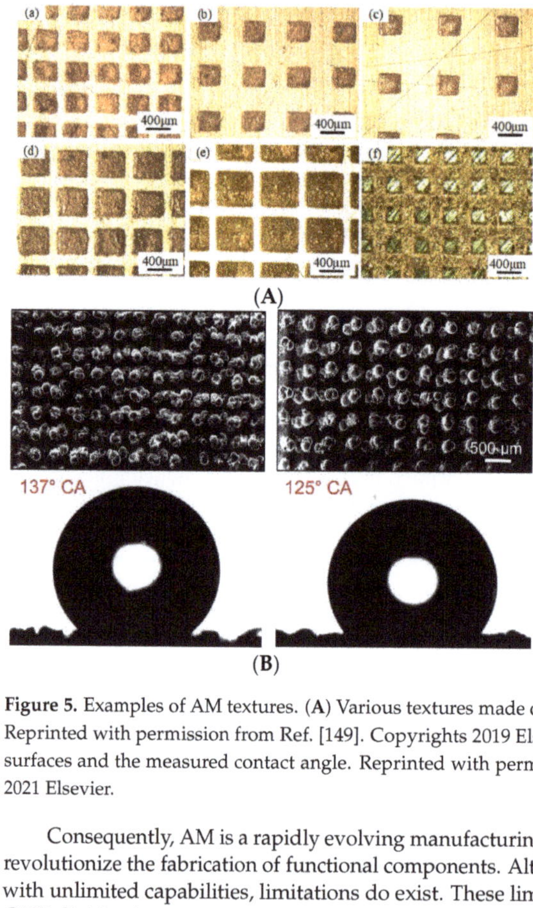

Figure 5. Examples of AM textures. (**A**) Various textures made of ink-printed copper nanoparticles. Reprinted with permission from Ref. [149]. Copyrights 2019 Elsevier. (**B**) Two cases of pillar array surfaces and the measured contact angle. Reprinted with permission from Ref. [150]. Copyrights 2021 Elsevier.

Consequently, AM is a rapidly evolving manufacturing process, with huge potential to revolutionize the fabrication of functional components. Although it is considered a method with unlimited capabilities, limitations do exist. These limitations take into consideration CAD digitization, process parameter optimization and the effect on material properties, the current capabilities of AM technologies, and the lifecycle of AM components, as well as metrology and quality control challenges [151].

Among the most popular technologies, such as SLS, stereolithography, and FDM, the need to expand the materials used for microscale AM is a major challenge. The adaptation of non-metallic materials, such as ceramics, polyamides, and composite powder-enriched resins, is necessary for the improvement of component functionality, as well as the expansion of the applications for which AM is used [26]. Another important limitation for most of the aforementioned processes is the fabrication of hollow, closed structures as it is difficult to remove excess material without invasive post-processing. Dimensional and geometrical deviations, linked with thermal history, heat-affected zones (HAZ), and material phase changes, impose great restrictions on AM processes when it comes to functional component fabrication. LOM, for example, is known to undergo severe shrinkage by as much as 18% in some cases. Especially in sintering, the feature size of the component is limited either by the particle-size limitations of the powders used or the technology's laser focus [152,153]. Moreover, increased surface roughness is also connected with powder size and HAZ during fabrication, further restricting the quality of AM components [154,155]. One of the most frequent defects in sintering processes, leading to poor mechanical properties and a decreased life cycle, is porosity. Being affected by both environmental and process parameters, along with thermal and oxidation effects, porosity is one of the most important constraints for both macro- and microscale AM [26,46].

When it comes to stereolithography, minimum layer thickness, as well as improved surface roughness, are challenging. Both are limited by the physical properties of the resins used. Surface tension and the viscosity of the resin are the limiting factors for layer thickness, and also affect the surface quality and post-processing needed to clean up the final component by removing all excess resin.

4.2. Selection of Processes and Materials

As previously described, AM consists of many different technologies (processes), each of them with unique capabilities and limitations. Every different technology uses specific materials or groups of materials. Thus, a critical point of the design process is the selection of a particular AM technology to utilize its advantages and obtain the best manufacturing quality, as well as achieve the optimal material based on the functionality of the fabricated component. For example, micro-stereolithography is a high-resolution method that uses photocurable resins capable of producing complicated components in large manufacturing volumes. Moreover, materials such as ceramics, metals (WC, Co, Al, Cu), and hydrogels can be used [156–159]. Micro-laser sintering is a powder-based method with a wide variety of materials, isotropic properties, and without the need for support structure during component fabrication. Among the most popular materials of MLS are 316 L stainless steel and a variety of metals. However, the technique requires post-processing and the components may suffer from porosity [160]. FDM is one of the most popular macroscale technologies that can also fabricate components in microscale. Besides thermoplastics, which are widely used, biomaterials are also available, making it possible to create medical and biological parts. Nevertheless, it is limited to low fabrication volumes, with high temperatures and poor repeatability [158,161]. Laminated object manufacturing (LOM), one of the oldest AM processes, uses metals such as 316 L stainless steel, zirconia, and ceramics for the fabrication of fully dense and high-mechanical-strength components. Its resolution, however, is limited to 80 μm; it lacks dimensional accuracy due to high shrinkage (12–18%), and the post-processing of parts is required [161,162]. The popular, 2D-based, Inkjet printing process uses a wide variety of materials and many biomaterials, making it ideal for biomedical applications and with fair repeatability, but the need for support is essential [26]. The other 2D-based method, FIBDW, uses mostly metals for the fabrication of high-resolution components but remains a slow method with poor repeatability [162,163]. Finally, one of the most widely used AM methods for micro-part fabrication, EFAB, is suitable for the high-resolution manufacturing of complex parts and assemblies, such as medical devices. Nevertheless, it has dimensional limitations and post-processing is necessary and sometimes even difficult [72,164,165].

4.3. Design Considerations

As discussed thoroughly in the previous paragraphs, AM, despite being a revolutionizing fabrication method with unique capabilities, still has limitations and restrictions when it comes to functional component manufacturing. The proposed actions during the process of design for AM are presented in Figure 6.

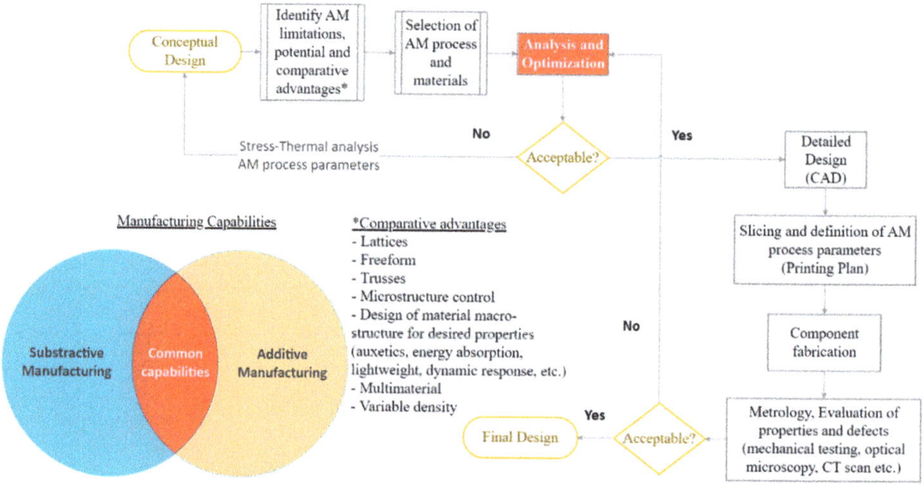

Figure 6. Design for Additive Manufacturing process.

The designer, on the one hand, needs to exploit all the comparative advantages of AM methods, such as freeform fabrication, the use of lattices and trusses, the ability to control the microstructure, and the use of metamaterials with advanced properties. On the other hand, the designer has to consider the limitations and weaknesses of each fabrication method. Therefore, the design process begins with the conceptual design of the component, based on functional and geometrical criteria. Then, the designer must select the most suitable AM method by taking into consideration the material, process limitations, and fabrication possibilities. After the material and process are selected, an optimization procedure is carried out, in order to minimize defects and increase repeatability and productivity. This optimization procedure involves coupled stress-thermal analysis to exploit the effects of the AM process on the final material properties (residual thermal stresses, density, material anisotropy, strength, fatigue life, etc.). Based on the large number of different fabrication methods and the limited computational tools that can accurately predict stress-thermal behavior during fabrication, many experimental methods are also popular for the optimization and evaluation procedures. Last, but not least, the optimization procedure also includes another critical aspect, the process parameter selection. Recent advances in the technology of AM and software development dedicated to AM fabrication allow the control of a variety of critical parameters for AM fabrication. This software uses dedicated algorithms that propose an optimized set of parameters and allow the user to manually intervene and control them. Finally, in the same way as in traditional subtractive manufacturing methods, metrological evaluation is necessary to ensure the quality of the fabricated component. On the microscale, the use of conventional metrological equipment is impossible. Therefore, other methods and procedures had to be created or adapted to cover the gap. Some of them are presented in the next paragraph.

4.4. Metrological and Performance Validation of AM Structures at the Microscale

The experimental evaluation of components produced with AM is critical for the optimization of the manufacturing process and the minimization of defects and deviations from nominal geometry. The micro-detail (mD) measurement of complex parts remains an open field of investigation. It is critical to experimentally establish the minimum dimensional and geometric limits to produce parts with mDs. These limits are highly dependent on the AM process, technology, and materials used [166,167].

One way of doing that is based on mathematical formulations, using voxelization (a process where the geometry of the part is represented with volumetric unitary elements).

The minimum feature size and obtainable tolerances of additive manufacturing processes are linked to the smallest volumetric elements (voxels) that can be created. These models can predict the behavior of the parts during fabrication. A test sample can then be used to evaluate the dimensional variation [168,169].

Dimensional restrictions in the microscale make the use of conventional metrological equipment impossible. One very popular alternative for the evaluation of AM components in microscale is X-ray computed tomography (XCT). The use of XCT for testing and analyzing AM components was established in the past few decades and is becoming extremely popular for dimensional evaluation, as well as structural integrity, density, and porosity analyses [170,171].

Another popular evaluation method is optical microscopy. This method can be non-destructive for the examination of surface quality and external geometrical characteristics, but it can also be used as a destructive method to obtain information about the internal porosities of a component [172,173].

5. Conclusions

This article presents advancements in the field of additive manufacturing at the microscale from a design perspective. The main ambition of this study is to highlight the critical aspects of the design process by combining generalized results, mainly regarding the effect of fabrication technologies and the use of materials. Even though microscale AM technologies have been extensively discussed in other recent studies, here, we conclude that the fabrication of microscale components using AM may be employed either by utilizing existing meso- or macroscale techniques or by developing new, dedicated methods at the micro- and nanoscales. Furthermore, this article collects recently presented applications of microscale AM in the field of the fabrication of MEMs, actuators, and soft robotics, as well as in the biomedical sector. The design process presented begins by considering the comparative advantages and limitations of microscale AM. Its advantages include freeform fabrication, the use of lattices and trusses, the ability to control the microstructure, and the use of metamaterials with advanced properties. The limitations are mainly associated with CAD digitization, the effects of fabrication process parameters on material properties, which must be assessed, and the lifecycle of AM components, as well as the current lack of metrology and quality control protocols. The design procedure proceeds with the selection of appropriate materials and processes, analysis and optimization using CAE, and detailed design, then concludes with fabrication and the post-evaluation.

Author Contributions: Conceptualization, N.R., C.V. and S.K.G.; methodology, N.R. and C.V.; validation, V.S., N.D.L. and S.K.G.; formal analysis, V.S. and S.K.G.; investigation, N.R., C.V. and N.D.L.; resources, N.R. and C.V.; data curation, V.S. and S.K.G.; writing—original draft preparation, N.R. and C.V.; writing—review and editing, S.K.G. and V.S.; visualization, V.S. and N.D.L.; supervision, S.K.G. and V.S. All authors have read and agreed to the published version of the manuscript.

Funding: This research received no external funding.

Acknowledgments: This research is co-financed by Greece and the European Union (European Social Fund—ESF) through the Operational Program "Human Resources Development, Education and Lifelong Learning" in the context of the project "Strengthening Human Re-sources Research Potential via Doctorate Research" (MIS-5000432), implemented by the State Scholarships Foundation (IKY).

Conflicts of Interest: The authors declare no conflict of interest.

References

1. Wiechert, L.; Wall, W.A. A nested dynamic multi-scale approach for 3D problems accounting for micro-scale multi-physics. *Comput. Methods Appl. Mech. Eng.* **2010**, *199*, 1342–1351. [CrossRef]
2. Kovachki, N.; Liu, B.; Sun, X.; Zhou, H.; Bhattacharya, K.; Ortiz, M.; Stuart, A. Multiscale modeling of materials: Computing, data science, uncertainty and goal-oriented optimization. *Mech. Mater.* **2022**, *165*, 104156. [CrossRef]
3. Van Der Giessen, E.; Schultz, P.A.; Bertin, N.; Bulatov, V.V.; Cai, W.; Csányi, G.; Foiles, S.M.; Geers, M.G.D.; González, C.; Hütter, M.; et al. Roadmap on multiscale materials modeling. *Model. Simul. Mater. Sci. Eng.* **2020**, *28*, 043001. [CrossRef]

4. Michihata, M. Surface-Sensing Principle of Microprobe System for Micro-Scale Coordinate Metrology: A Review. *Metrology* **2022**, *2*, 46–72. [CrossRef]
5. Lévy, M.; Ferrari, R.; Franks, P.J.S.; Martin, A.P.; Rivière, P. Bringing physics to life at the submesoscale. *Geophys. Res. Lett.* **2012**, *39*, 1–13. [CrossRef]
6. Nguyen, A.T.; Sathe, S.R.; Yim, E.K.F. From nano to micro: Topographical scale and its impact on cell adhesion, morphology and contact guidance. *J. Phys. Condens. Matter* **2016**, *28*, 183001. [CrossRef]
7. Chandel, V.S.; Wang, G.; Talha, M. Advances in modelling and analysis of nano structures: A review. *Nanotechnol. Rev.* **2020**, *9*, 230–258. [CrossRef]
8. Stoyanov, P.; Chromik, R.R. Scaling effects on materials tribology: From macro to micro scale. *Materials* **2017**, *10*, 550. [CrossRef]
9. Zeng, H.; Du, X.W.; Singh, S.C.; Kulinich, S.A.; Yang, S.; He, J.; Cai, W. Nanomaterials via laser ablation/irradiation in liquid: A review. *Adv. Funct. Mater.* **2012**, *22*, 1333–1353. [CrossRef]
10. Prasek, J.; Drbohlavova, J.; Chomoucka, J.; Hubalek, J.; Jasek, O.; Adam, V.; Kizek, R. Methods for carbon nanotubes synthesis - Review. *J. Mater. Chem.* **2011**, *21*, 15872–15884. [CrossRef]
11. Singh, T.; Dvivedi, A. Developments in electrochemical discharge machining: A review on electrochemical discharge machining, process variants and their hybrid methods. *Int. J. Mach. Tools Manuf.* **2016**, *105*, 1–13. [CrossRef]
12. Goud, M.; Sharma, A.K.; Jawalkar, C. A review on material removal mechanism in electrochemical discharge machining (ECDM) and possibilities to enhance the material removal rate. *Precis. Eng.* **2016**, *45*, 1–17. [CrossRef]
13. Unune, D.R.; Mali, H.S. Current status and applications of hybrid micro-machining processes: A review. *Proc. Inst. Mech. Eng. Part B J. Eng. Manuf.* **2015**, *229*, 1681–1693. [CrossRef]
14. Chalker, P.R. Photochemical atomic layer deposition and etching. *Surf. Coat. Technol.* **2016**, *291*, 258–263. [CrossRef]
15. Cai, Z.; Liu, B.; Zou, X.; Cheng, H.-M. Chemical Vapor Deposition Growth and Applications of Two-Dimensional Materials and Their Heterostructures. *Chem. Rev.* **2018**, *118*, 6091–6133. [CrossRef]
16. Baptista, A.; Silva, F.J.G.; Porteiro, J.; Míguez, J.L.; Pinto, G.; Fernandes, L. On the Physical Vapour Deposition (PVD): Evolution of Magnetron Sputtering Processes for Industrial Applications. *Procedia Manuf.* **2018**, *17*, 746–757. [CrossRef]
17. Leo Kumar, S.P.; Jerald, J.; Kumanan, S.; Prabakaran, R. A Review on Current Research Aspects in Tool-Based Micromachining Processes. *Mater. Manuf. Process.* **2014**, *29*, 1291–1337. [CrossRef]
18. El-Sayegh, S.; Romdhane, L.; Manjikian, S. A critical review of 3D printing in construction: Benefits, challenges, and risks. *Arch. Civ. Mech. Eng.* **2020**, *20*, 34. [CrossRef]
19. Moon, S.K.; Tan, Y.E.; Hwang, J.; Yoon, Y.-J. Application of 3D printing technology for designing light-weight unmanned aerial vehicle wing structures. *Int. J. Precis. Eng. Manuf. Technol.* **2014**, *1*, 223–228. [CrossRef]
20. Ntouanoglou, K.; Stavropoulos, P.; Mourtzis, D. 4D Printing Prospects for the Aerospace Industry: A critical review. *Procedia Manuf.* **2018**, *18*, 120–129. [CrossRef]
21. Manghnani, R. An exploratory study: The impact of additive manufacturing on the automobile industry. *Int. J. Curr. Eng. Technol.* **2015**, *5*, 3407–3410.
22. Tay, Y.W.D.; Panda, B.; Paul, S.C.; Noor Mohamed, N.A.; Tan, M.J.; Leong, K.F. 3D printing trends in building and construction industry: A review. *Virtual Phys. Prototyp.* **2017**, *12*, 261–276. [CrossRef]
23. Ventola, C.L. Medical Applications for 3D Printing: Current and Projected Uses. *Pharm. Ther.* **2014**, *39*, 704–711.
24. Ko, H.; Moon, S.K.; Hwang, J. Design for additive manufacturing in customized products. *Int. J. Precis. Eng. Manuf.* **2015**, *16*, 2369–2375. [CrossRef]
25. ASTM International. *Standard Terminology for Additive Manufacturing Technologies: Designation F2792-12a*; ASTM International: West Conshohocken, PA, USA, 2012.
26. Vaezi, M.; Seitz, H.; Yang, S. A review on 3D micro-additive manufacturing technologies. *Int. J. Adv. Manuf. Technol.* **2013**, *67*, 1721–1754. [CrossRef]
27. Behera, D.; Cullinan, M. Current challenges and potential directions towards precision microscale additive manufacturing – Part I: Direct ink writing/jetting processes. *Precis. Eng.* **2021**, *68*, 326–337. [CrossRef]
28. Paul, S.; Shibendu, S.R. Advances in Macro, Micro and Nano Additive Manufacturing Processes: A Review. In Proceedings of the National Conference on Advances in Research and Innovations in Mechanical Engineering, Material Science, Industrial Engineering and Management, Imphal, India, 12–13 December 2016.
29. Behera, D.; Chizari, S.; Shaw, L.A.; Porter, M.; Hensleigh, R.; Xu, Z.; Roy, N.K.; Connolly, L.G.; Zheng, X.; Saha, S.; et al. Current challenges and potential directions towards precision microscale additive manufacturing—Part II: Laser-based curing, heating, and trapping processes. *Precis. Eng.* **2021**, *68*, 301–318. [CrossRef]
30. Chizari, S.; Shaw, L.A.; Behera, D.; Roy, N.K.; Zheng, X.; Panas, R.M.; Hopkins, J.B.; Chen, S.-C.; Cullinan, M.A. Current challenges and potential directions towards precision microscale additive manufacturing—Part III: Energy induced deposition and hybrid electrochemical processes. *Precis. Eng.* **2021**, *68*, 174–186. [CrossRef]
31. Patel, D.K.; Sakhaei, A.H.; Layani, M.; Zhang, B.; Ge, Q.; Magdassi, S. Highly Stretchable and UV Curable Elastomers for Digital Light Processing Based 3D Printing. *Adv. Mater.* **2017**, *29*, 1606000. [CrossRef]
32. Kuang, X.; Zhao, Z.; Chen, K.; Fang, D.; Kang, G.; Qi, H.J. High-Speed 3D Printing of High-Performance Thermosetting Polymers via Two-Stage Curing. *Macromol. Rapid Commun.* **2018**, *39*, 1700809. [CrossRef]

33. Hegde, M.; Meenakshisundaram, V.; Chartrain, N.; Sekhar, S.; Tafti, D.; Williams, C.B.; Long, T.E. 3D Printing All-Aromatic Polyimides using Mask-Projection Stereolithography: Processing the Nonprocessable. *Adv. Mater.* **2017**, *29*, 1–7. [CrossRef] [PubMed]
34. Ge, Q.; Sakhaei, A.H.; Lee, H.; Dunn, C.K.; Fang, N.X.; Dunn, M.L. Multimaterial 4D Printing with Tailorable Shape Memory Polymers. *Sci. Rep.* **2016**, *6*, 31110. [CrossRef] [PubMed]
35. Bhattacharjee, N.; Parra-Cabrera, C.; Kim, Y.T.; Kuo, A.P.; Folch, A. Desktop-Stereolithography 3D-Printing of a Poly(dimethylsiloxane)-Based Material with Sylgard-184 Properties. *Adv. Mater.* **2018**, *30*, 1800001. [CrossRef]
36. Chen, D.; Zheng, X. Multi-material Additive Manufacturing of Metamaterials with Giant, Tailorable Negative Poisson's Ratios. *Sci. Rep.* **2018**, *8*, 9139. [CrossRef] [PubMed]
37. Vyatskikh, A.; Delalande, S.; Kudo, A.; Zhang, X.; Portela, C.M.; Greer, J.R. Additive manufacturing of 3D nano-architected metals. *Nat. Commun.* **2018**, *9*, 593. [CrossRef]
38. Harnisch, E.; Russew, M.; Klein, J.; König, N.; Crailsheim, H.; Schmitt, R. Optimization of hybrid polymer materials for 2PP and fabrication of individually designed hybrid microoptical elements thereof. *Opt. Mater. Express* **2015**, *5*, 456. [CrossRef]
39. Quick, A.S.; de los Santos Pereira, A.; Bruns, M.; Bückmann, T.; Rodriguez-Emmenegger, C.; Wegener, M.; Barner-Kowollik, C. Rapid Thiol-Yne-Mediated Fabrication and Dual Postfunctionalization of Micro-Resolved 3D Mesostructures. *Adv. Funct. Mater.* **2015**, *25*, 3735–3744. [CrossRef]
40. Jiang, L.; Xiong, W.; Zhou, Y.; Liu, Y.; Huang, X.; Li, D.; Baldacchini, T.; Jiang, L.; Lu, Y. Performance comparison of acrylic and thiol-acrylic resins in two-photon polymerization. *Opt. Express* **2016**, *24*, 13687. [CrossRef]
41. Saha, S.K.; Oakdale, J.S.; Cuadra, J.A.; Divin, C.; Ye, J.; Forien, J.-B.; Bayu Aji, L.B.; Biener, J.; Smith, W.L. Radiopaque Resists for Two-Photon Lithography To Enable Submicron 3D Imaging of Polymer Parts via X-ray Computed Tomography. *ACS Appl. Mater. Interfaces* **2018**, *10*, 1164–1172. [CrossRef]
42. Malinauskas, M.; Žukauskas, A.; Bičkauskaitė, G.; Gadonas, R.; Juodkazis, S. Mechanisms of three-dimensional structuring of photo-polymers by tightly focussed femtosecond laser pulses. *Opt. Express* **2010**, *18*, 10209. [CrossRef]
43. Maruo, S.; Nakamura, O.; Kawata, S. Three-dimensional microfabrication with two-photon-absorbed photopolymerization. *Opt. Lett.* **1997**, *22*, 132. [CrossRef] [PubMed]
44. Sun, H.-B.; Kawata, S. Two-Photon Photopolymerization and 3D Lithographic Microfabrication. *NMR 3D Anal. Photopolym.* **2006**, 169–273.
45. Uzan, N.E.; Ramati, S.; Shneck, R.; Frage, N.; Yeheskel, O. On the effect of shot-peening on fatigue resistance of AlSi10Mg specimens fabricated by additive manufacturing using selective laser melting (AM-SLM). *Addit. Manuf.* **2018**, *21*, 458–464. [CrossRef]
46. Hirt, L.; Reiser, A.; Spolenak, R.; Zambelli, T. Additive Manufacturing of Metal Structures at the Micrometer Scale. *Adv. Mater.* **2017**, *29*, 1604211. [CrossRef]
47. Godec, M.; Zaefferer, S.; Podgornik, B.; Šinko, M.; Tchernychova, E. Quantitative multiscale correlative microstructure analysis of additive manufacturing of stainless steel 316L processed by selective laser melting. *Mater. Charact.* **2020**, *160*, 110074. [CrossRef]
48. Lewis, J.A. Direct Ink Writing of 3D Functional Materials. *Adv. Funct. Mater.* **2006**, *16*, 2193–2204. [CrossRef]
49. Ahn, B.Y.; Duoss, E.B.; Motala, M.J.; Guo, X.; Park, S.-I.; Xiong, Y.; Yoon, J.; Nuzzo, R.G.; Rogers, J.A.; Lewis, J.A. Omnidirectional Printing of Flexible, Stretchable, and Spanning Silver Microelectrodes. *Science* **2009**, *323*, 1590–1593. [CrossRef]
50. Clendenning, S.B.; Aouba, S.; Rayat, M.S.; Grozea, D.; Sorge, J.B.; Brodersen, P.M.; Sodhi, R.N.S.; Lu, Z.-H.; Yip, C.M.; Freeman, M.R.; et al. Direct Writing of Patterned Ceramics Using Electron-Beam Lithography and Metallopolymer Resists. *Adv. Mater.* **2004**, *16*, 215–219. [CrossRef]
51. Bohandy, J.; Kim, B.F.; Adrian, F.J. Metal deposition from a supported metal film using an excimer laser. *J. Appl. Phys.* **1986**, *60*, 1538–1539. [CrossRef]
52. Florian, C.; Caballero-Lucas, F.; Fernández-Pradas, J.M.; Ogier, S.; Winchester, L.; Karnakis, D.; Geremia, R.; Artigas, R.; Serra, P. Printing of silver conductive lines through laser-induced forward transfer. *Appl. Surf. Sci.* **2016**, *374*, 265–270. [CrossRef]
53. Piqué, A.; Auyeung, R.C.Y.; Kim, H.; Charipar, N.A.; Mathews, S.A. Laser 3D micro-manufacturing. *J. Phys. D Appl. Phys.* **2016**, *49*, 223001. [CrossRef]
54. Kuznetsov, A.I.; Kiyan, R.; Chichkov, B.N. Laser fabrication of 2D and 3D metal nanoparticle structures and arrays. *Opt. Express* **2010**, *18*, 21198. [CrossRef] [PubMed]
55. Arnold, C.B.; Serra, P.; Piqué, A. Laser Direct-Write Techniques for Printing of Complex Materials. *MRS Bull.* **2007**, *32*, 23–31. [CrossRef]
56. Papakonstantinou, P.; Vainos, N.; Fotakis, C. Microfabrication by UV femtosecond laser ablation of Pt, Cr and indium oxide thin films. *Appl. Surf. Sci.* **1999**, *151*, 159–170. [CrossRef]
57. Zenou, M.; Sa'ar, A.; Kotler, Z. Digital laser printing of aluminum micro-structure on thermally sensitive substrates. *J. Phys. D Appl. Phys.* **2015**, *48*, 205303. [CrossRef]
58. Mattle, T.; Shaw-Stewart, J.; Schneider, C.W.; Lippert, T.; Wokaun, A. Laser induced forward transfer aluminum layers: Process investigation by time resolved imaging. *Appl. Surf. Sci.* **2012**, *258*, 9352–9354. [CrossRef]
59. Tóth, Z.; Szörényi, T. Pulsed laser processing of Ge/Se thin film structures. *Appl. Phys. A Solids Surfaces* **1991**, *52*, 273–279. [CrossRef]

60. Thomas, B.; Allonce, A.P.; Delaporte, P.; Sentis, M.; Sanaur, S.; Barret, M.; Collot, P. Experimental investigations of laser-induced forward transfer process of organic thin films. *Appl. Surf. Sci.* **2007**, *254*, 1206–1210. [CrossRef]
61. Fogarassy, E.; Fuchs, C.; de Unamuno, S.; Perriere, J.; Kerherve, F. High Tc Superconducting Thin Film Deposition by Laser Induced forward Transfer. *Mater. Manuf. Process.* **1992**, *7*, 31–51. [CrossRef]
62. Han, Y.; Dong, J. Electrohydrodynamic Printing for Advanced Micro/Nanomanufacturing: Current Progresses, Opportunities, and Challenges. *J. Micro Nano-Manuf.* **2018**, *6*. [CrossRef]
63. An, B.W.; Kim, K.; Lee, H.; Kim, S.-Y.; Shim, Y.; Lee, D.-Y.; Song, J.Y.; Park, J.-U. High-Resolution Printing of 3D Structures Using an Electrohydrodynamic Inkjet with Multiple Functional Inks. *Adv. Mater.* **2015**, *27*, 4322–4328. [CrossRef] [PubMed]
64. Han, Y.; Wei, C.; Dong, J. Super-resolution electrohydrodynamic (EHD) 3D printing of micro-structures using phase-change inks. *Manuf. Lett.* **2014**, *2*, 96–99. [CrossRef]
65. Jayasinghe, S.N.; Edirisinghe, M.J.; Wang, D.Z. Controlled deposition of nanoparticle clusters by electrohydrodynamic atomization. *Nanotechnology* **2004**, *15*, 1519–1523. [CrossRef]
66. Wei, C.; Dong, J. Direct fabrication of high-resolution three-dimensional polymeric scaffolds using electrohydrodynamic hot jet plotting. *J. Micromech. Microeng.* **2013**, *23*, 025017. [CrossRef]
67. Galliker, P.; Schneider, J.; Eghlidi, H.; Kress, S.; Sandoghdar, V.; Poulikakos, D. Direct printing of nanostructures by electrostatic autofocussing of ink nanodroplets. *Nat. Commun.* **2012**, *3*, 890. [CrossRef] [PubMed]
68. Jeong, Y.J.; Lee, X.; Bae, J.; Jang, J.; Joo, S.W.; Lim, S.; Kim, S.H.; Park, C.E. Direct patterning of conductive carbon nanotube/polystyrene sulfonate composites via electrohydrodynamic jet printing for use in organic field-effect transistors. *J. Mater. Chem. C* **2016**, *4*, 4912–4919. [CrossRef]
69. Kim, B.H.; Onses, M.S.; Lim, J.B.; Nam, S.; Oh, N.; Kim, H.; Yu, K.J.; Lee, J.W.; Kim, J.-H.; Kang, S.-K.; et al. High-Resolution Patterns of Quantum Dots Formed by Electrohydrodynamic Jet Printing for Light-Emitting Diodes. *Nano Lett.* **2015**, *15*, 969–973. [CrossRef]
70. Kim, M.; Yun, H.; Kim, G.H. Electric-field assisted 3D-fibrous bioceramic-based scaffolds for bone tissue regeneration: Fabrication, characterization, and in vitro cellular activities. *Sci. Rep.* **2017**, *7*, 3166. [CrossRef]
71. Han, Y.; Dong, J. High-resolution direct printing of molten-metal using electrohydrodynamic jet plotting. *Manuf. Lett.* **2017**, *12*, 6–9. [CrossRef]
72. Cohen, A.; Chen, R.; Frodis, U.; Wu, M.; Folk, C. Microscale metal additive manufacturing of multi-component medical devices. *Rapid Prototyp. J.* **2010**, *16*, 209–215. [CrossRef]
73. O'Donnell, J.; Kim, M.; Yoon, H.S. A Review on electromechanical devices fabricated by additive manufacturing. *J. Manuf. Sci. Eng. Trans. ASME* **2017**, *139*, 010801. [CrossRef]
74. Kumar, S.; Bhushan, P.; Pandey, M.; Bhattacharya, S. Additive manufacturing as an emerging technology for fabrication of microelectromechanical systems (MEMS). *J. Micromanuf.* **2019**, *2*, 175–197. [CrossRef]
75. Hines, L.; Petersen, K.; Lum, G.Z.; Sitti, M. Soft Actuators for Small-Scale Robotics. *Adv. Mater.* **2017**, *29*. [CrossRef] [PubMed]
76. Zolfagharian, A.; Kouzani, A.Z.; Khoo, S.Y.; Moghadam, A.A.A.; Gibson, I.; Kaynak, A. Evolution of 3D printed soft actuators. *Sens. Actuators A Phys.* **2016**, *250*, 258–272. [CrossRef]
77. Watson, B.; Friend, J.; Yeo, L. Piezoelectric ultrasonic micro/milli-scale actuators. *Sens. Actuators A Phys.* **2009**, *152*, 219–233. [CrossRef]
78. Chen, C.; Wang, X.; Wang, Y.; Yang, D.; Yao, F.; Zhang, W.; Wang, B.; Sewvandi, G.A.; Yang, D.; Hu, D. Additive Manufacturing of Piezoelectric Materials. *Adv. Funct. Mater.* **2020**, *30*, 2005141. [CrossRef]
79. Ricotti, L.; Trimmer, B.; Feinberg, A.W.; Raman, R.; Parker, K.K.; Bashir, R.; Sitti, M.; Martel, S.; Dario, P.; Menciassi, A. Biohybrid actuators for robotics: A review of devices actuated by living cells. *Sci. Robot.* **2017**, *2*, 1–18. [CrossRef]
80. Won, P.; Ko, S.H.; Majidi, C.; Feinberg, A.W.; Webster-Wood, V.A. Biohybrid Actuators for Soft Robotics: Challenges in Scaling Up. *Actuators* **2020**, *9*, 96. [CrossRef]
81. Shao, G.; Ware, H.O.T.; Li, L.; Sun, C. Rapid 3D Printing Magnetically Active Microstructures with High Solid Loading. *Adv. Eng. Mater.* **2020**, *22*, 3–9. [CrossRef]
82. Shao, G.; Ware, H.O.T.; Huang, J.; Hai, R.; Li, L.; Sun, C. 3D printed magnetically-actuating micro-gripper operates in air and water. *Addit. Manuf.* **2021**, *38*, 101834. [CrossRef]
83. Daniel, F.; Fontenot, J.; Radadia, A.D. Characterization of an electrothermal gripper fabricated via extrusion-based additive manufacturing. *Sens. Actuators A Phys.* **2022**, *333*, 113302. [CrossRef]
84. Tyagi, M.; Spinks, G.M.; Jager, E.W.H. Fully 3D printed soft microactuators for soft microrobotics. *Smart Mater. Struct.* **2020**, *29*, 085032. [CrossRef]
85. Lantada, A.D.; De Blas Romero, A.; Tanarro, E.C. Micro-vascular shape-memory polymer actuators with complex geometries obtained by laser stereolithography. *Smart Mater. Struct.* **2016**, *25*, 065018. [CrossRef]
86. Kozaki, S.; Moritoki, Y.; Furukawa, T.; Akieda, H.; Kageyama, T.; Fukuda, J.; Maruo, S. Additive manufacturing of micromanipulator mounted on a glass capillary for biological applications. *Micromachines* **2020**, *11*, 174. [CrossRef] [PubMed]
87. Kobayashi, Y.; Cordonier, C.E.J.; Noda, Y.; Nagase, F.; Enomoto, J.; Kageyama, T.; Honma, H.; Maruo, S.; Fukuda, J. Tailored cell sheet engineering using microstereolithography and electrochemical cell transfer. *Sci. Rep.* **2019**, *9*, 10415. [CrossRef]
88. Alblalaihid, K.; Overton, J.; Lawes, S.; Kinnell, P. A 3D-printed polymer micro-gripper with self-defined electrical tracks and thermal actuator. *J. Micromech. Microeng.* **2017**, *27*, 045019. [CrossRef]

89. Almeida, A.; Andrews, G.; Jaiswal, D.; Hoshino, K. The actuation mechanism of 3D printed flexure-based robotic microtweezers. *Micromachines* **2019**, *10*, 470. [CrossRef] [PubMed]
90. Bas, O.; Gorissen, B.; Luposchainsky, S.; Shabab, T.; Bertoldi, K.; Hutmacher, D.W. Ultrafast, miniature soft actuators. *Multifunct. Mater.* **2021**, *4*. [CrossRef]
91. Joyee, E.B.; Pan, Y. Multi-material additive manufacturing of functional soft robot. *Procedia Manuf.* **2019**, *34*, 566–573. [CrossRef]
92. Schaffner, M.; Faber, J.A.; Pianegonda, L.; Rühs, P.A.; Coulter, F.; Studart, A.R. 3D printing of robotic soft actuators with programmable bioinspired architectures. *Nat. Commun.* **2018**, *9*, 878. [CrossRef]
93. Sinatra, N.R.; Ranzani, T.; Vlassak, J.J.; Parker, K.K.; Wood, R.J. Nanofiber-reinforced soft fluidic micro-actuators. *J. Micromech. Microeng.* **2018**, *28*, 084002. [CrossRef]
94. Xavier, M.S.; Tawk, C.D.; Yong, Y.K.; Fleming, A.J. 3D-printed omnidirectional soft pneumatic actuators: Design, modeling and characterization. *Sens. Actuators A Phys.* **2021**, *332*, 113199. [CrossRef]
95. Zhang, Y.; Ng, C.J.; Chen, Z.; Zhang, W.; Panjwani, S. Miniature Pneumatic Actuators for Soft Robots by High-Resolution Multimaterial 3D Printing. *Adv. Mater. Technol.* **2019**, *4*, 1900427. [CrossRef]
96. Ge, L.; Dong, L.; Wang, D.; Ge, Q.; Gu, G. Sensors and Actuators A: Physical A digital light processing 3D printer for fast and high-precision fabrication of soft pneumatic actuators. *Sens. Actuators A Phys.* **2018**, *273*, 285–292. [CrossRef]
97. Ahangar, P.; Cooke, M.E.; Weber, M.H.; Rosenzweig, D.H. Current biomedical applications of 3D printing and additive manufacturing. *Appl. Sci.* **2019**, *9*, 1713. [CrossRef]
98. Aimar, A.; Palermo, A.; Innocenti, B. The Role of 3D Printing in Medical Applications: A State of the Art. *J. Healthc. Eng.* **2019**, *2019*, 5340616. [CrossRef]
99. Bozkurt, Y.; Karayel, E. 3D printing technology; methods, biomedical applications, future opportunities and trends. *J. Mater. Res. Technol.* **2021**, *14*, 1430–1450. [CrossRef]
100. Kotta, S.; Nair, A.; Alsabeelah, N. 3D Printing Technology in Drug Delivery: Recent Progress and Application. *Curr. Pharm. Des.* **2018**, *24*, 5039–5048. [CrossRef]
101. Wallis, M.; Al-Dulimi, Z.; Tan, D.K.; Maniruzzaman, M.; Nokhodchi, A. *3D Printing for Enhanced Drug Delivery: Current State-of-the-Art and Challenges*; Taylor & Francis: Abingdon, UK, 2020; Volume 46, ISBN 4412738728.
102. Prasad, L.K.; Smyth, H. 3D Printing technologies for drug delivery: A review. *Drug Dev. Ind. Pharm.* **2016**, *42*, 1019–1031. [CrossRef]
103. Dabbagh, S.R.; Sarabi, M.R.; Rahbarghazi, R.; Sokullu, E.; Yetisen, A.K.; Tasoglu, S. 3D-printed microneedles in biomedical applications. *iScience* **2021**, *24*, 102012. [CrossRef]
104. Huang, D.; Li, J.; Li, T.; Wang, Z.; Wang, Q.; Li, Z. Recent advances on fabrication of microneedles on the flexible substrate. *J. Micromech. Microeng.* **2021**, *31*. [CrossRef]
105. Hwang, H.H.; Zhu, W.; Victorine, G.; Lawrence, N.; Chen, S. 3D-Printing of Functional Biomedical Microdevices via Light- and Extrusion-Based Approaches. *Small Methods* **2018**, *2*, 1700277. [CrossRef] [PubMed]
106. Prabhakar, P.; Sen, R.K.; Dwivedi, N.; Khan, R.; Solanki, P.R.; Srivastava, A.K.; Dhand, C. 3D-Printed Microfluidics and Potential Biomedical Applications. *Front. Nanotechnol.* **2021**, *3*, 1–16. [CrossRef]
107. Kim, Y.; Son, K.; Lee, J. Auxetic structures for tissue engineering scaffolds and biomedical devices. *Materials* **2021**, *14*, 6821. [CrossRef] [PubMed]
108. Borovjagin, A.V.; Ogle, B.M.; Berry, J.L.; Zhang, J. From Microscale Devices to 3D Printing: Advances in Fabrication of 3D Cardiovascular Tissues. *Circ. Res.* **2017**, *120*, 150–165. [CrossRef] [PubMed]
109. Joyee, E.B.; Pan, Y. Additive manufacturing of multi material soft robot for on-demand drug delivery applications. *J. Manuf. Process.* **2020**, *56*, 1178–1184. [CrossRef]
110. Coltelli, M.A.; Catterlin, J.; Scherer, A.; Kartalov, E.P. Simulations of 3D-Printable biomimetic artificial muscles based on microfluidic microcapacitors for exoskeletal actuation and stealthy underwater propulsion. *Sens. Actuators A Phys.* **2021**, *325*, 112700. [CrossRef]
111. Vasilescu, S.A.; Bazaz, S.R.; Jin, D.; Shimoni, O.; Warkiani, M.E. 3D printing enables the rapid prototyping of modular microfluidic devices for particle conjugation. *Appl. Mater. Today* **2020**, *20*, 100726. [CrossRef]
112. Cesewski, E.; Haring, A.P.; Tong, Y.; Singh, M.; Thakur, R.; Laheri, S.; Read, K.A.; Powell, M.D.; Oestreich, K.J.; Johnson, B.N. Additive manufacturing of three-dimensional (3D) microfluidic-based microelectromechanical systems (MEMS) for acoustofluidic applications. *Lab Chip* **2018**, *18*, 2087–2098. [CrossRef]
113. Thomas, D.J.; Tehrani, Z.; Redfearn, B. 3-D printed composite microfluidic pump for wearable biomedical applications. *Addit. Manuf.* **2016**, *9*, 30–38. [CrossRef]
114. Taylor, A.P.; Velásquez-García, L.F. Miniaturized diaphragm vacuum pump by multi-material additive manufacturing. *J. Microelectromech. Syst.* **2017**, *26*, 1316–1326. [CrossRef]
115. Behrens, M.R.; Fuller, H.C.; Swist, E.R.; Wu, J.; Islam, M.M.; Long, Z.; Ruder, W.C.; Steward, R. Open-source, 3D-printed Peristaltic Pumps for Small Volume Point-of-Care Liquid Handling. *Sci. Rep.* **2020**, *10*, 1543. [CrossRef] [PubMed]
116. Economidou, S.N.; Uddin, M.J.; Marques, M.J.; Douroumis, D.; Sow, W.T.; Li, H.; Reid, A.; Windmill, J.F.C.; Podoleanu, A. A novel 3D printed hollow microneedle microelectromechanical system for controlled, personalized transdermal drug delivery. *Addit. Manuf.* **2021**, *38*, 101815. [CrossRef]

117. Rehmani, M.A.A.; Jaywant, S.A.; Arif, K.M. Study of microchannels fabricated using desktop fused deposition modeling systems. *Micromachines* **2021**, *12*, 14. [CrossRef]
118. Caudill, C.; Perry, J.L.; Iliadis, K.; Tessema, A.T.; Lee, B.J.; Mecham, B.S.; Tian, S.; DeSimone, J.M. Transdermal vaccination via 3D-printed microneedles induces potent humoral and cellular immunity. *Proc. Natl. Acad. Sci. USA* **2021**, *118*, e2102595118. [CrossRef]
119. Chen, Z.; Lin, Y.; Lee, W.; Ren, L.; Liu, B.; Liang, L.; Wang, Z.; Jiang, L. Additive Manufacturing of Honeybee-Inspired Microneedle for Easy Skin Insertion and Difficult Removal. *ACS Appl. Mater. Interfaces* **2018**, *10*, 29338–29346. [CrossRef]
120. Johnson, A.R.; Procopio, A.T. Low cost additive manufacturing of microneedle masters. *3D Print. Med.* **2019**, *5*, 1–10. [CrossRef]
121. Krieger, K.J.; Bertollo, N.; Dangol, M.; Sheridan, J.T.; Lowery, M.M.; O'Cearbhaill, E.D. Simple and customizable method for fabrication of high-aspect ratio microneedle molds using low-cost 3D printing. *Microsyst. Nanoeng.* **2019**, *5*, 42. [CrossRef]
122. He, Y.; Gao, Q.; Wu, W.; Bin, N.J.; Fu, J.Z. 3D printed paper-based microfluidic analytical devices. *Micromachines* **2016**, *7*, 108. [CrossRef]
123. Bégin-Drolet, A.; Dussault, M.A.; Fernandez, S.A.; Larose-Dutil, J.; Leask, R.L.; Hoesli, C.A.; Ruel, J. Design of a 3D printer head for additive manufacturing of sugar glass for tissue engineering applications. *Addit. Manuf.* **2017**, *15*, 29–39. [CrossRef]
124. Yeshmukhametov; Koganezawa; Yamamoto A Novel Discrete Wire-Driven Continuum Robot Arm with Passive Sliding Disc: Design, Kinematics and Passive Tension Control. *Robotics* **2019**, *8*, 51. [CrossRef]
125. Georgantzinos, S.K.; Giannopoulos, G.I.; Bakalis, P.A. Additive Manufacturing for Effective Smart Structures: The Idea of 6D Printing. *J. Compos. Sci.* **2021**, *5*, 119. [CrossRef]
126. Saleh, M.S.; Hu, C.; Brenneman, J.; Al Mutairi, A.M.; Panat, R. 3D printed three-dimensional metallic microlattices with controlled and tunable mechanical properties. *Addit. Manuf.* **2021**, *39*, 101856. [CrossRef]
127. Askari, M.; Hutchins, D.A.; Thomas, P.J.; Astolfi, L.; Watson, R.L.; Abdi, M.; Ricci, M.; Laureti, S.; Nie, L.; Freear, S.; et al. Additive manufacturing of metamaterials: A review. *Addit. Manuf.* **2020**, *36*, 101562. [CrossRef]
128. Zadpoor, A.A. Additively manufactured porous metallic biomaterials. *J. Mater. Chem. B* **2019**, *7*, 4088–4117. [CrossRef]
129. Fan, J.; Zhang, L.; Wei, S.; Zhang, Z.; Choi, S.-K.; Song, B.; Shi, Y. A review of additive manufacturing of metamaterials and developing trends. *Mater. Today* **2021**, *50*, 303–328. [CrossRef]
130. Savio, G.; Rosso, S.; Meneghello, R.; Concheri, G. Geometric Modeling of Cellular Materials for Additive Manufacturing in Biomedical Field: A Review. *Appl. Bionics Biomech.* **2018**, *2018*, 1654782. [CrossRef]
131. Singh, J.; Upadhyay, A.; Sehgal, S. A review on metallic micro lattice. *Mater. Today Proc.* **2020**, *33*, 1695–1700. [CrossRef]
132. Surjadi, J.U.; Gao, L.; Du, H.; Li, X.; Xiong, X.; Fang, N.X.; Lu, Y. Mechanical Metamaterials and Their Engineering Applications. *Adv. Eng. Mater.* **2019**, *21*, 1800864. [CrossRef]
133. Garcia-Taormina, A.R.; Alwen, A.; Schwaiger, R.; Hodge, A.M. A review of coated nano- and micro-lattice materials. *J. Mater. Res.* **2021**, *36*, 3607–3627. [CrossRef]
134. Obadimu, S.O.; Kourousis, K.I. Compressive Behaviour of Additively Manufactured Lattice Structures: A Review. *Aerospace* **2021**, *8*, 207. [CrossRef]
135. Zadpoor, A.A. Mechanical performance of additively manufactured meta-biomaterials. *Acta Biomater.* **2019**, *85*, 41–59. [CrossRef] [PubMed]
136. Noronha, J.; Qian, M.; Leary, M.; Kyriakou, E.; Brandt, M. Hollow-walled lattice materials by additive manufacturing: Design, manufacture, properties, applications and challenges. *Curr. Opin. Solid State Mater. Sci.* **2021**, *25*, 100940. [CrossRef]
137. Köhnen, P.; Ewald, S.; Schleifenbaum, J.H.; Belyakov, A.; Haase, C. Controlling microstructure and mechanical properties of additively manufactured high-strength steels by tailored solidification. *Addit. Manuf.* **2020**, *35*, 101389. [CrossRef]
138. Kumar, P.; Prakash, O.; Ramamurty, U. Micro-and meso-structures and their influence on mechanical properties of selectively laser melted Ti-6Al-4V. *Acta Mater.* **2018**, *154*, 246–260. [CrossRef]
139. Souza, J.; Großmann, A.; Mittelstedt, C. Micromechanical analysis of the effective properties of lattice structures in additive manufacturing. *Addit. Manuf.* **2018**, *23*, 53–69. [CrossRef]
140. Athanasiadis, A.E.F.; Dias, M.A.; Budzik, M.K. Can confined mechanical metamaterials replace adhesives? *Extrem. Mech. Lett.* **2021**, *48*, 101411. [CrossRef]
141. Kenel, C.; Casati, N.P.M.; Dunand, D.C. 3D ink-extrusion additive manufacturing of CoCrFeNi high-entropy alloy micro-lattices. *Nat. Commun.* **2019**, *10*, 904. [CrossRef]
142. Boulvert, J.; Costa-Baptista, J.; Cavalieri, T.; Perna, M.; Fotsing, E.R.; Romero-García, V.; Gabard, G.; Ross, A.; Mardjono, J.; Groby, J.-P. Acoustic modeling of micro-lattices obtained by additive manufacturing. *Appl. Acoust.* **2020**, *164*, 107244. [CrossRef]
143. Haubrich, J.; Gussone, J.; Barriobero-Vila, P.; Kürnsteiner, P.; Jägle, E.A.; Raabe, D.; Schell, N.; Requena, G. The role of lattice defects, element partitioning and intrinsic heat effects on the microstructure in selective laser melted Ti-6Al-4V. *Acta Mater.* **2019**, *167*, 136–148. [CrossRef]
144. Lozanovski, B.; Downing, D.; Tino, R.; du Plessis, A.; Tran, P.; Jakeman, J.; Shidid, D.; Emmelmann, C.; Qian, M.; Choong, P.; et al. Non-destructive simulation of node defects in additively manufactured lattice structures. *Addit. Manuf.* **2020**, *36*, 101593. [CrossRef]
145. Günther, J.; Brenne, F.; Droste, M.; Wendler, M.; Volkova, O.; Biermann, H.; Niendorf, T. Design of novel materials for additive manufacturing - Isotropic microstructure and high defect tolerance. *Sci. Rep.* **2018**, *8*, 1298. [CrossRef] [PubMed]

146. Echeta, I.; Dutton, B.; Leach, R.K.; Piano, S. Finite element modelling of defects in additively manufactured strut-based lattice structures. *Addit. Manuf.* **2021**, *47*, 102301. [CrossRef]
147. McGregor, D.J.; Tawfick, S.; King, W.P. Automated metrology and geometric analysis of additively manufactured lattice structures. *Addit. Manuf.* **2019**, *28*, 535–545. [CrossRef]
148. Yuan, L.; Ding, S.; Wen, C. Additive manufacturing technology for porous metal implant applications and triple minimal surface structures: A review. *Bioact. Mater.* **2019**, *4*, 56–70. [CrossRef]
149. Wang, M.; Wang, X.; Liu, J.; Wei, J.; Shen, Z.; Wang, Y. 3-Dimensional ink printing of friction-reducing surface textures from copper nanoparticles. *Surf. Coat. Technol.* **2019**, *364*, 57–62. [CrossRef]
150. Mekhiel, S.; Koshy, P.; Elbestawi, M.A. Additive texturing of metallic surfaces for wetting control. *Addit. Manuf.* **2021**, *37*, 101631. [CrossRef]
151. Thompson, M.K.; Moroni, G.; Vaneker, T.; Fadel, G.; Campbell, R.I.; Gibson, I.; Bernard, A.; Schulz, J.; Graf, P.; Ahuja, B.; et al. Design for Additive Manufacturing: Trends, opportunities, considerations, and constraints. *CIRP Ann.* **2016**, *65*, 737–760. [CrossRef]
152. Volkman, S.K.; Yin, S.; Bakhishev, T.; Puntambekar, K.; Subramanian, V.; Toney, M.F. Mechanistic Studies on Sintering of Silver Nanoparticles. *Chem. Mater.* **2011**, *23*, 4634–4640. [CrossRef]
153. Nanda, K.K.; Maisels, A.; Kruis, F.E.; Fissan, H.; Stappert, S. Higher Surface Energy of Free Nanoparticles. *Phys. Rev. Lett.* **2003**, *91*, 106102. [CrossRef]
154. Roy, N.K.; Behera, D.; Dibua, O.G.; Foong, C.S.; Cullinan, M.A. Single shot, large area metal sintering with micrometer level resolution. *Opt. Express* **2018**, *26*, 25534. [CrossRef] [PubMed]
155. Roy, N.K.; Behera, D.; Dibua, O.G.; Foong, C.S.; Cullinan, M.A. A novel microscale selective laser sintering (μ-SLS) process for the fabrication of microelectronic parts. *Microsyst. Nanoeng.* **2019**, *5*, 64. [CrossRef] [PubMed]
156. Bártolo, P.J. (Ed.) *Stereolithography*; Springer: Boston, MA, USA, 2011; ISBN 978-0-387-92903-3.
157. Crivello, J.V. The discovery and development of onium salt cationic photoinitiators. *J. Polym. Sci. Part A Polym. Chem.* **1999**, *37*, 4241–4254. [CrossRef]
158. Gibson, I.; Rosen, D.W.; Stucker, B. *Additive Manufacturing Technologies*; Springer: Berlin/Heidelberg, Germany, 2010; ISBN 9781441911193.
159. Lee, J.W.; Lee, I.H.; Cho, D.-W. Development of micro-stereolithography technology using metal powder. *Microelectron. Eng.* **2006**, *83*, 1253–1256. [CrossRef]
160. Regenfuß, P.; Ebert, R.; Exner, H. Laser Micro Sintering—A Versatile Instrument for the Generation of Microparts. *Laser Tech. J.* **2007**, *4*, 26–31. [CrossRef]
161. Woodfield, T.B.F.; Malda, J.; de Wijn, J.; Péters, F.; Riesle, J.; van Blitterswijk, C.A. Design of porous scaffolds for cartilage tissue engineering using a three-dimensional fiber-deposition technique. *Biomaterials* **2004**, *25*, 4149–4161. [CrossRef]
162. Yi, S.; Liu, F.; Zhang, J.; Xiong, S. Study of the key technologies of LOM for functional metal parts. *J. Mater. Process. Technol.* **2004**, *150*, 175–181. [CrossRef]
163. Hon, K.K.B.; Li, L.; Hutchings, I.M. Direct writing technology—Advances and developments. *CIRP Ann.* **2008**, *57*, 601–620. [CrossRef]
164. Butler, E.J.; Folk, C.; Cohen, A.; Vasilyev, N.V.; Chen, R.; del Nido, P.J.; Dupont, P.E. Metal MEMS tools for beating-heart tissue approximation. In Proceedings of the 2011 IEEE International Conference on Robotics and Automation, Shanghai, China, 9–13 May 2011; pp. 411–416.
165. Gad-el-Hak, M. (Ed.) *The MEMS Handbook-3 Volume Set*; CRC Press: Boca Raton, FL, USA, 2019; ISBN 9780429103872.
166. Stringer, J.; Derby, B. Limits to feature size and resolution in ink jet printing. *J. Eur. Ceram. Soc.* **2009**, *29*, 913–918. [CrossRef]
167. Dantan, J.-Y.; Huang, Z.; Goka, E.; Homri, L.; Etienne, A.; Bonnet, N.; Rivette, M. Geometrical variations management for additive manufactured product. *CIRP Ann.* **2017**, *66*, 161–164. [CrossRef]
168. Silva, M.R.; Pereira, A.M.; Sampaio, Á.M.; Pontes, A.J. Assessment of the Dimensional and Geometric Precision of Micro-Details Produced by Material Jetting. *Materials* **2021**, *14*, 1989. [CrossRef] [PubMed]
169. Thompson, M.K.; Mischkot, M. Design of Test Parts to Characterize Micro Additive Manufacturing Processes. *Procedia CIRP* **2015**, *34*, 223–228. [CrossRef]
170. Thompson, A.; Maskery, I.; Leach, R.K. X-ray computed tomography for additive manufacturing: A review. *Meas. Sci. Technol.* **2016**, *27*, 072001. [CrossRef]
171. du Plessis, A.; Yadroitsev, I.; Yadroitsava, I.; Le Roux, S.G. X-Ray Microcomputed Tomography in Additive Manufacturing: A Review of the Current Technology and Applications. *3D Print. Addit. Manuf.* **2018**, *5*, 227–247. [CrossRef]
172. Leach, R. Metrology for Additive Manufacturing. *Meas. Control* **2016**, *49*, 132–135. [CrossRef]
173. Lévesque, D.; Bescond, C.; Lord, M.; Cao, X.; Wanjara, P.; Monchalin, J.-P. Inspection of additive manufactured parts using laser ultrasonics. *AIP Conf. Proc.* **2016**, *1706*, 130003.

micromachines

Article

A Comparative Evaluation of Magnetorheological Micropump Designs

Sevki Cesmeci [1,*], Rubayet Hassan [1] and Mahmoud Baniasadi [2]

[1] Department of Mechanical Engineering, Georgia Southern University, Statesboro, GA 30460, USA; rh17100@georgiasouthern.edu
[2] Intel Corporation, Hillsboro, OR 97124, USA; mahmoud.baniasadi@intel.com
* Correspondence: scesmeci@georgiasouthern.edu

Abstract: In this study, we assessed the performance characteristics of five different magnetorheological micropump designs, two of which were our proposed designs, while others were from the existing designs in the literature. Comparisons have been performed based on physics-based simulations, and the fully coupled magneto-solid-fluid interaction simulations were carried out in COMSOL Multiphysics software. For a fair and meaningful comparison, both the material and geometric properties were kept the same, and the simulations were run for one complete pumping cycle. The results showed that the proposed flap and duckbill valve models could pump 1.09 μL and 1.16 μL respectively in 1 s, which was more than the rest of the existing micropump models. Moreover, at 0.5 s, when the magnetic flux density was maximum, the flap and duckbill valve models could pump almost twice as fluid as some of the existing valve models did. The results also demonstrated that the flap and duckbill valve models were nearly five times faster than some of existing models. In conclusion, the proposed two micropump models could propel more net fluid volume than the existing micropump designs, experienced low leakage during the contraction and expansion phase, and had faster response times. We believe that the present study provides valuable insights for future micropump designs, which have an extensive range of application areas, ranging from insulin dosing systems for T1D patients to artificial organs to transport blood and from organ-on-chip applications to micro-cooling systems.

Keywords: micropump; magnetorheological; MRE; magneto-solid-fluid interaction

Citation: Cesmeci, S.; Hassan, R.; Baniasadi, M. A Comparative Evaluation of Magnetorheological Micropump Designs. *Micromachines* **2022**, *13*, 764. https://doi.org/10.3390/mi13050764

Academic Editor: Stelios K. Georgantzinos

Received: 18 April 2022
Accepted: 10 May 2022
Published: 12 May 2022

1. Introduction

With the recent advancements in manufacturing technologies, researchers have begun to explore novel designs at the micro and nano levels. Various micropump designs have been proposed and studied in the literature, including AC magnetohydrodynamic, piezoelectric, electroosmotic, acoustic, shape memory, thermo-pneumatic and so on. These designs have their advantages and disadvantages over one another in terms of scalability, biocompatibility, complexity, accuracy, cost and reliability. In this study, we focus on the magnetorheological micropumps. Magnetorheological elastomer (MRE) is a type of semi-active material, consisting of a rubber-like matrix with micron-sized iron particles embedded in it. MREs change their mechanical behavior under the influence of magnetic field. Their stiffness and damping properties alter with the intensity of the magnetic field. Due to this unique feature, they were utilized in various applications, ranging from vibration isolation systems in buildings and bridges to sensors for structural health monitoring applications and actuation systems such as micropumps [1–4]. There have been various MR micropump designs proposed in the literature in the last decade [5–14]. These designs usually varied based on the valves used. For example, Behrooz and Gordaninejad's conceptual MR micropump design employed conical one-way valves to transport the flow unidirectionally [15,16], whereas Ehsani and Nejat investigated another conceptual design of an MR micropump, where they used one-way angled valves [17]. More recently, Xufeng et. al. proposed a magneto-active pulse pump (MAPP) with flexible check valves, resembling an eagle beak [18]. Researchers carried out fully coupled magneto-solid-fluid simulations to assess the performance of their proposed designs.

However, all of these designs were studied individually, and they are not compared against one another in terms of their pumping performances. In this study, we assessed the performance characteristics of five different MR micropump designs, two of which were our proposed designs, while others were from the existing designs in the literature. The models from the literature included the Behrooz, Ehsani and Xufeng [15–18] models. These were all the available magnetorheological micropumps with valves in the literature. Table 1 lists a brief overview of these pumps. All of these micropumps involved one-way valves with different designs. All included a flexible MRE top wall. While Behrooz and Ehsani models carried out fully coupled magneto-solid-fluid interaction simulations, Xufeng model did not. Behrooz and Ehsani models also carried out parametric studies to see the effects of important design parameters.

Table 1. Existing magnetorheological micropump designs in the literature.

Micropump Models	Number of Electromagnets	Valve Type	Upper Wall Material	Parametric Study	Simulation
Behrooz valve model	1	One-way conical valve	MRE	Yes	MFSI
Xufneg valve model	1	Check valve	MRE	No	No
Ehsani valve model	1	One-way angle valve	MRE	Yes	MFSI

The other two models are the ones that are proposed in this study, namely the duckbill valve and flap valve models. Comparisons have been performed based on physics-based simulations, and the fully coupled magneto-solid-fluid interaction (MFSI) simulations were carried out in COMSOL Multiphysics software. For a fair and meaningful comparison, both the material and geometric properties were kept the same, and the simulations were run for one complete pumping cycle. The details of the design will be discussed in Section 2. Highly coupled MFSI simulations were carried out in COMSOL Multiphysics software (v5.6). The paper is organized as follows: the proposed design is discussed in Section 2, simulation methodology of five micropump designs is presented in Section 3 and summary and conclusions are discussed in Section 4.

2. Proposed Design

Among the proposed designs, the flap valve model consists of a pump chamber, two one-way flap valves and an electromagnet, while the duckbill valve model is composed of a pump chamber, two one-way duckbill valves and an electromagnet (Figure 1). The top wall of the pump chamber is made of a semi-active smart material called magnetorheological elastomer (MRE), while the rest of the structure is made from a passive elastomer. MREs falling in the category of composite materials are composed of a rubber-like base matrix, such as silicone and micron-sized iron particles doped in it. Because of their ferromagnetic properties, the MREs deform or can be designed to resist the deformation in the presence of a magnetic field, depending on design intent. Since their mechanical properties alter by the induced magnetic field, they are properly categorized as semi-active materials. In the proposed micropump design, under an external magnetic field, the top wall contracts inwardly in the pumping chamber. The amount of contraction depends on the intensity of the applied magnetic field, which can be controlled via an external electromagnet. The deformation on the top wall creates a squeezing effect on the fluid inside the flow chamber, pushing the fluid through the one-way valve in the front end. While this is occurring, a one-way flap and duckbill valve in the rear end prevents the fluid from leaking backward, thus creating an effective unidirectional pumping mechanism.

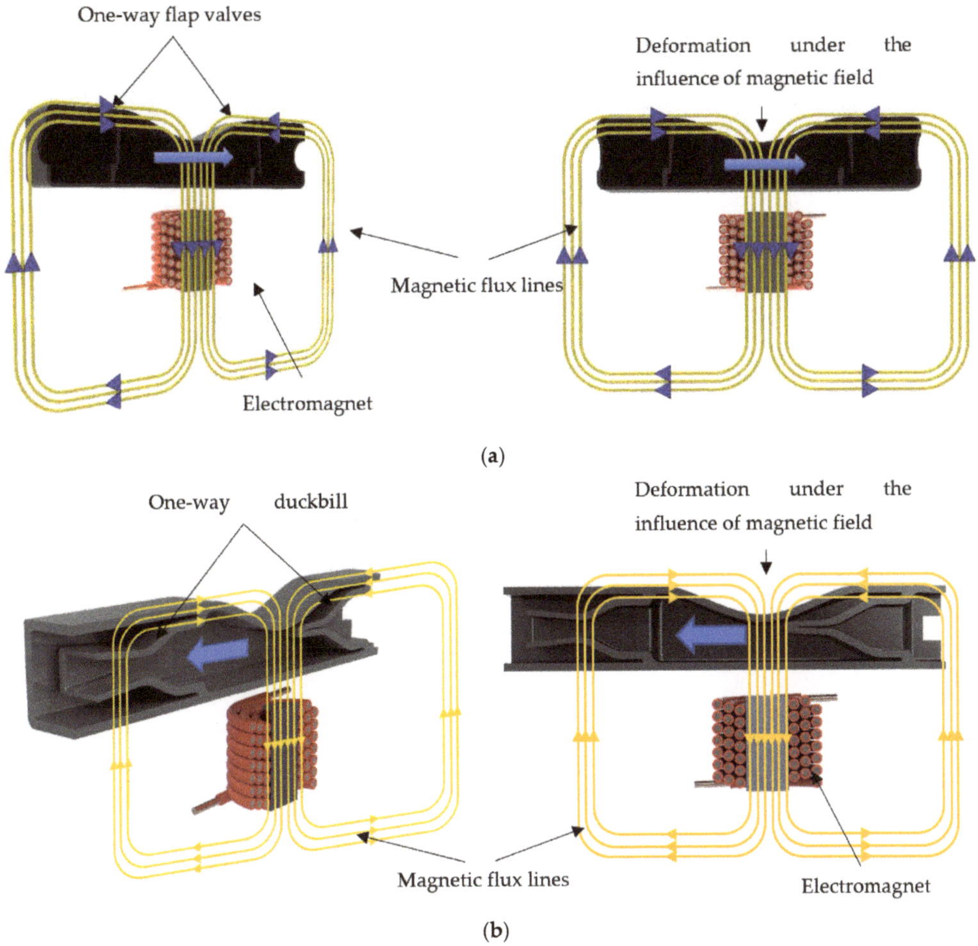

Figure 1. The proposed micropump designs with their main components: (**a**) flap valve model; and (**b**) duckbill valve model.

3. Simulation Methodology

3.1. Model Creation

The proposed design involves coupled magneto-solid-fluid interaction physics. Thus, the performance of the pump could best be predicted with the help of computer simulations rather than simplified 1D analytical models. In this study, simulations were carried out using COMSOL Multiphysics software (v5.6, Burlington, MA, USA). Figure 2 shows a schematic of the model created in COMSOL.

The geometric and material properties of the 2D models are given in Table 2, while the dimensions are shown in Figure 3 for convenience. The magnet is placed 0.35 mm below the pump chamber. The height and wall thickness of the pump chamber is 1.10 mm and 0.10 mm, respectively, for all models. However, to maintain the same pumping area the distance between two valves is varied on different valve models. The elastic modulus of the pump material and average magnetic flux density acting on the upper wall are also kept identical, which are 1.2 MPa and 0.018 T, respectively.

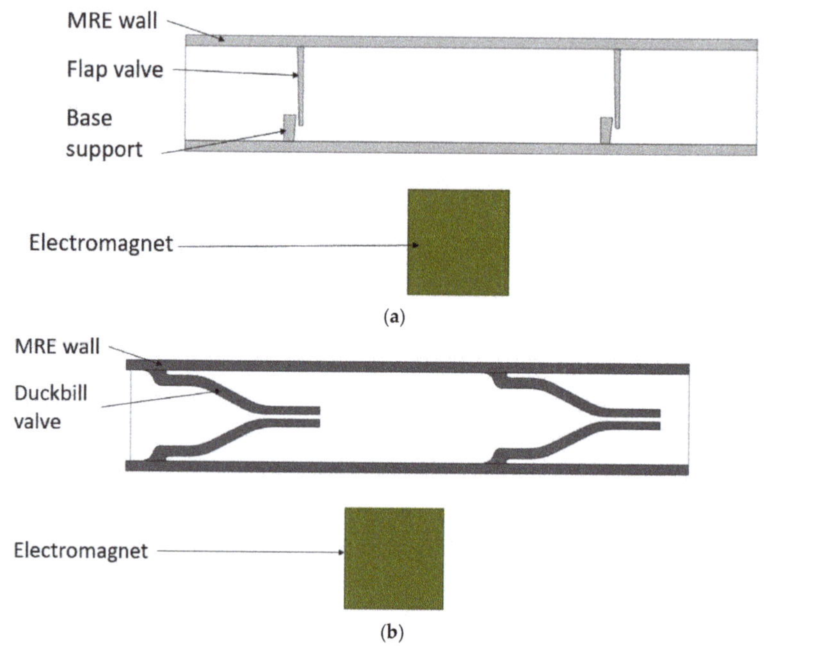

Figure 2. Micropump models with their main components: (**a**) flap valve model; and (**b**) duckbill valve model.

Table 2. Properties of microchannels.

Parameter	Symbol	Behrooz Valve Model	Xufeng Valve Model	Ehsani Valve Model	Duckbill Valve Model	Flap Valve Model
Height of the pump chamber	H	0.9 (mm)	0.9 (mm)	0.9 (mm)	0.9 (mm)	0.9 (mm)
Thickness of the upper wall	t	0.1 (mm)	0.1 (mm)	0.1 (mm)	0.1 (mm)	0.1 (mm)
Valve spacing distance	S	S1 = 3.079 (mm)	S2 = 3.238 (mm)	S3 = 3.069 (mm)	S4 = 3.204 (mm)	S5 = 3.102 (mm)
Length of micro channel	L	5.65 (mm)	5.65 (mm)	5.65 (mm)	5.65 (mm)	5.65 (mm)
Distance between the pump chamber and electromagnet	D	0.35 (mm)	0.35 (mm)	0.35 (mm)	0.35 (mm)	0.35 (mm)
Side of the electromagnet	A	1 (mm)	1 (mm)	1 (mm)	1 (mm)	1 (mm)
Magnetic flux density	B	0.018 (T)	0.018 (T)	0.018 (T)	0.018 (T)	0.018 (T)
Elastic modulus	E	1.2 (MPa)	1.2 (MPa)	1.2 (MPa)	1.2 (MPa)	1.2 (MPa)

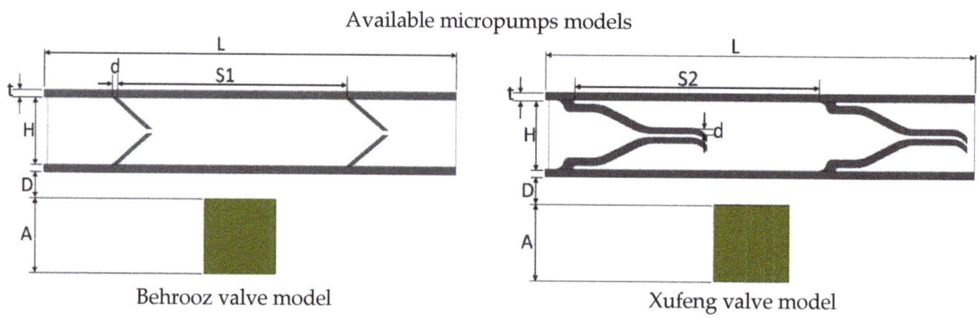

Figure 3. *Cont.*

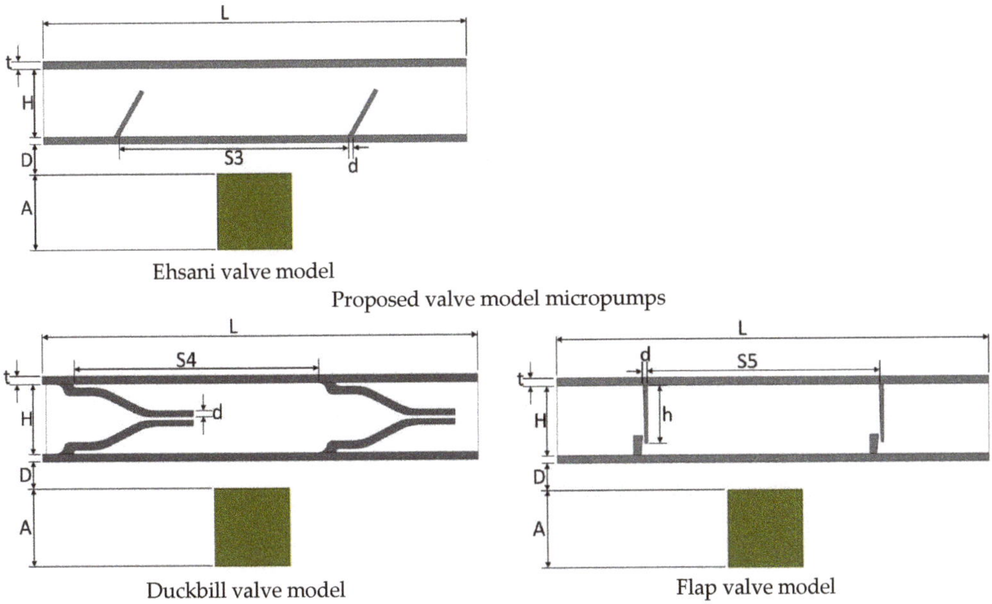

Figure 3. Valve models. Dimension values are given in Table 2.

The dimensions of the five 2D valve models are shown in the Figure 3.

In this study, MRE is assumed as a ferromagnetic material that has constant structural and magnetic properties. The wall at the bottom does not deform in the presence of the magnetic field. It is assumed to be sitting on a flat rigid surface so the deformation on the top wall can activate the fluid flow. The magnetic field analysis was conducted in an AC/DC module, whereas the structural deformation of the pump chamber, including the top wall and one-way valves and fluid flow through pump chamber and valves were carried out in Solid Mechanics and Laminar Flow modules, respectively (Figure 4).

3.2. Simulation Procedure

The simulation procedure is illustrated with the flowchart shown in Figure 5. First, the variables such as geometric and material properties were defined. Since the simulation involves the deformation of the pump chamber and one-way valves, the moving mesh schemes were defined next. Then the simulation geometry was created, which was followed by the material assignment to all solid and fluid domains. Next, the respective boundary conditions were assigned in each physics module, i.e., AC/DC module, solid mechanics module and laminar flow module. Then these different modules were communicated with each other to transfer data between different flow physics. For example, the upper wall of the pump chamber deforms under the influence of a magnetic field since it is a ferromagnetic material. To model this phenomenon, the AC/DC module was run to calculate the magnetic field over the entire domain, including all solid and fluid domains. Then this information was passed to the solid mechanics module to calculate the resultant deformation under the influence of the magnetic field. Similarly, the deformation data on the seal was transferred to the fluid domain via fluid-structure coupling between the solid mechanics and laminar flow modules. This reflected itself as a pressure load on the fluid domain at the fluid-solid interface, causing the fluid flow, and thus the pumping effect. While the fluid was pushed through the front valve, the pressure data on the valves were communicated, again, with the fluid-structure coupling. Note that the fluid-structure coupling was a two-way coupling, which provided convenience when the data was communicated back and forth between the solid and fluid domains. This was done by defining fluid-structure interaction Multiphysics at all interfaces between the fluid and solid domains.

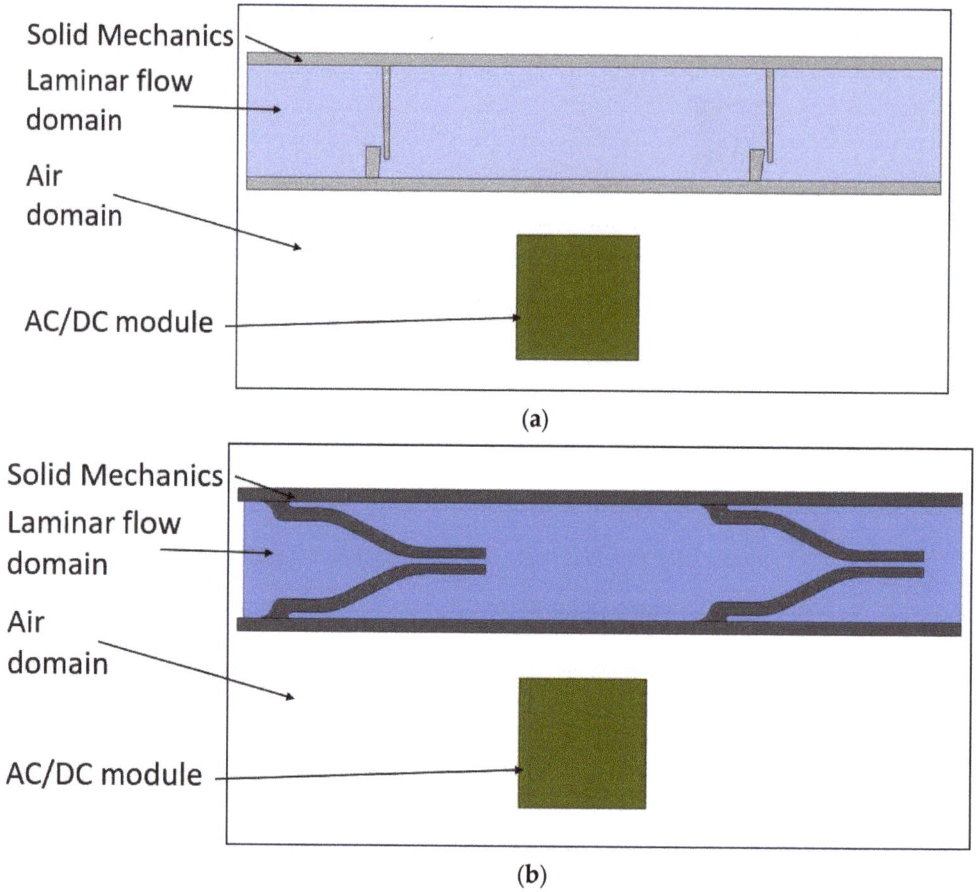

Figure 4. 2D simulation model geometries: (**a**) flap valve model; and (**b**) duckbill valve model.

3.3. Boundary Conditions

As for the boundary conditions, a current excitation was added to the coil in the AC/DC module, and the force calculation interface was selected for the top edge of the upper wall of the pump chamber. This was to ensure that the Maxwell forces were transferred to the top edge for the desired deformation, and thus the resulting pumping effect. A sinusoidal time function and an external current density with a specific amplitude were applied to the core of the electromagnet. In the solid mechanics module, fixed support was added at the bottom wall of the pump chamber and the Maxwell surface stress tensor (magnetic interaction force between the pump chamber and magnetic field) was selected in the boundary load interface. In the laminar flow module, the pressures at the left and right outlets were set to 0 Pa (gage). This was to allow the fluid freely to pass in and out from two terminals. In addition, a no-slip condition was applied to the walls surrounding the flow chamber.

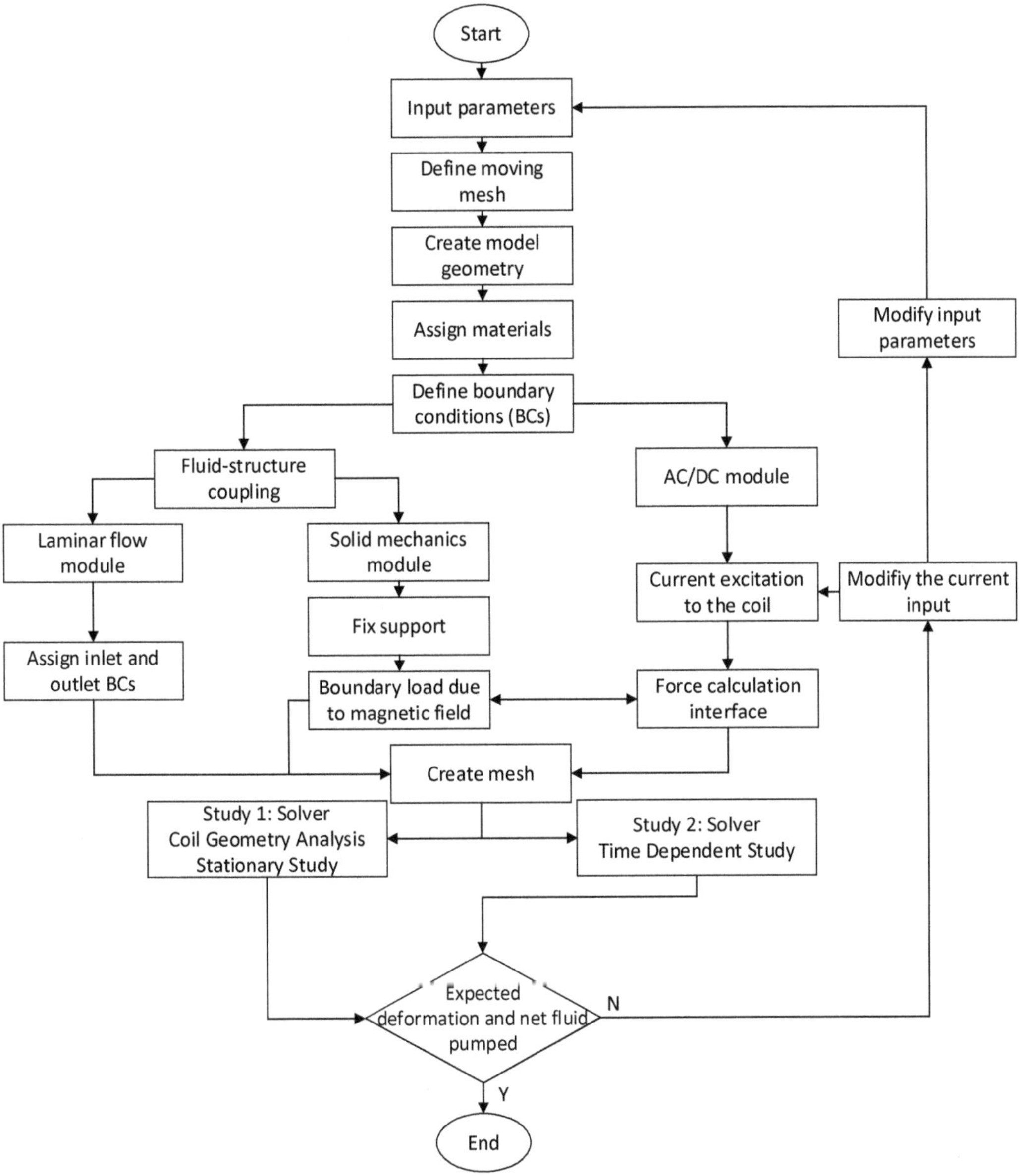

Figure 5. Flowchart of the simulation process in COMSOL.

3.4. Computing Equations

Next, we discuss the governing equations for each physics involved. The following equations were solved for the magnetic domain

$$\nabla \cdot J = 0 \tag{1}$$

$$\nabla \times H = J \tag{2}$$

$$B = \mu_0 \mu_r H \tag{3}$$

$$J = \nabla \times (\mu_r \mu_0)^{-1} B \tag{4}$$

where ∇ is the gradient operator, B is the magnetic flux density, J is the current density; μ_0 and μ_r are the permeability of the vacuum and the relative magnetic permeability, respectively. The magnetic flux density H is determined in Equation (1), which is calculated from Ampere's law. By combining Equations (1)–(3), the relation between J and B is obtained, which is given by Equation (4).

Moreover, Equation (5) shows the final equation of J

$$J = \sigma E + \sigma \mathrm{v} \times B + J_e \tag{5}$$

where σ is the electrical conductivity, E is the electric field, J_e is the external current density and v is the velocity of the conductor. The magnetic flux density B and external current density J_e can be calculated from Equations (6) and (7), respectively.

$$B = \nabla \times A \tag{6}$$

$$J_e = \frac{I \cdot n}{a \cdot b} \tag{7}$$

In Equations (6) and (7), A is the magnetic vector potential, n is the number of turns of the coil of the electromagnet, I is the input electrical current to the coil and a and b are the cross-sectional dimensions of the core. The magnetic field density B can be calculated from Equation (8), when the external current density J_e is known:

$$J_e = \nabla \times (\mu_r \mu_0)^{-1} B - \sigma E - \sigma \mathrm{v} \times B \tag{8}$$

The Maxwell stress is shown in Equations (9) and (10) the total stress, which is the result of magnetic stress and fluid pressure, is shown

$$n \cdot \sigma_{maxwell} = -0.5 n (H \cdot B) + (n \cdot H) B^T \tag{9}$$

$$\sigma_{total} = \sigma_{maxwell} + \sigma_p \tag{10}$$

where n is the normal vector, σ_{total} is the total stress, $\sigma_{Maxwell}$ is the Maxwell stress and σ_p is the stress due to fluid pressure.

In the fluid domain, the flow is set to laminar incompressible flow appropriately. The following equations were solved in the fluid domain:

Steady continuity:

$$\rho \nabla \cdot u_{fluid} = 0 \tag{11}$$

Navier-Stokes equations for stationary approach:

$$\rho (u_{fluid} \cdot \nabla) u_{fluid} = \nabla \cdot [-pI + K] + F \tag{12}$$

Navier-Stokes equation for the time-dependent approach

$$\rho \frac{\partial u_{fluid}}{\partial t} + \rho (u_{fluid} \cdot \nabla) u_{fluid} = \nabla \cdot [-pI + K] + F \tag{13}$$

where u_{fluid} is the velocity vector, ρ is the density, p is the pressure, K is the turbulent kinetic energy and F is the volume force vector. The turbulent kinetic energy K can be determined from Equation (14).

$$K = \mu (\nabla u_{fluid} + (\nabla u_{fluid})^T) \tag{14}$$

For the solid mechanics, the following equations were solved:

For stationary approach:

$$0 = \nabla \cdot (FS)^T + Fv,\ F = I + \nabla u_{solid} \tag{15}$$

For the time-dependent approach:

$$\rho \frac{\partial^2 u_{solid}}{\partial t^2} = \nabla \cdot (FS)^T + Fv, F = I + \nabla u_{solid} \tag{16}$$

In Equations (15) and (16), Fv is the volume force vector, u_{solid} is the displacement, I is the unit tensor and FS (F is the deformation gradient) is the first Piola-Kirchhoff stress tensor.

3.5. Grid Generation and Grid Independence Study

As is commonplace procedure for conducting simulations in both Finite Element Analysis (FEA) and Computational Fluid Dynamics (CFD), a mesh independency analysis was conducted. To begin this process, a course mesh was applied to an initial simulation. Then, the mesh size was reduced continually, and the net pumped volume was monitored after each run. The mesh size reduction continued until there was no significant change between two sequential cases. This process is shown in Table 3, which is also visually illustrated in Figure 6. From the recorded cases, it can be seen that the percent change in the target parameter was 0.14% between grid numbers 4 and 5. Thus, grid number 4 is selected for the rest of the simulations.

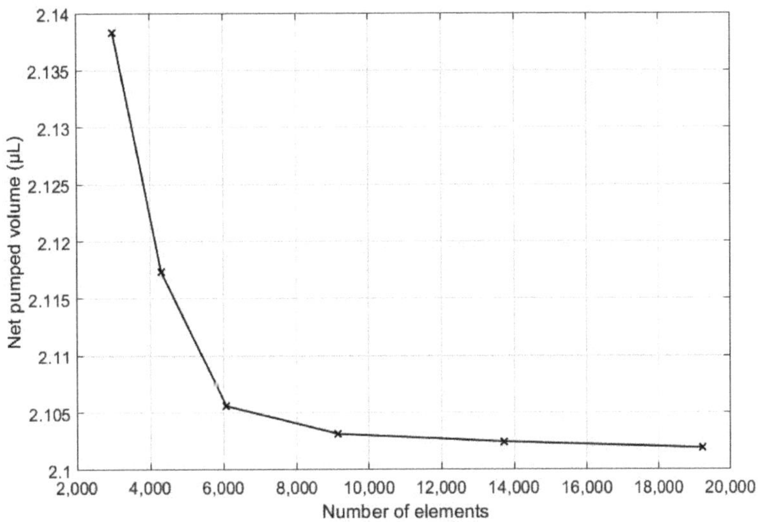

Figure 6. Net volume pumped is calculate for different number of elements.

Table 3. Net pumped volume for each grid number.

Grid Number	Number of Mesh Elements	Net Pumped Volume (µL)	The Difference in Net Pumped Volume (%)
1	3002	2.13832	-
2	4306	2.11735	0.98
3	6075	2.10558	0.56
4	9144	2.10314	0.51
5	13,708	2.10245	0.14
6	19,215	2.10191	0.11

3.6. Validity Study

Although COMSOL Multiphysics is a proven simulation tool and has been employed by thousands of scientific studies in the literature, it is always wise and scientifically required to validate the simulation approaches with the existing studies in literature when experimental data is not readily available. To this end, we selected the model presented in [18]. All parameters, boundary conditions, magnetic flux density and geometric dimensions were set to be the same. Figure 7 shows the model with the main components, as well as the dimensions.

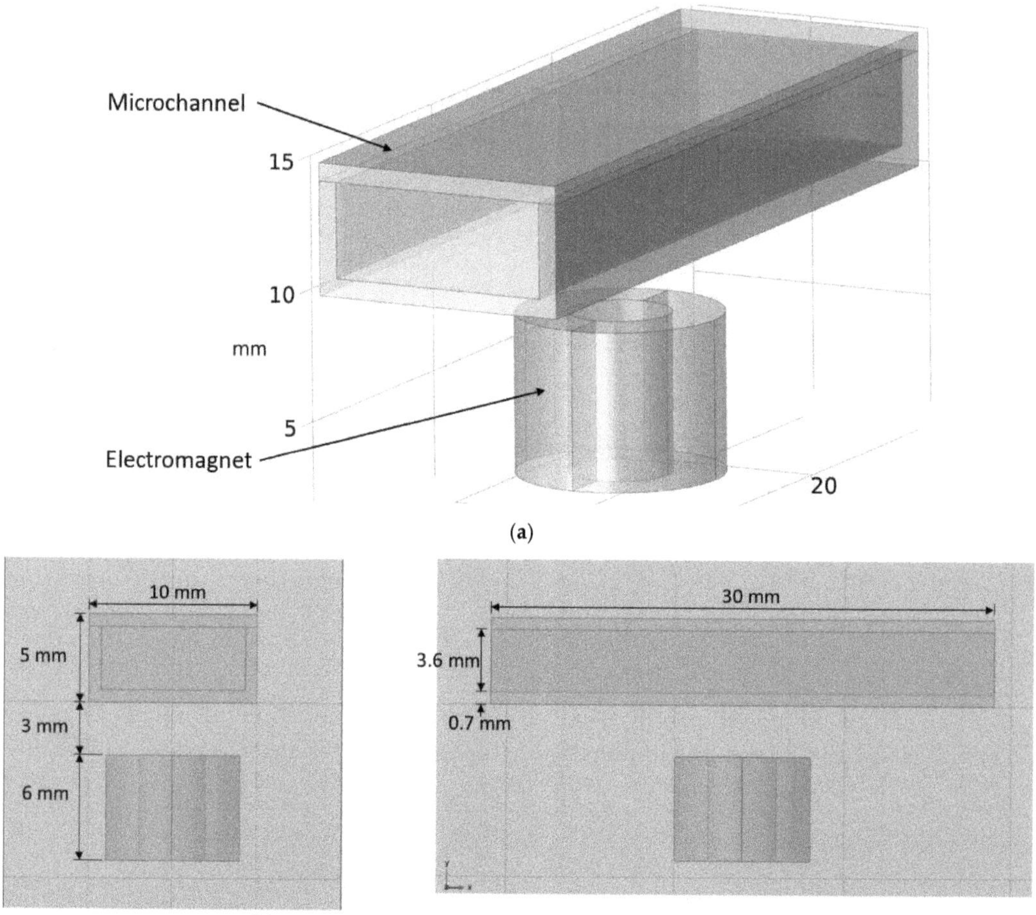

Figure 7. Schematic of the model studied in "Reprinted/adapted with permission from Ref. [18]. 2022, Rubayet Hassan": (**a**) 3D view; (**b**) front view; and (**c**) side view.

The simulations were carried out by following the procedure outlined in Figure 5. The comparisons between the benchmark study and this study are given in Table 4. From Table 4, it is seen that the upper wall displacement for both cases is the same for 75 mT, 145 mT and 175 mT magnetic flux densities. In addition to the deformations, volume flow rates were also compared. Figure 8 presents a comparison graph between the two cases. As seen from the figure, the volume flow rates matched closely between the two cases. Furthermore, the numerical data for the volume flow rates are presented in Table 5, with percent error margins. It can be seen that the average percent error between the two cases is about 1.6%, with minimum and maximum deviations being 0.00% and 4.16% at 150 mT and 175 mT, respectively. This validates our simulation methodology, allowing us to continue with the simulations of the pump designs.

Table 4. Comparisons of simulations between this study and benchmark study: (e)–(g) 3D shape of pipe under 75 mT, 145 mT, and 175 Mt during experimental study; (i)–(k) The deformation simulation corresponding to the experimental study in (e)–(g) "Reprinted/adapted with permission from Ref. [18]. 2022, Rubayet Hassan".

Magnetic Flux Density (mT)	Benchmark Study "Reprinted/Adapted with Permission from Ref. [18]. 2022, Rubayet Hassan".	This Study	Maximum Displacement of Upper Wall (mm)
75	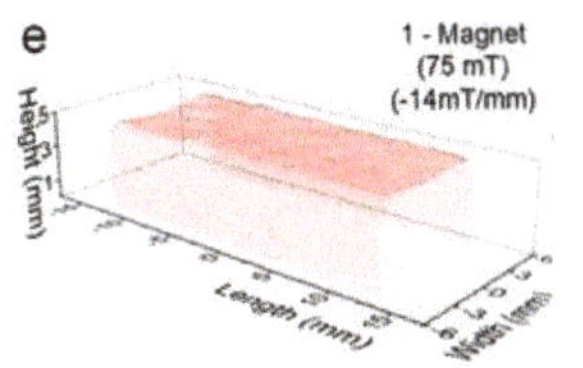 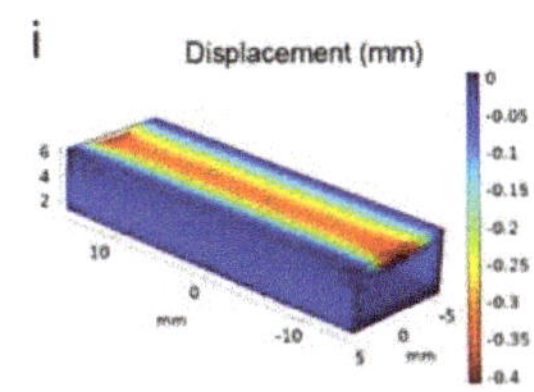	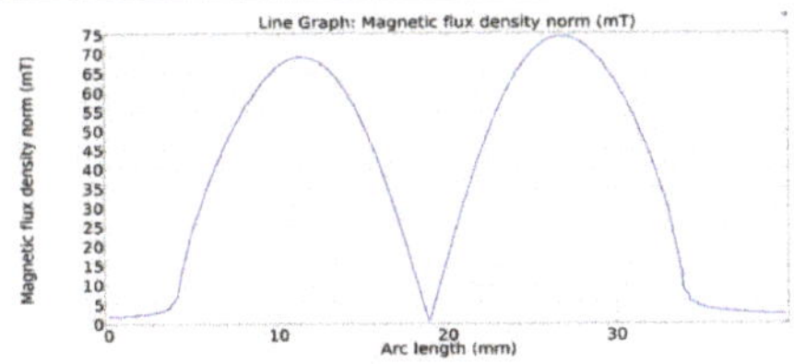 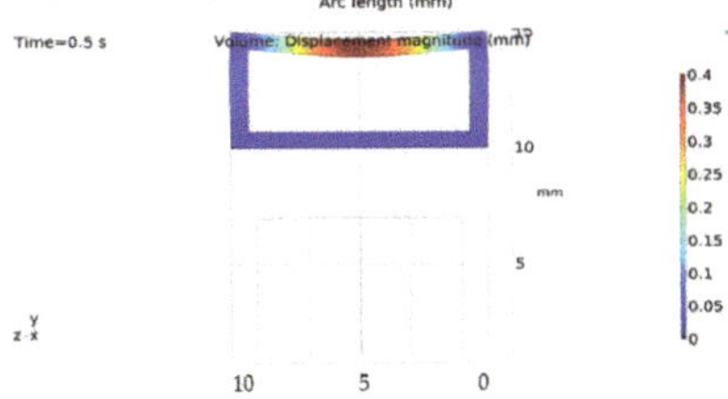	0.4
145	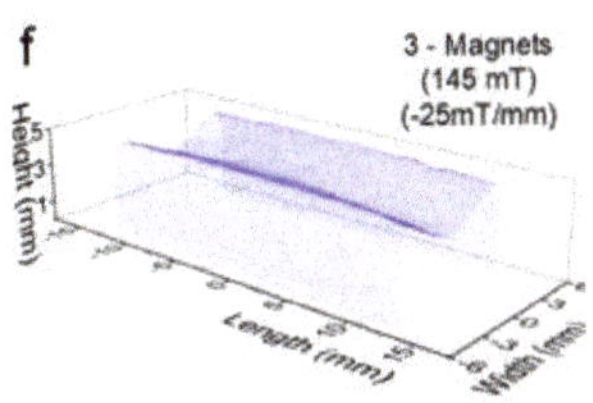 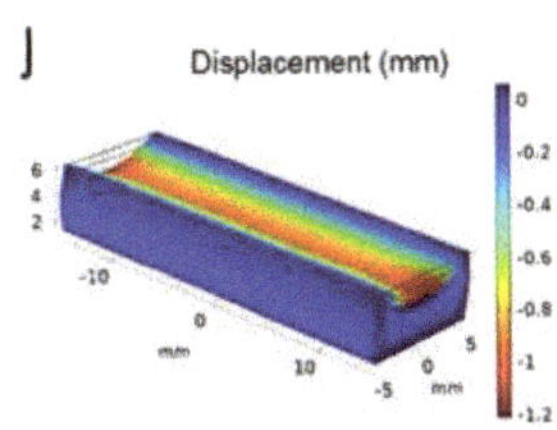	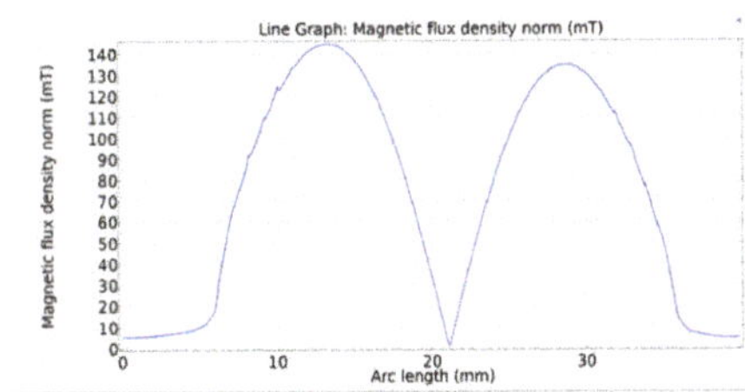 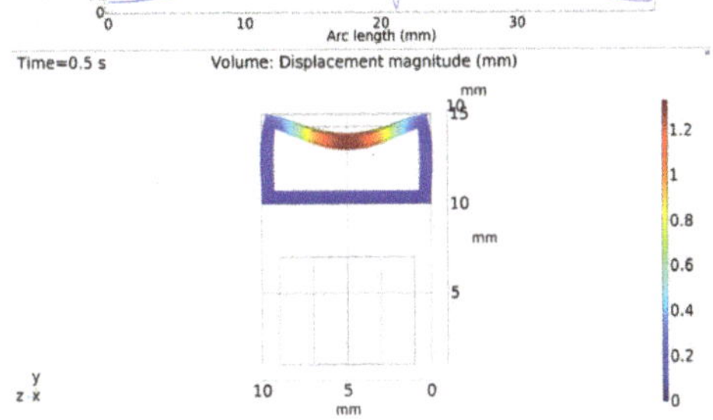	1.2

Table 4. *Cont.*

Magnetic Flux Density (mT)	Benchmark Study "Reprinted/Adapted with Permission from Ref. [18]. 2022, Rubayet Hassan".	This Study	Maximum Displacement of Upper Wall (mm)
175	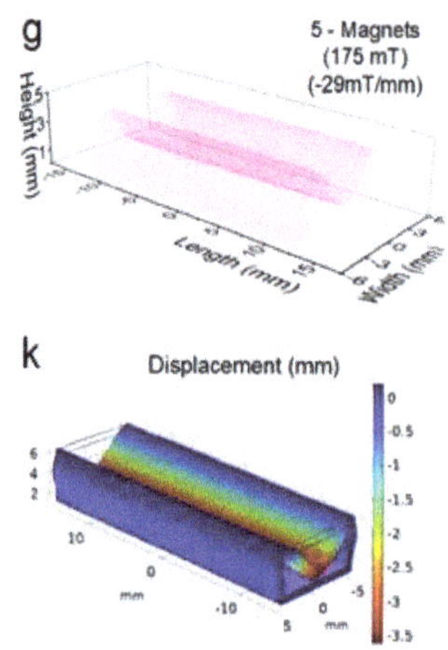	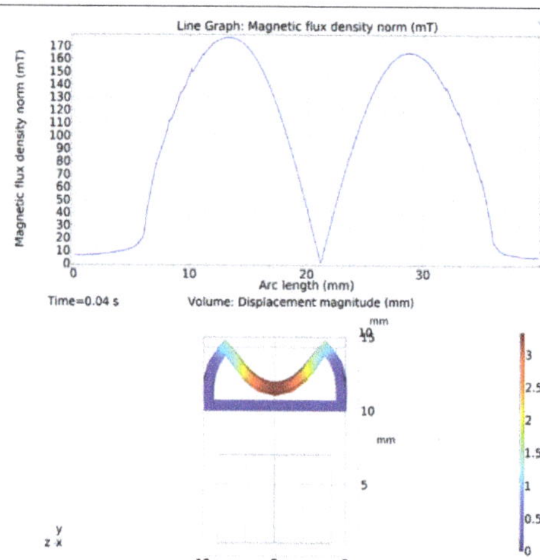	3.4

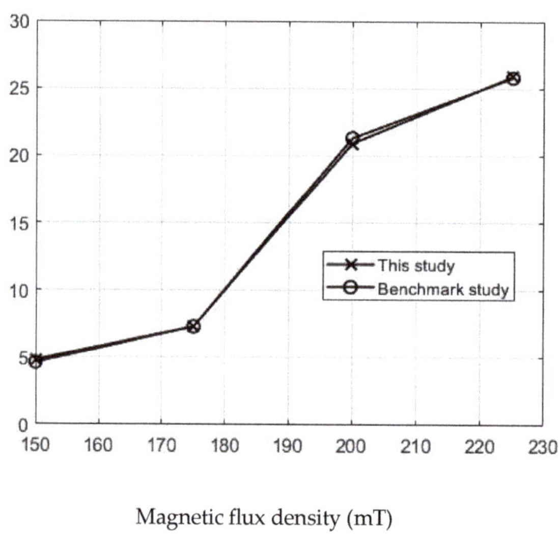

Figure 8. Net pumped volume comparisons between this study and benchmark study "Reprinted/adapted with permission from Ref. [18]. 2022, Rubayet Hassan".

Table 5. Percent error of the flow rates between this study and benchmark study "Reprinted/adapted with permission from Ref. [18]. 2022, Rubayet Hassan".

Magnetic Flux Density (mT)	Magnetic Flux Density (mT)	Flow Rate from Validation Case (μL/s)	Percentage of Error (%)
150	4.8	4.6	4.16
175	7.3	7.3	0
200	20.9	21.3	1.91
225	25.9	25.8	0.39

3.7. Results and Discussions

In this section, the pumping process is analyzed for one pumping cycle for the five micropump models. Comparisons are made in terms of volume flow rate and MRE wall response time. A full pumping cycle consists of a contraction phase, followed by an expansion phase. When the electromagnet is activated, the magnetic particles inside the MRE actuates and deforms the MRE wall downwards. The first phase of pumping is contraction. In this phase, the magnetic field over the MRE wall increase. Figure 9 represents the magnetic field of the upper MRE wall for all five models during the contraction period. The magnetic flux density is the same for the five models of micropump.

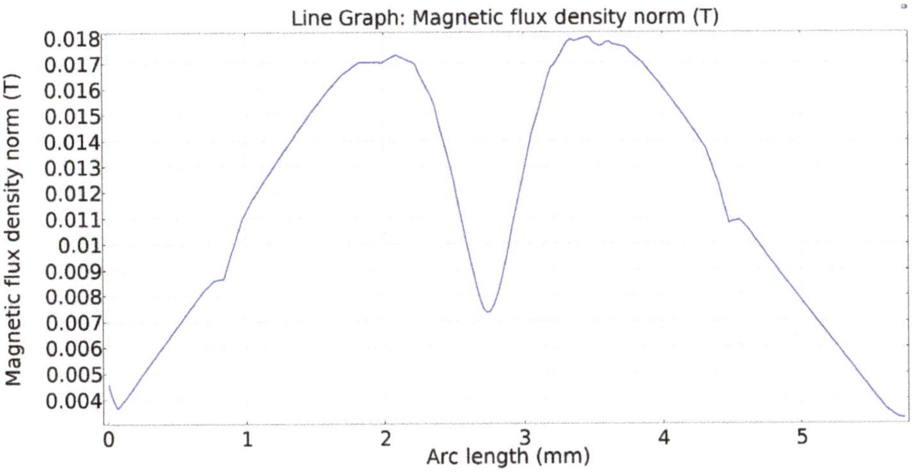

Figure 9. Magnetic flux density over the upper MRE wall under the magnetic field for the five valve models.

Figure 10 shows the vertical displacement of upper wall for five different valve models. The Ehsani valve model experienced the largest upper wall deformation. This is because the distance between the upper wall and the upper tip of the valves is larger than other valves. That's why the upper wall does not face any hindrance during deformation inwards. On the other hand, the MRE wall shows less deformation in the case of the Behrooz valve model. It is because the width of the Behrooz valve is more than the other valves, so the tips of the upper and lower valves sustain more resistance during the deformation of the MRE upper wall. The MRE wall bends due to the applied force. The deflection of MRE and the total applied force are reduced to zero when the expansion phase is completed. The fluid domain and solid parts are at rest initially for all five models.

When the upper wall begins to deform downwards it pushes the fluid inside the microchannel towards the two outlets. At this time, the fluid enclosed between the two valves deflects the left and right valves in a way that prevents the backflow, and propels the fluid to the right unidirectionally. This phenomenon is the same for all the five models.

Figure 11 shows that the right valve is fully open when the fluid velocity is maximum at the right outlet, and at the same time, the left valve remains closed to prevent the back flow. This is the same for all five models.

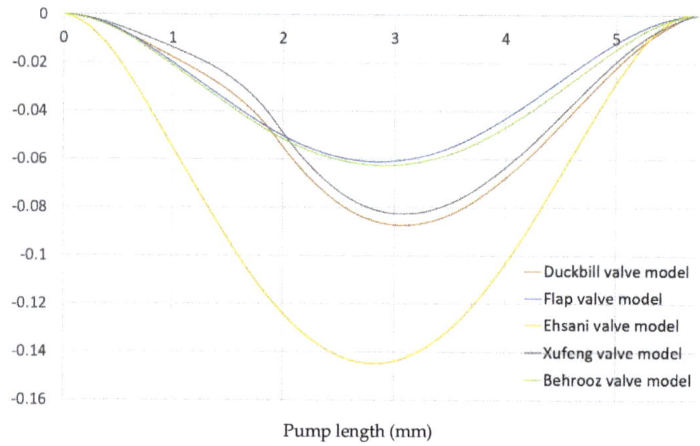

Figure 10. Vertical displacement of upper MRE wall for five different valve models.

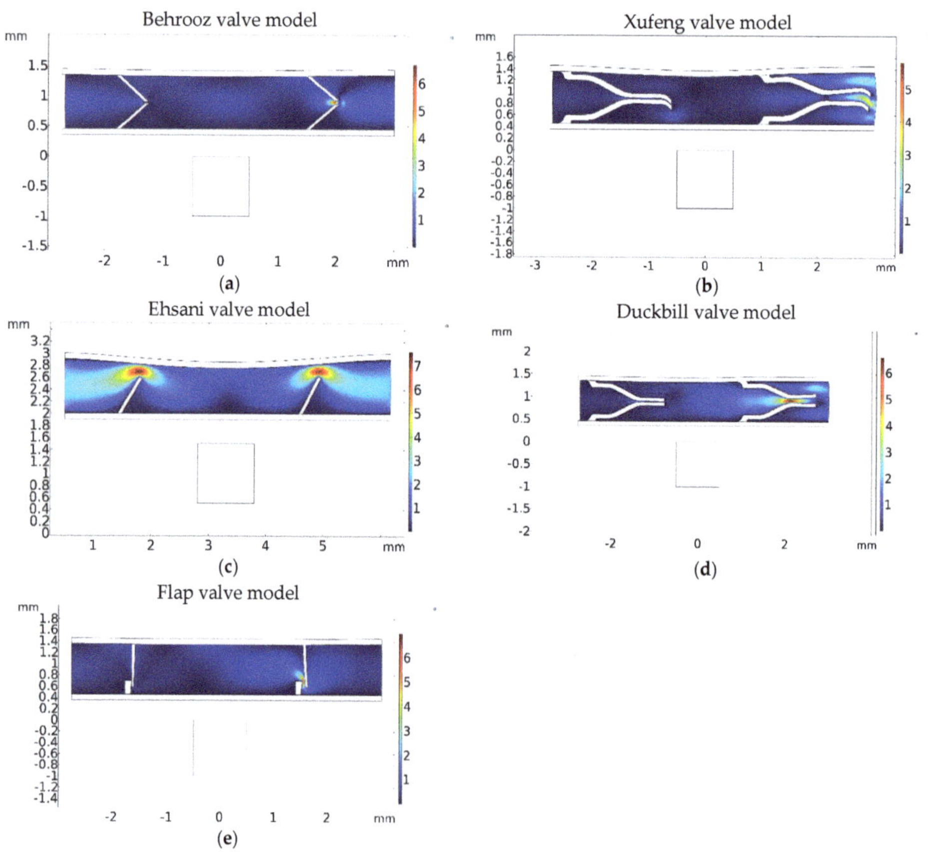

Figure 11. The velocity magnitude is shown for five different valve models. Maximum velocity magnitude at (**a**) $t = 0.1$ s for the Behrooz valve model, (**b**) $t = 0.5$ s for the Xufeng valve, (**c**) $t = 0.5$ s for the Ehsani valve, (**d**) $t = 0.1$ s for the Duckbill valve and (**e**) $t = 0.1$ s for Flap valve.

From Table 6 and Figure 12, it is seen that the maximum velocity occurs from 0.1 to 0.5 s for different valve models, which means that the magnetic flux acting on the upper MRE wall rapidly. Among all, the Flap, Duckbill and Behrooz valves have the fastest response time, which is an advantage. In addition, the maximum velocity is obtained for the Ehsani valve model, but has a slower response time. Thus, in terms of rapid deformation of the upper wall, the proposed flap and duckbill valves model show the best performance.

Table 6. Maximum velocity magnitude for all five valve models.

Valve Model	Maximum Velocity Occurred (s)	Maximum Velocity Magnitude at Outlet (mm/s)
Flap valve	0.1	1.10
Duckbill valve	0.1	0.91
Behrooz valve	0.1	0.59
Ehsani Valve	0.5	1.82
Xufeng valve	0.5	1.40

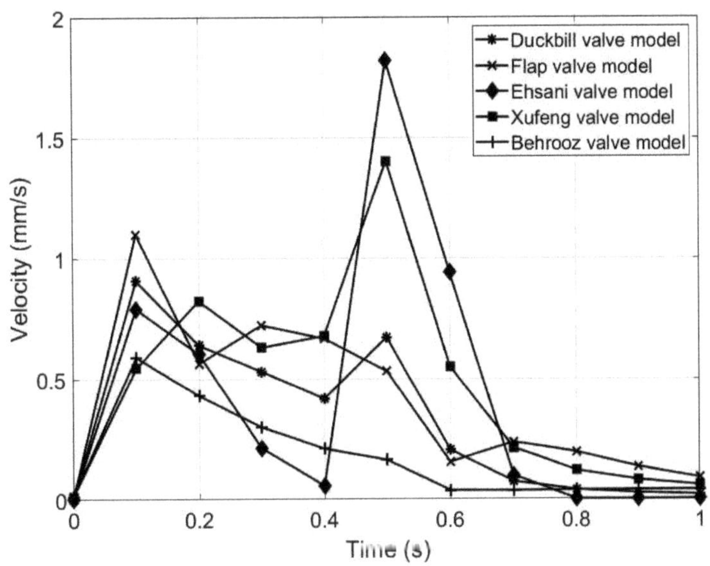

Figure 12. Graphical representation of velocity magnitude at the outlet of the five valve models.

Figure 13a represents the flow rate of the five valve models during the contraction phase. It is observed that the volume flow rate is high for the Ehsani valve model, which means that more fluid is propelled in the same amount of time compared to the other models. However, the Ehsani valve model is the only valve to have backflow during the contraction phase because of the large gap between the upper wall and the tip of the valves, which is a major disadvantage. Figure 13b represents the volume flow rate for both contraction and expansion phases. It is seen that all of the five models have backflow during the start of the expansion period (at t = 0.5 s), but the Ehsani valve model has the highest backflow compared to the other four models. Hence, the Ehsani model is the least advantageous among the five models. As seen from Figure 12, the Ehsani valve model shows the maximum velocity, but this maximum velocity is in the backward direction, which can also be confirmed by Figure 13b.

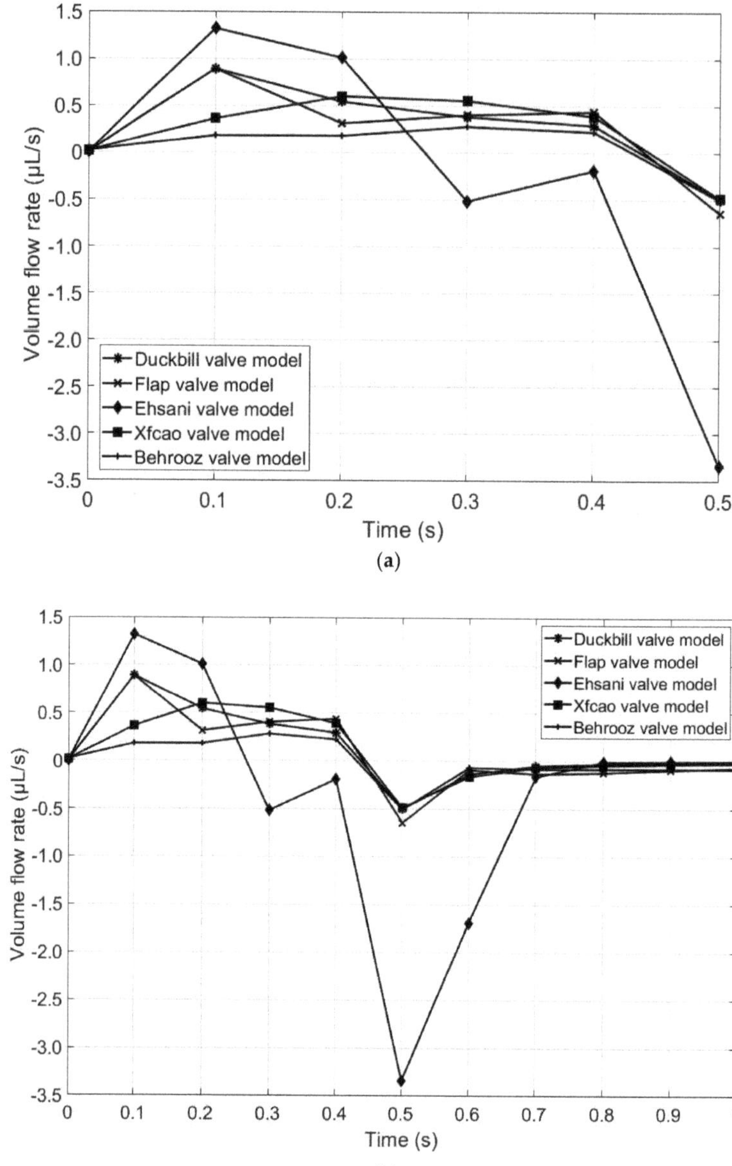

Figure 13. Comparison of volume flow rates for the five valve models: (**a**) contraction phase (until $t = 0.5$ s) and (**b**) both expansion and contraction phase (until $t = 1$ s).

Figure 14 and Table 7 demonstrate the net volume pumped by the five valve models during a 1 s pumping cycle. It is seen clearly from Table 7 that the duckbill valve model has the largest pumping capacity, followed by the flap valve model. It can thus be concluded that the proposed two valve models are superior to the other models in terms of net pumped volume during the same pumping cycle.

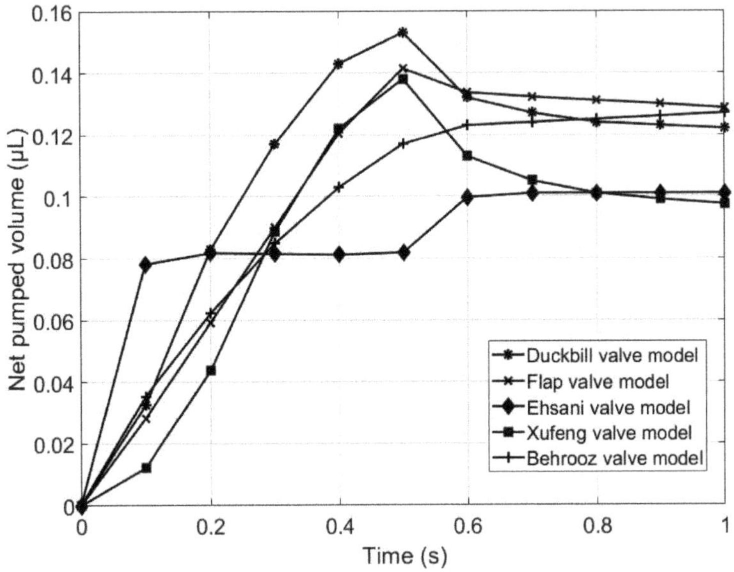

Figure 14. Comparison of net volume pumped.

Table 7. Net pumped volume of micropumps.

Micropump Model	Net Pumped Volume (μL)
Duckbill valve model	1.16
Flap valve model	1.09
Xufeng valve model	0.92
Ehsani valve model	0.91
Behrooz valve model	1.03

The material and the design of the chamber needed to be tuned to ensure the optimum deformation due to the electromagnetic force could be achieved, as stiff sidewalls would be a challenge for electromagnetic forces to be able to deform the chamber and increase the pressure to open the outlet valve. On the other hand, the return cycle needed to be as quick as possible to bring the chamber back to the initial shape ready for the next cycle, while generating enough suction to open the inlet valve and suck fluid into the chamber for the next pumping cycle. Both pressure and cycle time for pumping and suction cycles are directly related to the mechanical property of the elastic material and the design of the chamber. The thickness and design of the side walls has been tuned to optimize these pumping and suction performance of the pump. FEA analysis has been used to simulate the deformation and return of the chamber, to optimize the thickness and design of the chamber. While mechanical property of the chamber could be tuned by adjusting the UV-power density (power x exposure time) during the 3D printing process, a constant UV-power density (that was optimized for successful 3D printing of the thin wall for the specific elastic resin) has been used throughout all initial fabrication experiments.

The advantage of this fabrication is that geometry design and mechanical properties of the chamber could be easily modified and fine-tuned, with minimum limitation imposed from fabrication technique. The elastic resin used in this study is a new type of resin for 3D printing that has been developed by "Adaptive3D".

4. Summary and Conclusions

In this work, we presented a comparative evaluation of several magnetorheological micropump designs. All designs were positive displacement pumps, pushing the fluid in one direction via two flexible valves. The Behrooz, Ehsani and Xufeng valves were existing in the literature, while the flap and duckbill valve models were proposed in this study. The performances of the pumps were evaluated through physics-based multiphysics computer simulations. The comparisons of the performance characteristics of the five pumps were presented and discussed.

The major takeaways from this study can be listed as follows:

- The comparison study revealed that:
    - the proposed duckbill and flap valve micropumps were five times faster than the Ehsani and Xufeng valve micropumps;
    - the backflow was comparatively less for the duckbill and flap valve micropumps, while the Ehsani models showed the maximum backflow;
    - the proposed flap and duckbill valve model could pump 1.09 μL and 1.16 μL in 1 s of pumping cycle, respectively, which is more than each of the three existing micropump models.
- After examining the MRE wall deformation, velocity magnitude and volume flow rate, it is concluded that the proposed duckbill and flap valve models can propel the largest amount fluid in the same time interval, and both of them have short response time to apply the magnetic field. Thus, in terms of the performance measures, it can be concluded that the proposed two models provide better results.
- The proposed study could be used for future MR micropump design considerations.

Author Contributions: Conceptualization, S.C. and R.H.; methodology, S.C. and R.H.; software, R.H.; validation, R.H.; formal analysis, R.H.; investigation, R.H.; resources, S.C.; data curation, R.H.; writing—original draft preparation, R.H.; writing—review and editing, S.C. and M.B.; visualization, R.H.; supervision, S.C.; project administration, S.C.; and funding acquisition, S.C. All authors have read and agreed to the published version of the manuscript.

Funding: This research was funded by Faculty Seed Grants by Georgia Southern University.

Institutional Review Board Statement: Not applicable.

Informed Consent Statement: Not applicable.

Data Availability Statement: Not applicable.

Acknowledgments: The support from Georgia Southern University is gratefully acknowledged.

Conflicts of Interest: The authors declare no conflict of interest.

References

1. Behrooz, F.; Wang, M.; Gordaninejad, X. Seismic Control of Base Isolated Structures Using Novel Magnetorheological Elastomeric Bearing. In Proceedings of the SMASIS2013, Snowbird, UT, USA, 16–18 September 2013; pp. 1–9. [CrossRef]
2. Yarra, S.; Behrooz, M.; Pekcan, G.; Itani, A.; Gordaninejad, F. A large-scale adaptive magnetorheological elastomer-based bridge bearing. In Proceedings of the Active and Passive Smart Structures and Integrated Systems 2017, Portland, OR, USA, 11 April 2017; Volume 10164, p. 1016425. [CrossRef]
3. Behrooz, M.; Yarra, S.; Mar, D.; Pinuelas, N.; Muzinich, B.; Publicover, N.G.; Pekcan, G.; Itani, A.; Gordaninejad, F. A self-sensing magnetorheological elastomer-based adaptive bridge bearing with a wireless data monitoring system. In Proceedings of the Sensors and Smart Structures Technologies for Civil, Mechanical, and Aerospace Systems 2016, Las Vegas, NV, USA, 26 July 2016; Volume 9803, p. 98030D. [CrossRef]
4. Syam, T.M.I.; Muthalif, A.G.A. Magnetorheological Elastomer based torsional vibration isolator for application in a prototype drilling shaft. *J. Low Freq. Noise Vib. Act. Control* **2021**, *41*, 1–25. [CrossRef]
5. An, H.N.; Sun, B.; Picken, S.J.; Mendes, E. Long time response of soft magnetorheological gels. *J. Phys. Chem. B* **2012**, *116*, 4702–4711. [CrossRef] [PubMed]
6. Carlson, J.D.; Jolly, M. MR fluid, foam and elastomer devices. *Mechatronics* **2000**, *10*, 555–569. [CrossRef]
7. Davis, L.C. Model of magnetorheological elastomers. *J. Appl. Phys.* **1999**, *85*, 3348–3351. [CrossRef]

8. Ginder, J.M.; Davis, L.C. Shear stresses in magnetorheological fluids: Role of magnetic saturation. *Appl. Phys. Lett.* **1994**, *65*, 3410–3412. [CrossRef]
9. Gong, X.; Xu, Y.; Xuan, S.; Guo, C.; Zong, L. The investigation on the nonlinearity of plasticine-like magnetorheological material under oscillatory shear rheometry. *J. Rheol.* **2012**, *56*, 1375–1391. [CrossRef]
10. Oliveira, F.; Botto, M.A.; Morais, P.; Suleman, A. Semi-active structural vibration control of base-isolated buildings using magnetorheological dampers. *J. Low Freq. Noise Vib. Act. Control* **2018**, *37*, 565–576. [CrossRef]
11. Rabinow, J. The magnetic fluid clutch. *Electr. Eng.* **1948**, *67*, 1167. [CrossRef]
12. Rodríguez-López, J.; Elvira, L.; Resa, P.; Montero De Espinosa, F. Sound attenuation in magnetorheological fluids. *J. Phys. D Appl. Phys.* **2013**, *46*, 065001. [CrossRef]
13. Song, W.L.; Li, D.H.; Tao, Y.; Wang, N.; Xiu, S.C. Simulation and experimentation of a magnetorheological brake with adjustable gap. *J. Intell. Mater. Syst. Struct.* **2017**, *28*, 1614–1626. [CrossRef]
14. Sun, S.; Peng, X.; Guo, Z. Study on macroscopic and microscopic mechanical behavior of magnetorheological elastomers by representative volume element approach. *Adv. Condens. Matter Phys.* **2014**, *2014*, 232510. [CrossRef]
15. Behrooz, M.; Gordaninejad, F. A flexible micro fluid transport system featuring magnetorheological elastomer. *Smart Mater. Struct.* **2016**, *25*, 025011. [CrossRef]
16. Behrooz, M.; Gordaninejad, F. Three-dimensional study of a one-way, flexible magnetorheological elastomer-based micro fluid transport system. *Smart Mater. Struct.* **2016**, *25*, 095012. [CrossRef]
17. Ehsani, A.; Nejat, A. Conceptual design and performance analysis of a novel flexible-valve micropump using magneto-fluid-solid interaction. *Smart Mater. Struct.* **2017**, *26*, 055036. [CrossRef]
18. Cao, X.; Xuan, S.; Hu, T.; Gong, X. 3D printing-assistant method for magneto-active pulse pump: Experiment, simulation, and deformation theory. *Appl. Phys. Lett.* **2020**, *117*, 241901. [CrossRef]

micromachines

Article

Refractive Index Sensor Based on a Metal-Insulator-Metal Bus Waveguide Coupled with a U-Shaped Ring Resonator

Xiaoyu Zhang [1,2,3], Shubin Yan [1,2,*], Jilai Liu [1], Yifeng Ren [3], Yi Zhang [1,2] and Lifang Shen [1,2]

1 School of Electrical Engineering, Zhejiang University of Water Resources and Electric Power, Hangzhou 310018, China; zhangxiaoyu9725@163.com (X.Z.); liujl@zjut.edu.cn (J.L.); zhangyi@zjweu.edu.cn (Y.Z.); shenlf@zjweu.edu.cn (L.S.)
2 Joint Laboratory of Intelligent Equipment and System for Water Conservancy and Hydropower Safety Monitoring of Zhejiang Province and Belarus, Hangzhou 310018, China
3 School of Electrical and Control Engineering, North University of China, Taiyuan 030051, China; renyifeng126@126.com
* Correspondence: yanshb@zjweu.edu.cn; Tel.: +86-186-3611-2255

Abstract: In this study, a novel refractive index sensor structure was designed consisting of a metal-insulator-metal (MIM) waveguide with two rectangular baffles and a U-Shaped Ring Resonator (USRR). The finite element method was used to theoretically investigate the sensor's transmission characteristics. The simulation results show that Fano resonance is a sharp asymmetric resonance generated by the interaction between the discrete narrow-band mode and the successive wide-band mode. Next, the formation of broadband and narrowband is further studied, and finally the key factors affecting the performance of the sensor are obtained. The best sensitivity of this refractive-index sensor is 2020 nm/RIU and the figure of merit (FOM) is 53.16. The presented sensor has the potential to be useful in nanophotonic sensing applications.

Keywords: Fano resonance; SPPs; U-shaped ring resonator; sensor; metal-insulator-metal

Citation: Zhang, X.; Yan, S.; Liu, J.; Ren, Y.; Zhang, Y.; Shen, L. Refractive Index Sensor Based on a Metal-Insulator-Metal Bus Waveguide Coupled with a U-Shaped Ring Resonator. *Micromachines* **2022**, *13*, 750. https://doi.org/10.3390/mi13050750

Academic Editor: Stelios K. Georgantzinos

Received: 27 April 2022
Accepted: 7 May 2022
Published: 9 May 2022

1. Introduction

Surface plasmon polaritons (SPPs) are electromagnetic surface waves generated by the interaction between an external light field and free electrons in metal, which can reach the maximum field intensity on the surface and decay exponentially along the direction perpendicular to the interface [1,2]. SPPs can be excited in two ways, by electrons or light waves. When the size of a specific nanostructure is reached, SPPs can break through the limited conventional diffraction and control light on the nanoscale [3,4]. SPPs have three characteristics: low dimension and high intensity and subwavelength, which make them a good energy and information carrier, and their ability to combine subwavelengths can be used to make various optical devices [5], such as wavelength demultiplexers [6,7], plasmonic filters [8,9], logic gates [10], couplers [11], and sensors [12,13].

In recent years, refractive index sensors based on a metal-insulator-metal (MIM) waveguide have attracted wide attention because of their strong lateral confinement, easy manufacturability, low propagation loss, and shorter transmission length [14,15]. Sensors based on MIM waveguides can also produce nonlinear optical effects such as electromagnetically induced transparency (EIT) and Fano resonance. Fano resonance is a kind of scattering resonance phenomenon that can produce an asymmetrical line shape. The shape of Fano resonance is caused by interference in the scattering amplitude of the discrete narrow-band mode and successive wide-band mode. Due to the rapidly changing amplitude and phase, Fano resonance has a narrower full width at half maximum (FWHM) [16]. Because of the unique linear shape of Fano resonance, the smaller FWHM has the higher electromagnetic field binding ability, which is widely used to characterize the resolution of the instrument.

In addition, Fano resonance is very sensitive to structural parameters and refractive index changes, which is very helpful for the preparation and improvement of refractive index sensors [17]. It is an important method for designing refractive index sensors by using a waveguide coupling resonator to generate Fano resonance. Zhou et al. [18] proposed an MIM waveguide consisting of two baffling resonators and a ring resonator; this structure can increase the sensitivity to 1650 nm/RIU. Zhang et al. [19] put forward an MIM waveguide which can support Fano resonance with a sensitivity of 1268 nm/RIU. Liu et al. [20] designed a refractive-index nanosensor, which can reach 1510 nm/RIU sensitivities. Zhang et al. [21] proposed a tooth cavity-coupled ring-splitting cavity structure obtaining a sensitivity of 1200 nm/RIU. Herein, the proposed nanosensor, with a simple structure, can achieve a high sensitivity of 2020 nm/RIU with a FOM of 53.16.

In this article, a novel refractive index sensor structure consisting of a USRR and an MIM waveguide with two rectangular baffles is propounded theoretically. The USRR cavity can be represented with two parameters, which greatly reduces the difficulty of fabrication. Compared with other cavity structures, a USRR is more sensitive to the change in structural parameters, which is helpful for researching high sensitivity refractive index sensors. The standardized H_z field distributions and the propagation characteristics were theoretically proposed by the finite element method (FEM).

2. Materials and Methods

A schematic diagram of the designed structure is shown in Figure 1. A U-shaped cavity was chosen because it combines the advantages of circular and rectangular resonators. The rectangular resonator can change the vertical distance independently of the horizontal direction, and the straight waveguide is very beneficial to the transmission of light, but the coupling of the straight waveguide and MIM waveguide will cause a larger FWHM, and the perception performance will decrease. A circular resonator is the best choice, and displays a good performance in coupling with MIM waveguide. However, if the radius is increased, the increased cavity area in the vertical direction will lead to a decline in the sensor's performance and an increased loss, thus affecting the spread performance. In the end, we chose the USRR structure. The coupling mode between the resonator and bus waveguide can be divided into side coupling and shoulder coupling. The structure proposed in this work adopts the side coupling mode. The geometric parameters of the structure settings are as follows: R_1 and r_1 are the outer radius and inner radius of the USRR taken separately; the height of the USRR is described as d; g is the coupling gap between the USRR resonant cavity and the bus waveguide; the distance between the two rectangular baffles and the height of them are denoted as l and h. The input and output ports are P_1 and P_2, respectively, and ω represents the width of the MIM waveguide. ω is usually set to 50 nm to ensure only TM_0 mode exists in the waveguide [22].

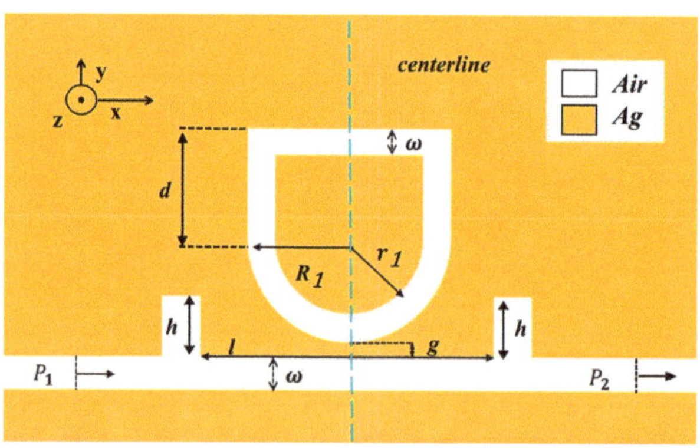

Figure 1. 2D schematic diagram of a USRR resonant cavity coupled with an MIM waveguide with two rectangular baffles.

The TM mode equation is [23,24]:

$$\tan h(k\omega) = -2kp\alpha_c / \left(k^2 + p^2k^2\right) \tag{1}$$

where $p = \varepsilon_{in}/\varepsilon_m$, k is the wave vector, and $\alpha_c = \left[{k_0}^2 \times (\varepsilon_{in} - \varepsilon_m) + k\right]^{1/2}$, ε_m and ε_{in} represent the permittivity of metal and dielectric, respectively. In free space, k can be depicted as $k_0 = 2\pi/\lambda_0$.

A 2D diagram can be used instead of a 3D diagram for simple calculation. The yellow and white parts were filled with silver and air, respectively. A sufficiently thick silver layer can be prepared on a silicon substrate by chemical vapor deposition (CVD). The desired structure can be obtained by electron beam etching on the silver layer. The silver was chosen as the filling metal, which is due to its higher electric field and lower power consumption. The relative permittivity of air is 1, and the relative permittivity of silver can be acquired using the Debye–Drude dispersion model:

$$\varepsilon_{Ag}(\omega) = \varepsilon_1(\omega) + \varepsilon_2(\omega)i = 1 - \frac{{\omega_p}^2\tau^2}{1 + \omega^2\tau^2} + \left(1 - \frac{\omega_p{\omega_p}^2\tau}{\omega(1 + \omega^2\tau^2)}\right)i \tag{2}$$

where ω represents the angular frequency of the light, ω_p $\left(\omega_p = 1.38 \times 10^{16}\right)$ can be expressed as the plasma frequency of silver, and $\tau\left(\tau = 7.35 \times 10^{15}\right)$ is the relaxation time [25,26].

The performances of the proposed structure can be evaluated by two parameters, sensitivity (S) and figure of merit (FOM), which can be reached as follows [27]:

$$S = \Delta\lambda/\Delta n \tag{3}$$

$$FOM = S/FWHM \tag{4}$$

where $\Delta\lambda$ and Δn represent the change in resonance wavelength and refractive indices, respectively.

Before the sensor was manufactured, the transmission spectrum of the structure was numerically simulated with COMSOL Multiphysics 5.4. Boundary conditions can be divided into absorptive boundary conditions and periodic boundary conditions. The absorbing boundary condition was established by a perfect matching layer to absorb spilled waves. By setting a high loss layer perfectly matching the dielectric impedance of the adjacent region in the boundary region, the electromagnetic wave decays rapidly without reflection until it is completely absorbed. The superfine mesh was selected to improve segmentation accuracy and guarantee perfect segmentation.

3. Results

Since Fano resonance plays a significant role in the sensitivity of the sensor, we will investigate the formation and characteristics of Fano resonance. The structural parameters are as follows: $R_1 = 210$ nm, $r_1 = R_1 - 50$, $d = 205$ nm, $h = 145$ nm, $l = 440$ nm, $g = 10$ nm. The method of Fano resonance excitation is divided into waveguide side coupled cavity excitation and symmetry breaking excitation. In this work, we used the waveguide side coupled resonator excitation mode. Fano resonance is caused by the interaction of the successive wide-band mode, which is excited by the bus waveguide with two rectangular baffles and the discrete narrow-band mode formed by the USRR resonant cavity. To evaluate the different transmission characteristics, three structures, with the whole system (red line), the single USRR resonant cavity (blue line), and two symmetrical rectangular baffles (yellow line), were chosen in Figure 2. The continuous solid yellow curve has a positive slope with ultra-high transmittance. Therefore, it can be considered as a continuous wideband mode. The SPPs, which are directly inspired by the TM beam of the incident light wave, enter the bus waveguide through P_1 and exit through P_2.

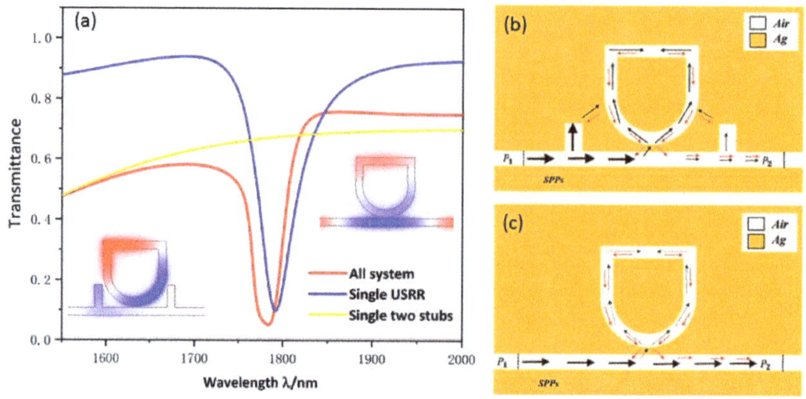

Figure 2. (**a**) Transmittance spectra of the single two baffles (yellow line), single USRR resonant cavity (blue line), and all system (red line); (**b**) The SPP pathway schematic of the system; (**c**) The SPP pathway schematic of the single USRR structure.

The blue line is a continuous curve with a narrow spectral width and low transmittance, and its transmission spectrum is similar to the shape of a Lorentz curve, showing a certain symmetry. Therefore, the transmission spectrum of single USRR can be considered as a discrete narrowband mode. As shown in Figure 2c, the SPPs indirectly motivated by the TM beam enter the bus waveguide from P_1 and most of them are trapped in the USRR resonant cavity. A few SPPs can return to the MIM waveguide and are propagated to P_2 to attain indirect excitation of the TM wave. The H_z field distribution indicates that the USSR structure has a certain field intensity distributed in the whole bus waveguide.

The red line is a continuous sharp asymmetric curve with narrow spectral width and low transmittance, and its transmission spectrum is the standard Fano resonance. As shown in Figure 2a, the red curve has the lowest transmittance, which indicates the device's ability to bind light is enhanced after the addition of double rectangular baffles to the MIM waveguide. As can be seen from Figure 2b, SPPs coupled to the USRR structure increase significantly, resulting in an increase in the field intensity of the USRR structure. The normalized H_z field suggests that the USRR resonant cavity and the left part of the bus waveguide have a relatively strong resonance. By comparison, when the bus waveguide is examined after adding the symmetric rectangular baffles, the ability of the whole structure to gather electric field and capture SPPs is significantly strengthened, resulting in the great modulation of the asymmetric line shape of Fano resonance.

The Fano resonance wavelength λ_1 can be calculated by:

$$\lambda_1 = 2\mathrm{Ln}_{\mathrm{eff}}/\mathrm{J} - \left(\varphi_{ref}/\pi\right) \tag{5}$$

where the effective resonant length of ring cavity is expressed as L; n_{eff} is the effective refractive index; the phase transition reflected by SPP at the MIM interface is represented as φ_{ref}; and J is the mode order (J = 1, 2, 3 . . .). According to Equation (5) and the analysis of the causes of Fano resonance, the parameters of the structure designed here will affect the line shape and the wavelength, thus affecting the performance of the refractive index sensor. The following changes will be made to the resonator and MIM waveguide parameters to analyze their influence on the sensitivity and figure of merit of the refractive index sensor, so as to achieve the best performance.

Based on the above analysis of the causes of Fano resonance, we know that the linear shape of Fano resonance can be changed by varying the broadband mode and narrow band mode, thus improving the performance of the sensor. By changing the parameters of USRR structure that generates narrowband mode, the influence of narrowband mode on the transmission performance of the whole structure was evaluated. The influence of USRR R_1 was further studied by setting R_1 to 215, 210, 205, 200 and 195 nm. In Figure 3a, as R_1 increases, the curves appear to undergo an obvious red shift with the transmittance decreased slightly. The reason for this is that due to the increase in the effective length of the USRR structure, the ability of the USRR to gather electric field increases, and eventually the resonant wavelength moves to a larger wavelength. Figure 3b shows the sensitivity fitting curve with a good linear relationship. Importantly, the change in sensitivity increases obviously from 1740 to 2000 nm/RIU as R_1 varies at 5 nm intervals. This indicates that R_1 is an important parameter to improve the sensitivity of the sensor. In practical application, proper parameters can be chosen according to the requirements of sensor manufacturing.

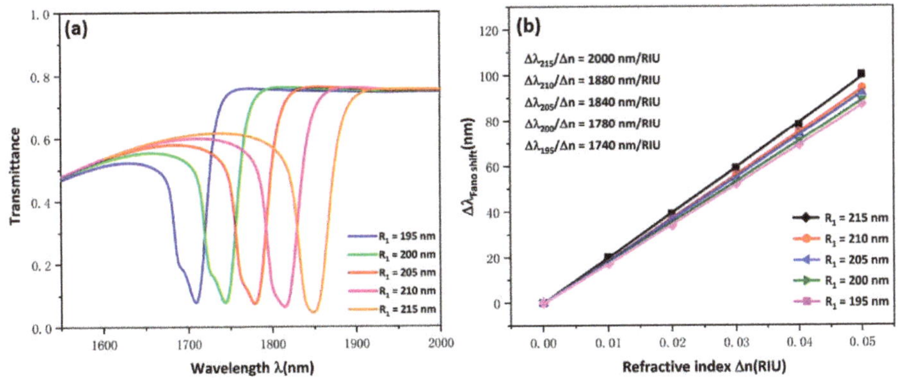

Figure 3. (**a**) Transmission spectra for USRR under various R_1 values; (**b**) Fitting lines of sensitivity at disparate values of R_1.

Subsequently, the influence of the changed d of USRR on the whole structure's transmission characteristics was evaluated, which were found to have increased from 175 to 215 nm. As Figure 4a shows, the transmission spectra showed an obvious redshift with d increased. The reason for this is also due to the increase in the effective length of the resonator. That the transmittance increases slightly indicates that the ability of USRR structure to gather electric field decreases slightly with the increase in cavity height. As Figure 4b shows, the sensitivity fitting curve has a good linear relationship and the sensitivity changes from 1720 to 1880. This indicates that the height of USRR structure has less influence on sensitivity than the radius; thus, changing d can improve sensitivity in a certain range.

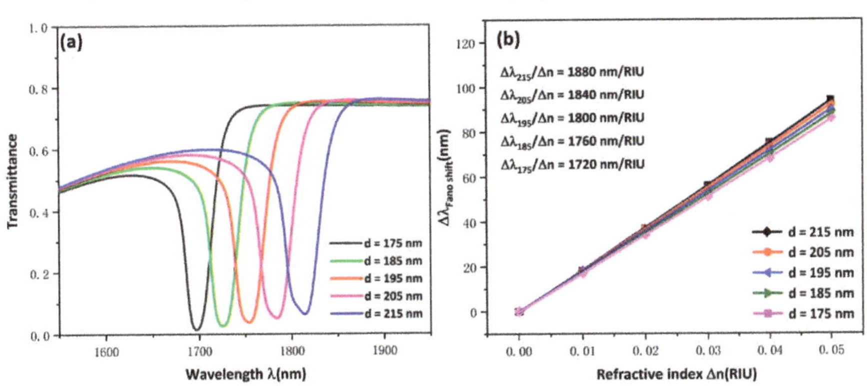

Figure 4. (**a**) Transmission spectra for USRR under various d values; (**b**) Fitting lines of sensitivity at disparate values of d.

Next, we investigated the broadband mode and evaluated the effect of broadband mode on the transmission performance of the whole structure by varying the parameters of the MIM waveguide structure that generates broadband mode. The MIM waveguide with two rectangular baffles structure and the MIM waveguide with single rectangular baffle structures were compared to explore the influence of different types of MIM waveguide on the transmission spectrum of the sensor. Both structures have the same parameters, which are set as: $R_1 = 210$ nm, $r_1 = R_1 - 50$, $d = 205$ nm, $l = 440$ nm, $g = 10$ nm, $h_1 = 145$ nm (two baffles), $h_2 = 70$ nm (sin gle baffle). The MIM waveguide with a single rectangular baffle structure is taken as the comparison structure. The contrast transmission spectrums of the two structures are shown in Figure 5a. The transmission spectrums show that the two structures will produce sharp curves of asymmetric Fano resonance and the wavelength remains almost unchanged. This indicates that the broadband mode has little effect on the refractive index of the sensor. It can be clearly seen that the blue curve has a high transmittance and narrow FWHM, which indicates that the ability of the whole structure to gather electric field and capture SPPs has been distinctly weakened. This can be interpreted as showing that the coupling ability of the double rectangular baffle structure to the USRR is much greater than that of the single rectangular baffle structure. As shown in Figure 5b,c, the normalized H_z field distribution shows that when the MIM waveguide contains two rectangular baffles, the USRR structure will gather a stronger electric field and exist to a stronger resonance. Thus the following studies of waveguide structure parameters are based on an MIM waveguide with double symmetrical baffle structures.

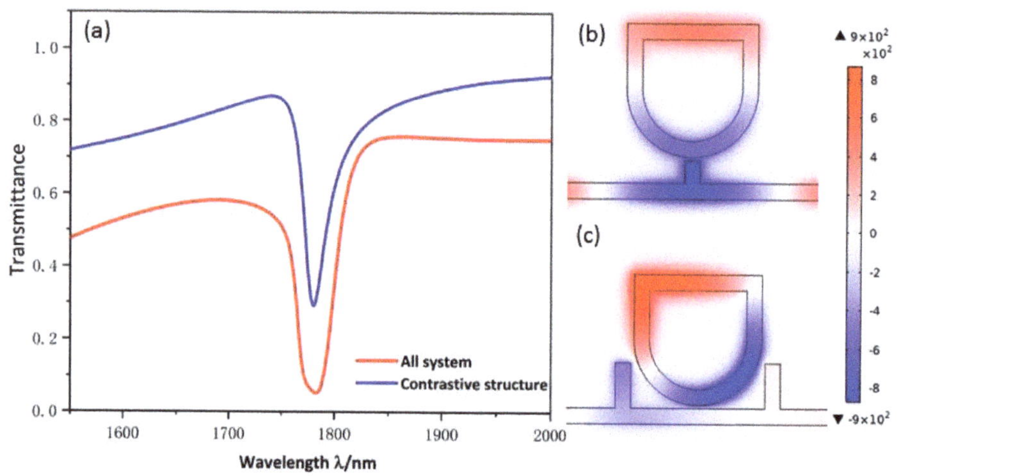

Figure 5. (**a**) Transmission spectra of two structures; (**b**) H_z field distributions of the contrastive structure; (**c**) H_z field distributions of the system.

The influence of MIM waveguide structural parameters on system characteristics were further researched; we changed the height of rectangular baffle h from 115 to 155 nm and increased the distance between the two rectangular baffles l from 420 to 460 nm with an increase of 10 nm. As shown in Figure 6a,b, the dip position remains constant no matter how h and l were changed, which indicates that the wavelength is basically unchanged and the two parameters are not sensitive to the refractive index sensor. However, it can be seen from Figure 6a that changing h will change the shape of the curves more obviously, which transformed an almost symmetrical shape into an entirely asymmetrical shape. When h increases, the transmittance will gradually decrease, which shows that as the rectangular baffles get closer to USRR, more electric fields can be accumulated on the USRR, which ensures the good performance of the sensor. The same phenomenon is also shown in Figure 6b. When l decreased, the transmittance will gradually increase, which shows that

USRR has a relatively strong resonance. Thus, the closer the rectangular baffles are to the resonator, the stronger the intensity of the field that will be gathered in the resonator and the performance of the device will be improved. The h is a more vital parameter than the distance between the two rectangular baffles in affecting the shape of the continuous broadband state. The above research indicates that transforming the h parameter can change the line's shape without affecting its dip wavelength.

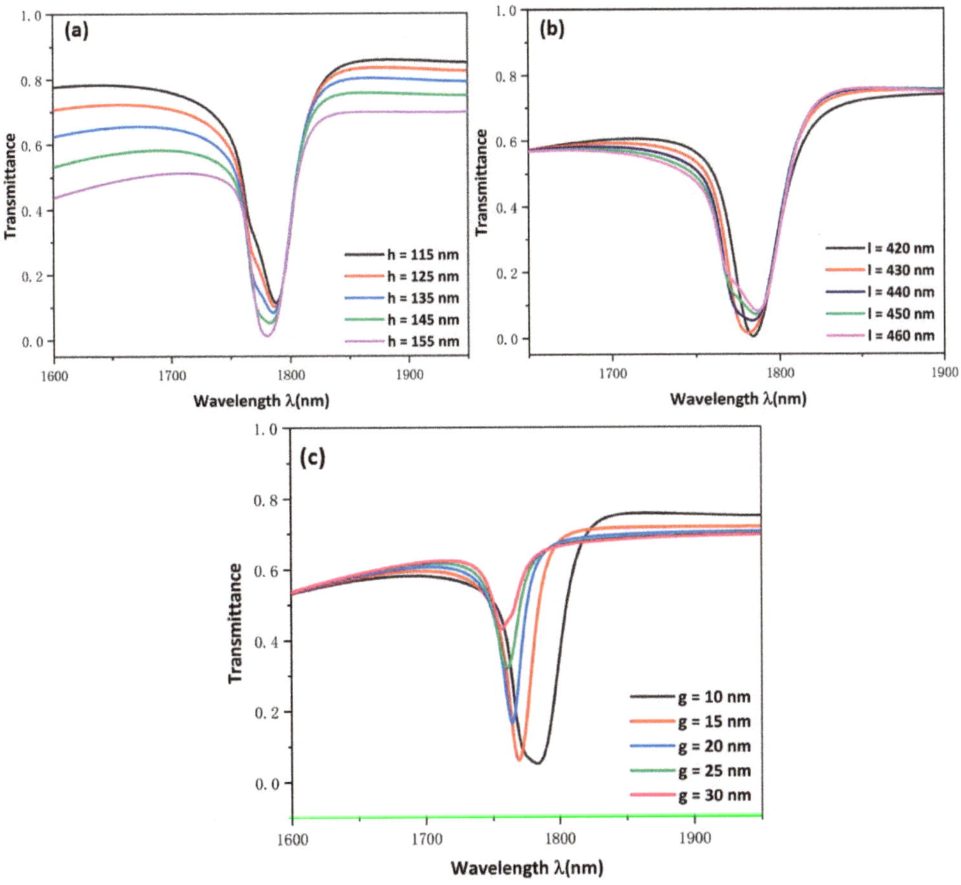

Figure 6. Transmission spectra at (**a**) disparate heights of rectangular baffles; (**b**) disparate distances between the two rectangular baffles; (**c**) various coupling gaps.

Finally, the influence of the coupling gap was evaluated by changing the g from 10 to 30 nm. Figure 6c indicates that the transmission spectrum has an obvious blue shift. The FWHM becomes narrower and the transmittance apparently increases. This indicates that increasing the coupling gap weakens the coupling of SPPs to USRR and the energy constraint of the cavity. This shows that the coupling gap g determines the low transmittance of the transmission spectrum. By balancing transmittance and FWHM, we chose 10 nm as the most optimal coupling distance for this sensor.

For optimal sensing performance, set the parameters to R_1 = 215 nm, d = 215 nm, h = 145 nm, l = 440 nm, g = 10 nm. The measuring theory of the refractive index sensing is that the resonance dip will change with the varying of the refractive index of surrounding materials. The refractive-index was set to 1.00, 1.01, 1.02, 1.03, 1.04, and 1.05. The transmission spectrum exhibited an obvious red shift is shown in Figure 7a. As Figure 7b shows, a maximum sensitivity and FOM of this structure were achieved, which are 2020 nm/RIU and 53.16. This is the optimal performance of the proposed structure, which is better than those which have been mentioned in the first part and are shown in Table 1. A comparison of the sensitivity of the structure indicates that the proposed structure provides better sensitivity to RI variation.

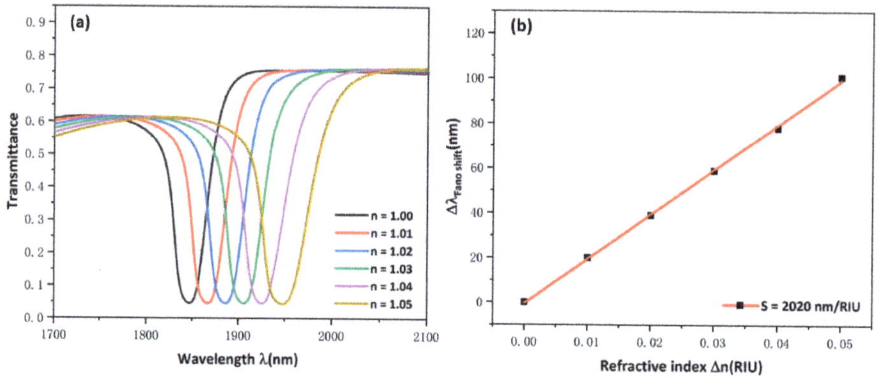

Figure 7. (**a**) Transmission spectra at various refractive indexes; (**b**) Fitting lines of sensitivity at disparate values of refractive indexes.

Table 1. Comparisons of the results with recent research.

Reference	Structure Type	Sensitivity (nm/RIU)
[18]	Two baffle resonators	1650
[19]	Ring resonator	1268
[20]	D-shaped cavity	1510
[21]	Tooth cavity-coupled ring splitting cavity	1200
This work	USRR structure	2020

4. Conclusions

In this study, a novel refractive index sensor structure was designed consisting of a metal-insulator-metal waveguide with two rectangular baffles, coupled with a U-shaped ring resonator. The results show that Fano resonance, which results from the interaction of the successive wide-band mode and the discrete narrow-band mode, plays a significant role in the performance of the refractive index sensors. It was also found that the broadband mode mainly affects the line shape and FWHM of Fano resonance, and the narrowband mode mainly affects the sensitivity of sensor. A maximum sensitivity and FOM of this structure were achieved, which are 2020 nm/RIU and 53.16. The presented structure has the potential to be useful in nanophotonic sensing applications.

Author Contributions: Conceptualization, S.Y. and X.Z.; methodology, X.Z. and J.L.; software, X.Z.; validation, S.Y., Y.Z. and L.S.; formal analysis, Y.R.; investigation, X.Z.; data curation, S.Y. and X.Z.; writing—original draft preparation, X.Z.; writing—review and editing, S.Y. and X.Z.; supervision, S.Y. All authors have read and agreed to the published version of the manuscript.

Funding: The work was supported in part by the National Natural Science Foundation of China under Grant No. 61875250 and Grant No. 61975189, in part by the Zhejiang Provincial Natural Science Foundation of China under Grant No. LD21F050001 and Grant No. Y21F040001, the Key Research Project by Department of Water Resources of Zhejiang Province under Grant No. RA2101, the Key Research and Development Project of Zhejiang Province under Grant No. 2021C03019, and the Scientific Research Foundation of Zhejiang University of Water Resources and Electric Power under Grant No. xky2022032.

Institutional Review Board Statement: Not applicable.

Informed Consent Statement: Not applicable.

Conflicts of Interest: The authors declare no conflict of interest.

References

1. Zheng, Z.P.; Luo, Y.; Yang, H.; Yi, Z.; Zhang, J.G.; Song, Q.J.; Yang, W.X.; Liu, C.; Wu, X.W.; Wu, P.H. Thermal tuning of terahertz met-amaterial properties based on phase change material vanadium dioxide. *Phys. Chem. Chem. Phys.* **2022**, 8846–8853. [CrossRef] [PubMed]
2. Meng, C.; Lu, F.; Zhang, W. Selective remote-excitation of gap mode in metallic nanowire-391 nanoparticle system using chiral surface plasmon polaritons. *IEEE J. Quantum Electron.* **2020**, *56*, 1–6. [CrossRef]
3. Chen, H.; Chen, Z.H.; Yang, H.; Wen, L.H.; Yi, Z.; Zhou, Z.G.; Dai, B.; Zhang, J.G.; Wu, X.W.; Wu, P.H. Multi-mode surface plasmon resonance absorber based on dart-type single-layer graphene. *RSC Adv.* **2022**, *12*, 7821–7829. [CrossRef] [PubMed]
4. Nishijima, Y.; Rosa, L.; Juodkazis, S. Surface plasmon resonances in periodic and random patterns of gold nano-disks for broadband light harvesting. *Opt. Express* **2012**, *20*, 11466–11477. [CrossRef] [PubMed]
5. Zhao, F.; Lin, J.C.; Lei, Z.H.; Yi, Z.; Qin, F.; Zhang, J.G.; Liu, L.; Wu, X.W.; Yang, W.X.; Wu, P.H. Realization of 18.97% theoretical efficiency of 0.9 μm Thick c-Si/ZnO Heterojunction Ultrathin-film Solar Cells via Surface Plasmon Resonance Enhancement. *Phys. Chem. Chem. Phys.* **2022**, *24*, 4871–4880. [CrossRef]
6. Chen, J.; Li, Z.; Li, J.; Gong, Q. Compact and high-resolution plasmonic wavelength demultiplexers based on Fano interference. *Opt. Express* **2011**, *19*, 9976–9985. [CrossRef]
7. Khani, S.; Danaie, M.; Rezaei, P. Double and triple-wavelength plasmonic demultiplexers based on improved circular nanodisk resonators. *Opt. Eng.* **2018**, *57*, 107102. [CrossRef]
8. Xia, X.; Wang, J.; Liang, X.; Tang, B.; Song, C.; Qu, S.; Liu, C. A dual-way directional surface-plasmon-polaritons launcher based on asymmetric slanted nanoslits. *J. Mod. Opt.* **2015**, *62*, 358–363. [CrossRef]
9. Zhang, H.; Shen, D.; Zhang, Y. Circular split-ring core resonators used in nanoscale metal-insulator-metal band-stop filters. *Laser Phys. Lett* **2014**, *11*, 115902. [CrossRef]
10. Li, Z.; Wen, K.; Chen, L.; Lei, L.; Zhou, J.; Zhou, D.; Fang, Y.; Wu, B. Refractive index sensor based on multiple Fano resonances in a plasmonic MIM structure. *Appl. Opt.* **2019**, *58*, 4878–4883. [CrossRef]
11. Zhang, L.; Wang, L.; Wu, Y.; Tai, R. Plasmonic Luneburg lens and plasmonic nano-coupler. *Chin. Opt. Lett.* **2020**, *18*, 092401. [CrossRef]
12. Lu, H.; Liu, X.M.; Mao, D.; Wang, G.X. Plasmonic nanosensor based on Fano resonance in waveguide-coupled resonators. *Opt. Lett.* **2012**, *37*, 3780–3782. [CrossRef] [PubMed]
13. Estevez, M.C.; Otte, M.A.; Sepulveda, B.; Lechuga, L.M. Trends and challenges of refractometric nanoplasmonic biosen-sors: A review. *Anal. Chim. Acta* **2014**, *806*, 55–73. [CrossRef] [PubMed]
14. Zheng, Z.P.; Zheng, Y.; Luo, Y.; Yi, Z.; Zhang, J.G.; Liu, Z.M.; Yang, W.X.; Yu, Y.; Wu, X.W.; Wu, P.H. Switchable terahertz device combining ultra-wideband absorption and ultra-wideband complete reflection. *Phys. Chem. Chem. Phys.* **2022**, *24*, 2527–2533. [CrossRef]
15. Zhou, W.; Li, K.; Song, C.; Hao, P.; Chi, M.; Yu, M.; Wu, Y. Polarization-independent and omnidirectional nearly perfect absorber with ultra-thin 2D subwavelength metal grating in the visible region. *Opt. Express* **2015**, *23*, 413–418. [CrossRef]
16. Fan, Z. A Tunable High-Sensitivity Refractive Index of Analyte Biosensor Based on Metal-Nanoscale Covered Photonic Crystal Fiber with Surface Plasmon Resonance. *IEEE Photonics J.* **2019**, *11*, 1–14. [CrossRef]
17. Wu, X.L.; Zheng, Y.; Luo, Y.; Zhang, J.G.; Yi, Z.; Wu, X.W.; Cheng, S.B.; Yang, W.X.; Yu, Y.; Wu, P.H. A four-band and polarization-independent BDS-based tunable absorber with high refractive index sensitivity. *Phys. Chem. Chem. Phys.* **2021**, *23*, 26864–26873. [CrossRef]
18. Zhou, C.; Huo, Y.; Guo, Y.; Niu, Q. Tunable Multiple Fano Resonances and Stable Plasmonic Band-Stop Filter Based on a Metal-Insulator-Metal Waveguide. *Plasmonics* **2021**, *3*, 1–9. [CrossRef]
19. Zhang, Y.; Cui, M. Refractive Index Sensor Based on the Symmetric MIM Waveguide Structure. *J. Electron. Mater.* **2019**, *48*, 1005–1010. [CrossRef]
20. Liu, X.; Li, J.; Chen, J.; Rohimah, S.; Tian, H.; Wang, J. Fano resonance based on D-shaped waveguide structure and its application for human hemoglobin detection. *Appl. Opt.* **2020**, *59*, 6424–6430. [CrossRef]
21. Zhang, Y.; Kuang, Y.; Zhang, Z.; Tang, Y.; Han, J.; Wang, R.; Cui, J.; Hou, Y.; Liu, W. High-sensitivity refractive index sensors based on Fano resonance in the plasmonic system of splitting ring cavity-coupled MIM waveguide with tooth cavity. *Appl. Phys. A* **2019**, *125*, 13. [CrossRef]
22. Phan, Q.H.; Lai, Y.R.; Xiao, W.Z.; Pham, T.T.; Lien, C.H. Surface plasmon resonance prism coupler for enhanced circular birefringence sensing and application to non-invasive glucose detection. *Opt. Express* **2020**, *28*, 24889–24899. [CrossRef] [PubMed]
23. Hassan, M.F.; Sagor, R.H.; Amin, M.R.; Islam, M.R.; Alam, M.S. Point of Care Detection of Blood Electrolytes and Glucose Utilizing Nano-Dot Enhanced Plasmonic Biosensor. *IEEE Sens. J.* **2021**, *21*, 17749–17757. [CrossRef]
24. Zhou, F.Q.; Qin, F.; Yi, Z.; Yao, W.T.; Liu, Z.M.; Wu, X.W.; Wu, P.H. Ultra-wideband and wide-angle perfect solar energy absorber based on Ti nanorings surface plasmon resonance. *Phys. Chem. Chem. Phys.* **2021**, *23*, 17041–17048. [CrossRef] [PubMed]

25. Chen, J.; Yang, H.; Fang, Z.; Zhao, M.; Xie, C. Refractive Index Sensing Based on Multiple Fano Resonances in a Split-Ring Cavity-Coupled MIM Waveguide. *Photonics* **2021**, *8*, 472. [CrossRef]
26. Mayer, K.M.; Hafner, J.H. Localized surface plasmon resonance sensors. *Chem. Rev.* **2011**, *111*, 3828–3857. [CrossRef]
27. Beijnum, F.V.; Veldhoven, P.; Geluk, E.J.; Dood, M.; Hooft, G.W.T.; Exter, M. Surface plasmon lasing observed in metal hole arrays. *Phys. Rev. Lett.* **2013**, *110*, 206802. [CrossRef]

micromachines

MDPI

Review

Latest Performance Improvement Strategies and Techniques Used in 5G Antenna Designing Technology, a Comprehensive Study

Iftikhar Ahmad, Wenhao Tan, Qasim Ali and Houjun Sun *

Beijing Key Laboratory of Millimeter Wave and Terahertz Techniques, School of Information and Electronics, Beijing Institute of Technology, Beijing 100081, China; iftikhar.roghani@bit.edu.cn (I.A.); tanwenhao@bit.edu.cn (W.T.); qasimali@bit.edu.cn (Q.A.)
* Correspondence: sunhoujun@bit.edu.cn

Abstract: In the recent era, fifth-generation technology (5G) has not been fully implemented in the realm of wireless communication. To have excellent accessible bandwidth feasibility, and in order to achieve the aims of 5G standards, such as higher data rates and ultrahigh-definition video streaming, the millimeter wave (mmWave) band must be employed. Services with minimal latency and many other features are feasible only in the mmWave spectrum. To avoid numerous communication complexities such as high connection losses, short wavelength, and restricted bandwidth, as well as path-loss challenges in the mmWave range, an antenna with wide bandwidth, high gain, narrow steerable beam, high isolation, low side-lobe levels, and multiband features is required to alleviate these difficulties and meet 5G communication standards. To overcome these challenges, specific strategies and techniques should be employed in the traditional antenna designing procedure to excellently improve the performance of the antenna in terms of bandwidth, gain, and efficiency and to reduce the mutual coupling effect between the closely colocated antenna elements in MIMOs and arrays. The researchers reported on a variety of bandwidth and gain improvement approaches. To gain broader coverage, traditional antenna design techniques must be modified. In this study, the latest state-of-the-art work is reviewed, such as the role of the metamaterials (MMTs), parasitic patches, hybrid feeding, EBG structure, impact of the slots with different geometrical shapes in the radiator to achieve the goal of wide bandwidth, boosted gain, reduced side-lobes level, as well as stable radiation properties. Mutual coupling reduction techniques are also briefly reported. The role of reconfigurability is focused on in this study, and at the end, the future challenges in the field of antenna design and possible remedies to such issues are reviewed.

Keywords: 5G, mmWave, SIW, hybrid feeding, SIC, wide bandwidth, ECC, stable high gain, isolation, MEG; DIG; SAR; SSL; metamaterials (MMTs); DRAs

Citation: Ahmad, I.; Tan, W.; Ali, Q.; Sun, H. Latest Performance Improvement Strategies and Techniques Used in 5G Antenna Designing Technology, a Comprehensive Study. *Micromachines* **2022**, *13*, 717. https://doi.org/10.3390/mi13050717

Academic Editor: Stelios K. Georgantzinos

Received: 28 March 2022
Accepted: 26 April 2022
Published: 30 April 2022

1. Introduction

In mobile communication, 5G is the latest technical advancement [1–10]. Facilitating users with delivery of high data rates, HD video broadcasting with ultrahigh speed, flawless traffic capacity, as well as an improved network and spectrum efficiency or efficient frequency sweep [2,3,11–22]. In addition, 5G Technology would not only facilitate excellent communication services and accessibility to the Internet of Things but also be an excellent feasible solution for device-to-device connectivity. An automation in industrial manufacturing, such as tele surgery, automatic fracture detection application, remote sensing, telemedicine, and much more would be possible with advanced fashion and zeal. Due to heavy traffic and more interferences offered by different services, the Sub 1 GHz and the sub-6 GHz older spectrum bandwidths have become limited [11]. The term "bandwidth" refers to the most important property of an antenna system being responsible for the improvement of capacity to meet the aforementioned 5G design goals. Ultralow latency with higher data rates can easily be achieved and can be utilized, but we are still faced with the much higher ratio of challenges and limitations such as link losses, multipath effects, small-scale fading as well as short coverage [11]. To fully overcome and alleviate these challenges and limitations, different valuable solutions are available which might handle these issues, such as advanced beam-forming techniques, antennas with high gain and narrow beamwidth, small-cell technology in advanced format, massive MIMO antennas with excellent steerable characteristics of low scanning losses and reconfigurable characteristics [2,15,16], and bandwidth enhancement techniques [23–28].

For mitigating these challenges, antennas with high performance parameters such as antennas with small beam width, small side-lobe level, with characteristics of low scanning losses, are required, which might be able for two- and three-dimensional steering capability at higher levels of frequency scanning. It is a challenging and difficult task to maintain narrow beam width with low scanning levels. For flawless reliable communication, broadband antennas with excellent steerable characteristics, that are small in size, and have narrow beam width and high gain are required at mobile base stations.

The key challenges for designers is to maintain a suitable position of the antenna at base station and to maintain stability in the radiation characteristics with other communication components such as Wi Fi, 4G LTE, and LCD; besides these challenges, the back casing composed of metal is an enormous challenge for designers and researchers.

Specific absorption losses in terms of the effect of the user's body on antenna performance are required to be inspected. Between each two antenna elements, the mutual coupling should be reduced or isolation between each two elements should be high [29], especially in the case of MIMO antennas. Some other performance parameters such as channel capacity loss (CCL), diversity gain (DG), mean effective gain (MG), and total active reflection coefficient also need to be analyzed [29,30].

The microstrip antennas are mostly used due to being lower in cost, simple in fabrication, and easy in installation, but they have some limitations such as limited level of gain, narrow in bandwidth, lower efficiency, and low power-handling capability. Various strategies and techniques are used to enhance the bandwidth, such as to create the slots in the radiator of microstrip patch antennas in the ground plane of antennas, as well as impedance matching techniques, increasing the height, and stacking arrangements of patch [2,3,15–18].

For microstrip patch antennas, these are conventional performance enhancement techniques in terms of bandwidth, but they effect some additional performance parameters such as gain, radiation pattern, side-lobe level, and cross-polarization level [25,31].

Simultaneous use of the coalescence of numerous technologies is needed in the designing of an antenna for enhancement of bandwidth, gain, and efficiency as well as reduction in side-lobe level and achieving a high level of isolation among interelements of a MIMO or array antenna. The above-mentioned techniques are famous conventional techniques for enhancement of performance, especially in bandwidth

Microstrip antennas are known for their narrow bandwidth, low gain, and low efficiency [31,32]. By using new technology, researchers are able to increase bandwidth by 70% and gain 30–35 dBi and various structural alterations. There are several methods for increasing bandwidth which have been employed by researchers in the evolution and development of 5G antenna systems for a flawless communication system.

The main focus is on the latest key techniques responsible for the improvement in bandwidth, gain, and efficiency, and in the case of MIMO, isolation enhancement is the main target studied in the literature [33].

First, we discuss in short the global fequency spectrum used by different regions for 5G application [32], as shown in Figure 1. The global snapshot elaborates the frequency ranges for each region where work for 5G communication is in progress. Under the umbrella of (ITU-R), the master international frequency register (MIFR) is the key element of international frequency management, which is a permanent database that contains the spectrum characteristics for radio stations' operations throughout the world. The international recognition and protection against interference is conferred by these stations. This database is managed by BR and currently contains 2.6 million frequency assignments for terrestrial services, and over 200,000 are added every year.

	<1 GHz	3GHz	4GHz	5 GHz	24–28 GHz	37–40 GHz	64–71GHz
	600 MHz	3.5GHz		5.9–7.1GHz	27.5–28.35GHz	37–37.6GHz	64–71GHz
	600 MHz	3.5GHz		5.9–7.1GHz	27.5–28.35GHz	37.6–40GHz	64–71GHz
	700 MHz	3.4–3.8GHz		5.9–6.4GHz	24.5–27.5GHz	37–37.6GHz	
		3.4–3.8GHz			26, 28GHz	37–40GHz	
		3.4–3.7GHz			26, 28GHz		
		3.46–3.8GHz			26GHz		
		3.6–3.8GHz					
		3.3–3.6GHz			24.5–27.5GHz	37.5–42.5GHz	
		3.4–3.7GHz	4.8–5GHz		26.5–29.5GHz		
		3.6–4.2GHz			27.5–29.5GHz		
		3.4–3.7GHz	4.4–4.9GHz		28 GHz	39GHz	

Figure 1. Snapshot of global frequency spectrum.

More than 1.1 million assigned frequencies are contained in this database. In addition, about 350,000 assigned frequencies for the broadcasting-satellite service and 25,000 allotted frequencies for the fixed-satellite service are planned for future uses.

In [32], the authors discussed the global frequency spectrum as shown in Figure 1. A detailed summary of frequency bands both licensed and unlicensed are shown. Furthermore, the paper is split out into six sections. The first section of "Introduction" covers 5G technology's goals as well as a summarized package of the literature review consisting of the latest state-of-the-art work focused on the latest performance improvement strategies and techniques in performance parameters. There are three main categories of frequencies in the assigned global frequency spectrum. The first one is assigned as high bands above 24 GHz, the midband frequencies range is from 1 GHz to 7 GHz, and the low band is below 1 GHz. Furthermore, these bands are divided into licensed bands, over 40 bands globally for LTE, which remains the industry's top priority. the unlicensed spectrum is 2.4 GHz/5.9–7.1 GHz globwise.

After introduction, the second section of the paper is focused on 5G antenna trends for base stations or access points and for mobile terminals.

The third section of the paper summarizes the latest performance enhancement techniques discussed in detail in the literature review.

The fourth section is about the beam steer ability and beam forming techniques.

In section five, the reconfigurability for different environmental scenarios are discussed.

The sixth section presents a deep focus on the MIMO antenna performance-enhancement techniques such as mutual coupling reduction/decoupling techniques, which might effectively play a vital role in mutual coupling reduction.

The seventh section is the future challenges and opportunities.

Section eight is the conclusion, which concludes the paper in a summarized package.

There are a variety of strategies that must be used in conjunction with traditional antenna design techniques, data rate, signal-to-noise ratio and a lot more.

2. 5G Antenna Development Trends for Base Station (Access Point) and Mobile Terminal

The rapid exponential growth of bandwidth-hungry applications of current smart phone users have evoked researchers across the globe to redesign and update the existing commercial communication system to fulfill the near-future high data rate and wide bandwidth applications. The sub-6 GHz mmWave spectrum being overcrowded, researchers and engineers have focused to proceed toward efficient designing technologies and techniques to achieve higher bands in mmWaves instead of sub-6 GHz to overcome these issues [1–3]. For the near-future 5G communication systems, the feasibility of mmWaves being used as a carrier frequency is coming closer to the reality. The issue at the primary level is the inherent high path losses at mmWave frequencies, which need to be mitigated by high-gain antennas used on both sides: at the mobile terminal and at the base stations or access points. Before targeting the antenna structure for 5G applications for a base station, we focus on the basic scenario of 5G communication in a coexisting multiband scenario of fixed-earth stations and 5G base stations which would help us in selecting the techniques to be implemented in the designing of antennas for base stations or access points and mobile terminals. Base-station antenna technology (BSAs) and evolution details are as follows.

2.1. Evolution of (BSAs) Technology and Capacity Enhancement Techniques

First-generation networks are based on omnidirectional cells focused mainly on coverage and not on capacity. With each exponential increase in the number of users, the operators of the second-generation network started to search for more effective ways to overcome the main issue of capacity [4]. The main technique used for capacity improvement was sectorization. In 2G, the capacity-enhancement techniques used was to divide the existing omnidirectional cells into three sectors with a coverage of 1200 each. Each antenna has the capability of a 10 dB beam width of 1200 each. Polarization diversity in this era was a second techniques for capacity improvement at +450, with the rapid increase in the number of users moving from 2G to 3G as the mobile data services were introduced in 3G. A key factor for an effective increase in the capacity was to further subdivide the existing sectors into narrower sectors [5]. Half-power beam widths such as 65° or even 33° are shown in the Figure 2.

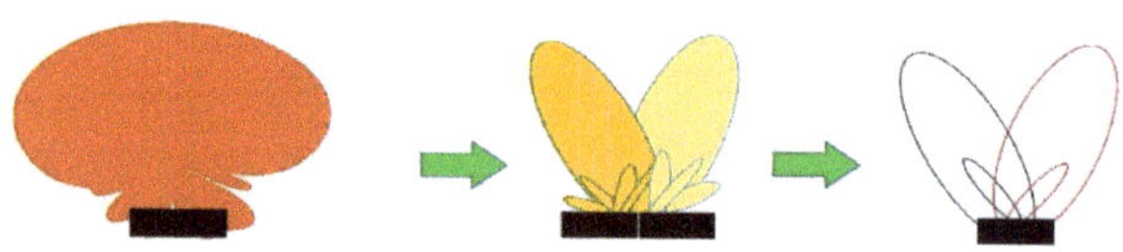

Figure 2. Single-beam BSA, two narrow-beam BSAs, and a twin-beam BSA [5].

The main drawback was the increasing number of the antenna elements leading to overload in the tower. One solution was initiated to avoid the overloading of antennas by introducing the multibeam panel antennas techniques. Hybrid couplers were used for achieving multibeam capabilities, but they need comprehensive planning for managing the network to reduce cell-edge interferences, and operators need to adjust cell/sector coverage.

The use of phase shifters played a key role in network optimization. Two bands of 800 MHz and 900 MHz were introduced in this scheme, and 4G LTE with additional spectrum was introduced with up to 2.6 GHz.

2.2. 5G Antennas for Base-Station Communication Scenario

A conceptual scenario based on a practical cellular communication system is presented with a clear aim of understanding the transmission and reception of wireless communication signals at 900, 1800, and 2700 compared with a 450 beam shutoff range, and as a result, minor improvement in the exclusion zone is achieved [6]; as the beam shutoff range increases to 900, 1800, and 2700, the restriction zone then decreases significantly. Both zones can be achieved by loading the 5G BSs and ES actual parameters.

In this study, an angular protection scheme for fixed-earth stations (FEs) and 5G base stations is analyzed. Relative location, relative distance, as well as angular changes are analyzed. The angular protection scenario is the effective way to solve the coexistence-related issues of FSS and 5G BSs. The implementation of both the angular protection and distance protection should lead to more strengthening in the angular protection. The whole scenario is shown in Figure 3. The problem caused by interference can be solved effectively by a distance protection scheme. Four types of interferences are reported in [7] between FSS and 5G system. The types of interferences reported are as follows.

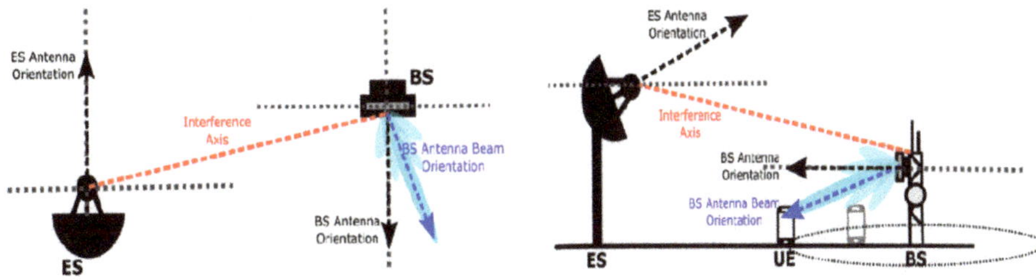

(a) (b)

Figure 3. Carrier frequency set to the center frequency [7] in the (**a**) azimuthal and (**b**) elevation planes.

From the FSS earth station to 5G base station.
From the FSS Es to the 5G user.
From 5G BS to FSS devices.
From 5G users to FSS devices.

In the work previously discussed, we have focused on the calculation of interferences in the elevation, and azimuth planes are presented by the formulas reported in [2,16] as follows.

$$I_{FSS\ ES \rightarrow BS} = P_{BS\ tx} + G_{FSS\ ES} + G_{BS} - PL$$

where I_{FSS} is the power density (dB/MHz) of the transmit station.

The gain of the antenna of the ES in the direction of the transmit station (in dBi) is represented by $G_{FSS\ ES}$, where G_{BS} represents the gain of 5G BS by using MIMO antenna array in the direction of ES considering the antenna pattern(dBi) with beam-forming capability.

To properly compute the range of the beam shut-off angle ϕ_{off} (i.e., the size of the beam angle that must be shut off) the BS interference was calculated in each step. If the I/N interference threshold value remains less than that of the interference with the current beam sweep angle ϕs, then the beam angle must need to be shut off for protection of ES. The shut-off angle ϕ_{off} of the beam increases by 1°, as shown in Figure 4.

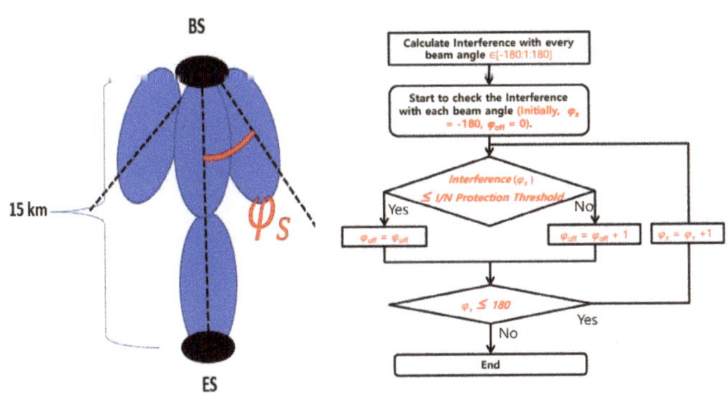

(a) (b)

Figure 4. Fixed base station—earth station: (**a**) relative azimuth and (**b**) flux gram [7].

2.3. 5G Antenna Applications in Base Station or Access Point

Several researchers have explored antennas for base station or access points. For suppression of side-lobe levels, different techniques have been proposed in [1–3]. For phase error correction, the parasitic patches have been used with an excellent approach for aperture designing employed. Design with circular polarization is proposed in [9,10]. In [34], a standard dielectric substrate is used with integration of EBG unit cells periodically for suppression of the surface waves to achieve better radiation efficiency.

A planar array with high-density elementary radiators for next-generation 5G base stations is described in [35], yielding a compact size; the antenna is designed as densely interconnected in a stacked structure with high integral capability. According to the specific application-driven requirements, the optimization is performed in such a way throughout the entire planner arrangement to meet these requirements. Detail is shown in Figure 5.

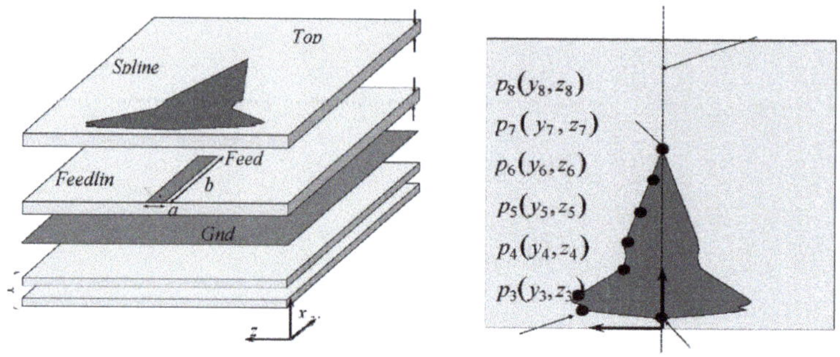

(a) (b)

Figure 5. Geometry of the HDI—based filtering 5G patch radiator [35] (**a**) 3D and (**b**) top views.

At a resonance of 28 GHz, suitable performance in terms of realized gain, polarization purity, impedance matching, and inter element isolation is achieved. Out-of-band filtering capability and wide noncontiguous stop bands without any additional circuitry is obtained. The array is shown in Figure 5. Reference coordinates are shown in Figure 6.

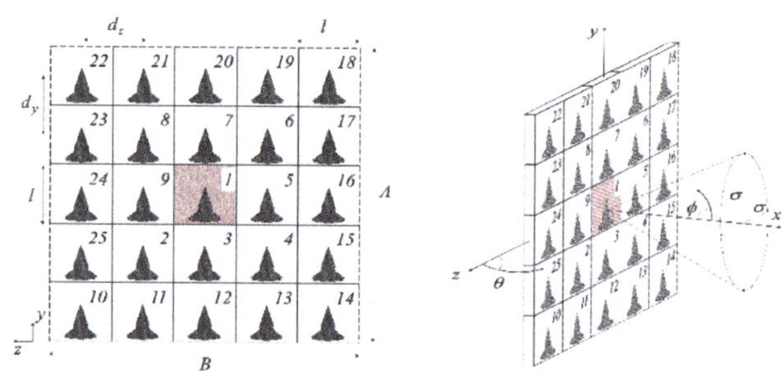

(a) (b)

Figure 6. Planar array for 5G BSAs. (**a**) Front view and (**b**) the central embedded element identifying the referenced coordinate [35].

2.4. 5G Antenna Applications in Mobile Terminal (Shared Aperture)

The realization of a shared-aperture sub-6 GHz and 5 GmmWave antenna system is proposed in [36] for application in mobile terminals or handheld devices. The antenna is designed in such a way that the integration for both bands of sub-6 GHz and mmWave operation is excellently achieved with the realization of both the simulated and measured results. The integrated antenna structure is composed up of a dipole and a tapered slot operating at multibands i.e., 3.6 GHz in sub-6 GHz and also the validation of 28 GHz in mmWave is performed. The tapered slot itself operates at 28 GHz and is also used to excite the dipole at 3.6 GHz. The simulated results of the tapered slot, rectangular slot, as well as the results for using the stub and without stub, are also shown in the figure. The dipole arms work as an antenna footprint. The designed structure has an overall size of 75 × 25 × 0.254 mm^3 using Ro-5880 substrate. The validity is confirmed by matching the simulated and measured results. Mobile phone antenna designing is always the art of how to compromise between size, appearance, and performance. In recent years, metal casing for smart phones is gaining significant interest in the mobile industry. The main reason for using metal casing is their improved mechanical strength and attractive appearance and performance, as well as better thermal conductivity. The structure with results is shown in Figure 7.

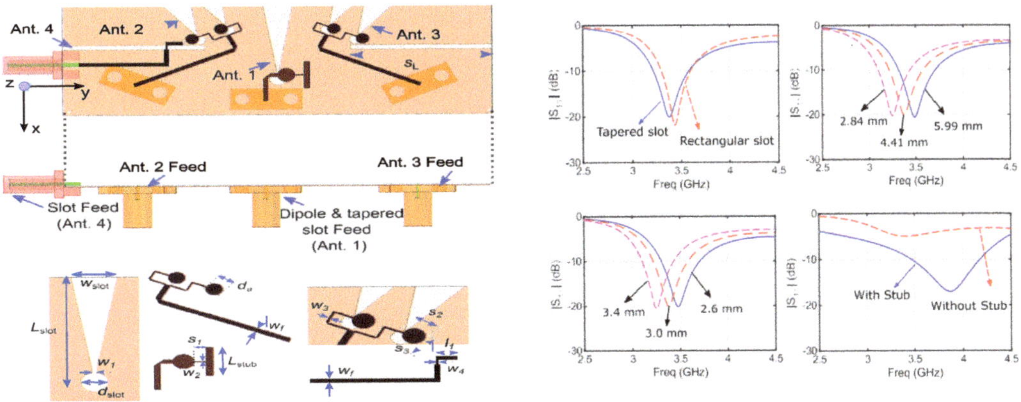

(a) (b)

Figure 7. (**a**) The design geometry of sub-6 GHz and mm wave common aperture 5G antenna system [36] and (**b**) results.

2.5. Shared Aperture Using Reconfigurability Techniques

To design a common aperture for both mmWave and sub-6 GHz, the authors have used reconfigurability techniques for 5G smart phone applications. In [11], a microstrip patch through a PIN diode is linked up with a meanderline structure to achieve the reconfigurability between the two desired frequency bands. An excellent MIMO performance is observed across the entire bandwidth for both proposed bands of mmWaves and sub-6 GHz to fulfill the requirements of an efficient MIMO antenna system. The antenna geometry and results are shown in Figure 8.

2.6. Shared Aperture Using Reconfigurability Techniques

Integration of sub-6 GHz and mmWave band is a challenging task to maintain the compactness of 5G smart phones owing to large frequency ratios. In [12], a microstrip patch radiator loaded with an inverted u-shaped slot to ensure dual-band operation in the mmWave and sub-6-GHz frequencies, i.e., 2.8 GHz and 38 GHz. In the proposed design, a patch radiator is linked up with a meander radiating structure through a compact microstrip resonant cell (CMRC) as a low-pass filter and a 12.9%, 5.8%, and 2.4% wide decoupled bandwidth are achieved.

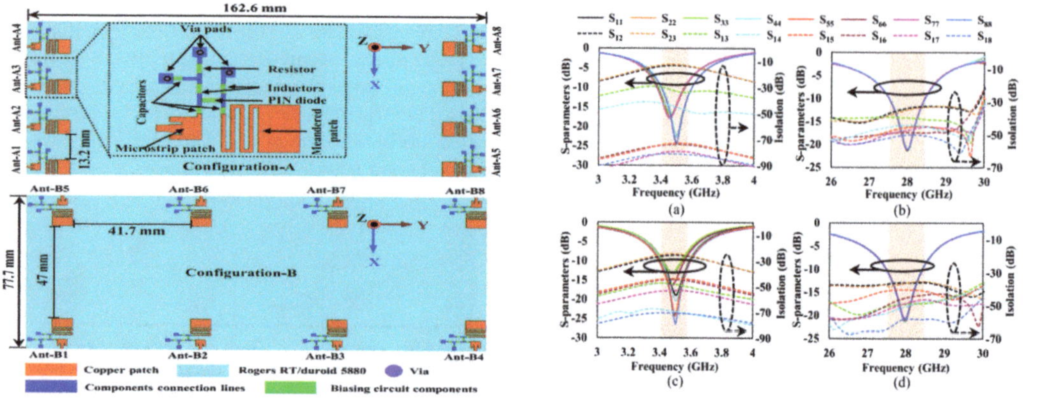

(a) (b)

Figure 8. Antenna configurations (**a**) reconfigurable 8 × 8 MIMO system and (**b**) S11 parameters and isolation [11].

3. Antenna Performance Enhancement Techniques

Antennas are one of the pivotal parts of 5G devices and a main part of a communication system. Due to the natural phenomena of using microstrip technology, especially in 5G, plenty of radiation losses as well as issues in the key performance parameters are presented, such as narrow bandwidth, lower efficiency, and less gain. To overcome all these burning issues, we need to design such antennas to fulfill the very basic requirements in terms of wide or broad bandwidth, high gain, and high efficiency, which would provide an excellent provision for low latency, ultrareliable flawless communication with better spectral efficiency. The one and only solution is to exercise such excellent techniques which might be able to overcome these issues and make the design as the best solution. The main focus is on the latest key techniques responsible for improvement in bandwidth, gain, and efficiency, and in case of MIMO isolation enhancement is the main target discussed in the literature. First, we elaborate the layout of all key performance parameters to summarize our study and make a provision for better understanding other researchers' working in the same field. Researchers have also worked to find out the human body effects on the performance of antenna functionality. It is a challenging task for research industry to practically implement the techniques to minimize these effects and make functionality and efficiency of antennas better. Some other challenges, such as the effects of high-frequency signals on the human tissues are also key challenges to be addressed.

3.1. Effect of Substrate Choice on Performance Enhancement

The appropriate selection of a substrate is the main requirement of an antenna implementation for valuable and fruitful results, as in [37]. For antenna fabrication, various substrates with different permittivity and loss tangents are available to be used based on choice and requirements of a researcher. For enhancement of gain and reduction in power loss, the selection of a substrate could play a vital role. Substrates with less relative permittivity and low loss tangents must be selected to achieve the goal of enhanced value of gain and a decreased value of power loss.

3.2. Effect of Corrugation on Bandwidth and Front-to-Back Ratio

Removing the metal part, which might be in any geometrical shape such as square, rectangular, triangular, sine or any other), from the radiator's edge has a great impact on improving the most important performance parameters in terms of bandwidth, gain, and front-to-back ratio [38].

3.3. *Impact of Dielectric Lens on Directivity and Gain*

Electrostatic radiation transmitted in one direction by the dielectric lens has the capability to divert the direction of the electrostatic radiation in one direction, which leads to an increase in the directivity as well as gain of an antenna, as in [39]. Different geometrical shapes of dielectric lens are available, and they are designed by alignment with the same or different substrate materials.

3.4. *Multielements*

The gain of an antenna can further be increased by using the antenna with multi-element. It also improves the bandwidth and efficiency of an antenna, as in [40]. In applications where a single element antenna cannot fulfill the high gain and wide bandwidth requirements, a multielements antenna is the best solution and is more effective in this case.

3.5. *Effect of Dielectric Resonator on Gain and Bandwidth Enhancement*

How to achieve a high gain and wide bandwidth a dielectric resonator antenna (DRA) for 5G wireless communication is proposed in [18]. In a sequence, to achieve a high value of gain, the DRA antenna is operational at higher order mode TEx δ_{15}, while to improve the bandwidth, the quality factor is reduced by employing a hollow cylinder at the center of the DRA, as shown in Figure 9. A 50 Ω micro strip line with a narrow aperture slot was used for excitation of DRA. For the TEx δmm mode, the resonance frequencies can be extracted from the dielectric waveguide model (DWM); the three waves K_X, K_Y, and Kz could be extracted by solving the transcendental equation as follows.

$$k_x \tan\left(\frac{k_x a}{2}\right) = \sqrt{(\varepsilon_r - 1)k_0^2 - k_x^2} \tag{1}$$

$$k_x^2 + k_y^2 + k_z^2 = \varepsilon_r k_0^2 \tag{2}$$

$$k_0 = \frac{2\pi f_0}{a}, k_y = \frac{m\pi}{b}, k_z = \frac{n\pi}{d} \tag{3}$$

$$f_0 = \frac{c}{2\pi\sqrt{\varepsilon_r}}\sqrt{k_x^2 + k_y^2 + k_z^2} \tag{4}$$

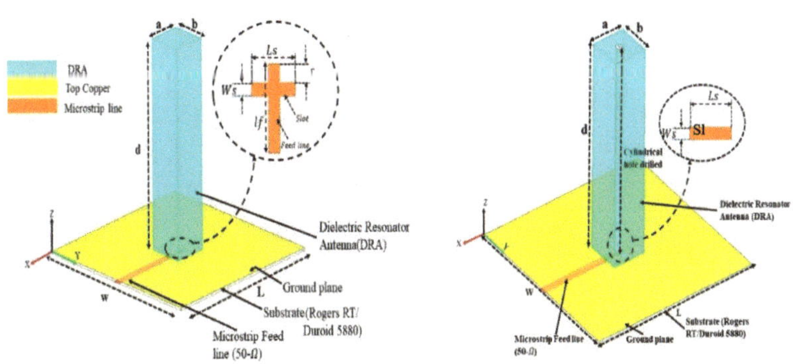

(a) (b)

Figure 9. Configuration of (DRAs): (**a**) (DRA1) without hole TEδ15 mode; (**b**) (DRA2) with cylindrical hole TEδ15 mode [18].

3.6. *Substrate-Integrated Waveguide Feeding Techniques (SIW)*

Substrate-integrated waveguide (SIW) techniques have so many excellent features, on the basis of which one can easily chose such a cheaper and effective way, especially in the designing structure of antenna for 5G application. In such techniques, the radiation losses are comparatively lower as in the conventional feeding mechanism. There is much possibility of radiation losses, especially in 5G antennas, so the power manipulating capability is higher than the one in conventional feeding ways. Various RF components which use SIW techniques are much cheaper in cost. In the process of mounting, the discrete components on SIW structures have a high density of integration.

A substrate-integrated waveguide broadband, with high gain and low-cost features, is presented in [19]. For 60 GHz bands, a single-layered SIW feeding network for achieving high gain with wide bandwidth is employed with a combined structure of cavity-backed patch antenna and wide-band T-junction. Although the antenna has a multilayered structure, it can still be easily printed with single-layered PCB techniques. There is a unidirectional symmetrical radiation pattern with higher gain of 19.6 dBi with lower cross-polarization characteristics securing 27.5% impedance bandwidth. In [20], a substrate-integrated waveguide feeding structure is used for wide-band characteristics, the SIW antenna offers a wideband (25.2–30.2) GHz with a peak value of stable gain 16.4 dBi for orthogonal polarization discrimination above 30 dB with high efficiency above 30 dB.

To suppress the cross Polarization in the single Element a coupler with four way broad wall for 2*2 subarray is applied a 4*4 main Array exhibiting 26.7% Bandwidth is designed on the basis of two sub Array having a peak Gain value of 26.7 dB are shown in Figure 9. A substrate-integrated waveguide SIW-fed array antenna for 5G application with broadband and high gain characteristics is demonstrated in [21]. A two-layer SIW feeding network of the aperture coupling method is adopted. For the purpose of broadening the bandwidth, the heights of the posts are designed as lower than the height of the cavity. Moreover, for in-band duplex communication scenarios, this SIW technique is employed. The peak self-interference cancellation value of >36 dBi is achieved. The full duplex communication scenario is shown in Figure 8. SIW provide the benefits of planner integration. The proposed design integrates a dual linear polarized three port differential antenna, a three-port SIW common-mode power combiner, and a 180° phase shifter at 28 GHz. The system operation is shown in Figure 10.

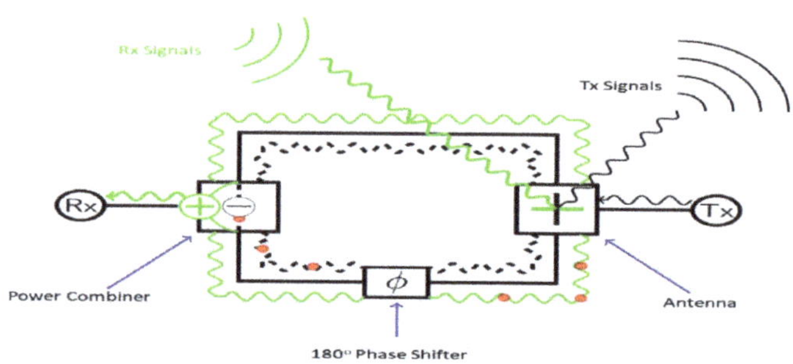

Figure 10. Proposed SIW antenna system operation (red dot represent the out-of-phase component) [21].

A stacked microstrip patch antenna array with SIW feeding structure is used for wideband application. In [22], an antenna element with linear polarization and wide bandwidth of 83.75 from (50–74) GHz with 8.7 dBi value of average gain is demonstrated. In [22,37,38,41–44], with different strategies, the SIW techniques have been used and effective results in terms of gain, bandwidth, and efficiency were achieved. Advantages and disadvantages of different techniques are shown in Table 1.

Table 1. Performance Enhancement Techniques: Advantages and Disadvantages.

Reference Antennas	Performance Enhancement Techniques	Advantages	Disadvantages
[45,46]	Substrate Choice	Substrate with low permittivity characteristics have a significant impact on the performance of an antenna and provide enhanced gain, wide bandwidth and high efficiency. Having high permittivity characteristics improves the value of return loss.	Substrate with low permittivity is costly and not easily available.
[47–49]	Mutual Coupling Reduction /Decoupling	Excellently improves the impedance matching and directly enhances both the gain and efficiency. Mutual coupling reduction techniques also reduce the size of antenna.	Mutual reduction has an impact on antenna designing and increases the complexity.
[50,51]	Multielement	It significantly improves the return loss, bandwidth and radiation efficiency, besides these properties, it also effectively reduces the side- and back-lobe levels.	For such techniques, the feeding network is a difficult task to design and makes complexity to some extent.
[52–54]	Corrugation	Improvement in the gain, efficiency, and bandwidth as well as return loss.	Significantly reduces the input impedance.
[55,56]	Dielectric Lens	Gain enhancement improvement in front-to-back ratio with stable radiation pattern. Enhancement in the gain, improvement in the front-to-back ration, stability in the radiation pattern, and radiation in the front-side direction.	The size of antenna definitely increases.

3.7. Effect of Slots in Radiator with Different Geometrical Shapes

In [31], a patch antenna in a rectangular shape with a triangular slot loaded on the surface of the radiator on the right side from top to bottom is presented for wide-band applications. The combined effect of the triangular slot and partial ground plane excellently increases the impedance bandwidth more than 90% efficiency and 12 dBi gain is achieved. The impact on bandwidth is shown in Figure 11.

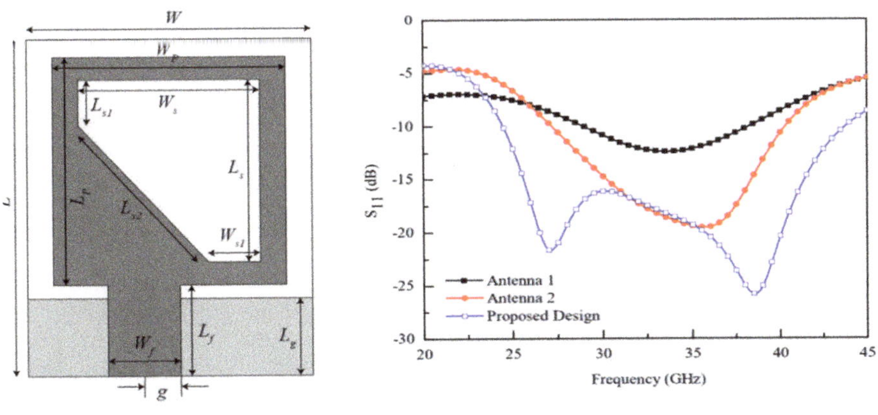

(a) (b)

Figure 11. Proposed patch antenna with triangular slot loaded. (**a**) Top view. (**b**) Enhancement in bandwidth due to slot in blue [31].

An ultra-wideband antenna with a truncated "U"-shaped slot in the patch is presented in [32]. A compact size 2.9–23.5 GHz and a LTY-5 Taconic εr = 2.2 substrate is used; the lower corner of the patch is modified with a semicircle. For the wide-band characteristics' realization, a slot in U shape is etched at the center of the radiator to ensure achievement of the notch band. A Structure in the semi ring shape in [26] has excellent impact on bandwidth enhancement.

In [24], for enhancement of impedance bandwidth, leaky wave half width microstrip patch antenna is presented. The enhancement of bandwidth is realized by etching four circular slots on the surface of the radiating patch. The main beam with a wide scanning range of +12 degrees and +70 degrees is achieved where range frequency sweep is 4.28 GHz to 7.13 GHz with peak gain of 10.31 dBi at 5 GHz. A high radiation efficiency with main-beam continuous scanning capabilities in forward direction only by changing the operating frequency is achieved. Due to the circular slots in the radiation element, the level of cross-polarization decreases.

In [57], a multimode characteristic E-shaped patch antenna for 5G mm wave application is presented. A slot is introduced in the proposed design and a 45.4% wide impedance with low loss polarization and stable radiation pattern is achieved.

3.8. Role of Parasitic Patches in the Band and Gain Enhancement

The use of parasitic patches has shown an outstanding role in the enhancement of performance parameters, with especially effective behavior in the bandwidth enhancement. In [25], the authors presented a simple low-profile spiral monopole antenna upgraded with vias in the ground plane in the inverted-"U"-shaped sequence. A 11.5 dBi peak gain with more than 83% radiation efficiency is achieved. The back side of the antenna is introduced with parasitic patches in the hexagonal shape to restrain the flow of surface waves and to minimize the interelement mutual coupling. Bandwidth ranging from 23.76 to 42.15 GHz is achieved. In the geometry of the antenna, two spiral arms are lengths L1 and L2, respectively. These arms are then connected to the feed line. A rectangular slot is created in the back side of the antenna. The width of the slot is the same as in the feed line. RT/Duriod Rogers 5880 is used with dielectric constant εr = 2.2, and the thickness of substrate is 0.254 mm. A low-profile parasitic antenna in G shape for 5G is presented in [27]. The antenna offers the impedance bandwidth with two separate ranges. For the result-oriented geometry of the parasitic design, the antenna is optimized and a wide bandwidth due to the parasitic patch was observed, as in Step 1, and as 7.46 to 8.42 GHz in the second step from 4.43 to 8.82 GHz.

The authors of [27] have cross-verified the results by cross-checking on both software CST microwave studio and high-frequency structured simulator HFSS results from both software are well agreed. For converting the end fire direction and pattern of beam to the broadside, the extension in the U-shape have a key role.

Similarly, in [28] the parasitic elements are placed on either side of the monopole radiator to improve the bore-sight gain. The mutual coupling between of two adjacent elements is reduced by the arrangement of radiating patches and their feed lines with a rotation of 180 degrees out-of-phase alternating sequence discouraging the flow of surface waves, which directly results in improving the symmetry of the radiation pattern, and mutual coupling of 20 dB with uniform distribution of amplitude and a 19.8 dBi maximum broadside gain is achieved with a side-lobe level of −12.1 dB. These characteristics rendering the antenna would be an excellent choice for 5G wireless applications. The geometry of the parasitic patch is shown in Figure 12.

The coupling of 20 dB with uniform distribution of amplitude and a 2 dBi maximum broadside gain is achieved and a side-lobe level of −12.1 dB is achieved. All simulated parameters well agree with the measured one.

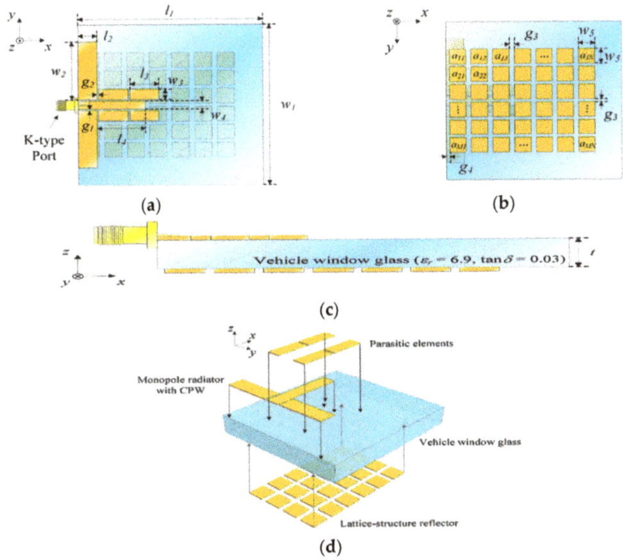

Figure 12. Geometry of the proposed printed antenna: (**a**) top view; (**b**) bottom view; (**c**) side view; (**d**) isometric view [28].

3.9. Role of Hybrid Feeding Structure in Performance Enhancement

A hybrid-feed wide-band and high-efficiency antenna for mmWave 5G communication is proposed in [58]. The proposed designed antenna consists of a novel hybrid-feed structure in a planar antenna array with excellent performance in terms of high efficiency and wide bandwidth is presented. To achieve the goal of high-performance key parameters such as wide bandwidth and high efficiency, the conventional high-loss ridged feeding network and substrate-based feeding network are replaced by an effective low-loss ridge-gap waveguide feeding network. The measured bandwidth ranges from 26.05 to 31.15 GHz (16%), and the maximum achieved gain is 25.15 dBi. The proposed design might be an excellent choice for 5G application. In comparison with [38], the results are much improved due to the use of ridge gap waveguide (RGWG) instead of gap waveguide GWG. Structrual detail with different layers of the design is shown in Figure 13.

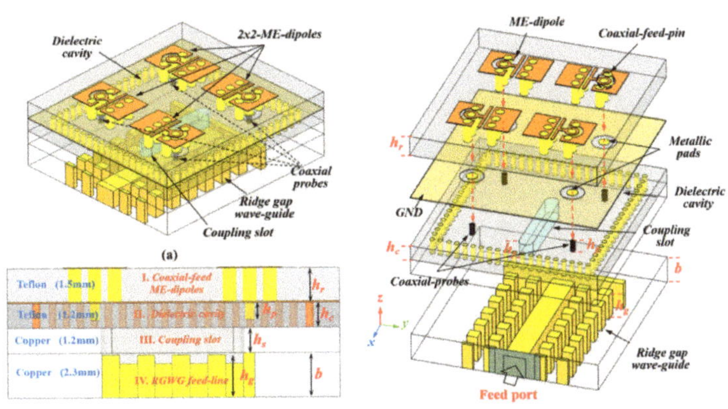

Figure 13. Hybrid feed geometry of a 2 × 2 unit for ME dipole subarray. (**a**) Perspective view. (**b**) Side view. (**c**) Exploded view [57].

3.10. Differential Feeding Network (Specific Feeding)

In [39], a patch antenna fed differentially with wide-band and dual polarization capability with excellent filtering response is proposed. For excitation of square patch antenna, a feeding structure with novel E-shaped is proposed. One end of the feeding network is shortened to the ground for achieving the wider operation bandwidth, also an extra resonant mode is introduced for generation of a third resonance mode. A stepped.

Resonator with symmetric short/open-circuit is proposed to further enhance the bandwidth. The same resonator not only makes feasible the enhancement in bandwidth but effectively enhances the gain suppression level outside the operation band. For an active VSWR value of fewer than 1.5, a wide operation bandwidth (3.12–3.9 GHz at 22.2%) is achieved with in-hand stable gain suppression level for out-band is better than 18.5 dB. The geometry of the differential feeding network is shown in Figure 7 as follows. The radiating patch and E-shaped feeding structure are printed on the top layer of the upper substrate. One end of each E-shaped feeding structure is shorted to the ground with metal posts. The lower layer is a F4B substrate with a dielectric permittivity of 2.65 and thickness of 0.8 mm. In addition, four pairs of O/SCSIRs are printed on the top layer of the lower substrate. The four ends that close to the center of the substrate are connected to the E-shaped feeding structures with metal posts. The other four ends of the O/SCSIRs are connected with 50 Ohm coaxial cables. A metal ground is printed at the bottom layer of the lower substrate with four U-shaped slots. When port1 and port2 are excited with differential signals, a 45° polarization is generated. Similarly, a 45° polarization can be obtained when port3 and port4 are excited with differential signals. When the shorting posts are introduced (the parallel inductance L2 is added), the overall inductance decreases, and the first resonant mode moves to lower frequencies. Thus, the two resonant modes are separated by the shorting posts, and a wide operation band is obtained. The proposed structure with metal post and upper substrate is shown in Figure 14.

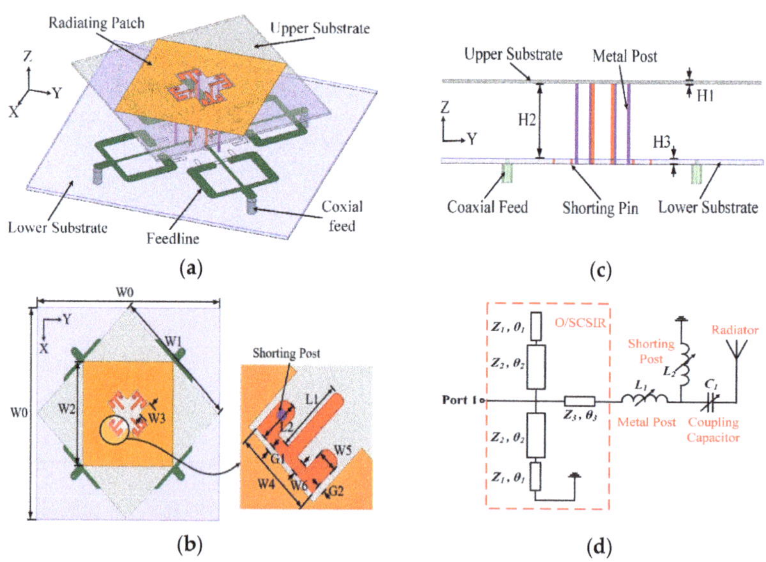

Figure 14. Antenna configuration: (**a**) 3D view, (**b**) top view, (**c**) front view, (**d**) Equivalent circuit [39].

3.11. Performance Enhancement Using Metamaterials

Artificial materials are composed up of periodic material structures and specific dielectric whose electromagnetic properties in terms of electric permittivity as well as magnetic permeability could be redesign to a level higher than the nature of the available materials. Materials having negative permittivity and negative permeability do not exist in nature, and metamaterial is the one and only source. If we replace conventional patch antenna with MMTs, which are commonly used with DNG double-negative characteristics, it will lead to an excellent shift to higher frequencies, and thus will definitely increase bandwidth [59,60]. In [61], a modified patch with diagonal slot and a layer of metasurface is demonstrated in low profile with circular polarization. A patch with a diagonally positioned slot in the center of the patch to achieve wide bandwidth as well as circular polarization is proposed. The metasurface is designed without generating any gap on the patch surface to make provision for miniaturization. A wide-band low-profile antenna with circularly polarized metasurface-based antenna for realization of performance enhancement in terms of high gain, wide, bandwidth, high efficiency, and circular polarization is presented, as well as a wide bandwidth of 37.4% (24–34.1 GHZ) for the value of S11 < −10 dB with a high value of gain 11 dBic. The gain response is stable at 9.5–11 dBic. CP in the left hand is achieved. At 11 dBic the same Antenna is tuned. Large axial ratio as well as high bandwidth and high efficiency is achieved. Similarly, in the antenna in [62], a circularly polarized structure with wide band and bidirectional metasurface-based antenna is proposed. A metasurface composed up of two identical periodic metallic layers as a metal surface is used. In this case, the metasurfaces convert the linear polarized radiation to circular polarized radiation which was produced by the rectangular slot. In total, 14.3% bandwidth is achieved (5.2–6.0 GHz). The antenna offers state-of-the-art operating bandwidths compare with simple planar antennas, shown in Figure 15.

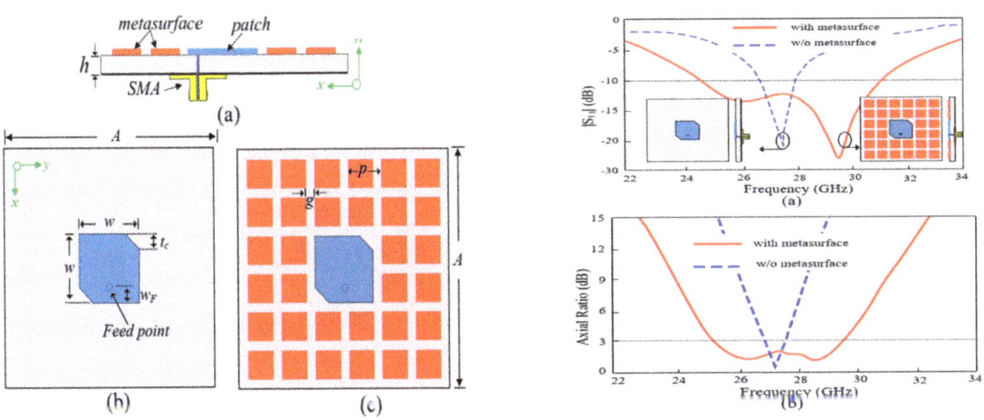

Figure 15. Antenna geometry: (**a**, **left**) side view (**b**,**c**, **left**) top and bottom view, and (**a**,**b**) (**right**) impact of MS on S_{11} and axial ratio [61].

4. Beam Steerability and Beam Forming

4.1. Dielectric Resonator with Beemsteerability

A microstrip patch antenna is presented in this research work. The front and back side of the dielectric slab is arranged periodically to properly adjust the VCRM position. A good radiation pattern is attained with good matching of numerical and experimental results. A high value of 11.9 dB gain is achieved in [62]. A dielectric resonator antenna with parasitic elements is used to achieve high-order mode steerability. The beam steering was successfully achieved by switching the termination capacitor on the parasitic element and a narrow aperture in the ground plane to achieve a wider bandwidth, which can be potentially applied for device–device (D2D) communication in 5G Internet of Things (IOTs), which can interconnect, as shown in Figure 16. For a significant performance achievement, the size of the resonator might be less than the device antenna [63], and 36.78% impedance bandwidth is achieved. The entire design mechanism is shown in Figure 16.

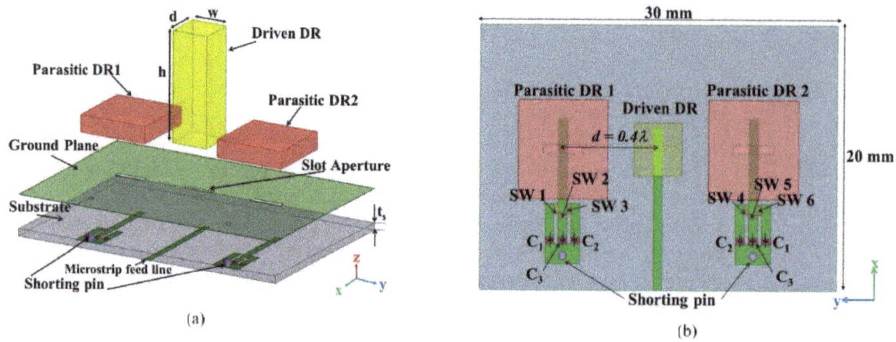

Figure 16. The geometrical configuration of the proposed DRA array. (**a**) 3D view. (**b**) Top view [62].

A 193.16% in [57,64] is achieved and four CSRRs in the ground plane are used. A total of 5.71% impedance bandwidth while the simple patch Antenna without the use of CSRRs offer an impedance bandwidth of 4.28% at 28 GHz.

4.2. Impact of Fabrication Techniques

Antennas need to be robust mechanically and should be efficient with wide bandwidth as well as having desirable radiation characteristics [65–72] which might ensure high simulation accuracy [31]. Radiation pattern and gain of the antenna are required to be tested for distortion or degradation. In the light of the literature studies, here is the summarized comparison on the bases of advantages and disadvantages in terms of bandwidth, efficiency, and gain. A brief summery has been studied in the literature about how different fabrication techniques affect the performance of an antenna. A summary of different antenna performances along with fabrication techniques is presented in Table 2.

Table 2. A Summary of different Antennas Performances with Different Fabrication Techniques.

References Antennas	Number of Unit Cells		Total Size Array (λ0)		Effective Bandwidth (S11 < −10 dB)		Maximum Gain		Radiation Efficiency (%)	Fabrication Techniques
[65]	8 × 8	150 × 75 × 7	6.4 × 6.4 × 0.17	Ground Slot	34–40.1 Hz	3.3–6.0	24 dBi	>18	41%	PCB
[66]	5 × 6	124 × 74 × 6	5.07 × 3.82 × 0.1	Ground Slot	33.95–34.86 GHz	3.3–3.6	17.09 dBi	>15	22%	PCB
[67]	8 × 8	150 × 80 × 0.8	6.12 × 6.8 × 0.5	No	55.4–66.5 GHz	3.4–3.6	26.1 dBi	>17.5	70%	Multi-Layer PCB
[68]	4 × 4	150 × 75 × 0.8	2 × 2 × 0.12	Orthogonal Polarization	25.5–40.2 GHz	3.3–3.8	16.1 dBi	>15	83%	LTCC
[69]	16 × 16	150 × 75 × 8	15.7 × 16 × 0.8	Orthogonal Polarization	71–86 GHz	3.4–3.6	32.9 dBi	>17	86.60%	Diffusion Bonding
[70]	4 × 4	145 × 75 × 6	5.3 × 5.3 × 1.1	No	29.6–30.7 GHz	3.4–3.6	22.4 dBi	>15	99%	Machining
[71]	4 × 4	150 × 75 × 7	3.5 × 3.4 × 0.3	No	28.8–34 GHz	3.4–3.6	21.2 dBi	>12.7	70%	PCB
[72]	4 × 8	150 × 75 × 7	11.8 × 11.4 × 2.2	Neutralization Line	86.7–102.2 GHz	3.4–3.6	23 dBi	>11.5	N/A%	PCB + Machining
[58]	8 × 8	140 × 70 × 1	5.4 × 5.4 × 0.6	No	26.05–31.15 GHz	3.4–3.6	25 dBi	>11.2	85%	
						3.4–3.8		>15.5		

5. Reconfigurability

Since past decades, reconfigurable antennas have gained considerable attention [33,73,74]. Unlike wideband and multiband antennas, the reconfigurable antennas have excellent feasibility for customers to select a frequency band on their own [74]. For facing different environments with different complexities, the response of the antenna is a key asset for flawless, smooth, and reliable communication. For several applications such as searching, rescue, and tracking, the antenna needs to have effective switching capabilities as well as performance-boosting technologies. To handle different applications scenarios, different switching techniques, phased arrays with adaptive nulling characteristics, and multiple beams with low side lobe are required. For achieving such multiple tasks, different kinds of signal processing techniques can excellently provide effective solutions. A high degree of reconfigurable antenna is required for reliable communication. In [33], a low-profile simple reconfigurable antenna is proposed. Two pin diodes are used, as shown in Figure 17. These pin diodes connect the two additional stubs and the triangular monopole. The same techniques are used in Figure 18.

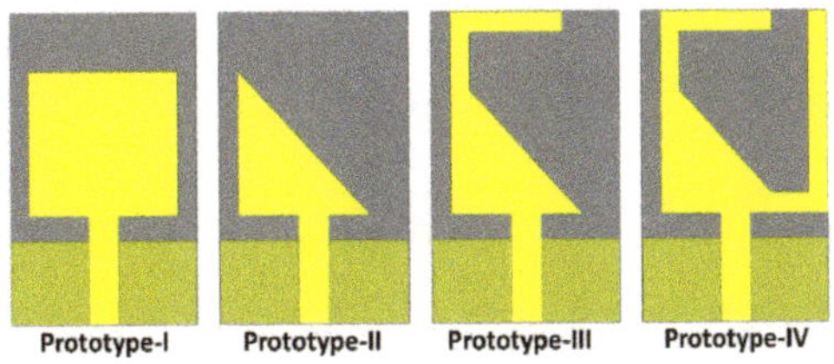

(a) (b) (c) (d)

Figure 17. Evolution steps of antenna presented in Figure 17. Different geometrical shapes (**a**–**d**).

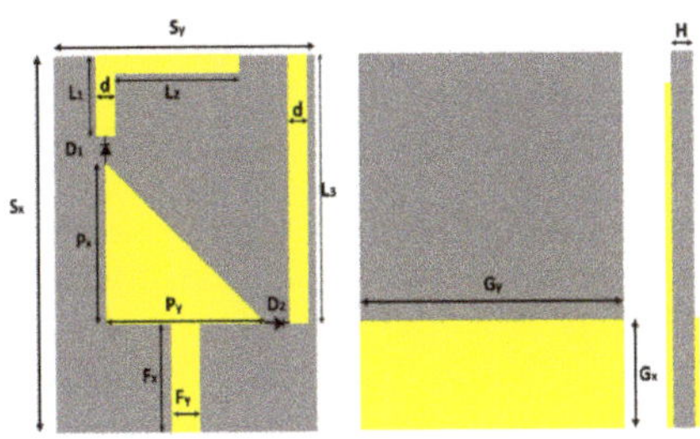

(a) (b) (c)

Figure 18. Final geometry of antenna element. (**a**) Top view, (**b**) bottom view, and (**c**) side view [73].

The antenna has a feasibility in two different bands and when the two switching states are ON, thus the antenna operates in a single band of 3.3 to 4.2 GHz. In the case of only one switch being in the ON position, the antenna then offers the dual band ranging from 3.3 to 4.2 and 5.8 to 7.2. The antenna with low profile and compact size of 0.27λ0 × 0.16λ0 × 0.017λ0 can be used for several applications, such as for wireless systems in vehicular systems and wireless local area networks (WLAN). A bidirectional beam characteristic is achieved.

In [33], in the first stage, the simple patch antennas is modified to a right-angle triangle and then two stubs are introduced on the top angle of the triangle in an inverted L-shape for enhancement of bandwidth. Two diodes, D1 and D2, were placed between the two stubs and the radiator. The antenna offers the broadside radiation pattern when both the diodes D1 and D2 are in the OFF state in H-plane while in the E-plane a small tilt is observed in case D1 is OFF and D2 is ON. On the other hand, when D1 is ON and D2 is OFF, it provides pattern radiation in 30 GHz and 32 GHz, respectively. When both diodes (D1 and D2) are in the ON state, then a little tilted beam is achieved by the antenna. The tilt scenario is due to the truncated ground. When both diodes are in the ON state, the impedance bandwidth offered by the antenna is 6.3 GHz, ranging from 29.95 to 32.25, which covers the potential spectrum from 26–29 GHz for 5G applications. When both diodes are in the OFF state, there is a bandwidth of 6.75 GHz ranging from 27.81–34.560 GHz, where S11 < 10 dB. In case (01), when one diode D1 is ON and the second one is OFF, the impedance bandwidth is of 6.42 GHz (27.81–42 GHz), where S11 is < −10 dB.

The design is shown in Figure 19.

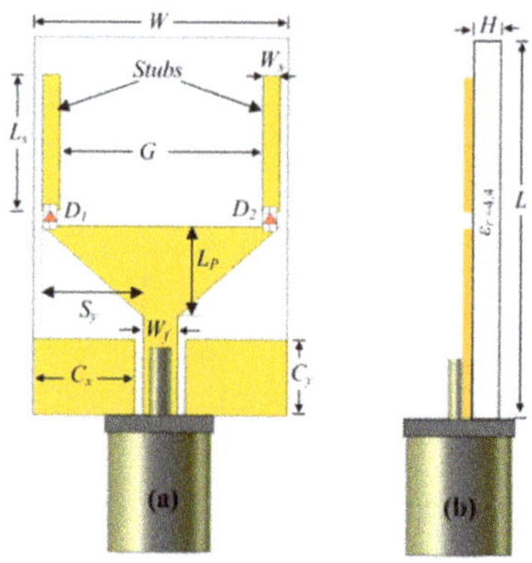

Figure 19. A reconfigurable antenna [33]. (a) Front view and (b) side view.

After generalizing the view of the reconfigurability scenario, the frequency and pattern reconfigurability must be explored. In [27], lumped elements as a switching component for hexaband switching capability is presented. Lumped elements are used to achieve tunable capacitance responsible for frequency reconfigurability of hexaband frequency spectrum, i.e., 2.10 GHz, 4.11 GHz, and 2.4 GHz and 5.2 GHz on affordable, compact geometry capable to be integrated with ease in modern communication systems.

Pattern reconfigurable antenna is presented in [72]. For 5G, there are new radio frequency bands –N77(3.3 GHz–3.8 GHz). The structure of the radiating element is modified by a swastika-shaped structure. Around the radiator, four parasitic arc-shaped elements are loaded and each parasitic element is loaded with a pin diode to control the antenna radiation field pattern. Table 3 is the comparison of different antennas using pin diodes for switching.

Table 3. Comparison of different antennas using pin diode as a witching component.

References Antennas	Antenna Size (λ0 × λ0)	Number of Pin Diodes	Single Band	Multi Band	Wide Band	Bandwidth (%)
[75]	1.00 × 0.41	2	4	4	8	26.4%, 37.4%
[76]	0.62 × 0.41	1	4	4	8	13.5%, 35.72%, 9.94%
[77]	0.36 × 0.33	6	4	8	4	123.5%, 28.5%
[78]	0.60 × 0.28	2	4	8	4	44.89%, 10.55%
[79]	0.55 × 0.59	4	8	4	4	74%, 8.2%, 9.79%, 15.4%
[33]	150 × 0.16	2	4	4	4	64.40%, 24%, 25.5%

A UWB antenna [80] with improved gain and omega structure loaded in the radiating patch is presented. In omega structure, a lumped capacitor is inserted for tuning the notched frequency spectrum 5.7 GHZ to 3.8 GHz for both WLAN and WIMAX bands, respectively. For gain improvement, a 3 × 2 single layer array with a unit cell of metamaterial in L-shape is used for UWB and gain improvement. Lumped capacitors have an effective role for the mitigation of interference of WiMAX and WLAN spectrum. Antennas with reconfigurable techniques through electronic switching components such as PIN diodes, Varacter diodes, and RF MEMS are reported in [33,73,74,81,82]. Each technique has some supremacy or advantages as well as some weaknesses/disadvantages. A comprehensive study on many kinds of electronics switching components used for gaining reconfigurability are also reported in [33,73,74]. Photoconductive switches are composed up of the famous semiconductor materials silicon and gallium arsenide [29,30]. Biasing lines are used instead of electrical wire, obviating large sizes and offering high isolation along with fewer interference. The details of different switching techniques are summarized in Table 4.

Table 4. A comparison of different switch components [83–87].

Switching Types	Supremacy/ Advantages	Weaknesses/ Disadvantages
PIN Diode	Extremely reliable Very low in cost Most probable choice for reconfiguration	Power handling capability is high The tuning speed is very high High DC biasing capability in ON state
MEMS	Feasible for small flow of current continuous tuning Integration is easy	A nonlinear source Lower range dynamically Biasing circuitry is complex
Varacters	Linear impedance bandwidth with high isolation Comparatively less noise figure and low power losses	Voltage control level is high. The switching speed is slow Limited life cycle is limited compared with other components.

6. MIMO Antenna Performance Enhancement Techniques

In recent years in communication systems, the topic of multiinput and multioutput (MIMO) systems has received a considerable attraction, especially the performance in terms of the channel capacity of a communication system which can badly be affected by the mutual coupling of closely correlated interelements of a MIMO system. The capacity gain of a communication system has a significantly deep impact on the overall efficiency of an antenna system. So, a good communication system needs to have good isolation between the interelement or closely collocated elements.

A summary of MIMO antenna designing is reported in [81]. Most authors have used one or more techniques for designing antennas for mutual coupling reduction in MIMO antennas described in the literature for 5G and wide-band applications, these designing and decoupling techniques can effectively be used and better results could be achieved. The detailed layout is shown in Figure 19. The most prominent techniques of MIMO antenna designing as well as mutual coupling reduction techniques are presented to better explain the idea of mutual reduction or decoupling methods and possibilities discussed in the literature from [29,30,73,74,81,82]. In multielement antennas, either in arrays or especially in the case of MIMO antenna design, the collocated antenna elements on a single PCB might affect the performance of each other. To avoid or reduce this effect, researchers have implemented different strategies and techniques for mutual coupling reduction or decupling of these collocated elements n the case of MIMO antennas, which is the main focus in our study, named as isolation or decoupling strategies and techniques in the literature. Some state of the art techniques are explained in the next section. The decoupling techniques play an excellent role to achieve the optimum performance in case of MIMO antennas. These techniques are a predetermined and unavoidable part in the designing procedure of the MIMO antenna. These techniques are explained below.

6.1. Mutual Coupling Reduction Using CDRA in Ground

A CDRA-based antenna is presented in [81,82]. In [82] the CDRA is used back to back on the opposite side of the substrate. For excitation of Port no.1 and Port no.2, a pair of coplanar waveguides is used with conformal microstrip lines. CDRA is placed on yop with a height of 6.5 mm and on the bottom with 6.0 mm. The excitation of Port no.3 and port no. 4 is conducted by microstrip line conformal fed to stripline. An improved interelement isolation is realized by generating two modes HEx and in each CDRA. Two different polarizations are produced due to the two different modes HEx and HEy. Excitation from opposite directions and two orthogonal modes in opposite directions make it feasible to excellently improve the isolation of closely colocated elements. DRA have shown better results in terms of bandwidth and gain [82]. In [30,62], monopole antenna with inverted "L"-shaped geometry and its MIMO design with eight elements is demonstrated. Good efforts have been carried out in [29]. In [30] a MIMO antenna the size of 136 × 68 mm^2 with eight elements covering 450 MHz bandwidth is presented.

Isolation with value of 15 dB as well as high Gain of 4 dBi is achieved. In the structure of the antenna, a parasitic patch is extended from the ground plane. Excellent impedance matching in the desired frequency spectrum. MIMO antenna designing and mutual coupling reduction or decoupling techniques are shown in Figure 20.

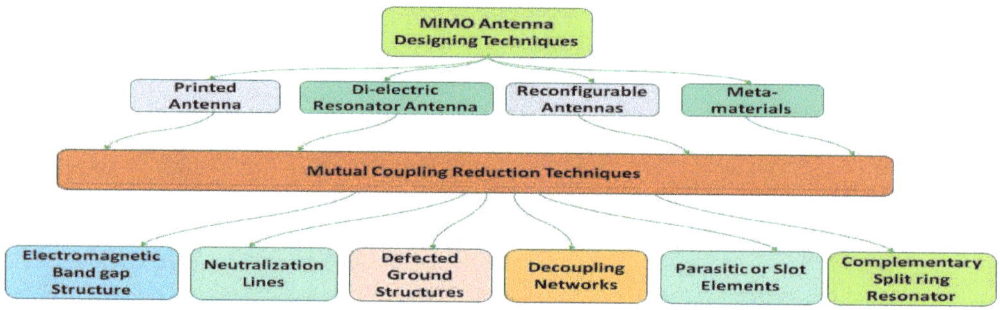

Figure 20. Designing and mutual coupling reduction technique of MIMO antennas, a summarized generic view.

6.2. Neutralization Lines

Having capability to pass electromagnetic waves between antenna elements to completely decouple or reduce the effect of mutual coupling by using lumped elements or metallic slits. It improves the antenna bandwidth and reduces the occupying space of antennas. The improvement in bandwidth is realized

6.3. Decoupling Network

Cross admittance becomes transformed to purely imaginary value in decoupling network by adding a discrete component or transmission lines. To reduce mutual coupling, this technique employs a plane decoupling network which acts like a resonator. The pattern diversity for multielements could also be achieved through decoupling by implementation of dummy load and coupled resonators technique. To improve isolation, it is a cost effective way to be implemented [88].

6.4. Electromagnetic Bandgap (EBG) Structure

For the transmission of electromagnetic waves, it acts like a medium. EBG structure provides capability of low mutual coupling and excellent efficiency, as in [88]. An improved scattering parameter with exhibiting wideband gap characteristics is achieved. In the desired direction far field gain pattern is achieved.

6.5. Dielectric Resonator

An antenna that contains a dielectric resonator is called as dielectric resonator antenna (DRA). DRAs are capable of providing low loss, high gain, and high radiation efficiency. Dual-band property as well as high isolation value can be achieved using DRAs [88].

6.6. Defected Ground Structure (DGS)

Different geometrical shapes, defects or slots are consolidated in the ground plane of the antenna. Low mutual coupling and wide bandwidth with maximum efficiency is the structure with defected ground in [88].

6.7. Metamaterial

Metamaterials are materials which can be designed manually by using two or more than two different materials which contain electromagnetic characteristics. Different types of metamaterials with the characteristics of single negative, electromagnetic, electromagnetic band gap, double negative, anisotropic, isotropic, terahertz, chiral, tunable, photonic, frequency selective surface-based, nonlinear, and tunable metamaterial. Using metamaterial, it is possible to have an antenna with low mutual coupling, high gain, bandwidth, and be compact in size [37,88]. Slot elements are used to achieve enhancement in the impedance bandwidth using the coupling method in the radiator or patch or in the ground plane. The slot antenna have the capability to provide wide bandwidth, high gain, and high efficiency but have a high mutual coupling value [89]. Complementary split-ring resonators (CSRR) are used for isolation improvement, to perform iterating function, and to provide lower mutual coupling. CSRRs are also used to provide high efficiency and miniaturize the size of the antenna [88]. CSRR is made up of two concentric ring structures with slots opposite to each other. Frequency reconfiguration is based on switching techniques by using varactor diodes and MEMS switches, and PIN diode reconfigurability can be achieved to increase frequency range and enhance the value of the envelop correlation coefficient.

6.8. Recnfigurability

The reconfigurable antenna structure can provide lower mutual coupling as well as a high value of diversity gain, and high efficiency, wide bandwidth, and front-to-back ratio can also be achieved [30,33,73,74,81,82,88]. These antennas are either a single element or multielement. The main focus of these enhancement techniques is to improve the antenna performance throughout in terms of mean effective gain (MEG), envelop correlation coefficient ECC, and 38.1 bps/Hz channel capacity.

A comprehensive and summarized detail of MIMO antenna designing techniques as well as performance enhancement techniques in terms of mutual coupling reduction is shown in Table 5. Table 6 elaborates the comparison of different techniques employed for mutual coupling reduction/decoupling. in this summery, the comparison of different techniques employed for mutual coupling reduction/decoupling is presented.

Table 5. Mutual Coupling Reduction/Decoupling Techniques Advantages and Disadvantages.

Reference Antennas	Performance Enhancement Techniques	Advantages	Disadvantages
[90]	Neutralization Lines	This technique is mostly used in compact-size antenna to decouple the closed interelement coupling.	Complexities in structure
[91,92]	Dielectric Resonator Antenna	Bandwidth Gain and Efficiency Enhancement	Complexities in structure
[93,94]	Neutralization Lines	This technique is mostly used in compact-size antenna to decouple the closed interelement coupling.	Complexities in structure
[95,96]	Frequency Reconfigurable Antenna	It provides feasibility for antenna to be in compact size and excellent provision for compactness dimensionally.	External component provision is not an easy task.
[97–100]	Slot or Parasitic element Metamaterials Decoupling Network	Diversity gain, bandwidth, and efficiency could excellently be improved. Enhancement in the diversity gain, bandwidth and envelop correlation coefficient (ECC) Diversity gain and impedance matching could be improved effectively.	Designing and decision about position is difficult and a time-consuming activity. Decisions about position and designing are not an easy job. Complexity in design as well as low in gain

Table 6. Comparison of Different Techniques employed for Mutual Coupling Reduction/Decoupling.

Reference Antennas	Geometry	Decoupling Techniques	Effective Bandwidth (GHz)	Isolation(dB)	Channel Capacity Loss	ECC
[101]	Non Planar	Decoupling based on FSS	3–11	20	<0.20	<0.20
[102]	Non Planar	Parasitic structure for decoupling	3.1–10.6	20	<0.70	<0.1
[103]	Non Planar	Configuration based on 3D element distance	3–11	20	-	<0.5
[104]	Planar	Decoupling structure based on grounded slits	2–11	15	-	<0.2
[89]	Planar	Orthogonal polarization	2.5–2.7	12	-	<0.12
[105]	Planar	Decoupling structure of ENG-NZI metamaterial	3.4–3.6, 4.8–5.0	12	<0.08	<0.15
[106]	Planar	DN (secoupling network)	3.4–3.6	14	-	<0.2
[107]	Planar	Element positioning + geometrical slotting	3.27–5.92	14.5	-	<0.1
[108]	Planar	Port distance plus positioning of elements	3.2–6.1	18	-	<0.21
[109]	Planar	Parasitic structure for decoupling	3–11	20	<0.35	<0.0025
[110]	Planar	Decoupling structure of ENG-NZI metamaterial	3–11	28	<0.30	<0.1

Some techniques are used not only for gain and bandwidth enhancement in single element and arrays but also used for mutual coupling reduction or decoupling the correlated closed elements' place on the same single PCB. Like metamaterials, researchers have used metasuface structure for bandwidth enhancement in [31].

The same metameterial plays a vital role in MIMO as it is used for mutual coupling reduction due to its unique double-negative (DNG) characteristics. Different geometrical shapes are used by different researcher for achieving the goal of wide bandwidth, high gain and mutual coupling reduction. The role of metamaterials could not be ignored as they have excellent characteristics.

The Antenna in [82] offers 2.6 times a 4 × 4 MIMO antenna. All parameters effecting the performance of antenna such as SAR in terms of user hand effect is also taken in consideration.

7. Future Challenges and Opportunities

Novel circuit antenna co-integration needs to be a solution for active beamforming antenna arrays for 5G telecommunication applications in the sub 6 GHz (FR1) and near-30 GHz (FR2) mmWave frequency bands. Active integrated antennas can be designed with nontraditional functionalities (e.g., tailored power combining, active load modulation for power amplifiers [111], active source modulation for low-noise amplifiers, and RF filtering [112]) that help to minimize power loss, particularly at mmWave frequencies.

Antenna arrays with high directivity are implemented for fixed-beam communication at frequencies beyond 100 GHz. Most high-efficiency mmWave array antennas are waveguide (WG) structures that are typically designed with multilayer H-plane split blocks that are galvanically connected. This method results in a hollowed WG multilayer structure formed by stacking machined metal plates.

Due to crucial connectivity power dissipation and physical constraints, conventional systems incorporating phase shifter ICs attached to each element may not be practical at 100G+ frequency. Other systems, such as those based on reconfigurable antenna elements equipped by varactor or PIN diodes, suffer from comparable problems, as well as design complexity, high cost, and time-consuming assembly.

Innovative array antenna designs enabling two-dimensional (fullspace) beam control and effective beam steering are ideal for integration and hybrid packaging alongside active IC modules. Antenna subarray modularization (ASM) is one of the important technologies in this context for making the system practical and robust [112]. ASM will be investigated in order to determine optimally sparse topologies to relax the physical restrictions of antennas while meeting beamforming objectives with a lower number of amplitude-phase controls per element and remaining design compatible with semiconductor technology. One specific goal is to reduce signal latency while operating several phase shifters along the same ASM module.

The micromachined THz system platform also allows for the implementation of MEMS (micro-electromechanical systems) switches, allowing for reconfigurable antenna front ends. There is significant potential to realized silicon-micro-machined antenna systems with low-order (1 to 2 bit) MEMS-based phase control to support beamforming abilities of large-scale array antennas, with the benefits of low insertion loss, high compactness, convenience of control circuitry, and high-power packing abilities. The disadvantage associated with the limited number of degrees of freedom in beamforming can be overcomes by combining the above-mentioned array antenna topologies.

8. Conclusions

In communication systems, antenna plays a vital role and could be considered the important asset of a communication system. The main focus in this research is the study of state-of-the-art performance improvement strategies and techniques used for bandwidth, gain, and efficiency enhancement, as well as the mutual coupling reduction techniques used in the array of multiple elements or MIMO antenna elements. Different strategies and techniques used for excellent enhancement in the performance of antennas used for 5G and ultrawide band applications has been reviewed. Metamaterials play a multidimensional role in the performance enhancement of an antenna, especially for 5G and wideband application. MMTs not only play a role in the enhancement of bandwidth but have the capability to reduce mutual coupling between the closely collocated antenna elements in MIMOs and the arrays of multiple elements. The role of the EBG structure slots in the radiators with different geometrical shapes in the antenna performance enhancement in terms of bandwidth, gain, and efficiency is included. Reconfigurability is focused as having an important role and solution for different environmental complexities such as surveillance, tracking, and smart-cities scenarios. Reconfigurability techniques for sharing the same aperture for MMwave and sub-6 GHz application is also included. The role of beam steerability and beam forming is studied in detail.

Author Contributions: Conceptualization, I.A. and Q.A.; methodology, I.A.; software, I.A.; validation, H.S. and W.T.; formal analysis, I.A. investigation, H.S.; resources, H.S.; data curation, Q.A.; writing—original draft preparation, I.A.; writing—review and editing H.S. and W.T.; supervision, H.S.; project administration, H.S.; funding acquisition, H.S. All authors have read and agreed to the published version of the manuscript.

Funding: This research was funded by the Beijing Key Laboratory of Millimeter Wave and Terahertz Techniques, School of Information and Electronics Beijing Institute of Technology, Beijing 100081, China.

Acknowledgments: I am highly indebted to Houjun Sun for all the support extended to me in paving the research ideas and shaping my academic career. I must pay gratitude to him for the technical and financial support and boost throughout all academic and research work. It would not have been possible to obtain my research work here without his lasting and sustained cooperation as and when needed.

Conflicts of Interest: The authors declare no conflict of interest.

References

1. Khalily, M.; Tafazolli, R.; Rahman, T.A.; Kamarudin, M.R. Design of Phased Arrays of Series-Fed Patch Antennas with Reduced Number of the Controllers for 28-GHz mm-Wave Applications. *IEEE Antennas Wirel. Propag. Lett.* **2016**, *15*, 1305–1308. [CrossRef]
2. Khalily, M.; Tafazolli, R.; Xiao, P.; Kishk, A.A. Broadband mm-Wave Microstrip Array Antenna with Improved Radiation Characteristics for Different 5G Applications. *IEEE Trans. Antennas Propag.* **2018**, *66*, 4641–4647. [CrossRef]
3. Mao, C.-X.; Khalily, M.; Xiao, P.; Brown, T.W.C.; Gao, S. Planar Sub-Millimeter-Wave Array Antenna with Enhanced Gain and Reduced Sidelobes for 5G Broadcast Applications. *IEEE Trans. Antennas Propag.* **2019**, *67*, 160–168. [CrossRef]
4. Koul, S.K.; Karthikeya, G.S. *Millimetre Wave Antennas for 5G Mobile Terminals and Base Stations*, 1st ed.; CRC Press: Boca Raton, FL, USA, 2020. [CrossRef]
5. Liu, S.; Wei, Y.; Hwang, S.-H. Hybrid of Angular and Distance Protection for Coexistence of 5G Base Stations and Satellite Earth Stations. *Electronics* **2022**, *11*, 623. [CrossRef]
6. Farasat, M.; Thalakotuna, D.N.; Hu, Z.; Yang, Y. A Review on 5G Sub-6 GHz Base Station Antenna Design Challenges. *Electronics* **2021**, *10*, 2000. [CrossRef]
7. Wei, Y.; Liu, S.; Hwang, S.-H. Distance Protection for Coexistence of 5G Base Station and Satellite Earth Station. *Electronics* **2021**, *10*, 1481. [CrossRef]
8. Jiang, T.; Zhang, J.; Tang, P.; Tian, L.; Zheng, Y.; Dou, J.; Asplund, H.; Raschkowski, L.; D'Errico, R.; Jämsä, T. 3GPP Standardized 5G Channel Model for IIoT Scenarios: A Survey. *IEEE Internet Things J.* **2021**, *8*, 8799–8815. [CrossRef]
9. Aliakbari, H.; Abdipour, A.; Mirzavand, R.; Costanzo, A.; Mousavi, P. A single feed dual-band circularly polarized millimeter-wave antenna for 5G communication. In Proceedings of the 10th European Conference on Antennas and Propagation (EuCAP), Davos, Switzerland, 10–15 April 2016; pp. 1–5. [CrossRef]
10. Liu, C.; Guo, Y.; Bao, X.; Xiao, S.-Q. 60-GHz LTCC Integrated Circularly Polarized Helical Antenna Array. *IEEE Trans. Antennas Propag.* **2012**, *60*, 1329–1335. [CrossRef]
11. Zada, M.; Shah, I.A.; Yoo, H. Integration of Sub-6-GHz and mm-Wave Bands with a Large Frequency Ratio for Future 5G MIMO Applications. *IEEE Access* **2021**, *9*, 11241–11251. [CrossRef]
12. Islam, S.; Zada, M.; Yoo, H. Low-Pass Filter Based Integrated 5G Smartphone Antenna for Sub-6-GHz and mm-Wave Bands. *IEEE Trans. Antennas Propag.* **2021**, *69*, 5424–5436. [CrossRef]
13. Andrews, J.G.; Buzzi, S.; Choi, W.; Hanly, S.V.; Lozano, A.; Soong, A.C.; Zhang, J.C. What will 5 G Be? *IEEE J. Sel. Areas Commun.* **2014**, *32*, 1065–1082. [CrossRef]
14. Hong, W.; Baek, K.-H.; Lee, Y.; Kim, Y.; Ko, S.-T. Study and prototyping of practically large-scale mmWave antenna systems for 5G cellular devices. *IEEE Commun. Mag.* **2014**, *52*, 63–69. [CrossRef]
15. Aryanfar, F. Millimeter-wave beam forming as an enabling technology for 5 G cellular communications: Theoretical feasibility and prototype results. *IEEE Commun. Mag.* **2014**, *52*, 106–113.
16. Gandhi, O.; Riazi, A. Absorption of Millimeter Waves by Human Beings and its Biological Implications. *IEEE Trans. Microw. Theory Tech.* **1986**, *34*, 228–235. [CrossRef]
17. Yu, B.; Yang, K.; Sim, C.-Y.; Yang, G. A Novel 28 GHz Beam Steering Array for 5G Mobile Device with Metallic Casing Application. *IEEE Trans. Antennas Propag.* **2018**, *66*, 462–466. [CrossRef]
18. Ali, I.; Jamaluddin, M.H.; Gaya, A.; Rahim, H.A. A Dielectric Resonator Antenna with Enhanced Gain and Bandwidth for 5G Applications. *Sensors* **2020**, *20*, 675. [CrossRef]
19. Ishteyaq, I.; Masoodi, I.S.; Muzaffar, K. A compact double-band planar printed slot antenna for sub-6 GHz 5G wireless applications. *Int. J. Microw. Wirel. Technol.* **2020**, *13*, 469–477. [CrossRef]
20. Xu, J.; Hong, W.; Jiang, Z.H.; Zhang, H. Wideband, low-profile patch array antenna with corporate stacked micro strip and substrate integrated waveguide feeding structure. *IEEE Trans. Antennas Propag.* **2019**, *67*, 1368–1373. [CrossRef]
21. Masaud, S.; Cheema, H.M.; Abbasi, Q.H. Substrate Integrated Waveguide Antenna System for 5G In-Band Full Duplex Applications. *Electronics* **2021**, *10*, 2456.
22. Zhu, Q.; Bo, K.; Chi, N.; Chan, H.; Luk, K.-M. Substrate-Integrated-Waveguide-Fed Array Antenna Covering 57–71 GHz Band for 5G Applications. *IEEE Trans. Antennas Propag.* **2017**, *65*, 6298–6306. [CrossRef]
23. Yin, J.; Wu, Q.; Yu, C.; Wang, H.; Hong, W. Broadband symmetrica E-shaped patch antenna with multimode resonance for 5G millimeter-wave applications. *IEEE Trans. Antennas Propag.* **2019**, *67*, 4474–4483. [CrossRef]
24. Mohsen, M.K.; Isa, M.S.; Isa, A.A.; Abdulhameed, M.K.; Attiah, M.L.; Dinar, A.M. Enhancement bandwidth of half width-microstrip leaky wave antenna using circular slots. *Prog. Electromagn. Res.* **2019**, *94*, 59–74. [CrossRef]
25. Ullah, H.; Tahir, F.A. A Novel Snowflake Fractal Antenna for Dual-Beam Applications in 28 GHz Band. *IEEE Access* **2020**, *8*, 19873–19879. [CrossRef]
26. Rafique, U.; Din, S.; Khalil, H. Compact CPW-fed super wideband planar elliptical antenna. *Int. J. Microw. Wirel. Technol.* **2021**, *13*, 407–414. [CrossRef]
27. Shah, I.; Hayat, S.; Basir, A.; Zada, M.; Shah, S.; Ullah, S. Design and analysis of a hexa-band frequency reconfigurable antenna for wireless communication. *AEU-Int. J. Electron. Commun.* **2018**, *98*, 80–88. [CrossRef]
28. Im, C.; Lim, T.-H.; Jang, D.; Kong, N.-K.; Choo, H. Design of a Printed 5G Monopole Antenna on Vehicle Window Glass Using Parasitic Elements and a Lattice-Structure Reflector for Gain Enhancement. *Appl. Sci.* **2021**, *11*, 9953. [CrossRef]

29. Kiani, S.H.; Altaf, A.; Abdullah, M.; Muhammad, F.; Shoaib, N.; Anjum, M.R.; Damaševičius, R.; Blažauskas, T. Eight Element Side Edged Framed MIMO Antenna Array for Future 5G Smart Phones. *Micromachines* **2020**, *11*, 956. [CrossRef]
30. Abdullah, M.; Altaf, A.; Anjum, M.R.; Arain, Z.A.; Jamali, A.A.; Alibakhshikenari, M.; Falcone, F.; Limiti, E. Future Smartphone: MIMO Antenna System for 5G Mobile Terminals. *IEEE Access* **2021**, *9*, 91593–91603. [CrossRef]
31. Ahmad, I.; Sun, H.; Rafique, U.; Yi, Z. Triangular Slot-Loaded Wideband Planar Rectangular Antenna Array for Millimeter-Wave 5G Applications. *Electronics* **2021**, *10*, 778. [CrossRef]
32. Hussain, N.; Jeong, M.; Park, J.; Rhee, S.; Kim, P.; Kim, N. A compact size 2.9–23.5 GHz micro strip patch antenna with WLAN band-rejection. *Microw. Opt. Technol. Lett.* **2019**, *61*, 1307–1313. [CrossRef]
33. Awan, W.; Naqvi, S.; Ali, W.; Hussain, N.; Iqbal, A.; Tran, H.; Alibakhshikenari, M.; Limiti, E. Design and Realization of a Frequency Reconfigurable Antenna with Wide, Dual, and Single-Band Operations for Compact Sized Wireless Applications. *Electronics* **2021**, *10*, 1321. [CrossRef]
34. Haraz, O.M.; Elboushi, A.; Alshebeili, S.A.; Sebak, A.-R. Dense Dielectric Patch Array Antenna with Improved Radiation Characteristics Using EBG Ground Structure and Dielectric Superstrate for Future 5G Cellular Networks. *IEEE Access* **2014**, *2*, 909–913. [CrossRef]
35. Salucci, M.; Castlunger, C.; Marcantonio, D.; Oliveri, G.; Robol, F.; Rosatti, P.; Tosato, L.; Zardi, F.; Massa, A. High Density Interconnect Microstrip Patch Antenna for 5G Base Stations with Integrated Filtering Performance. *Technologies* **2018**, *6*, 45. [CrossRef]
36. Ikram, M.; Nguyen-Trong, N.; Abbosh, A.M. Common-Aperture Sub-6 GHz and Millimeter-Wave 5G Antenna System. *IEEE Access* **2020**, *8*, 199415–199423. [CrossRef]
37. Park, J.; Jeong, M.; Hussain, N.; Rhee, S.; Park, S.; Kim, N. A lowprofile high-gain filtering antenna for fifth generation systems based on nonuniform metasurface. *Microw. Opt. Technol. Lett.* **2019**, *61*, 2513–2519. [CrossRef]
38. Dixit, A.S.; Kumar, S. A Miniaturized Antipodal Vivaldi Antenna for 5G Communication Applications. In Proceedings of the 7th International Conference on Signal Processing and Integrated Networks (SPIN), Noida, India, 27–28 February 2020; pp. 800–803. [CrossRef]
39. Li, Y.; Zhao, Z.; Tang, Z.; Yin, Y. Differentially-Fed, Wideband Dual-Polarized Filtering Antenna with Novel Feeding Structure for 5G Sub-6 GHz Base Station Applications. *IEEE Access* **2019**, *7*, 184718–184725. [CrossRef]
40. Zarifi, D.; Farahbakhsh, A.; Zaman, A.U.; Kildal, P.-S. Design and Fabrication of a High-Gain 60-GHz Corrugated Slot Antenna Array with Ridge Gap Waveguide Distribution Layer. *IEEE Trans. Antennas Propag.* **2016**, *64*, 2905–2913. [CrossRef]
41. Chen, Z.; Liu, H.; Yu, J.; Chen, X. High Gain, Broadband and Dual-Polarized Substrate Integrated Waveguide Cavity-Backed Slot Antenna Array for 60 GHz Band. *IEEE Access* **2018**, *6*, 31012–31022. [CrossRef]
42. Li, Y.; Wang, J.; Luk, K.M. Millimeter-wave multi beam aperture-coupled magneto electric dipole array with planar substrate integrated beam forming network for 5G applications. *IEEE Trans. Antennas Propag.* **2017**, *65*, 6422–6431. [CrossRef]
43. Tian, Y.; Ouyang, J.; Hu, P.F.; Pan, Y. Millimeter-Wave Wideband Circularly Polarized End fire Planar Magneto-Electric Dipole Antenna Based on Substrate Integrated Waveguide. *IEEE Antennas Wirel. Propag. Lett.* **2021**, *21*, 49–53. [CrossRef]
44. Kumar, K.B.; Shanmuganantham, T. 3-Port T-junction SIW power divider for 60 GHz applications. In Proceedings of the 2017 IEEE International Conference on Antenna Innovations & Modern Technologies for Ground, Aircraft and Satellite Applications (iAIM), Bangalore, India, 24–26 November 2017; pp. 1–4.
45. Xi, L. A wideband planar filtering dipole antenna for 5G communication applications. *Microw. Opt. Technol. Lett.* **2019**, *61*, 2746–2751. [CrossRef]
46. Niu, Z.; Zhang, H.; Chen, Q.; Zhong, T. Isolation Enhancement for 1 × 3 Closely Spaced E-Plane Patch Antenna Array Using Defect Ground Structure and Metal-Vias. *IEEE Access* **2019**, *7*, 119375. [CrossRef]
47. Li, M.Y.; Ban, Y.L.; Xu, Z.Q.; Wu, G.; Sim, C.Y.D.; Kang, K.; Yu, Z.F. Eight-Port Orthogonally Dual-Polarized Antenna Array for 5G Smartphone Applications. *IEEE Trans. Antennas Propag.* **2016**, *64*, 3820–3830. [CrossRef]
48. Vallappil, A.K.; Rahim, M.K.A.; Khawaja, B.A.; Iqbal, M.N. Compact Metamaterial Based 4 × 4 Butler Matrix with Improved Bandwidth for 5G Applications. *IEEE Access* **2020**, *8*, 13573–13583. [CrossRef]
49. Ullah, H.; Tahir, F.A. Broadband planar antenna array for future 5G communication standards. *IET Microw. Antennas Propag.* **2019**, *13*, 2661–2668. [CrossRef]
50. Dzagbletey, P.A.; Jung, Y.B. Stacked micro strip linear array for millimeter-wave 5G baseband communication. *IEEE Antennas Wirel. Propag. Lett.* **2018**, *17*, 780–783. [CrossRef]
51. Liu, H.; Yang, W.; Zhang, A.; Zhu, S.; Wang, Z.; Huang, T. A Miniaturized Gain-Enhanced Antipodal Vivaldi Antenna and Its Array for 5G Communication Applications. *IEEE Access* **2018**, *6*, 76282–76288. [CrossRef]
52. Goel, T.; Patnaik, A. Novel Broadband Antennas for Future Mobile Communications. *IEEE Trans. Antennas Propag.* **2018**, *66*, 2299–2308. [CrossRef]
53. Dixit, A.S.; Kumar, S. A Survey of Performance Enhancement Techniques of Antipodal Vivaldi Antenna. *IEEE Access* **2020**, *8*, 45774–45796. [CrossRef]
54. Dadgarpour, A.; Sorkherizi, M.S.; Kishk, A.A. High-Efficient Circularly Polarized Magnetoelectric Dipole Antenna for 5G Applications Using Dual-Polarized Split-Ring Resonator Lens. *IEEE Trans. Antennas Propag.* **2017**, *65*, 4263–4267. [CrossRef]
55. Kim, E.; Ko, S.-T.; Lee, Y.J.; Oh, J. Millimeter-Wave Tiny Lens Antenna Employing U-Shaped Filter Arrays for 5G. *IEEE Antennas Wirel. Propag. Lett.* **2018**, *17*, 845–848. [CrossRef]

56. Alizadeh, F.; Ghobadi, C.; Nourinia, J.; Zayer, R. Band width enhancement of patch antennas loaded with complementary split-ring resonators. In Proceedings of the 7th International Symposium on Telecommunications (IST'2014), Tehran, Iran, 9–11 September 2014.
57. Tan, W.; Xiao, Y.; Li, C.; Zhu, K.; Luo, H.; Sun, H. A Wide-Band High-Efficiency Hybrid-Feed Antenna Array for mm-Wave Wireless Systems. *Electronics* **2021**, *10*, 2383. [CrossRef]
58. Griguer, H.; Marzolf, E.; Lalj, H.; Riouch, F.; Drissi, M. Patch antenna bandwidth enhancement through the use of meta materials. In Proceedings of the International Conference on Telecommunications, Marrakech, Morroco, 25–27 May 2009; pp. 323–327.
59. Li, T.; Chen, Z.N. Shared-Surface Dual-Band Antenna for 5G Applications. *IEEE Trans. Antennas Propag.* **2019**, *68*, 1128–1133. [CrossRef]
60. Hussain, N.; Jeong, M.; Abbas, A.; Kim, T.; Kim, N. A Meta surface-Based Low-Profile Wideband Circularly Polarized Patch Antenna for 5G Millimeter-Wave Systems. *IEEE Access* **2020**, *8*, 22127–22135. [CrossRef]
61. Hussain, N.; Jeong, M.; Abbas, A.; Kim, N. Metasurface-Based Single-Layer Wideband Circularly Polarized MIMO Antenna for 5G Millimeter-Wave Systems. *IEEE Access* **2020**, *8*, 130293–130304. [CrossRef]
62. Shahadan, N.H.; Jamaluddin, M.H.; Kamarudin, M.R.; Yamada, Y.; Khalily, M.; Jusoh, M.; Dahlan, S.H. Steerable Higher Order Mode Dielectric Resonator Antenna with Parasitic Elements for 5G Applications. *IEEE Access* **2017**, *5*, 22234–22243. [CrossRef]
63. Yassin, M.E.; Mohamed, H.A.; Abdallah, E.A.F.; El-Hennawy, H.S. Single-fed 4G/5G multiband 2.4/5.5/28 GHz antenna. *IET Microw. Antennas Propag.* **2019**, *13*, 286–290. [CrossRef]
64. Jiang, X.; Jia, F.; Cao, Y.; Huang, P.; Yu, J.; Wang, X.; Shi, Y. Ka-Band 8 × 8 Low-Side lobe Slot Antenna Array using a 1-to-64 High-Efficiency Network Designed by New Printed RGW Technology. *IEEE Antennas Wirel. Propag. Lett.* **2019**, *18*, 1248–1252. [CrossRef]
65. Liu, B.; Zhao, R.; Ma, Y.; Guo, Z.; Wei, X.; Xing, W.; Wang, Y. A 45° Linearly Polarized Slot Array Antenna with Substrate Integrated Coaxial Line Technique. *IEEE Antennas Wirel. Propag. Lett.* **2018**, *17*, 339–342. [CrossRef]
66. Li, Y.; Luk, K.M. A 60-GHz Wideband Circularly Polarized Aperture-Coupled Magneto-Electric Dipole Antenna Array. *IEEE Trans. Antennas Propag.* **2016**, *64*, 1325–1333. [CrossRef]
67. Li, Y.; Wang, C.; Guo, Y.X. A Ka-Band Wideband Dual-Polarized Magnetoelectric Dipole Antenna Array on LTCC. *IEEE Trans. Antennas Propag.* **2019**, *68*, 4985–4990. [CrossRef]
68. Tomura, T.; Hirokawa, J.; Hirano, T.; Ando, M. A 45 degrees Linearly Polarized Hollow-Waveguide 16 × 16-Slot Array Antenna Covering 71–86 GHz Band. *IEEE Trans. Antennas Propag.* **2014**, *62*, 5061–5067. [CrossRef]
69. Ferrando-Rocher, M.; Herranz-Herruzo, J.I.; Valero-Nogueira, A.; Vila-Jimenez, A. Single-Layer Circularly-Polarized Ka-Band Antenna Using Gap Waveguide Technology. *IEEE Trans. Antennas Propag.* **2018**, *66*, 3837–3845. [CrossRef]
70. Sorkherizi, M.S.; Dadgarpour, A.; Kishk, A. Planar High-efficiency Antenna Array Using New Printed Ridge Gap Waveguide Technology. *IEEE Trans. Antennas Propag.* **2017**, *65*, 3772–3776. [CrossRef]
71. Cao, J.; Wang, H.; Mou, S.; Soothar, P.; Zhou, J. An Air Cavity-Fed Circularly Polarized Magneto-Electric Dipole Antenna Array With Gap Waveguide Technology for mm-Wave Applications. *IEEE Trans. Antennas Propag.* **2019**, *67*, 6211–6216. [CrossRef]
72. Han, T.-Y.; Hsieh, Z.-K.; Lo, J.; Sim, C.-Y. A Pattern Reconfigurable Antenna Design for 5G Communication System. In Proceedings of the 2021 International Symposium on Antennas and Propagation (ISAP), Taipei, Taiwan, 19–22 October 2021; pp. 1–2. [CrossRef]
73. Ahmad, I.; Sun, H.; Zhang, Y.; Ali, Q. Low Profile, Compact Size Frequency Reconfigurable Antenna for 5G mm-Wave Wireless Communication. In Proceedings of the 5th International Conference on Computer and Communication Systems (ICCCS), Shanghai, China, 15–18 May 2020; pp. 712–716. [CrossRef]
74. Afzal, M.U.; Matekovits, L.; Esselle, K.P.; Lalbakhsh, A. Beam-Scanning Antenna Based on Near-Electric Field Phase Transformation and Refraction of Electromagnetic Wave through Dielectric Structures. *IEEE Access* **2020**, *8*, 199242–199253. [CrossRef]
75. Gonçalves, R.; Pinho, P.; Carvalho, N.B. Compact, frequency reconfigurable, printed monopole antenna. *Int. J. Antennas Propag.* **2012**, *2012*, 602780. [CrossRef]
76. Majid, H.A.; Rahim, M.K.A.; Hamid, M.R.; Ismail, M.F. Frequency reconfigurable micro strip patch-slot antenna with directional radiation pattern. *Prog. Electromagn. Res.* **2014**, *144*, 319–328. [CrossRef]
77. Li, T.; Zhai, H.; Wang, X.; Li, L.; Liang, C. Frequency-Reconfigurable Bow-Tie Antenna for Bluetooth, WiMAX, and WLAN Applications. *IEEE Antennas Wirel. Propag. Lett.* **2014**, *14*, 171–174. [CrossRef]
78. Ullah, S.; Hayat, S.; Umar, A.; Ali, U.; Tahir, F.A.; Flint, J.A. Design, fabrication and measurement of triple band frequency reconfigurable antennas for portable wireless communications. *AEU-Int. J. Electron. Commun.* **2017**, *81*, 236–242. [CrossRef]
79. Ali, W.A.E.; Mohamed, H.A.; Ibrahim, A.A.; Hamdalla, M.Z.M. Gain improvement of tunable band-notched UWB antenna using metamaterial lens for high speed wireless communications. *Microsyst. Technol.* **2019**, *25*, 4111–4117. [CrossRef]
80. Feng, S.; Zhang, L.; Yu, H.W.; Zhang, Y.X.; Jiao, Y.C. A SingleLayer Wideband Differential-Fed Microstrip Patch Antenna with Complementary Split-Ring Resonators Loaded. *IEEE Access* **2019**, *7*, 132041–132048. [CrossRef]
81. Hussain, S.; Qu, S.W.; Zhou, W.L.; Zhang, P.; Yang, S. Design and Fabrication of Wideband Dual-Polarized Dipole Array for 5G Wireless Systems. *IEEE Access* **2020**, *8*, 65155–65163. [CrossRef]
82. Parchin, N.O.; Al-Yasir, Y.; Abdulkhaleq, A.M.; Elfergani, I.; Rayit, A.; Noras, J.M.; Rodriguez, J.; Abd-Alhameed, R.A. Frequency reconfigurable antenna array for mm-wave 5G mobile handsets. In Proceedings of the International Conference on Broadband Communications, Networks and Systems, Faro, Portugal, 19–20 September 2018.

83. Rutschlin, M.; Sokol, V. Reconfigurable Antenna Simulation: Design of Reconfigurable Antennas with Electromagnetic Simulation. *IEEE Microw. Mag.* **2013**, *14*, 92–101. [CrossRef]
84. Thummaluru, S.R.; Kumar, R.; Chaudhary, R.K. Isolation and frequency reconfigurable compact MIMO antenna for WLAN applications. *Microw. Antennas Propag.* **2019**, *13*, 519–525. [CrossRef]
85. Ding, Z.; Jin, R.; Geng, J.; Zhu, W.; Liang, X. Varactor Loaded Pattern Reconfigurable Patch Antenna with Shorting Pins. *IEEE Trans. Antennas Propag.* **2019**, *67*, 6267–6277. [CrossRef]
86. Alazemi, A.J. A tunable single-feed triple-band LTE antenna with harmonic suppression. *IEEE Access* **2019**, *7*, 104667–104672. [CrossRef]
87. Ullah, S.; Yeo, W.-H.; Kim, H.; Yoo, H. Development of 60-GHz millimeter wave, electromagnetic bandgap ground planes for multiple-input multiple-output antenna applications. *Sci. Rep.* **2020**, *10*, 8541. [CrossRef]
88. Li, M.Y.; Xu, Z.Q.; Ban, Y.L.; Sim, C.Y.D.; Yu, Z.F. Eight-port orthogonally dual-polarised MIMO antennas using loop structures for 5G smartphone. *IET Microw. Antennas Propag.* **2017**, *11*, 1810–1816. [CrossRef]
89. Nadeem, I.; Choi, D.-Y. Study on Mutual Coupling Reduction Technique for MIMO Antennas. *IEEE Access* **2018**, *7*, 563–586. [CrossRef]
90. Sharawi, M.S.; Podilchak, S.K.; Khan, M.U.; Antar, Y.M. Dual-frequency DRA-based MIMO antenna system for wireless access points. *IET Microw. Antennas Propag.* **2017**, *11*, 1174–1182. [CrossRef]
91. Kowalewski, J.; Eisenbeis, J.; Jauch, A.; Mayer, J.; Kretschmann, M.; Zwick, T. A mmW Broadband Dual-Polarized Dielectric Resonator Antenna Based on Hybrid Modes. *IEEE Antennas Wirel. Propag. Lett.* **2020**, *19*, 1068–1072. [CrossRef]
92. Ali, W.A.; Ibrahim, A.A. A compact double-sided MIMO antenna with an improved isolation for UWB applications. *AEU-Int. J. Electron. Commun.* **2017**, *82*, 7–13. [CrossRef]
93. Jiang, W.; Liu, B.; Cui, Y.; Hu, W. High-Isolation Eight-Element MIMO Array for 5G Smartphone Applications. *IEEE Access* **2019**, *7*, 34104–34112. [CrossRef]
94. Liu, R.; An, X.; Zheng, H.; Wang, M.; Gao, Z.; Li, E. Neutralization Line Decoupling Tri-Band Multiple-Input Multiple-Output Antenna Design. *IEEE Access* **2020**, *8*, 27018–27026. [CrossRef]
95. Nie, Z.; Zhai, H.; Liu, L.; Li, J.; Hu, D.; Shi, J. A Dual-Polarized Frequency-Reconfigurable Low-Profile Antenna with Harmonic Suppression for 5G Application. *IEEE Antennas Wirel. Propag. Lett.* **2019**, *18*, 1228–1232. [CrossRef]
96. Ikram, M.; Nguyen-Trong, N.; Abbosh, A. A Simple Single-Layered Continuous Frequency and Polarization-Reconfigurable Patch Antenna Array. *IEEE Trans. Antennas Propag.* **2019**, *68*, 4991–4996. [CrossRef]
97. Jin, G.; Deng, C.; Xu, Y.; Yang, J.; Liao, S. Differential Frequency-Reconfigurable Antenna Based on Dipoles for Sub-6 GHz 5G and WLAN Applications. *IEEE Antennas Wirel. Propag. Lett.* **2020**, *19*, 472–476. [CrossRef]
98. Shen, X.; Liu, Y.; Zhao, L.; Huang, G.L.; Shi, X.; Huang, Q. A Miniaturized Micro strip Antenna Array at 5G Millimeter-Wave Band. *IEEE Antennas Wirel. Propag. Lett.* **2019**, *18*, 1671–1675. [CrossRef]
99. Zhang, S.; Syrytsin, I.; Pedersen, G.F. Compact Beam-Steerable Antenna Array With Two Passive Parasitic Elements for 5G Mobile Terminals at 28 GHz. *IEEE Trans. Antennas Propag.* **2018**, *66*, 5193–5203. [CrossRef]
100. Bilal, M.; Saleem, R.; Abbasi, H.H.; Shafique, M.F.; Brown, A.K. An FSS-Based Nonplanar Quad-Element UWB-MIMO Antenna System. *IEEE Antennas Wirel. Propag. Lett.* **2017**, *16*, 987–990. [CrossRef]
101. Jafri, S.I.; Saleem, R.; Shafique, M.F.; Brown, A.K. Compact reconfigurable multiple-input-multiple-output antenna for ultra-wideband applications. *IET Microw. Antennas Propag.* **2016**, *10*, 413–419. [CrossRef]
102. Khan, M.S.; Rigobello, F.; Ijaz, B.; Autizi, E.; Capobianco, A.D.; Shubair, R.; Khan, S.A. Compact 3-D eight elements UWB-MIMO array. *Microw. Opt. Technol. Lett.* **2018**, *60*, 1967–1971. [CrossRef]
103. Mathur, R.; Dwari, S. 8-port multibeam planar UWB-MIMO antenna with pattern and polarisation diversity. *IET Microw. Antennas Propag.* **2019**, *13*, 2297–2302. [CrossRef]
104. Hu, W.; Liu, X.; Gao, S.; Wen, L.-H.; Qian, L.; Feng, T.; Xu, R.; Fei, P.; Liu, Y. Dual-Band Ten-Element MIMO Array Based on Dual-Mode IFAs for 5G Terminal Applications. *IEEE Access* **2019**, *7*, 178476–178485. [CrossRef]
105. Li, M.-Y.; Ban, Y.-L.; Xu, Z.-Q.; Guo, J.; Yu, Z.-F. Tri-Polarized 12-Antenna MIMO Array for Future 5G Smartphone Applications. *IEEE Access* **2017**, *6*, 6160–6170. [CrossRef]
106. Chen, H.-D.; Tsai, Y.-C.; Sim, C.-Y.; Kuo, C. Antenna Array Design for Sub-6 GHz 5G NR Bands Metal-Frame Smartphone Applications. *IEEE Antennas Wirel. Propag. Lett.* **2020**, *19*, 1078–1082. [CrossRef]
107. Singh, A.; Saavedra, C.E. Wide-bandwidth Inverted-F stub fed hybrid loop antenna for 5G sub-6 GHz massive MIMO enabled handsets. *IET Microw. Antennas Propag.* **2020**, *14*, 677–683. [CrossRef]
108. Shabbir, T.; Saleem, R.; Al-Bawri, S.S.; Shafique, M.F.; Islam, M.T. Eight-Port Metamaterial Loaded UWB-MIMO Antenna System for 3D System-in-Package Applications. *IEEE Access* **2020**, *8*, 106982–106992. [CrossRef]
109. Shabbir, T.; Islam, M.T.; Al-Bawri, S.S.; Aldhaheri, R.W.; Alharbi, K.H.; Aljohani, A.J.; Saleem, R. 16-Port Non-Planar MIMO Antenna System With Near-Zero-Index (NZI) Metamaterial Decoupling Structure for 5G Applications. *IEEE Access* **2020**, *8*, 157946–157958. [CrossRef]
110. Iupikov, O.A.; Hallberg, W.; Maaskant, R.; Fager, C.; Rehammar, R.; Buisman, K.; Ivashina, M.V. A Dual-Fed PIFA Antenna Element with nonsymmetric Impedance Matrix for High-Efficiency Doherty Transmitters: Integrated Design and OTA-characterization. *IEEE Trans. Antennas Propagat.* **2020**, *68*, 21–32. [CrossRef]

111. Iupikov, O.A.; Perez-Cisneros, J.-R.; Meyer, P.; Akesson, D.; Maaskant, R.; Buisman, K.; Rehammar, R.; Fager, C.; Ivashina, M.V. A Cavity-Backed Patch Antenna with Distributed Multi-Port Feeding, Enabling Efficient Integration with Doherty Power Amplifier and Band-Pass Filter. *IEEE Trans. Antennas Propag.* **2021**, *69*, 4412–4422. [CrossRef]
112. Chou, H.-T.; Huang, H.-J. Multi-Level Subarray Modularization to Construct Hierarchical Beamforming Networks for Phased Array of Antennas with Low Complexity. *IEEE Trans. Antennas Propag.* **2017**, *65*, 5819–5828. [CrossRef]

micromachines

MDPI

Article

Design of Deployable Structures by Using Bistable Compliant Mechanisms

Tinghao Liu and **Guangbo Hao** *

Electrical and Electronic Engineering, School of Engineering and Architecture, University College Cork, T12 K8AF Cork, Ireland; 120220250@umail.ucc.ie
* Correspondence: g.hao@ucc.ie

Abstract: A deployable structure can significantly change its geometric shape by switching lattice configurations. Using compliant mechanisms as the lattice units can prevent wear and friction among multi-part mechanisms. This work presents two distinctive deployable structures based on a programmable compliant bistable lattice. Several novel parameters are introduced into the bistable mechanism to better control the behaviour of bistable mechanisms. By adjusting the defined geometry parameters, the programmable bistable lattices can be optimized for specific targets such as a larger deformation range or higher stability. The first structure is designed to perform 1D deployable movement. This structure consists of multi-series-connected bistable lattices. In order to explore the 3D bistable characteristic, a cylindrical deployable mechanism is designed based on the curved double tensural bistable lattice. The investigation of bistable lattices mainly involves four types of bistable mechanisms. These bistable mechanisms are obtained by dividing the long segment of traditional compliant bistable mechanisms into two equal parts and setting a series of angle data to them, respectively. The experiment and FEA simulation results confirm the feasibility of the compliant deployable structures.

Keywords: deployable structure; compliant mechanism; bistable mechanism; programmable behaviour

Citation: Liu, T.; Hao, G. Design of Deployable Structures by Using Bistable Compliant Mechanisms. *Micromachines* **2022**, *13*, 651. https://doi.org/10.3390/mi13050651

Academic Editor: Stelios K. Georgantzinos

Received: 10 March 2022
Accepted: 15 April 2022
Published: 19 April 2022

1. Introduction

In recent years, deployable structures have been widely used in many fields, including but not limited to intelligent aerospace, architecture and the medical field due to the enormous dimensional deformability [1–3]. A novel type of closed-looped deployable structure was developed as a supporting frame for the large-diameter antenna in space engineering [4]. A negative pressure room based on deployable structures can effectively reduce the spread of the virus while ensuring the air circulation [5]. In addition, the Wren parallel mechanisms, which undergo one degree-of-freedom (DOF) Borel–Bricard motion, are designed as lattices of deployable structures as well [6].

This paragraph lists four common deployable mechanisms, which consist of different lattices [7]. The first category is scissor-like deployable structures. The deployable structures, comprised of scissor-like elements (SLEs), are widely applied because of their high loading capacity and excellent equilibrium stability [8–10]. The second type of deployable structure is the tensegrity structure, where tensional and compressional members co-exist. These members are connected to form a stable system [11]. In addition, another branch of deployable structures is derived from origami structures. These structures typically consist of thin membranes and slender elements [12–14]. In this list, the last approach to making structures deployable is to use the compliant multi-stable mechanisms as the lattices.

Compliant mechanisms can transfer or transform motion, force and energy via the deformation of flexible members [15]. Therefore, using compliant mechanisms can reduce costs and part-counts, save assembly time and also simplify manufacturing processes [16–18]. The wear among multi-part mechanisms can be avoided as well. Another benefit of utilizing compliant mechanisms is increasing the performance. The working precision of mechanisms can be significantly increased if replacing the traditional hinged mechanisms with compliant mechanisms [19–21]. The compliant mechanisms are demanded in many fields that require higher precision. A compact large-range XY compliant parallel manipulator was designed based on the characteristic of high performance [22]. In addition to the high performance, the dynamic characteristic of compliant mechanisms attracts considerable attention due to their high reliability and wide range of applications.

A compliant bistable mechanism has two stable equilibrium positions. The bistable mechanisms will tend to one of two positions if no external forces are acting on it [23]. Each stable position represents the local potential energy minimum, called the potential energy wells. A specific amount of external energy is required if the compliant bistable mechanism jumps from one stable position to another. This energy is called an energy barrier between two potential wells [24]. Additionally, there is a certain distance between two equilibrium positions. This distance can change the overall dimensions of the structure, accordingly making the structure deployable. In addition, this distance determines the deformation ability of the deployable structure.

Compliant bistable mechanisms can be categorized into two groups: spatial- and planar-compliant bistable mechanisms. The representative of the spatial bistable mechanism is the composite Laminates structure and torsion structure. The composite Laminates can change their configurations between saddle shape and cylindrical shape, thus getting its bistable characteristic [25]. Another tunable bistable component using shape memory polymers can perform twisting and rotational bistability [26]. Compared with spatial mechanisms, the planar mechanisms have the advantage of easy manufacturing and lower cost. The planar-compliant bistable mechanisms are primarily composed of straight and curved compliant segments [27–29]. The compliant mechanisms with straight segments have been applied extensively. A miniature latching accelerometer that does not require electrical power was designed based on compliant bistable mechanisms with straight segments [30]. When the load applied to this miniature accelerometer reaches a certain threshold, the structure will switch to a second stable state and lock. The advantage of using compliant mechanisms with straight segments is that the manufacturing process is more straightforward, and geometry parameters are easier to control. In addition to the straight segments, two curved centrally clamped parallel segments are introduced into the compliant bistable mechanisms as well [31]. This research proved the feasibility of using curved beams as segments of compliant bistable mechanisms.

Several indicators are used to measure the bistability of bistable structures, such as the critical forces and second stable position. These indicators are mainly proposed based on the force–displacement relationship. When reviewing the existed structures, it is found that these indicators of the bistable mechanism are almost impossible to adjust individually. These indicators will be influenced mutually even if changing only one geometry parameter. However, some design targets aim to optimize an individual indicator without affecting other indicators. For traditional bistable mechanisms, it is difficult to control the bistable behaviour because of the limited amount of geometry parameters. Hence, a novel analysis method is necessary. We introduced several new parameters to control the bistable behaviour using the generalized bistable model. Then, these bistable indicators can be better controlled by adjusting the geometric parameters. This work investigated how each geometry parameter influences bistable behaviour.

The aim of this work mainly involves designing deployable structures based on a programmable compliant bistable lattice with straight segments. The programmable feature refers to the highly controllable ability of bistable behaviour. The programmable feature is obtained by using the generalized bistable model. The structures are designed as both planar and spatial. Firstly, the novel type of 1D bistable lattice is proposed. The bistable lattice is tessellated into a plane to investigate the multi-stable characteristics in a single dimension. Subsequently, the lattice is curved to fit the surface of the cylinder to investigate whether the bistable characteristic will disappear when bending the lattice.

This paper is organized as below. Section 2 describes the methodologies of designing a bistable lattice with programmable behaviour. The bistable mechanism is designed from a novel perspective. The segments of the traditional compliant bistable mechanism are equally divided into two segments. The segments are defined with a series of angles. After the finite element analysis, four different types of bistable lattices are summarized. Then, the bistable lattice is optimized by revising the geometric parameters, such as the thicknesses and length of segments. The objective of optimizing the mechanism is designing a bistable lattice with larger deformation range. Section 3 mainly introduces several planar and spatial tessellations both in spatial and planar bistable lattices. Lastly, conclusions are finally drawn in Section 4.

2. Design of a Bistable Lattice with Programmable Behaviour

The main content of this section is to propose a method of designing a proper bistable lattice. The target bistable feature depends on the specific design aim. For instance, designing nonexplosive release mechanisms aims to make more potential energy be stored in the deformed segments [32]. Accordingly, the release mechanism can perform with higher reliability. The expected result in this work is to maximise the distance between two stable configurations while trying to increase its stability in a second stable position. Firstly, the traditional compliant bistable mechanism is introduced, and the bistable curve is also explained. Understanding the principle of the traditional bistable mechanism can contribute to the innovation of the designing method. Furthermore, four types of compliant mechanisms with compresural or tensural segments are listed by modifying geometry shapes. The double-tensural compliant mechanism is chosen to be the final mechanism. After determining the shape, the geometric parameters are analysed to optimize the bistable mechanism. Finally, the planar bistable lattice is designed and prototyped. In the experimental result, the force–displacement relationship proves that the planar compliant mechanism has the bistable feature.

The traditional compliant bistable mechanism can be seen in Figure 1a. Two compliant segments under the constraint of the frame are connected to the centre shuttle. When applying a displacement input to the shuttle, the segments will bend. Due to the symmetry characteristic, the shuttle will move along the centre line, which is just like moving in a guide. The bistable curve, which is shown in Figure 1b, can represent the typical force–displacement relationship of the bistable mechanism. The displacement u is defined as the distance that the shuttle has moved along its centre axis. The positive direction of displacement u is the same as the input direction. The reaction force F represents the reaction force of the shuttle along the centre axis. Some points in the bistable curve are defined to explain their significance. The diagram starts from the original point, i.e., the first stable point. No external force or displacement is input to the system at that status. Therefore, the structure is in stable status at that position. When the displacement increases, the reaction force will become larger. This process lasts until the first extreme point, called the first critical point. At that point, the displacement and reaction force are defined as u_{c1} and F_{c1}, respectively. If we continue increasing the displacement u, the bistable mechanism will arrive at its snap-through point u_s. When the displacement of the shuttle reaches the snap-through position, the shuttle starts to act opposite to the first stable point and quickly shifts to the second stable configuration. The displacement of the second critical point is defined as u_{c2}. Correspondingly, the reaction force at the second critical point is F_{c2}. If the second critical force F_{c2} is above zero, the structure has only one equilibrium position. The second stable position is located at u_2. The structure can keep its configuration at that position without adding any external force. The parametric analysis is mainly based on the above-defined parameters since they represent the performance of bistable mechanisms.

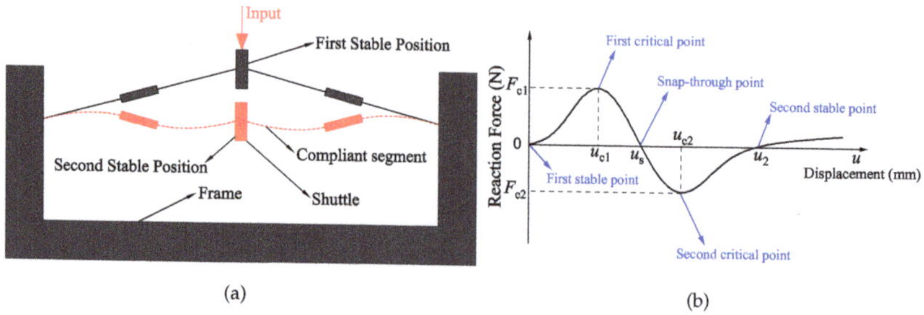

Figure 1. Traditional compliant bistable mechanism: (**a**) the design; (**b**) the force–displacement curve.

A novel method of designing a compliant bistable lattice is proposed in this work. The schematic of the generalized compliant bistable mechanism can be seen in Figure 2. The traditional compliant segments are divided into two equal parts. These two parts are fixed by a rigid connection. The aim is to investigate whether the segments will perform bistable behaviour when changing the angle of each half-segment. In this method, the middle point of each half-segment is fixed with a relative location. The relative position of the two middle points (Mp_1 and Mp_2) is defined by two parameters, i.e., γ and d. The half-segments are set in a series of angles. The angle range varies from 0° to 360° to make an investigation with relatively more extensive coverage. The angle intervals are initially set as 60° to explore the existence of bistable features. These structures are designed in three layers. The middle layer is the compliant half-segments. The problem of interference, when changing the angles of half-segments, is solved by adding out-plane rigid connections. These connections are designed with enough rigidity to resist the inner force. In addition, the deformation of these connections is small enough to be ignored.

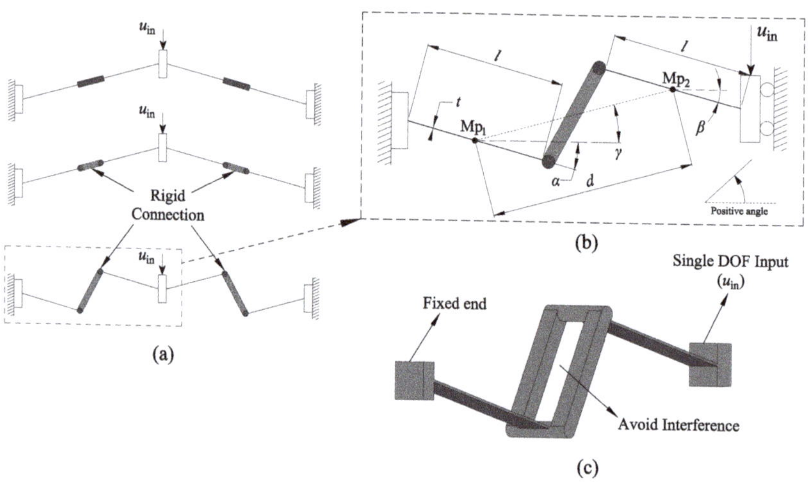

Figure 2. The generalized compliant bistable mechanism: (**a**) the derivation process from traditional compliant bistable mechanism to generalized mechanism; (**b**) The defined parameters; (**c**) The multi-layer analysis model.

Four types of compliant bistable mechanisms are derived from the above generalized model. They can be obtained from Figure 3. Each of them consists of compresural or tensural half-segments. The compresural segment refers to the compliant part that will be under compressive stress during the flexure. The naming of tensural segments is based on the same principle as compresural segments [33,34]. Type one is the traditional compliant bistable mechanism. Both of the two half-segments are under compresural stress when displacement input exists. The second type of bistable mechanism has both compresural and tensural half-segments. The compresural half-segment, which is fixed to the frame, is placed on the left-hand side. The tensural half-segment is connected to the shuttle. The shuttle is represented by a roller on a guide in the schematic diagram due to its symmetric feature. After that, another bistable mechanism, whose compresural half-segment is fixed to the frame, is proposed. The tensural half-segment of this mechanism is connected to the shuttle. The mechanical properties of a compliant compresural–tensural bistable mechanism have been investigated by previous researchers [34]. The last type of compliant bistable mechanism is designed with a double tensural half-segment. The angle parameters of these compliant bistable mechanisms are shown in Table 1. This work mainly focuses on the bistable mechanisms with double tensural segments.

Table 1. The angle parameters of compliant bistable mechanisms.

	Type I	Type II	Type III	Type IV
α	15°	0°	180°	180°
β	15°	180°	0°	180°
γ	15°	15°	15°	15°

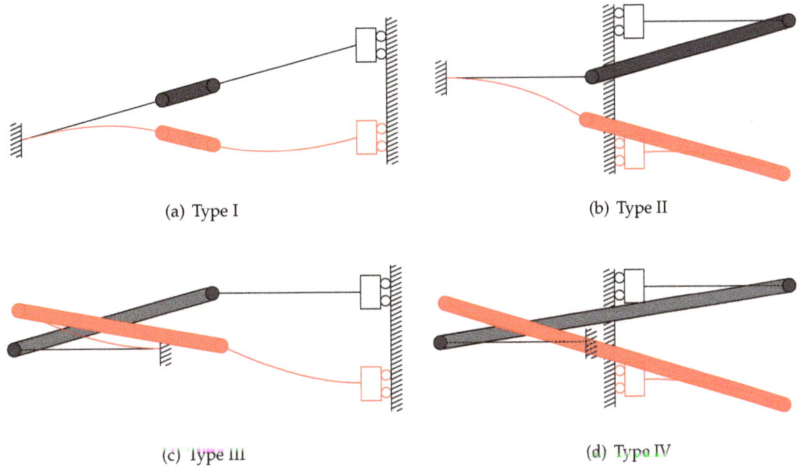

Figure 3. Compliant bistable mechanisms derived from the generalized model when the parameters α and β are set at different angles. These mechanisms are designed based on the multi-layer analysis model. The left half-segments are anchored to the frame, and the right half-segments are connected to the shuttle: (**a**) compliant compresural–compresural bistable mechanism (traditional compliant bistable mechanism); (**b**) compliant tensural–compresural bistable mechanism; (**c**) compliant compresural–tensural bistable mechanism; (**d**) compliant tensural–tensural bistable mechanism.

Finite element methods are used to analyse static or dynamic objects and systems. This method decomposes an object or system into a geometric model composed of multiple interconnected, simple, independent points. Linear static analysis is one type of method in which the displacement and force applied to the structure have a linear relationship. The stress is regarded as still in the elastic range during the linear analysis process, and its stiffness matrix is constant. However, the linear static analysis is insufficient for the mechanism with large deformation since it cannot reflect the actual relationships between force and displacement. Hence, in this paper, nonlinear FEA is used to analyse the bistable behaviour of lattices. The software used for FEA analysis in this paper is Strand7. During the simulation, a square mesh is applied to subdivide the structure. The maximum edge length of each mesh is set as 3% of the overall geometry size. Both geometry and material nonlinearity are considered. The Strand7 nonlinear static solver uses an algorithm based on the generalized Newton–Raphson method. This algorithm has the following features:

- The load increments can be defined easily.
- Multiple freedom cases can be used.
- The stiffness matrix need not be updated in every iteration.
- Restart can be used to continue a previously completed or aborted solution.

The two interrelated parameters α and β are both 180° in the preliminary analysis process. In order to perform the analysis in higher resolution, the angle interval is decreased to 5°. The range of angles varies from 165° to 195°. The reaction forces at the second critical point and the displacement of the second stable points are listed in Figure 4. In Figure 4a, these values represent the magnitudes of the forces at the second critical point. The maximum force located at the point where $\alpha = \beta = 185°$. It is found that in Figure 4b, the maximum displacement of the second stable status also happens in that angle pair. Hence, the angles α and β are decided to be 185° to get a larger deformation range and higher stability in the second stable status. A larger force at the second critical point means the mechanism is more stable in the second equilibrium position. Increasing the stroke of the shuttle of the deployable lattice can expand the range of deformation. Hence, these two parameters are the key points of designing a bistable lattice. After the determination of α and β, the next step of designing is to determine the other four parameters: angle γ, distance d, segment length l and segment thickness t.

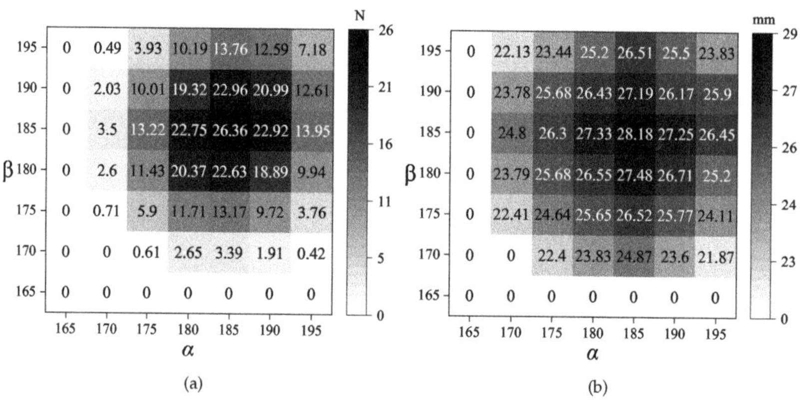

Figure 4. The switch-back forces and second stable positions when α and β are close to 180°. When the switch-back force is 0, it represents the structure is absent of bistable characteristics: (**a**) the forces at the second critical points; (**b**) the displacements at the second stable points.

The first parameter that needs to be investigated is the angle γ, which is defined between the line connecting two middle points and the horizontal line. The γ is set in a series of angles, varying from 5° to 25°. As can be seen in Figure 5a, the forces at the first critical point F_{c1} of the bistable mechanism increase with the increment of angle γ. For example, when the angle is set as 25°, the force F_{c1} is around 150 N. The magnitude of the force F_{c1} represents the resistance of the changing configuration. In addition, the distance between two configurations (u_2) will be larger when increasing the angle γ. However, if the switch force is too big, it will be difficult for the mechanism to change its configuration. In terms of designing the deployable structure in this work, the relatively large forces at critical points are not necessary. Hence, the angle γ is finally decided to be 25°.

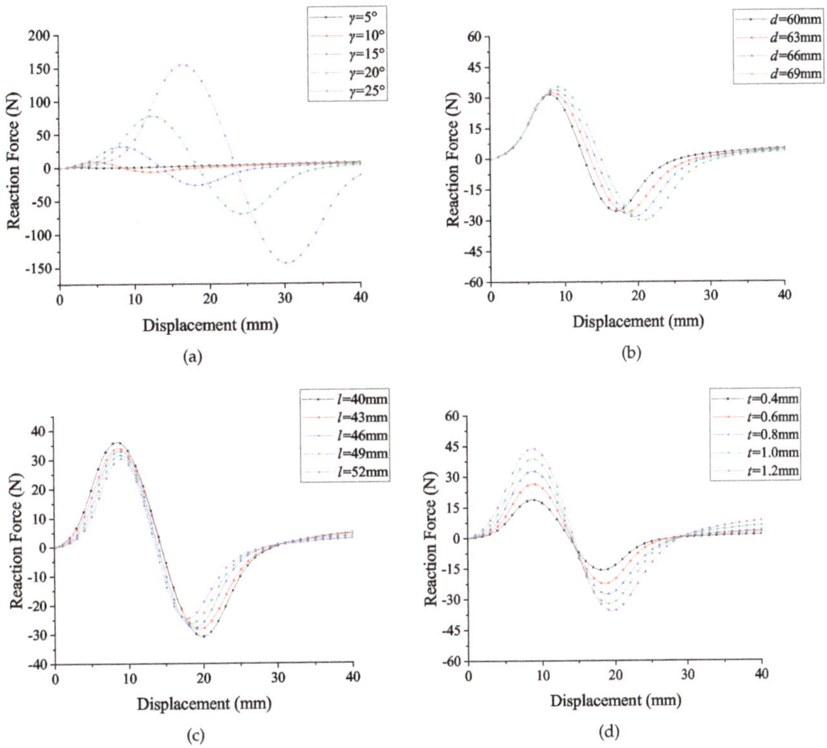

Figure 5. The bistable curves under different variables: (**a**) the angle γ as variable; (**b**) the distance between middle points d as variable; (**c**) the segment length l as variable; (**d**) the thickness of segments t as variable.

The following parameter that needs to be determined in the design process is the distance d between two middle points. The FEA result is shown in Figure 5b, the distance is set in four variables that vary from 60 mm to 69 mm. The change of d causes a tiny influence on the magnitude of the forces at the critical points. When the distance d becomes larger, the magnitude of the force F_{c2} will increase. Moreover, the position of these two critical points, u_{c1} and u_{c2}, are influenced as well. The position u_{c2} will tend to increase when setting a longer point distance. The switch position u_{c1} will increase as well, although the magnitude of change is not as large as u_{c2}. If the distance d is designed to be too large, it will affect the overall size of the compliant bistable lattice. The overall size of the bistable lattice is limited to the manufacturing feasibility. In contrast, the distance d has little effect on the bistable behaviour. Hence, the distance d is determined as 63mm, which is a moderate value in the range.

Conversely, in terms of the segment length l, the reaction forces at the critical points will become smaller when increasing the length. The maximum switch force in Figure 5c occurs when setting 40 mm as the segment length, which is the shortest segment in that range. The position of the second equilibrium point u_2 keeps stable when changing the lengths of segments. Changing the segment length has almost no influence on the equilibrium positions. The valid range of the segment lengths is limited by the geometric shape. The segment length is determined as 46 mm since the reaction forces are moderate. In addition to the segment length, if the thickness of segments t is set as the variable during the analysis process, the variation of thickness will only affect the magnitude of forces. The position of the critical points will not change as well. Another consideration is that the thickness of segments (t) is limited to the manufacturing precision. A very thin segment is difficult to manufacture, especially for CNC manufacturing methods, since the precision of thin membranes will be influenced by the milling or cutting stress. The thickness of the segments t highly depends on the manufacturing quality. Therefore, increasing the thickness of segments contributes to easier manufacturing. The thicknesses of segments are determined as 1.2 mm to maximise the feasibility of prototyping.

In summary, when changing the previous defined geometry parameters, the bistable curve will be affected to some extent. The adjustment of the bistable curve is mainly based on changing these parameters. And the design purpose determines which specific geometry should be adjusted. The adjustment method depends on the particular demand of adjustment. If the only need is changing the magnitude of forces, the thickness of segments (t) might be the only parameter that needs to be adjusted. If both the second stable positions and force magnitudes need to be adjusted, changing the angle (γ) is a good solution.

Through the above process of adjusting the geometry parameters, the final planar-compliant bistable lattice is determined. The parameters that are used in the planar mechanism can be found in Table 2. The rigid connections in the multi-layer analysis model are replaced by another type of connection. The bistable lattice is designed in plane.

Table 2. The determined geometry parameters of the planar bistable lattice.

α	β	γ	d	l	t
185°	185°	25°	63 mm	46 mm	1.2 mm

The planar bistable lattice is shown in Figure 6a. The final determined lattice is simulated by using FEA software. The mesh result and deformation result is shown in Figure 6b. A triangular mesh is selected since it is better fit to the curve profiles. In this design, the h_0 is designed as 210 mm, the shrunken length of the lattice Δh is 42.61 mm. The lattice is prototyped in half scale due to the limitations of the 3D printer. The 3D printer used is an Ultimaker S5. The build volume of this 3D printer is 330$\times$ 240 $\times$ 300 mm. The 3D printing software used to slice the model is Ultimaker Cura. During the prototyping process, the print core used in the printer is AA 0.4. The infill of the prototype is set as 100% to achieve the best performance. The material used for prototyping is Polylactide (PLA). The material properties are shown in Table 3. The centre shuttle of the 3D manufactured model can be pushed to its second stable configuration. The two distinct configurations can be seen in Figure 6d.

Table 3. The mechanical properties of Polylactide (PLA).

Property	**Tensile Modulus**	**Flexural Moudulus**	**Yield Tensile Stress**	**Yield Elongation**
Value	2346.5 MPa	3150.0 MPa	49.5 MPa	3.3%

A compress experiment is performed in this work. The comparison between the experimental result and the FEA simulation result is shown in Figure 6c. The setup of the experiment is shown in Figure 6e. The bistable lattice is fixed by two fixtures on the above and bottom, respectively. The shuttle moves along the direction of input displacement. This experiment was performed on the texture analyser (TA. Hd plus texture static test system), which is a load–displacement measurement system that moves in one single direction to compress or stretch the mechanism. A load cell of 5 (kg) was selected with a force resolution of 0.1 (g). The displacement was applied as the input to the system. The upper fixture moved downward at a shallow speed (0.1mm/s) to minimize the dynamic effect. Due to the limitation of devices, the only experiment performed in this work was investigating the bistable behaviour of planar mechanism. As can be seen in the comparison, some error exists. The error might be caused by the following reasons:

- The orthotropic functional properties of additively manufacturing.

- The uneven thickness of the segments.
- The frame lacks rigidity.
- The plastic deformation during the deformation process.

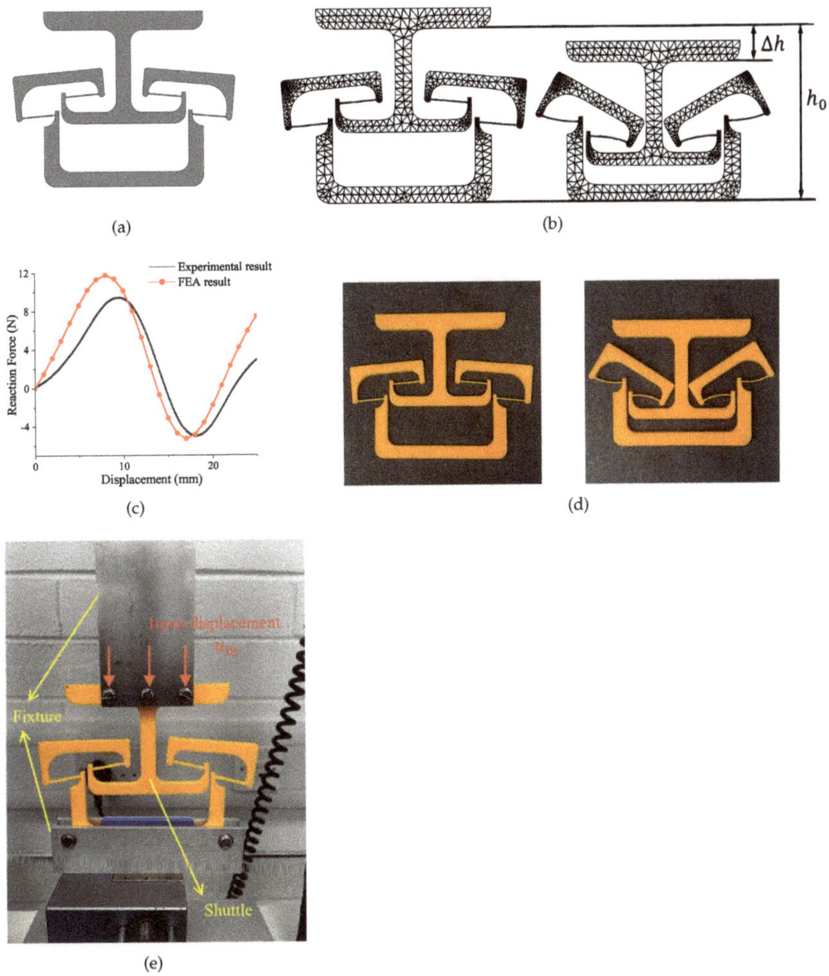

Figure 6. The planar compliant bistable mechanism: (**a**) the design of planar bistable mechanism; (**b**) the deformation result of FEA simulation; (**c**) the comparison between FEA simulation and experiment; (**d**) the 3D-printed prototype using PLA as material; (**e**) the experimental setup.

A comparison between the generalized planar-compliant bistable mechanism with double tensural segments and the traditional bistable mechanism is performed in this work. The comparison results are shown in Figure 7. These two mechanisms are designed with the same parameters except the angle α and β. As can be seen in Figure 7a, the generalized mechanism has the benefit of a relatively larger distance between two stable configurations. By utilizing the generalized mechanism as the lattice of deployable structure, the structure can increase its deployment range. In addition, the critical point forces F_{c1} and F_{c2} of the generalized mechanism are larger. This advantage can help to increase the stability of the deployment. When designing a deployable structure, consideration must be given to the ability of the structure to resist shock and load. The harder the switching process performs, the more energy it takes for the lattice to change its configuration. In other words, the structure has a higher ability to resist shock and load in its first and second stable state. Therefore, the larger these forces are, the more stable the deployable structure is. In Figure 7b, the results show that the generalized mechanism stored more potential energy in its second stable status compared to the traditional bistable mechanism. This benefit can increase the efficiency of energy conversion and harvest. In addition, a ratio of potential energy is introduced to the analysis process which is shown in Figure 7c. This ratio is defined as the result of the magnitudes of potential energy divided by the local maximum energy during the deformation process. As mentioned previously, the energy well is located at the position of the second stable point. At this position, the potential energy arrives at the position of local minimum potential energy. Therefore, in the position of the energy well, if there is no external force adding to the mechanism, the structure will keep in its stable status. The advances of the generalized mechanism are summarized as follows:

- Larger deformation range between two stable configurations.
- Better stability before/after deployment.
- More potential energy stored in the second stable status.

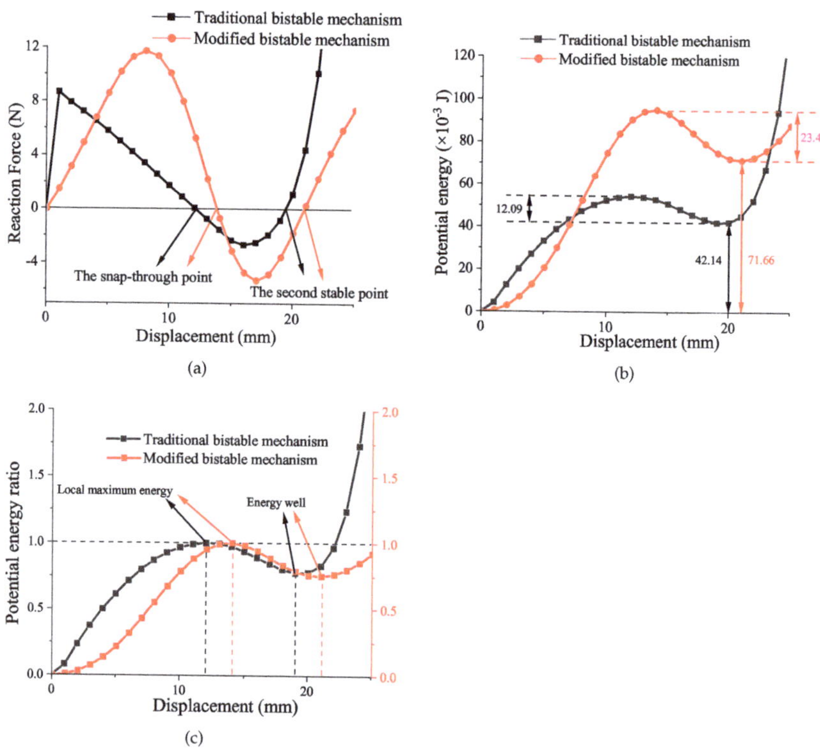

Figure 7. The comparison between the generalized bsitable mechanism and the traditional bistable mechanism: (**a**) the relationship between reaction forces and displacement; (**b**) the potential energy of mechanisms; (**c**) the ratio of potential energy compared to its local maximum energy.

3. Tessellated Deployable Structure

In the designing process of a structure with multi lattices, the tessellation method has various types. This section mainly discusses two modes of tessellating the bistable lattices into 1D and cylindrical configurations. In the 1D configuration, the bistable lattice is series connected to derive a multistable deployable structure. In the cylindrical configuration, the lattices are bent to form a cylindrical structure to investigate whether the bistable characteristic will disappear when using the curved bistable lattice.

3.1. 1D Deployable Structure

A multi-stable mechanism can be derived from tessellating the bistable lattices. In this part, the lattice is designed to be fitted into a single dimension structure. As mentioned previously, the shuttle of the bistable lattice moves along its centre axis. A theoretical multi-stable 1D deployable structure is obtained by overlapping the bistable lattice centre line. The series-connected mechanism can be seen in Figure 8. As can be seen, there are n lattices series-connected to form an entire 1D deployable structure. The overall height of the mechanism before deployment is $n \cdot h_0$. The deformation range of this 1D structure is $n \cdot \Delta h$.

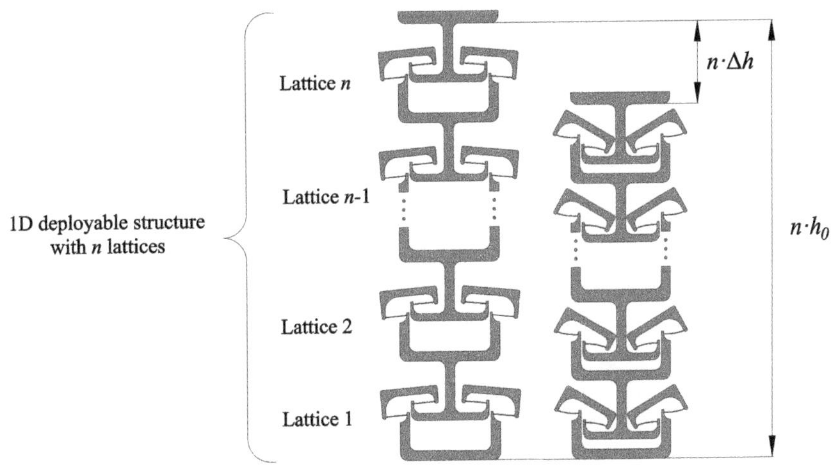

Figure 8. The multi-stable mechanism derived by tessellating bistable lattices.

In this work, the key content of the investigation is the three-connected multi-stable mechanism. This is because three lattices connected can perform a more generalized deformation compared to the two-connected structures. Each lattice of the deployable structure has two distinct stable configurations. The first stable position of them is when there is no deformation.

To better explain the process of the deformation, several parameters are introduced to the three-connected structure. The definition of these parameters can be seen in Figure 9. The FEA simulation operates based on a constant input displacement on the top of lattice three. The input displacement is named as u_{in}. In this mode, the displacement u_{in} increases linearly. The deformation u_{in} is relative to the base which is fixed. To better explore the deformation process of each lattice, the deformation distance Δh_n is introduced to the model (n is the order of lattices). The deformation distance Δh_n is relative to the corresponding coordinate system $O_n - x_n y_n$. The coordinate system is fixed on the frame centre of each lattice. In addition, the magnitude of u_{in} is equal to the sum of Δh_n.

The determination of multi-stable characteristics is based on measuring the relative displacement between frame and centre shuttle. When the structure is under load, the bistable lattice will change its configuration when the displacement of the shuttle reach and pass the snap-through point. It means that when the displacement u reaches the snap-through position u_s, the bistable lattice will change its configuration to the second stable equilibrium position. In addition, these lattices are series-connected but do not change their configuration at the same time. The order of change depends on the specific load situation.

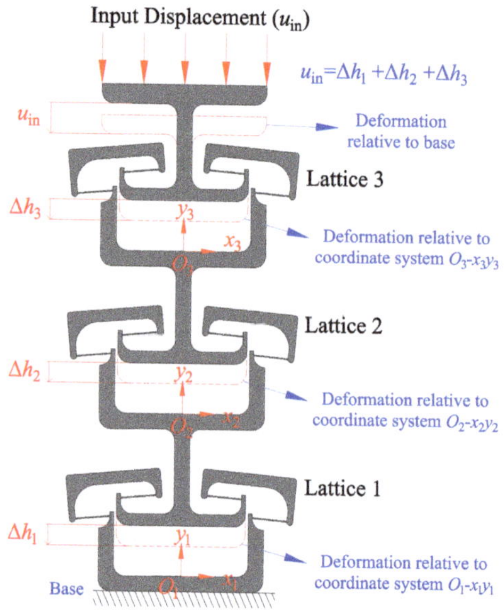

Figure 9. The parameters defined to describe the process of deformation.

As can be seen from the single bistable curve in Figure 1b, when the input displacement is before the snap-through point, the elastic deformation of the structure produces a force opposite the direction of motion. As the displacement increases, the reactive force of the structure tends to bring it back to its initial position. However, after passing the snap-through point, the reaction force of the structure becomes negative. The force after this point promotes the structure to switch to the next stable state. Without constraints, this reactive force would cause the displacement of shuttles to abruptly switch to a second stable state. For multiple series-connected bistable structures, when only the uppermost end is loaded with an overall input displacement, the relative position between the shuttle and the frame is not constrained. When a bistable structure switches stable configurations, its shuttle must pass through its snap-through position. The most intuitive phenomenon of the switching configuration of a bistable lattice is that there is a mutation in displacement and will not return to the first stable state. Therefore, we judge the sequence of lattice switching configuration in the structure by observing the sequence of generating displacement mutation. If three springs are connected in series, and the same load is applied to them, the springs will deform uniformly and linearly because the stiffness of the springs is constant. However, for the series-connected multistable mechanisms proposed in this paper, their nonlinear behaviour in the force and displacement relationship results in them not moving uniformly like the springs. When the overall displacement u_{in} is input, each lattice does not produce a uniform displacement because the force between them makes it difficult to achieve equilibrium.

In Figure 10, the overall displacement of the three-connected structure is set as increasing at a constant rate. The sum displacement of these three lattices is equal to the overall displacement. As mentioned, when the displacement of each shuttle is larger than u_s, the lattice will change to its second stable configuration. Due to the restraint of displacement, when a lattice changes its configuration, the other two lattices will also be influenced. The figure marks the snap-through position as a red dashed line. When the displacement Δh_n is larger than the switch position, the lattice will change its configuration to the second stable status. Based on the FEA simulation result, the first lattice that changes to the second stable status is Lattice 2. From the figure, the first intersection between lines happened in the field of Lattice 2. The second lattice that changes its configuration is Lattice 1. The last one is the

Lattice 3. The displacement result shows that three lattices change their configuration in total. Therefore, the result proves that the structure in this situation has four stable equilibrium status. The above diagram proves that the structure with three series-connected bistable lattices has four distinct equilibrium positions. When the displacement is set as input, lattices will affect each other during the configuration change process.

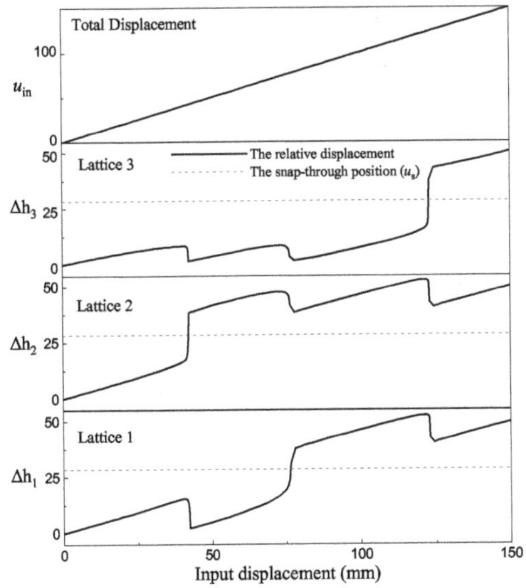

Figure 10. The displacement of each element under the load of input displacement.

3.2. *Cylindrical Deployable Structure*

Designing the cylindrical structure provides a possible solution to the cylindrical bistable mechanism with programmable behaviour. Compared with the cylindrical bistable mechanism composed of traditional bistable lattice, this generalized mechanism can better be targeted optimized due to several more geometric parameters being introduced. Hence, the method of optimizing the bistable behaviour can be applied to the cylindrical structure as well. Therefore, the proposal of cylindrical structure is not only designing a bistable structure but also introducing the programmable characteristic into it. This part proposes a cylindrical deployable structure based on the curved programmable bistable lattice. If the cylindrical deployable lattice is designed in a multi-layer structure, it will become a long telescopic tube with no diameter change during the deformation. This mechanism can be used as an engagement aid for transport pipelines.

The structure and its second stable configuration can be seen in Figure 11a. The cylindrical structure is prototyped by 3D printing technology as well. Then, an FEA simulation was performed. The result of the force–displacement relationship is shown in Figure 11b. This curve has three intersection points with the horizon line. This curve meets all the features of the bistable curve. Hence, the cylindrical structure has bistable characteristics. In addition, the 3D model can be found in Figure 11c. The 3D model proves that the structure has two stable configurations, i.e., it will keep in two stable statuses when there is no force acting on it. Therefore, this result proves that the bistable characteristic will not disappear when bending the lattices into a cylinder shape.

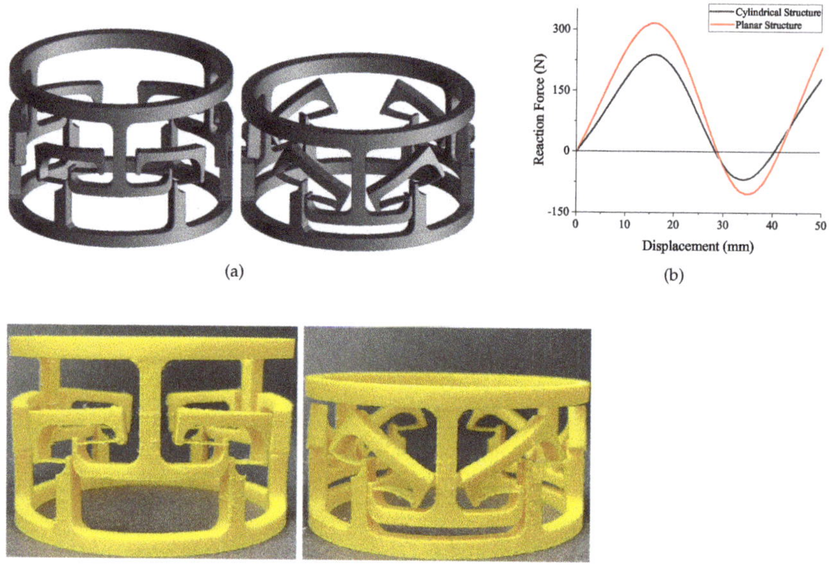

(c)

Figure 11. The cylindrical deployable structure: (**a**) the design; (**b**) the FEA simulation result; (**c**) two stable configurations of the cylindrical prototype.

A comparison between planar and cylindrical structures was performed. Both of these two structures consist of three bistable lattices. The lattices are arranged horizontally. The result illustrates that the cylindrical structure has relatively lower critical forces. This might be caused by the out-of-cylinder motion during the deformation process. Furthermore, the cylindrical structure design demonstrates the feasibility of translating the concept into a potential practical application. The aim of converting the flat 2D model to a 3D cylindrical model is to investigate the bistable behaviour in a higher dimension.

4. Conclusions

In this work, we presented two types of deployable structures including a 1D and a cylindrical model. These structures are designed based on the programmable bistable lattices. We performed a series of FEA simulations and parametric analysis to obtain a proper bistable lattice. Four types of bistable lattices derived from traditional compliant bistable mechanisms are listed in this work as well. The experimental results and FEA simulations prove that the designed structure has more than one stable status when no external force is acting on it. The breakthrough in this work provides more options for designing the deployable structure and bistable mechanisms.

Author Contributions: Conceptualization, G.H. and T.L.; methodology, G.H. and T.L.; software, T.L.; validation, T.L.; formal analysis, T.L.; investigation, T.L.; resources, G.H. and T.L.; data curation, T.L.; writing—original draft preparation, T.L.; writing—review and editing, G.H.; visualization, G.H. and T.L.; supervision, G.H.; project administration, G.H.; funding acquisition, G.H. All authors have read and agreed to the published version of the manuscript.

Funding: This work is partially funded by the EnABLES TA project (No.122). EnABLES (http://www.enables-project.eu/) has received funding from the EU Horizon 2020 research & innovation programme, under Grant Agreement No. 730957.

Acknowledgments: The authors would like to acknowledge Timothy Power, Mike O'Shea and Paul Conway for their assistance on the prototyping and experiments. The authors would also like to thank Denis Kelliher for his help on the FEA simulation.

Conflicts of Interest: The authors declare no conflict of interest.

References

1. Santo, L.; Quadrini, F.; Accettura, A.; Villadei, W. Shape memory composites for self-deployable structures in aerospace applications. *Procedia Eng.* **2014**, *88*, 42–47. [CrossRef]
2. Doroftei, I.; Doroftei, I.A. Deployable Structures for Architectural Applications—A Short Review. *Appl. Mech. Mater.* **2014**, *658*, 233. [CrossRef]
3. Taylor, A.; Miller, M.; Fok, M.; Nilsson, K.; Tsz Ho Tse, Z. Intracardiac magnetic resonance imaging catheter with origami deployable mechanisms. *J. Med. Devices* **2016**, *10*, 020957. [CrossRef]
4. Zhao, P.; Liu, J.; Wu, C.; Li, Y.; Chen, K. Novel surface design of deployable reflector antenna based on polar scissor structures. *Chin. J. Mech. Eng.* **2020**, *33*, 1–15. [CrossRef]
5. Chahal, A.M.; Van Dewark, K.; Gooch, R.; Fukushima, E.; Hudson, Z.M. A rapidly deployable negative pressure enclosure for aerosol-generating medical procedures. *MedRxiv* **2020**. [CrossRef]
6. Wang, J.; Kong, X.; Yu, J. Design of deployable mechanisms based on Wren parallel mechanism units. *J. Mech. Des.* **2022**, *144*, 063302. [CrossRef]
7. Dragoljevic, M.; Viscuso, S.; Zanelli, A. Data-driven design of deployable structures: Literature review and multi-criteria optimization approach. *Curved Layer. Struct.* **2021**, *8*, 241–258. [CrossRef]
8. Peng, Q.; Wang, S.; Zhi, C.; Li, B. A new flexible multibody dynamics analysis methodology of deployable structures with scissor-like elements. *Chin. J. Mech. Eng.* **2019**, *32*, 1–10. [CrossRef]
9. Zhao, J.S.; Chu, F.; Feng, Z.J. The mechanism theory and application of deployable structures based on SLE. *Mech. Mach. Theory* **2009**, *44*, 324–335. [CrossRef]
10. Akgün, Y.; Gantes, C.J.; Sobek, W.; Korkmaz, K.; Kalochairetis, K. A novel adaptive spatial scissor-hinge structural mechanism for convertible roofs. *Eng. Struct.* **2011**, *33*, 1365–1376. [CrossRef]
11. Ma, S.; Chen, M.; Skelton, R.E. Design of a new tensegrity cantilever structure. *Compos. Struct.* **2020**, *243*, 112188. [CrossRef]
12. Deleo, A.A.; O'Neil, J.; Yasuda, H.; Salviato, M.; Yang, J. Origami-based deployable structures made of carbon fiber reinforced polymer composites. *Compos. Sci. Technol.* **2020**, *191*, 108060. [CrossRef]
13. Zhai, Z.; Wang, Y.; Jiang, H. Origami-inspired, on-demand deployable and collapsible mechanical metamaterials with tunable stiffness. *Proc. Natl. Acad. Sci. USA* **2018**, *115*, 2032–2037. [CrossRef] [PubMed]
14. Shah, S.I.H.; Tentzeris, M.M.; Lim, S. A deployable quasi-Yagi monopole antenna using three origami magic spiral cubes. *IEEE Antennas Wirel. Propag. Lett.* **2018**, *18*, 147–151. [CrossRef]
15. Howell, L.L.; Magleby, S.P.; Olsen, B.M. *Handbook of Compliant Mechanisms*; John Wiley & Sons: Hoboken, NJ, USA, 2013.
16. Culpepper, M.L.; Anderson, G. Design of a low-cost nano-manipulator which utilizes a monolithic, spatial compliant mechanism. *Precis. Eng.* **2004**, *28*, 469–482. [CrossRef]
17. Kota, S.; Ananthasuresh, G. Designing compliant mechanisms. *Mech. Eng.-CIME* **1995**, *117*, 93–97.
18. Wegner, J.; Frey, M.; Busch, R.; Kleszczynski, S. Additive manufacturing of a compliant mechanism using Zr-based bulk metallic glass. *Addit. Manuf. Lett.* **2021**, *1*, 100019. [CrossRef]
19. Hao, G.; Zhu, J. Design of a monolithic double-slider based compliant gripper with large displacement and anti-buckling ability. *Micromachines* **2019**, *10*, 665. [CrossRef]
20. Zhao, L.; Yu, X.; Li, P.; Qiao, Y. High-precision compliant mechanism for lens XY micro-adjustment. *Rev. Sci. Instrum.* **2020**, *91*, 035004. [CrossRef]
21. Zhu, W.L.; Zhu, Z.; Shi, Y.; Wang, X.; Guan, K.; Ju, B.F. Design, modeling, analysis and testing of a novel piezo-actuated XY compliant mechanism for large workspace nano-positioning. *Smart Mater. Struct.* **2016**, *25*, 115033. [CrossRef]
22. Hao, G.; Kong, X. A novel large-range XY compliant parallel manipulator with enhanced out-of-plane stiffness. *J. Mech. Des. Trans. ASME* **2012**, *134*, 061009. [CrossRef]
23. Howell, L.L. *Compliant Mechanisms*; John Wiley & Sons: Hoboken, NJ, USA, 2001.
24. Liang, H.; Hao, G.; Olszewski, O.Z. A review on vibration-based piezoelectric energy harvesting from the aspect of compliant mechanisms. *Sens. Actuators A Phys.* **2021**, *331*, 112743. [CrossRef]
25. Emam, S.A.; Inman, D.J. A review on bistable composite laminates for morphing and energy harvesting. *Appl. Mech. Rev.* **2015**, *67*, 060803. [CrossRef]
26. Jeong, H.Y.; An, S.C.; Seo, I.C.; Lee, E.; Ha, S.; Kim, N.; Jun, Y.C. 3D printing of twisting and rotational bistable structures with tuning elements. *Sci. Rep.* **2019**, *9*, 324. [CrossRef] [PubMed]
27. Hussein, H.; Le Moal, P.; Younes, R.; Bourbon, G.; Haddab, Y.; Lutz, P. On the design of a preshaped curved beam bistable mechanism. *Mech. Mach. Theory* **2019**, *131*, 204–217. [CrossRef]
28. Pane, I.Z.; Asano, T. Investigation on bistability and fabrication of bistable prestressed curved beam. *Jpn. J. Appl. Phys.* **2008**, *47*, 5291. [CrossRef]

29. Wu, C.C.; Lin, M.J.; Chen, R. The derivation of a bistable criterion for double V-beam mechanisms. *J. Micromech. Microeng.* **2013**, *23*, 115005. [CrossRef]
30. Hansen, B.; Carron, C.; Jensen, B.; Hawkins, A.; Schultz, S. Plastic latching accelerometer based on bistable compliant mechanisms. *Smart Mater. Struct.* **2007**, *16*, 1967. [CrossRef]
31. Qiu, J.; Lang, J.H.; Slocum, A.H. A curved-beam bistable mechanism. *J. Microelectromech. Syst.* **2004**, *13*, 137–146. [CrossRef]
32. Zirbel, S.A.; Tolman, K.A.; Trease, B.P.; Howell, L.L. Bistable mechanisms for space applications. *PLoS ONE* **2016**, *11*, e0168218. [CrossRef]
33. Wu, K.; Hao, G. Design and nonlinear modeling of a novel planar compliant parallelogram mechanism with general tensural-compresural beams. *Mech. Mach. Theory* **2020**, *152*, 103950. [CrossRef]
34. Han, Q.; Jin, K.; Chen, G.; Shao, X. A novel fully compliant tensural-compresural bistable mechanism. *Sens. Actuators A Phys.* **2017**, *268*, 72–82. [CrossRef]

MDPI
St. Alban-Anlage 66
4052 Basel
Switzerland
Tel. +41 61 683 77 34
Fax +41 61 302 89 18
www.mdpi.com

Micromachines Editorial Office
E-mail: micromachines@mdpi.com
www.mdpi.com/journal/micromachines